International Series in Operations Research & Management Science

Volume 290

More information about this series at http://www.springer.com/series/6161

Vincent Charles · Juan Aparicio ·
Joe Zhu
Editors

Data Science and Productivity Analytics

Springer

Editors
Vincent Charles
School of Management
University of Bradford
Bradford, UK

Juan Aparicio
Center of Operations Research
University Miguel Hernandez
Elche, Alicante, Spain

Joe Zhu
Foisie Business School
Worcester Polytechnic Institute
Worcester, MA, USA

ISSN 0884-8289 ISSN 2214-7934 (electronic)
International Series in Operations Research & Management Science
ISBN 978-3-030-43383-3 ISBN 978-3-030-43384-0 (eBook)
https://doi.org/10.1007/978-3-030-43384-0

This Springer imprint is published by the registered company Springer Nature Switzerland AG
The registered company address is: Gewerbestrasse 11, 6330 Cham, Switzerland

Preface

Data science is defined as the collection of scientific methods, processes, and systems dedicated to extracting knowledge or insights from data to support more intelligent data-driven decision-making. It develops on concepts from various domains, containing mathematics and statistical methods, operational research, machine learning, computer programming, pattern recognition, and data visualization, among others. The use of data science techniques has spread over many fields, from manufacturing to retail, public services, telecommunications, banking and financial services, health care and education, just to name a few. The art of data science is not a new phenomenon; but in recent years, its prospect has been exponentially heightened by the explosion of Big Data, which has brought with it both challenges and opportunities. More than 90% of the data that currently exists was created in just the last few years; analyzing these new and expanding data streams has become a key basis of competition and innovation for any organization, underpinning new waves of productivity growth. As a matter of fact, the capabilities and affordances of data science and big data now fuel almost every new business initiative or product development process.

The present book *Data Science and Productivity Analytics* brings a fresh look onto the various ways that data science techniques could unleash value and drive productivity from these mountains of data. This book should be of interest to data scientists, researchers, and practitioners alike. The book is organized in 15 chapters, contributed by authors from all around the globe: Canada, China, Ecuador, France, India, Iran, Ireland, Spain, Turkey, The United Kingdom, and the United States.

Chapter 1 by Dariush Khezrimotlagh and Joe Zhu builds upon the study by Khezrimotlagh, Zhu, Cook, and Toloo (2019), whose framework proposed the fastest available technique in the literature to deal with large-scale data envelopment analysis (DEA). The examples provided show the applicability of the framework and its superiority to the existing methodologies when dealing with large-scale DEA problems.

DEA has had a distinguished career as an analytical tool derived to analyze efficiency and productivity, but it is now ready to be unfastened from its economics/efficiency/productivity moorings and generalized into a machine learning, big data, analytics tool. In this sense, Chap. 2 by José H. Dulá discusses the history of DEA algorithms, computations, and geometry, and how what we have learned designing algorithms can be used in machine learning.

Data Science and decision-making are intertwined, as data science enables in-depth analysis of data to support data-driven decision-making. Chapter 3 by Alex Rabasa and Ciara Heavin explores data science and its applications, specifically focusing on classification rules to enable better decision-making. Four case studies are presented for better understanding.

Identifying the presence of congestion can help improve productivity. Chapter 4 by Mahmood Mehdiloo, Biresh Sahoo, and Joe Zhu is concerned with the precise identification of both weak and strong forms of congestion of production units. The authors develop three computational algorithms for identifying the congestion statuses of all decision-making units (DMUs) in any finite-size sample and provide numerical examples to illustrate their superiority over existing approaches.

Chapter 5 by Gabriel Villa and Sebastián Lozano further explores the envelopment side of the DEA methodology by considering the nonparametric derivation of the production possibility set, the multiplicity of DEA models, and how to handle different types of situations and data.

Chapter 6 by Luis Orea serves as a guide to efficiency evaluation from an econometric perspective. To this aim, the chapter summarizes the main features of the standard econometric approach to measuring firms' inefficiency and productivity. The authors conclude by emphasizing the importance of choosing a suitable analytical framework that is in accordance with the industry characteristics and the restrictions faced by the firm.

Systems with a two-stage structure are very common in production or service organizations. Chapter 7 by Qingxian An, Haoxun Chen, Beibei Xiong, Jie Wu, and Liang Liang addresses the fair setting of the target intermediate products for such systems. To this aim, the authors propose an approach based on a new DEA model and a Nash bargaining game to determine the target intermediate products and further the production frontier projections of all DMUs, with an application to insurance companies.

Chapter 8 by Tao Ding, Feng Li, and Liang Liang further presents a new approach to dealing with fixed cost and resource allocation issues, under a centralized decision environment, in a two-stage network production system by considering the factor of technology heterogeneity. Two examples are provided to show the feasibility of the proposed models.

The following seven chapters not only make methodological contributions, but also deal with the application of DEA in various fields, such as education, ports, banking, health care, environmental issues, pension funds and mutual funds, stocks, and portfolios.

In Chap. 9, Jose Manuel Cordero, Cristina Polo, and Rosa Simancas apply some of the most recent nonparametric methods to assess and compare the education production efficiency of a sample of secondary schools operating in 67 different countries using data from the OECD Programme for International Student Assessment (PISA 2015).

Ports have become one of the main funnels to enhance the competitiveness in emerging markets of Latin America. In Chap. 10, Emilio J. Morales-Núñez, Xavier R. Seminario-Vergara, Sonia Valeria Avilés-Sacoto, and Galo Eduardo Mosquera-Recalde use DEA to evaluate and compare the performance of the Ecuadorian Guayaquil Contecon Port with 14 major container seaports in Latin America and the Caribbean.

Chapter 11 by Cong Xu, Guo-liang Yang, Jian-bo Yang, Yu-wang Chen, and Hua-ying Zhu proposes a DEA model to test whether employees under different types of Loci of Control will react differently to a positive (or negative) policy. The authors further investigate how the Locus of Control impacts a bank's policies through a case study of a Chinese state-owned bank.

Chapter 12 by Babak Daneshvar Rouyendegh (B. Erdebilli), Asil Oztekin, Joseph Ekong, and Ali Dag develops a holistic data analytic approach to measure and improve hospital productivity. The DEA methodology is integrated with the Fuzzy Analytic Hierarchy Process to develop a hybrid fuzzy logic-based multi-criteria decision-making model, which is then applied to a dataset of public healthcare institutions in Turkey.

Chapter 13 by Anyu Yu, Simon Rudkin, and Jianxin You proposes a meta-frontier DEA allocation model that reflects the potential technology heterogeneity of DMUs to analyze the division of carbon abatement tasks by considering corresponding regional-level collaboration and the dual optimization of carbon reduction and output maximization.

Chapter 14 by Maryam Badrizadeh, and Joseph C. Paradi introduces a new Mixed Variable DEA model that provides an approach to performance measurement when dealing with moderately different cultures and rules, but in the same industry, with an application to pension funds and mutual funds in Canada. The proposed methodology could be used in various other business areas.

Finally, Chap. 15 by Mercedes Landete, Juan F. Monge, José L. Ruiz, and José V. Segura proposes a new Sharpe ratio portfolio selection strategy based on a cross-efficiency evaluation, under the assumption that the risk-free asset is unknown, but still within a given interval. The proposed approach is compared with other classical solutions through the study of two cases, namely Eurostock50 and USA industry portfolios.

January 2020

The Editors
Vincent Charles
School of Management
University of Bradford
Bradford, UK

Juan Aparicio
Center of Operations Research
University Miguel Hernandez
Elche, Spain

Joe Zhu
Foisie Business School
Worcester Polytechnic Institute
Worcester, USA

Contents

Chapter 1
Data Envelopment Analysis and Big Data: Revisit with a Faster Method

Dariush Khezrimotlagh and Joe Zhu

Abstract Khezrimotlagh et al. (Eur J Oper Res 274(3):1047–1054, 2019) propose a new framework to deal with large-scale data envelopment analysis (DEA). The framework provides the fastest available technique in the DEA literature to deal with big data. It is well known that as the number of decision-making units (DMUs) or the number of inputs–outputs increases, the size of DEA linear programming problems increases; and thus, the elapsed time to evaluate the performance of DMUs sharply increases. The framework selects a subsample of DMUs and identifies the set of all efficient DMUs. After that, users can apply DEA models with known efficient DMUs to evaluate the performance of inefficient DMUs or benchmark them. In this study, we elucidate their proposed method with transparent examples and illustrate how the framework is applied. Additional simulation exercises are designed to evaluate the performance of the framework in comparison with the performance of the two former methods: build hull (BH) and hierarchical decomposition (DH). The disadvantages of BH and HD are transparently demonstrated. A single computer with two different CPUs is used to run the methods. For the first time in the literature, we consider the cardinalities, 200,000, 500,000 and 1,000,000 DMUs.

Keywords Performance evaluation · Big data · Data envelopment analysis (DEA) · Simulation

D. Khezrimotlagh (✉)
School of Science, Engineering and Technology, Pennsylvania State University - Harrisburg, Pennsylvania, PA 17057, USA
e-mail: dk@psu.edu; dzk349@psu.edu

J. Zhu
Foisie Business School, Worcester Polytechnic Institute, Worcester, MA 01609, USA
e-mail: jzhu@wpi.edu

V. Charles et al. (eds.), *Data Science and Productivity Analytics*,
International Series in Operations Research & Management Science 290,
https://doi.org/10.1007/978-3-030-43384-0_1

1.1 Introduction

Charnes et al. (1978) developed data envelopment analysis (DEA) to evaluate the performance of a set of homogeneous decision-making units (DMUs), with multiple factors that are classified as inputs and outputs. Using linear programming, DEA measures the performance of each DMU and identifies efficient and inefficient DMUs. The standard DEA models, such as the standard variable returns to scale (VRS) (Banker et al. 1984), which have been substantially used in the literature, should be solved n times (once for each DMU) to evaluate the performance of n DMUs. As n increases, the elapsed time to evaluate DMUs sharply increases. In other words, as the number of decision variables or constraints in a DEA model increases, the elapsed time to run the DEA model increases.

Several frameworks are developed in the DEA literature to reduce the elapsed time of applying DEA models and they can be found in Ali (1993), Dulá and Helgason (1996), Barr and Durchholz (1997), Dulá and Thrall (2001), Dulá (2008), Chen and Cho (2009), Dulá and López (2013), Chen and Lai (2017), and Khezrimotlagh et al. (2019). For example, in one of the most recent studies, Zhu et al. (2018) use 10 computers in parallel to decrease 57% of the elapsed time by the traditional DEA approach, when there were 20,000 DMUs with $2+2$ inputs–outputs. They used the hierarchical decomposition (HD) procedure proposed by Barr and Durchholz (1997).

HD is designed to partition a given sample of n DMUs into q blocks. Three parameters, b, β and γ, should be defined for HD by the user. The parameter b is used to find the size of each block, β to check the proportion of best-practice DMUs in the blocks and γ to increase the size of each block in each iteration. HD first partitions n DMUs into q blocks in which each block includes b DMUs at most. Then, it finds best-practice DMUs of each block. Here, we say best-practice DMUs instead of efficient DMUs because best-practice DMUs are efficient in a particular block and might be inefficient if they were in another block. Now, HD aggregates the found best practices in one single block. If the proportion of best practices within all blocks is greater than the parameter β, then HD evaluates the DMUs in that single block to find the efficient DMUs. Otherwise, HD repeats the steps above by partitioning that single block into q' blocks each with size $b\gamma$ at most. The procedure usually stops after a few iterations and efficient DMUs are found. The HD framework is very useful for decreasing the elapsed time to evaluate big data, using several processors in parallel.

Fourteen years later, Dulá (2011) proposed a framework, called build hull (BH), and showed that BH performs much faster than HD does. In BH framework, the size of the first linear programming is 2, and it is gradually increased until all required efficient DMUs to build the hull are determined. Indeed, BH uses a minimal number of DMUs to envelop all DMUs and finds efficient DMUs (one at a time). Each BH's linear programming is dependent on the other linear programming.

Khezrimotlagh et al. (2019) show the disadvantage of BH: using the dependent models from one iteration to another to find the efficient DMUs and to enhance the HD framework to substantially decrease the elapsed time of evaluating the performance of DMUs. They just used one single computer. In other words, they showed that the elapsed time of BH (as well as the traditional DEA approach) can be decreased up to 99.9% by the new framework using a single computer.

In this study, we elucidate the framework of Khezrimotlagh et al. (2019) by several practical examples. We provide a very basic example to show how the algorithms in their framework are applied. In addition, a theoretical example is designed with a simulation experiment to clarify the disadvantages of BH framework. The example also represents the situations that HD performs weaker than BH. We add the performance of HD for all the simulation experiences.

Khezrimotlagh et al. (2019) used the square root of the number of DMUs for selecting a subsample (as well as the size of blocks) to apply their framework versus BH. In this study, we increase the size of blocks to square root of the product of the number of DMUs to the number of inputs–outputs. We show that such a way to partition DMUs increases the power of HD in comparison with BH. The same increase in the size of the subsample is also considered in applying the Khezrimotlagh et al. (2019) framework.

Moreover, Khezrimotlagh et al. (2019) used Matlab2017a (student version) and a single laptop with an Intel® Core™ i7-7820HK CPU @2.90 GHz, 16 GB memory and a 64-bit Windows 10 operating system to develop their framework. In this study, we use the same device as well as a computer with an Intel® Core™ i9-7980XE CPU @3.10 GHz to run our simulations. Using a single computer with an Intel® Core™ i9-7980XE CPU, we show that when the number of DMUs is 1,000,000 with 1 + 1 input–output, BH uses 34.56 h (almost a day and a half) to find efficient DMUs, whereas that of Khezrimotlagh et al. (2019) only uses 8 min (using a single computer). The elapsed time of BH is decreased by 99.6% using a single computer. We note that BH can decrease the elapsed time of applying the traditional approach (to find efficient DMUs) by 99%, meanwhile the framework of Khezrimotlagh et al. (2019) can also decrease the elapsed time of BH by 99%, using a single computer. As a result, it is obvious that using two computers in parallel, almost 100% of the elapsed time of BH is decreased.

The study is organized into eight sections. Section 1.2 describes the framework of Khezrimotlagh et al. (2019). Section 1.3 illustrates how the framework is applied by a basic example. A theoretical example is given in Sect. 1.4 to discuss the pros and cons of each method. In Sects. 1.5–1.7 several different datasets are considered to compare the elapsed time of the methods with different cardinalities and dimensions. Conclusions are given in Sect. 1.8.

In this study, "the framework of Khezrimotlagh et al. (2019)", "the proposed method by Barr and Durchholz (1997)" and "the proposed method by Dual (2011)" are shortened to "the Framework", HD and BH, respectively.

1.2 The Framework

Suppose that a set of n observations DMU_j, $(j = 1, \ldots, n)$ is given, where each DMU_j has m non-negative inputs $\boldsymbol{x}_j = (x_{1j}, \ldots, x_{mj})$ and s non-negative outputs $\boldsymbol{y}_j = (y_{1j}, \ldots, y_{sj})$. Khezrimotlagh et al. (2019) proposed the Framework as follows:

1. Start,
2. Get a sample of DMUs,
3. Select a subsample of size b by the following algorithm,
   3.1. Use the theorems proposed by Ali (1993) to select $m + s$ DMUs (or more); such that, DMU_l is located on the VRS frontier if one of the following equations holds: $x_{il} = \min\{x_{ij} | j = 1, \ldots, n\}$, for $i = 1, \ldots, m$, and $y_{rl} = \max\{y_{rj} | j = 1, \ldots, n\}$, for $r = 1, \ldots, s$.
   3.2. For each unselected DMU_j assign a pre-score as follows,
   {$u = 0$,
   {For each i and k_i
   If $x_{ij} \leq k_i$-th percentile of $\boldsymbol{x}^i$ then
   $u = u + 1$}
   {For each r and k'_r
   If $y_{rj} \geq k'_r$-th percentile of $\boldsymbol{y}^r$ then
   $u = u + 1$}
   pre-score_$j = u$ }
   3.3. Sort DMUs in descending order by the assigned pre-scores, and the remaining DMUs (to construct a subsample with size b) are selected as those having the greatest pre-scores.
4. Find the best-practice DMUs,
5. Find exterior DMUs according to the hull of best-practice DMUs in Step 4,
6. If the found set in Step 5 is empty, go to 7 (as the efficient DMUs are found and the inefficient DMUs are already evaluated).
   Otherwise,
   6.1. Add the found set in Step 5 to the found set in Step 4, and find the best-practice DMUs, (in this step all required efficient DMUs to build the hull are found.)
   6.2. Evaluate the rest of the DMUs with respect to the DMUs hull in Step 6.1,
7. End.

In the above pseudocode, $\boldsymbol{x}^i$ and $\boldsymbol{y}^r$ represent the ith inputs and the rth outputs of all DMUs, $i = 1, \ldots, m$ and $r = 1, \ldots, s$; for instance, $\boldsymbol{x}^1 = (x_{11}, x_{12}, \ldots, x_{1n})$. The indexes, k_i and k'_r, are changed from 0 to 100, and they refer to the percentiles of $\boldsymbol{x}^i$ and $\boldsymbol{y}^r$, respectively.

As quoted by Khezrimotlagh et al. (2019), the Framework can be introduced as an HD procedure with the difference being that instead of partitioning DMUs into several blocks and evaluating each block, only a subsample of DMUs (i.e., one block) is used to evaluate all other DMUs (i.e., DMUs in other blocks) based upon the selected subsample. In this study, the aggregated set in Step 6.1 is partitioned into several blocks similar to that of HD. In other words, we only apply the Framework for the first iteration of HD in this study, and after that, the procedure is continued, similar to HD. We also consider the same parameters, b, β and γ for both HD and the Framework in this study.

1.3 A Basic Example

1.3.1 Applying the Framework Without Step 3

In this section, we illustrate a very basic example to show the structure of the Framework. We do not apply the Framework completely, but we design the example such that all possible outcomes (when a subsample is selected) occur. After that, we apply the Framework completely and show the differences.

Suppose that we have a sample of 14 DMUs, labeled $A - N$, where each DMU has one input and one output, as shown in Table 1.1 (Fig. 1.1).

In a basic view, the Framework is started by selecting a subsample of DMUs. Suppose that a random subsample of size 4 is given, such as: $\{B, D, H, N\}$. As discussed, we intentionally select this subsample to illustrate all possible outcomes when a subsample is selected. In other words, in this subsection, we do not apply Step 3 of the Framework to select the subsample, but we suppose that the subsample is given.

Figure 1.2 illustrates the efficient frontier which is measured by the subsample, $\{B, D, H, N\}$. As can be seen, it is possible that some DMUs within the subsample are inefficient, like N. It is also possible that some DMUs are not enveloped by the selected subsample. As a result, three different situations are possible when a subsample is selected: (1) DMUs which build the hull, like $\{B, D, H\}$, (2) interior DMUs, like $\{A$, F, I, J, K, L, M, N$\}$ and (3) exterior DMUs, like $\{C, E, G\}$.

Step 4 of the Framework explains how the best-practice frontier in Fig. 1.2 is known. Indeed, the corresponding standard VRS model in output-oriented with respect to the subsample, $\{B, D, H, N\}$, is formulated as follows to evaluate $\text{DMU}_l(x_l, y_l)$:

Table 1.1 A sample of 14 DMUs

DMU	*A*	*B*	*C*	*D*	*E*	*F*	*G*	*H*	*I*	*J*	*K*	*L*	*M*	*N*
x	5.51	4.70	9.00	7.00	4.00	6.30	9.00	5.00	7.20	7.00	7.50	10.00	8.20	5.70
y	1.12	1.10	2.14	1.90	0.69	0.80	1.93	1.35	1.67	1.03	1.48	1.59	1.00	1.40

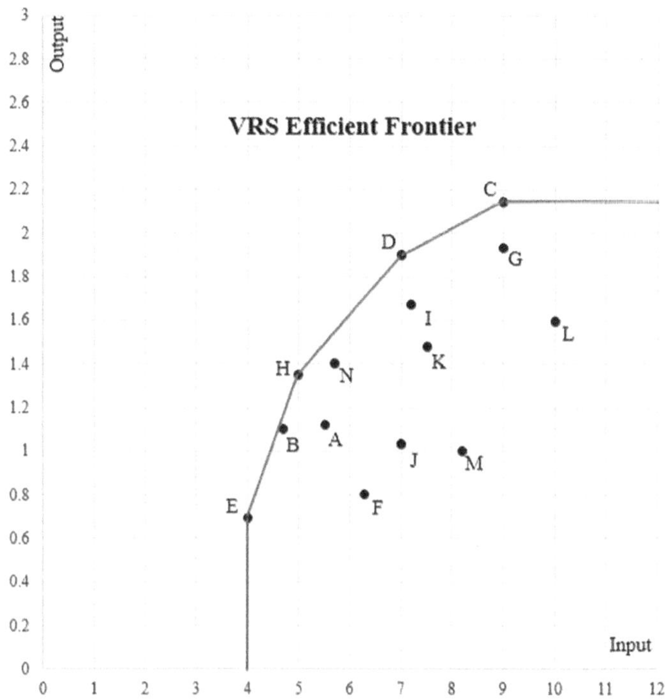

Fig. 1.1 An example of 14 DMUs

$$
\begin{aligned}
&\max \varphi_l \\
&\text{subject to} \\
&\lambda_B x_B + \lambda_D x_D + \lambda_H x_H + \lambda_N x_N \le x_l, \\
&\lambda_B y_B + \lambda_D y_D + \lambda_H y_H + \lambda_N y_N \ge \varphi_l y_l, \\
&\lambda_B + \lambda_D + \lambda_H + \lambda_N = 1, \\
&\lambda_B \ge 0,\ \lambda_D \ge 0,\ \lambda_H \ge 0,\ \lambda_N \ge 0.
\end{aligned} \tag{1.1}
$$

By running model (1.1), N is known as inefficient and the best-practice DMUs in the subsample, that is, $\{B, D, H\}$, are known, and we will have the following model with a smaller number of decision variables:

$$
\begin{aligned}
&\max \varphi_l \\
&\text{subject to} \\
&\lambda_B x_B + \lambda_D x_D + \lambda_H x_H \le x_l, \\
&\lambda_B y_B + \lambda_D y_D + \lambda_H y_H \ge \varphi_l y_l, \\
&\lambda_B + \lambda_D + \lambda_H = 1, \\
&\lambda_B \ge 0,\ \lambda_D \ge 0,\ \lambda_H \ge 0.
\end{aligned} \tag{1.2}
$$

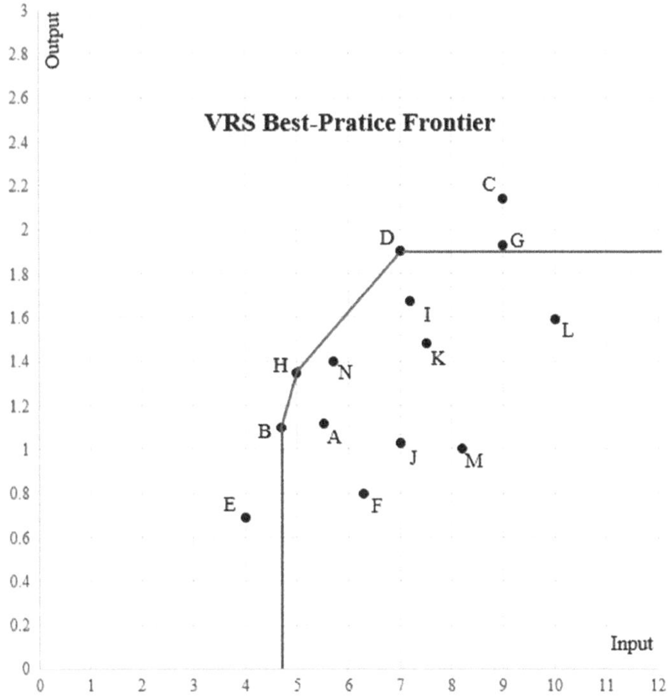

Fig. 1.2 The best-practice frontier according to the selected subsample

In Step 5, each DMU except DMUs in $\{B, D, H, N\}$ is evaluated to find the exterior DMUs to the hull of $\{B, D, H\}$. In other words, DMUs in $\{A, C, E, F, G, I, J, K, L, M\}$ are evaluated by model (1.2). For instance, when A is evaluated, we have the following model:

$$\begin{aligned}
&\max \varphi_A \\
&\text{subject to} \\
&\lambda_B x_B + \lambda_D x_D + \lambda_H x_H \leq x_A, \\
&\lambda_B y_B + \lambda_D y_D + \lambda_H y_H \geq \varphi_A y_A, \\
&\lambda_B + \lambda_D + \lambda_H = 1, \\
&\lambda_B \geq 0,\ \lambda_D \geq 0, \lambda_H \geq 0.
\end{aligned} \tag{1.3}$$

As a result, four different situations might occur, in particular: (1) the optimal objective is greater than 1, (2) the optimal objective is 1, (3) the optimal objective is between 0 and 1 and (4) the standard output-oriented VRS model is infeasible. The optimal objective in model (1.2) is greater than 1, when a DMU under evaluation is inefficient, that is, φ^* is greater than 1 for DMUs in $\{A, F, I, J, K, L, M, N\}$. There are no more DMUs with the unit score in this example. The optimal objective for

DMUs $\{C, G\}$ is between 0 and 1, and the model is infeasible for evaluating E. We assume a 0 score for such a DMU. As a result, the range of the optimal objective when model (1.2) is applied is within $[0, \infty)$.

If there were no exterior points to the hull of the subsample $\{B, D, H, N\}$, all DMUs would be already evaluated and the evaluation process would be finished. This is mentioned in Step 6.1 of the Framework.

However, in this example, there are three exterior DMUs to the hull of the subsample. Thus, the best-practice DMUs are aggregated with the exterior DMUs, that is, $\{B, D, H\} \cup \{C, E, G\}$, and model (1.4) is considered.

$$\begin{aligned}
&\max \varphi_A \\
&\text{subject to} \\
&\lambda_B x_B + \lambda_D x_D + \lambda_H x_H + \lambda_C x_C + \lambda_E x_E + \lambda_G x_G \le x_l, \\
&\lambda_B y_B + \lambda_D y_D + \lambda_H y_H + \lambda_C y_C + \lambda_E y_E + \lambda_G y_G \ge \varphi_l y_l, \\
&\lambda_B + \lambda_D + \lambda_H + \lambda_C + \lambda_E + \lambda_G = 1, \\
&\lambda_B \ge 0,\ \lambda_D \ge 0,\ \lambda_H \ge 0,\ \lambda_C \ge 0,\ \lambda_E \ge 0,\ \lambda_G \ge 0.
\end{aligned} \tag{1.4}$$

Similarly, model (1.4) evaluates DMUs $\{B, D, H, C, E, G\}$. In this case, the optimal value is either greater than 1 or equal to 1. Thus, the efficient DMUs are found to be $\{C, D, E, H\}$.

Note that, when the number of DMUs is large, we partition the aggregated DMUs in Step 6.1 into several blocks similar to those of HD. We can also apply Step 3 on the set of exterior DMUs and rerun Steps 4, 5 and 6.

From model (1.4) the efficiency scores of DMUs B and G are found; thus model (1.5) is considered in measuring the efficiency scores of DMUs $\{A, F, I, J, K, L, M, N\}$.

$$\begin{aligned}
&\max \varphi_l \\
&\text{subject to} \\
&\lambda_C x_C + \lambda_D x_D + \lambda_E x_E + \lambda_H x_H \le x_l, \\
&\lambda_C y_C + \lambda_D y_D + \lambda_E y_E + \lambda_H y_H \ge \varphi_l y_l, \\
&\lambda_C + \lambda_D + \lambda_E + \lambda_H = 1, \\
&\lambda_C \ge 0, \lambda_D \ge 0, \lambda_E \ge 0, \lambda_H \ge 0.
\end{aligned} \tag{1.5}$$

As a result, all DMUs are evaluated and the procedure ends.

1.3.2 Applying the Framework with Step 3

Suppose that the DMUs in Table 1.1 are given. We now want to select a subsample of size 4 by applying Step 3.

Since E has the minimum amount of inputs among DMUs, and C has the maximum amount of outputs among DMUs, these two DMUs are on the efficient frontier (Ali 1993) and selected. We now need two more DMUs to construct the subsample.

Step 3 also assigns a pre-score to the rest of DMUs. For example, the pre-score of D is calculated as follows:

- The input of D (=7) is the 47th percentile of $\boldsymbol{x}^1$ (DMUs' input). Thus, 54 points (regarding the 47th–100th percentiles of $\boldsymbol{x}^1$) are assigned to D.
- The output of D (=1.9) is between the 84th and 85th percentiles of $\boldsymbol{y}^1$ (DMUs' output). Thus, 85 points (regarding the 0th–84th percentiles of $\boldsymbol{y}^1$) are assigned to D.
- As a result, the pre-score for D is $54 + 84 = 139$.

Similarly, the input of B (=4.7) is between the 7th and the 8th percentiles of $\boldsymbol{x}^1$ and the output of B (=1.1) is between the 30th and 31st percentiles of $\boldsymbol{y}^1$; thus, the pre-score for B is $93 + 31 = 124$. Table 1.2 illustrates the pre-scores of DMUs in Table 1.1.

The DMUs are sorted in descending order by the assigned pre-scores, and the remaining two DMUs (to construct a subsample with size 4) are selected as those having the greatest pre-scores, and they are DMUs D and H. As a result, the selected subsample is $\{C, D, E, H\}$. Until here, the Framework does not know whether all efficient DMUs are found. Thus, model (1.5) is formulated and the scores of all DMUs $\{C, D, E, H\}$ are 1, that is, none of the DMUs is eliminated from the subsample. Now, model (1.5) is used to evaluate all DMUs except $\{C, D, E, H\}$. Since there are no exterior DMUs to the hull of $\{C, D, E, H\}$, the procedure ends, and the efficient DMUs are known and all DMUs are evaluated.

1.4 A Theoretical Example

In this section, we design a theoretical dataset to transparently show the disadvantages of BH and HD designs. From the example, users can clearly see the limitations in the structures of BH and HD and their weaknesses in evaluating large-scale DEA.

We note that BH uses a minimal number of efficient DMUs to build the hull. This is, of course, an advantage of using BH; however, linear programming models in BH are not independent of each other. This fact does not allow parallel processing and results in a huge elapsed time.

On the other hand, HD uses parallel processing, but it uses the maximum number of efficient DMUs to build the hull. Thus, the size of the linear programming in HD can be very large; and this results in a huge elapsed time in comparison with BH.

In contrast, the Framework removes the naïve structure of HD to use the parallel processing and substantially decreases the elapsed time of BH to apply large-scale DEA.

Here, we use (1) a single computer with a CPU i7 with 8 processors and (2) a single computer with a CPU i9 with 16 processors. Thus, the only way to apply

Table 1.2 The pre-scores of DMUs in Table 1.1

DMU	*A*	*B*	***D***	*F*	*G*	***H***	*I*	*J*	*K*	*L*	*M*	*N*
Pre-score	116	124	**139**	70	109	**132**	116	78	93	71	40	124

parallel processing in this situation is to use the available power of the CPU (either with 8 or 16 processors). In other words, while the Framework can use all of the power of a single computer, BH is not able to use more than 90% of the available power of a computer. Here HD uses all of the power of the computer, but it is totally ineffective. Of course, if we use two or more computers in parallel, HD may perform better than BH, but it is always weaker than the Framework.

1.4.1 Generating Data

We generate several samples of size 1,000, 5,000, 10,000, 15,000, 25,000 and 50,000 with 5 + 5 dimensions, where $\boldsymbol{y} = 2\boldsymbol{x}$. Here, $\boldsymbol{x}$ and $\boldsymbol{y}$ denote the vector of inputs' and outputs' data. This represents 100% density in the samples as well as 100% correlation between outputs and inputs.

1.4.2 The Outcomes of BH

In this example, BH only needs two efficient DMUs to build the hull, and after that, it measures the radial VRS scores of all the other efficient DMUs (which are all equal to 1).

This is the most convenient sample for BH because, regardless the cardinality, BH builds the hull with only two DMUs and the size of its LP is 2 + 1 in the entire execution, that is, two lambdas and an alpha (except for the first LP, which has a lambda and an alpha).

The advantage of BH can be clearly seen in the size of LP, which is 3, but the disadvantage is that BH can only solve one problem at a time to find this minimal number of sizes. In other words, although two efficient DMUs are enough to build the hull, BH needs to check all DMUs, one at a time, to find out the fact that the first two DMUs were enough to build the hull. When the number of DMUs increases, this disadvantage substantially increases the elapsed time to evaluate DMUs. We show this fact in Sect. 4.4.

Of course, after the first stage (finding the efficient DMUs), we use parallel processing for BH to measure the efficiency scores of the rest of DMUs. This means, we expect that the difference between the elapsed times of BH in Stage 1 and the total elapsed time is not too large.

1.4.3 The Outcomes of HD

On the other hand, HD partitions DMUs into q blocks, each of size b. For example, b can be considered as $\lfloor\sqrt{n}\rfloor$, the nearest integer value to the square root of the number of DMUs. Here, we assume that $\beta = 0.95$ and $\gamma = 1.5$, as suggested by Dulá (2011). Since all DMUs are efficient, HD builds the hull using all DMUs.

As discussed, HD needs the maximum number of efficient DMUs to build the hull. Indeed, at the first step, HD divides DMUs into several blocks, say q blocks each of size b. Then, HD evaluates DMUs within each block, solving n LPs with size $b + 1$. After that, HD counts the number of best practices, which is the same as the number of DMUs in this example, within blocks. In other words, in the first iteration of partitioning, HD finds that the proportion of best practices within blocks is 100%. Since the ratio of the number of best practices over the number of DMUs is greater than $\beta = 0.95$, HD finishes categorizing DMUs and aggregates all best practices in one block. From now on, HD totally performs as poorly as the traditional DEA approach does, that is, solving n LPs of size $n + 1$.

As a result, the differences between the elapsed times of BH versus HD are very large, although BH solves one problem at a time. This fact clearly shows the naïve implementation of parallel processing by HD. Hence, we do not record the elapsed time of HD in this example.

1.4.4 Comparison Between BH and the Framework

In contrast with BH and HD, the Framework is finished in Step 6 (without executing Steps 6.1 and 6.2) and it identifies that all DMUs are efficient. The only parameter that affects the elapsed time of the Framework is the size of the selected subsample. Khezrimotlagh et al. (2019) considered the size of the selected subsample as $\lfloor\sqrt{n}\rfloor$. Table 1.3 shows the elapsed time of BH and the Framework in Stage 1 (the elapsed time to find the efficient DMUs), Total (the total elapsed time to measure the efficiency scores of all DMUs) and R#DB (the required number of DMUs to build the hull).

It becomes apparent that when cardinality is 50,000 using the computer with a CPU i7, BH takes 871.88 s to complete Stage 1 with Total of 946.60 s. For the subsample of size $\lfloor\sqrt{50,000}\rfloor = 223$, the framework completes Stage 1 and Total in 95.74 s. In other words, the Framework finishes the tasks without executing Steps

Table 1.3 Comparing BH and the framework, using CPU i7 and $b = \sqrt{n}$

Dim.	Card.	Eff.	BH			The framework		
			S.1	Total	R#DB	S.1	Total	R#DB
5 + 5	1,000	1,000	7.06	8.54	2	2.42	2.42	31
	5,000	5,000	36.87	44.23	2	10.84	10.84	70
	10,000	10,000	84.77	99.90	2	18.85	18.85	100
	15,000	15,000	136.01	158.11	2	27.80	27.80	122
	25,000	25,000	281.63	318.60	2	47.47	47.47	158
	50,000	50,000	871.88	946.60	2	95.74	95.74	223

Table 1.4 The elapsed time of the framework with two different subsample sizes, using CPU i7

Dim.	Card.	Eff.	Size: $\min(\lfloor\sqrt{(m+s)n}\rfloor, \lfloor n/2\rfloor)$			Size: $\lfloor 10\sqrt{n}\rfloor$		
			S.1	Total	R#DB	S.1	Total	R#DB
5 + 5	1,000	1,000	2.56	2.56	100	2.63	2.63	310
	5,000	5,000	11.08	11.08	223	16.22	16.22	700
	10,000	10,000	23.03	23.03	316	43.87	43.87	1,000
	15,000	15,000	34.38	34.38	387	81.57	81.58	1,220
	25,000	25,000	61.51	61.51	500	192.08	192.09	1,580
	50,000	50,000	151.92	151.92	707	651.63	651.64	2,230

6.1 and 6.2 because in the first iteration the Framework identifies that all DMUs are efficient and stops the calculation. As a result, the Framework decreases the elapsed time of BH by 90%. This phenomenon transparently elucidates the power of the Framework using one single computer in comparison with BH which solves one problem at a time to find the minimum number of DMUs to build the hull.

Table 1.4 also illustrates the elapsed time for the Framework when subsamples of size $\min(\lfloor\sqrt{(m+s)n}\rfloor, \lfloor n/2\rfloor)$ and $\lfloor 10\sqrt{n}\rfloor$ are selected. As can be seen, when cardinality is 50,000 and the size of the selected subsample to build the hull is 707 (2,230), the Framework still completes the evaluation of DMUs faster than BH (solving an LP of size 3, one at a time) does.

Even if we select a subsample of size 2,500 to build the hull of 50,000 DMUs with 5 + 5 dimensions, the total elapsed time of the Framework is 793.96 s, which is still less than that of the first stage of BH (Stage 1 = 871.88 and Total = 946.60 s).

We note that, using the Framework, a subsample of size 2,500 is selected when there are 6,250,000 DMUs. Here, we only increased the size of subsamples to clarify that even if we build the hull with a subsample of size 2,500 DMUs, the Framework is still faster than BH, which solves one LP with size 3 at a time. This is when we use only one single computer and if we use several computers in parallel, the elapsed time of the Framework decreases substantially.

We now use a computer with a CPU i9 to show the advantage of the Framework versus BH. We also consider the number of DMUs 1,000, 5,000, 10,000, 15,000, 25,000, 50,000, 100,000, 200,000, 500,000 and 1,000,000 DMUs with 5 + 5 dimensions. Table 1.5 illustrates the results.

As can be seen, when the cardinality is 50,000, the elapsed time for the Framework is 40.29 s using CPU i9, in comparison with 95.74 s using CPU i7. This shows a decrease in the elapsed time of the Framework by 58%, using CPU i9 instead of CPU i7. It is obvious that when two computers are used in parallel, the elapsed time of the Framework can be decreased substantially.

When the cardinality is 100,000, BH uses 2,759.93 s (46 min) to measure the efficiency scores of DMUs (using an LP with size 3 only). In contrast, the Framework decreases the elapsed time of BH by 97.24% and measures the efficiency scores of DMUs in 76.13 s.

Table 1.5 Comparing BH and the framework, using CPU i9 and $b = \sqrt{n}$

Dim.	Card.	Eff.	BH			The framework		
			S.1	Total	R#DB	S.1	Total	R#DB
5 + 5	1,000	1,000	6.32	6.93	2	2.30	2.30	32
	5,000	5,000	31.22	33.50	2	6.83	6.83	71
	10,000	10,000	66.72	71.12	2	11.14	11.14	100
	15,000	15,000	116.80	123.44	2	15.13	15.13	122
	25,000	25,000	250.80	261.73	2	19.44	19.44	158
	50,000	50,000	775.80	797.62	2	40.29	40.29	224
	100,000	100,000	2,716.40	2,759.93	2	76.13	76.13	316
	200,000	200,000	10,749.18	10,835.97	2	151.33	151.33	447
	500,000	500,000	66,419.60	66,637.23	2	448.61	448.61	707
	1,000,000	1,000,000	257,634.32	258,072.30	2	1,196.48	1,196.48	1,000

The power of the Framework over BH can transparently be seen when the number of DMUs is 1,000,000. In this case, the elapsed time for BH in Stage 1 is 257,634.32 s (3 days) in comparison with 1,196.48 s (19.94 min) for the Framework. This shows 99.54% decrease in the elapsed time of BH using a single computer.

1.5 Uniform and Cobb-Douglas Approaches

We now use two different ways to generate data and compare the results to illustrate that different ways of generating datasets do not affect the performance of the Framework versus BH. In other words, the pros of the Framework versus BH are independent of the way of generating DMUs. When data are generated by the Cobb-Douglas approach (in comparison with a uniform approach), the output is positively correlated to inputs, and we expect a higher number of VRS efficient DMUs in the samples. We only consider a few samples for each case, because the standard deviation of elapsed time is negligible and the differences between the methods are notable and obvious.

1.5.1 Generating Data

We generate 30 random samples by a uniform approach. For each sample, we consider 1,000, 5,000, 10,000, 15,000, 25,000 and 50,000 cardinalities and 1 + 1 to 10 + 1 dimensions. We also generate 30 random samples by the Cobb-Douglas approach with the same cardinalities and dimensions for each sample.

For example, when the number of inputs is 3, we generate 30 samples of size n with a uniform approach by randomly generating 3 + 1 numbers between [10,

20] for each DMU. To generate samples using the Cobb-Douglas approach, for each sample, we first uniformly generate 1–3 values in [0, 1], β_i, in which the sum of these values is between 0.4 and 0.9, to satisfy the VRS condition for the Cobb-Douglas function. We then uniformly generate 1–3 input values in [10, 20], x_{ij}, corresponding to the numbers of β_i, for $j = 1, 2, \ldots, n$. For each j, we calculate $\prod_i x_{ij}^{\beta_i}$ to find the output value, called the true output value. We then uniformly generate a value in [0, 0.2], σ^2, calculate the exponential of the inverse of the half-normal distribution $\left|N\left(0, \sigma^2\right)\right|$, and call it the true inefficiency score for that DMU. Finally, we divide the true output value over the corresponding true inefficiency score for that DMU and find the corresponding output value, y_i, for x_{ij}. It is shown by Banker and Chang (2006) that, after generating the true inefficiencies (using exponential-half-normal for samples with 2 + 1 dimensions and cardinalities 50–150), the linear correlation between the measured scores by the radial VRS model and the true inefficiency scores is about 92%. For this experiment, we select the size of blocks using $\min\left(\left\lfloor\sqrt{(m+s)n}\right\rfloor, \lfloor n/2 \rfloor\right)$, and $\beta = 1.5$ and $\gamma = 0.9$.

1.5.2 The Outcome

In this section, we only show the results when the number of inputs is 3 and the cardinality is 10,000. The same results are observed when the number of inputs is changed from 1 to 10.

As discussed, we compare HD, BH and the Framework to show that, for the purpose of our study, using the uniform distribution to generate data provides results equivalent to those of using a parametric approach to generate data or using a real-life dataset. In other words, regardless of the possible correlation between inputs and outputs, the Framework performs faster than BH using a single computer with a small number of 8 or 16 processors. Obviously, we can simply increase the number of computers in parallel or increase the number of processors of the CPUs and have a better result than BH.

Moreover, as noted above, the Framework is implemented similar to how HD is implemented, and any improvement for HD can be applied to the Framework, as well. The only difference between HD and the Framework is in the first iteration of the HD. In other words, the Framework uses one representative block and evaluates all other blocks based upon the selected subsample. Since the elapsed time to select a subsample is very negligible, we clearly expect that the elapsed time of HD is always less than or equal to that of the Framework.

Figures 1.3 and 1.4 show the differences between elapsed times of the Framework and HD when data is generated using the Cobb-Douglas approach versus when data is generated using a uniform approach. As dimension and cardinality increases, the advantages of the Framework versus HD are clearer.

Figures 1.5 and 1.6 also represent the comparison between the elapsed times of the Framework and BH. As can be seen, the Framework performs faster than BH, and

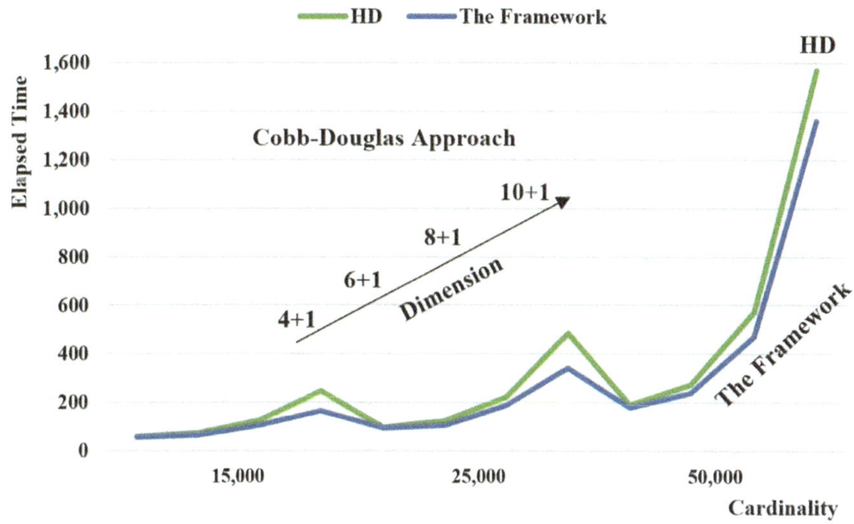

Fig. 1.3 Comparing the framework and HD, using the Cobb-Douglas approach, using CPU i7

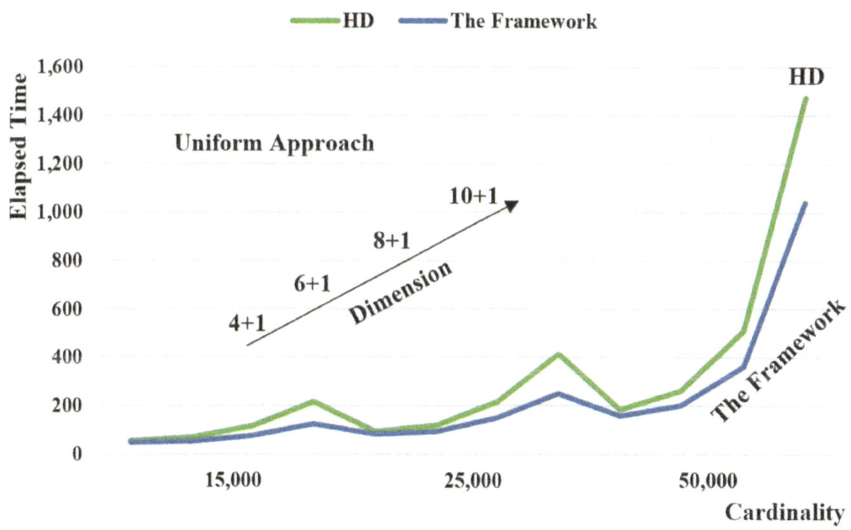

Fig. 1.4 Comparing the framework and HD, using a uniform approach, using CPU i7

when cardinality (dimension) increases, BH becomes weaker in comparison with the Framework, regardless of the way of data generation.

From the figures, it can be seen that the average elapsed times of the methods in the Cobb-Douglas approach is more than that when a uniform approach is considered. Nevertheless, using a single computer, the Framework on the uniform generated

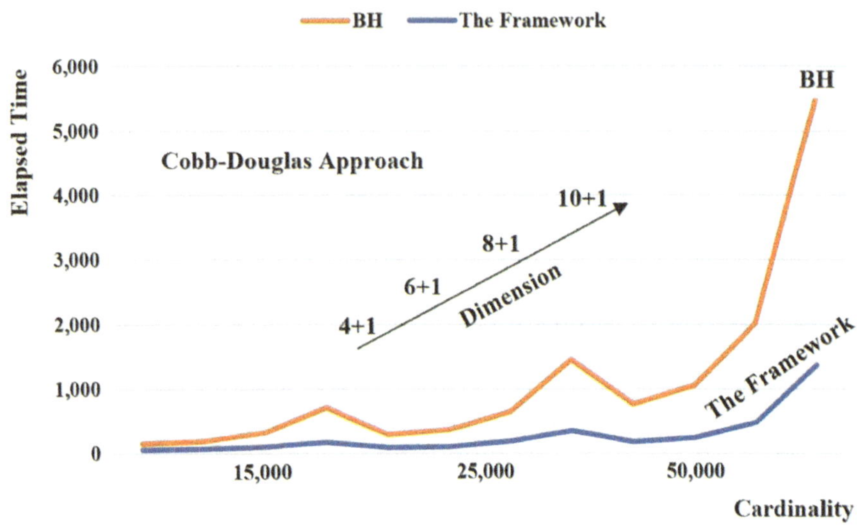

Fig. 1.5 Comparing the framework and BH, using the Cobb-Douglas approach, using CPU i7

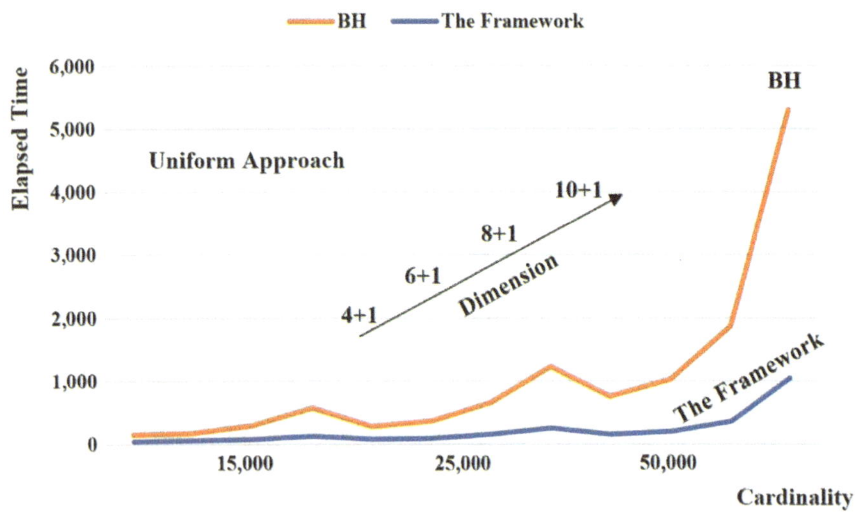

Fig. 1.6 Comparing the framework and BH, using a uniform approach, using CPU i7

dataset decreases the average elapsed times of BH by 79%, and on the Cobb-Douglas generated dataset, decreases the average elapsed times of BH by 74%.

We note that in each experiment, the Framework in the worst situation (considering its maximum elapsed times) still performs better than the best situations of BH (considering its minimum elapsed times). For all of the other cardinalities and dimensions, the same results are concluded, that is, the maximum elapsed time of the Framework is smaller than the minimum elapsed time of BH using a CPU i7.

Figure 1.7 also shows the differences between the elapsed times of BH and the Framework on both datasets, when the elapsed times of 30 samples are sorted from the least to the most elapsed times of the Framework. As can be seen, BH cannot perform better than the Framework whether data is generated by a uniform distribution or by a parametric approach.

When cardinality is 50,000 and dimensions are 10 + 1 with 10% density using CPU i7, the Framework completes Stage 1 in 624.62 s with Total of 1,359.5 s, whereas BH completes Stage 1 in 4,737.5 s with Total of 5,476.4 s. This means the Framework decreases the elapsed time of BH by 75%. Clearly, the difference between the elapsed time of the Framework and BH is remarkable, and the Framework dominates BH on both datasets.

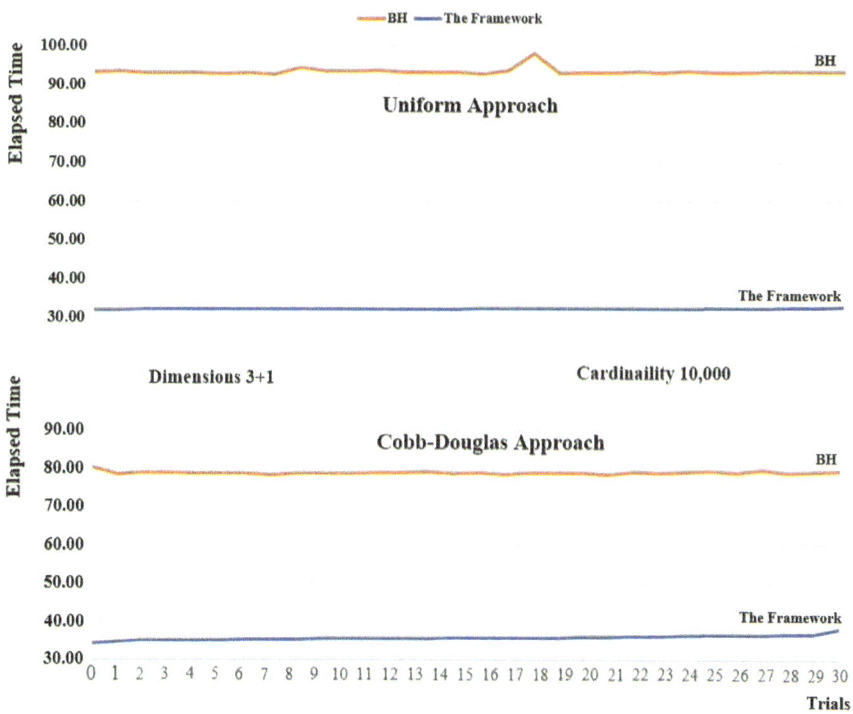

Fig. 1.7 The elapsed times of BH and the framework for 30 samples, using CPU i7

1.6 Changing the Cardinality and Dimension

1.6.1 Generating Data

We now compare HD, BH and the Framework to illustrate the effects of dimension and cardinality on elapsed times. We generate random samples by a uniform approach, when cardinalities are 1,000, 5,000, 10,000, 15,000, 25,000 and 50,000, and dimensions are $1 + 1$ to $10 + 10$, using the computer with CPU i7. We then use a computer with CPU i9 and compare the elapsed time when cardinalities are 10,000, 50,000, 100,000, 200,000, 500,000 and 1,000,000 and dimensions are $1 + 1$, $2 + 2$ and $5 + 5$.

1.6.2 Selecting the Parameters

Khezrimotlagh et al. (2019) used the size of blocks as $\lfloor\sqrt{n}\rfloor$; we select the size of blocks by the formula $\min\left(\lfloor\sqrt{(m+s)n}\rfloor, \lfloor n/2\rfloor\right)$ and consider $\beta = 1.5$ and $\gamma = 0.9$. It is already shown by Dulá (2011) that as dimension (and cardinality) increases, HD becomes weaker in comparison with BH. Dulá (2011) selects $b = 512$, $\beta = 1.5$ and $\gamma = 0.9$ to run HD. We do not suggest selecting a constant value for b (the size of blocks). We show that when the size of blocks is selected based upon the number of DMUs and the numbers of inputs and outputs, that is, $b = \min\left(\lfloor\sqrt{(m+s)n}\rfloor, \lfloor n/2\rfloor\right)$, HD performs much faster than BH in most cases. In other words, it is clear that HD performs well when cardinality is small, or dimension is small. We show that selecting an appropriate size of blocks as well as the parameters β and γ can decrease the elapsed times of HD substantially in comparison with BH. Of course, when dimensions increase, the power of HD in comparison with BH is descending, as is shown in this section. In contrast with HD's performance, we show that increasing the dimension or cardinality does not affect the performance of the Framework versus BH, using a single computer.

We note that both HD and the Framework can perform much faster than BH using several computers in parallel, however, here only a single computer is used. The Framework (similar to HD) allows us to use all of the power of the computer, whereas BH solves one problem at a time. As a result, when HD is implemented correctly and all of the power of the computer is used, HD can perform much faster than BH. Although, in these experiments, we select the size of blocks using $\min\left(\lfloor\sqrt{(m+s)n}\rfloor, \lfloor n/2\rfloor\right)$ for both HD and the Framework, the same results can also be concluded for the Framework versus BH when the size of blocks is selected using $\lfloor\sqrt{n}\rfloor$ (see Khezrimotlagh et al. 2019).

1.6.3 Outcomes Using CPU I7

Figure 1.8 illustrates the elapsed times of HD, BH and the Framework in Stage 1 (the time for finding efficient DMUs), when dimensions are 1 + 1 to 10 + 10 and cardinality is 50,000. The elapsed times for HD when dimensions are less than 8 + 8 are substantially less than that of BH. However, HD performs weaker than BH when dimensions are 9 + 9 and 10 + 10. This is due to the fact that HD builds the hull using 36,749 and 40,605 DMUs when dimensions are 9 + 9 and 10 + 10, respectively (to find the efficient DMUs). In contrast, the Framework builds the hull with 18,282 and 22,759 DMUs for dimensions 9 + 9 and 10 + 10, respectively.

Tables 1.6 and 1.7 show the pros of the Framework versus BH and HD as well as the pros of HD (using parallel processing) versus BH (solving one LP at a time).

Table 1.6 illustrates the differences between the elapsed times to apply the Framework, HD and BH. The first column represents the dimensions followed by the cardinalities and the numbers of efficient DMUs in the samples. For each method, the elapsed times in Stage 1 and Total as well as the required number of DMUs to

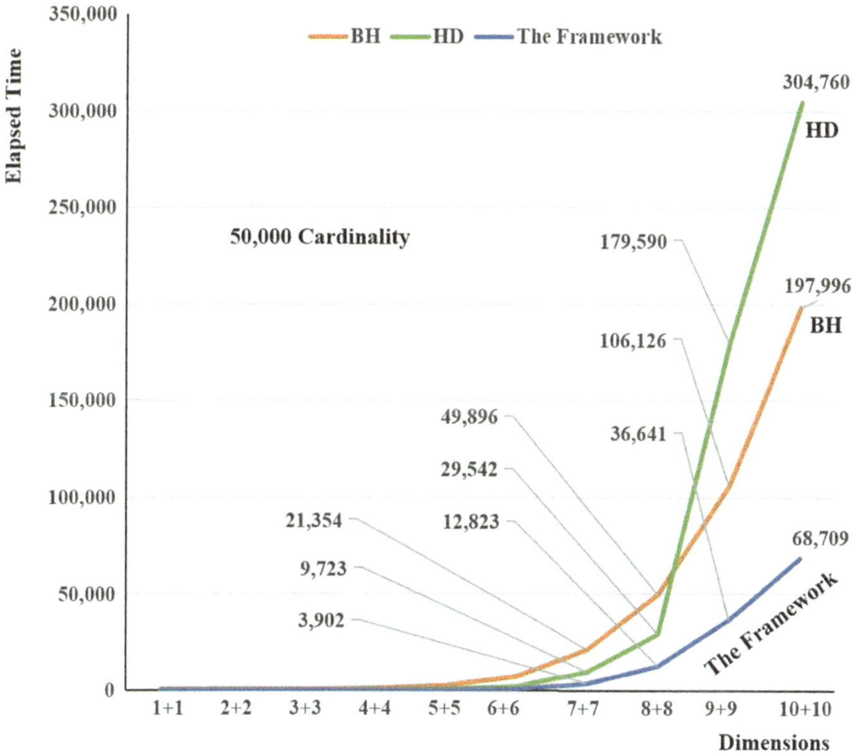

Fig. 1.8 The elapsed times in Stage 1 when dimension is changed, using CPU i7

Table 1.6 Comparing the elapsed times when dimension is 1 + 1 to 5 + 5, using CPU i7

Dim.	Card.	Eff.	BH			HD			The framework		
			S.1	Total	R#DB	S.1	Total	R#DB	S.1	Total	R#DB
1 + 1	1,000	5	7.02	8.50	5	2.65	4.16	5	2.41	3.92	5
	5,000	6	36.35	43.74	6	10.31	17.75	6	9.67	9.67	6
	10,000	7	75.62	90.39	6	18.85	33.80	7	18.03	18.03	6
	15,000	10	117.41	139.64	8	28.06	50.23	10	25.96	25.96	8
	25,000	11	209.34	246.41	8	45.20	82.03	12	42.71	79.74	8
	50,000	14	537.36	611.77	10	89.15	162.59	13	81.25	154.76	10
2 + 2	1,000	30	7.44	8.91	30	2.94	4.47	30	2.18	3.64	30
	5,000	49	37.43	44.90	47	10.63	18.10	49	8.83	16.28	48
	10,000	66	77.72	92.76	64	20.09	35.23	67	16.88	31.91	67
	15,000	65	121.30	143.70	63	29.91	52.63	65	24.95	47.38	65
	25,000	89	227.91	266.06	87	50.00	88.05	89	40.95	79.27	88
	50,000	95	626.20	702.71	90	102.72	178.52	95	80.43	156.06	92
3 + 3	1,000	111	8.46	9.87	110	3.83	5.24	111	2.30	3.68	111
	5,000	196	40.44	48.23	191	13.39	21.23	196	9.18	16.87	195
	10,000	248	85.34	101.64	244	25.24	41.61	248	17.62	33.64	248
	15,000	258	132.89	157.59	256	37.02	61.80	258	25.71	50.26	256
	25,000	405	276.26	320.62	395	63.24	108.45	405	43.11	88.06	405
	50,000	484	757.10	852.88	466	132.62	230.11	484	84.98	180.57	479
4 + 4	1,000	209	9.38	10.72	209	5.07	6.40	209	2.46	3.74	209
	5,000	509	50.44	59.78	507	21.01	30.27	509	10.37	19.49	509

(continued)

Table 1.6 (continued)

Dim.	Card.	Eff.	BH			HD			The framework		
			S.1	Total	R#DB	S.1	Total	R#DB	S.1	Total	R#DB
	10,000	629	105.96	127.23	624	38.11	59.37	629	19.90	41.21	626
	15,000	822	186.05	224.63	816	59.48	97.83	822	31.28	69.01	822
	25,000	1,015	394.73	472.85	1,008	102.23	181.20	1,015	52.27	129.22	1,014
	50,000	1,324	1,127.4	1,335.9	1,305	224.70	432.91	1,324	107.03	316.55	1,322
5 + 5	1,000	346	11.15	12.40	346	6.78	8.00	361	2.70	3.89	346
	5,000	906	71.17	84.03	903	34.02	46.80	906	13.58	25.59	906
	10,000	1,213	166.17	203.07	1,206	66.60	103.67	1,213	27.62	63.44	1,212
	15,000	1,720	358.91	445.76	1,709	131.00	218.42	1,720	51.09	134.50	1,718
	25,000	2,320	863.06	1,096.3	2,311	261.89	494.56	2,320	107.73	332.36	2,319
	50,000	2,926	2,443.8	3,140.8	2,888	610.02	1,323.10	2,926	218.64	910.82	2,921

Table 1.7 Comparing the elapsed times when dimension is 6 + 6 to 10 + 10, using CPU i7

Dim.	Card.	Eff.	BH			HD			The framework		
			S.1	Total	R#DB	S.1	Total	R#DB	S.1	Total	R#DB
6 + 6	1,000	474	13.49	14.65	473	8.86	10.00	487	3.15	4.21	474
	5,000	1,473	123.63	143.77	1,471	66.66	86.82	1,473	22.61	41.14	1,473
	10,000	2,057	327.53	400.14	2,052	156.19	228.66	2,057	49.99	119.55	2,057
	15,000	2,714	718.44	893.66	2,703	293.49	470.38	2,714	97.64	266.02	2,712
	25,000	3,750	1,904.26	2,436.83	3,734	737.69	1,275.00	3,750	244.94	754.34	3,750
	50,000	5,554	7,037.53	9,325.95	5,536	2,263.58	4,566.86	5,554	824.64	3,048.77	5,550
7 + 7	1,000	607	16.63	17.68	606	6.01	6.57	793	3.77	4.70	607
	5,000	2,065	218.80	248.45	2,064	79.93	102.94	2,743	40.72	66.90	2,065
	10,000	3,198	756.96	898.81	3,192	413.12	555.40	3,198	126.15	256.22	3,198
	15,000	4,321	1,817.01	2,199.13	4,312	896.25	1,279.02	4,321	305.49	652.58	4,321
	25,000	5,791	4,610.20	5,801.59	5,766	2,079.47	3,284.51	5,791	688.17	1,815.95	5,787
	50,000	9,384	21,354	28,204	9,362	9,723	16,629	9,385	3,901.54	10,317	9,384
8 + 8	1,000	723	20.45	21.34	722	5.41	5.69	920	4.38	5.15	723
	5,000	2,540	333.19	370.81	2,539	140.89	158.04	3,865	62.18	95.09	2,540
	10,000	4,286	1,443.99	1,663.01	4,277	687.30	812.80	6,734	270.64	462.87	4,286
	15,000	5,632	3,306.78	3,911.61	5,622	2,040.24	2,390.88	9,591	627.37	1,172.30	5,631
	25,000	8,182	10,383	12,782	8,164	6,133.33	8,516.77	8,182	2,130.51	4,279.25	8,181
	50,000	13,475	49,896	66,267	13,448	29,542	45,884	13,475	12,823	27,759	13,474
9 + 9	1,000	796	23.19	23.92	796	5.78	6.00	949	4.85	5.49	796
	5,000	3,121	535.19	578.47	3,120	215.23	226.07	4,530	98.97	136.02	3,121

(continued)

Table 1.7 (continued)

Dim.	Card.	Eff.	BH			HD			The framework		
			S.1	Total	R#DB	S.1	Total	R#DB	S.1	Total	R#DB
	10,000	5,396	2,538.95	2,833.30	5,393	1,371.38	1,488.26	8,195	526.85	777.89	5,396
	15,000	7,364	6,470.02	7,429.71	7,350	4,759.64	5,138.56	12,002	1,403.14	2,257.65	7,364
	25,000	10,714	20,191	24,466	10,694	22,774	24,479	19,328	5,394.19	9,152.77	10,714
	50,000	18,281	106,126	138,876	18,230	179,590	193,345	36,749	36,641	65,270	18,282
10 + 10	1,000	874	26.76	27.30	874	6.23	6.37	973	5.66	6.09	874
	5,000	3,467	698.05	745.07	3,464	267.70	276.87	4,703	137.38	176.90	3,467
	10,000	6,286	3,781.34	4,161.10	6,278	2,216.21	2,299.45	9,190	862.39	1,186.40	6,286
	15,000	8,829	10,423	11,785	8,821	8,031.18	8,334.53	13,625	2,671.44	3,840.67	8,829
	25,000	13,590	38,366	45,007	13,584	33,410	35,650	21,171	12,350	17,868	13,590
	50,000	22,759	197,996	248,399	22,732	304,760	322,159	40,605	68,709	112,159	22,759

build the hull (R#DB) are illustrated. Note that we eliminate the decimals when the measured elapsed times are large, so data can fit into the table.

In this situation, the larger the cardinality, the smaller the density; and the larger the dimension, the larger the density. As shown in Table 1.6, the result summarizes that regardless of the cardinality, dimension or density, the Framework performs faster than both HD and BH.

As can be seen, using CPU i7, the Framework at least decreases the elapsed times of BH by 53–81.4%.

Correspondingly, when cardinality is 50,000 with 5 + 5 dimensions and the density 6%, the Framework decreases the elapsed times of BH by 71%. Similarly, the Framework decreases the elapsed times of HD from 2.8 to 66% on this dataset, using CPU i7. For instance, when cardinality is 50,000 with 5 + 5 dimensions and the density 6%, the Framework decreases the elapsed times of HD by 31%.

Figures 1.9 and 1.10 also show the comparison between the total elapsed times when the dimension is fixed to 5 + 5 and 9 + 9 and the cardinality is changed from 1000 to 50,000. Similarly, Figs. 1.11 and 1.12 illustrate the differences between the total elapsed times when cardinality is 50,000 and dimension is changed from 1 + 1 to 10 + 10. As can be seen, the Framework always performs faster than HD and BH. Figure 1.13 also indicates that as the dimension increases, HD becomes weaker in

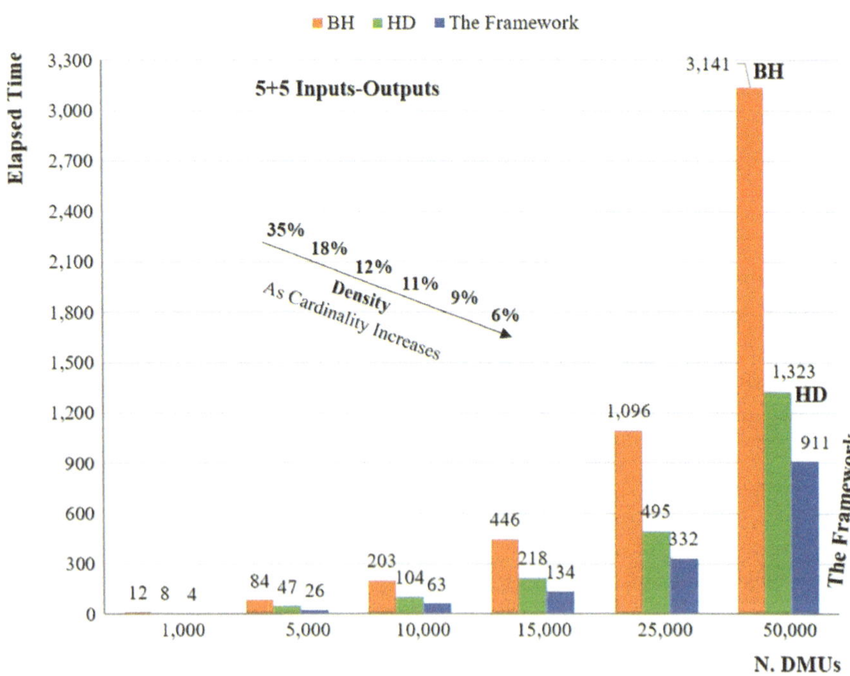

Fig. 1.9 The total elapsed times when dimension is 5 + 5, using CPU i7

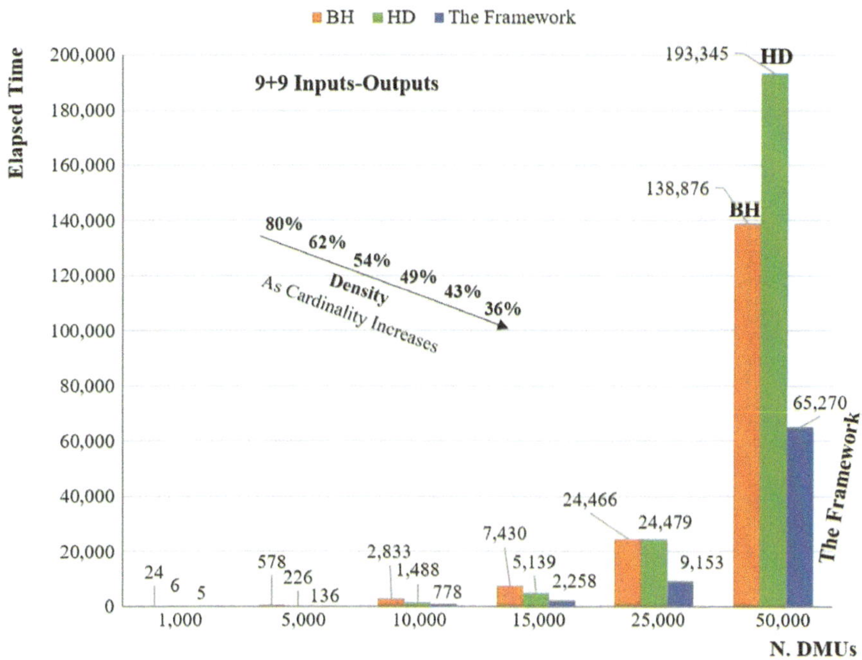

Fig. 1.10 The total elapsed times when dimension is 9 + 9, using CPU i7

comparison with the Framework. It also shows that when dimension and cardinality are small enough, HD always performs faster than BH.

Figures 1.9 to 1.12 also illustrate that when dimensions increase from 1 + 1 to 10 + 10, the densities increase and when cardinalities increase from 1,000 to 50,000, the densities decrease. Regardless of the change in cardinalities or dimensions, the Framework performs better than both HD and BH. We emphasize that these results are measured when a single computer is used, and if we use two computers (or more) in parallel, the existing methods are not comparable with the Framework. We also note that these results are recorded when only the first iteration of HD is improved, that is, Step 3 is used once in the Framework.

For instance, when dimensions are 6 + 6 and the cardinality is 50,000 with 11.1% density, the elapsed time of the Framework in Stage 1 is 824.64 s with Total of 3,048.77 s, whereas that of BH is 7,037.53 s in Stage 1 with Total of 9,325.95 s. This means that the Framework decreases the total elapsed time of BH by 67.3%.

Similarly, when dimension is 9 + 9 with 25,000 cardinality, the elapsed time of the Framework in Stage 1 is 5,394.19 s and Total is 9,152.77 s (2.5 h), whereas those of BH is 20,191 and 24,466 (6.8 h), respectively. In this case, the Framework decreases the total elapsed time of BH by 63%, where density is 43%.

In addition, when dimension is 10 + 10 with 50,000 cardinalities, the elapsed time of the Framework in Stage 1 is 68,709 s and Total is 112,159 s (31.15 h or 1.3 days), whereas those of BH is 197,996 and 248,399 (69 h or almost 3 days), respectively.

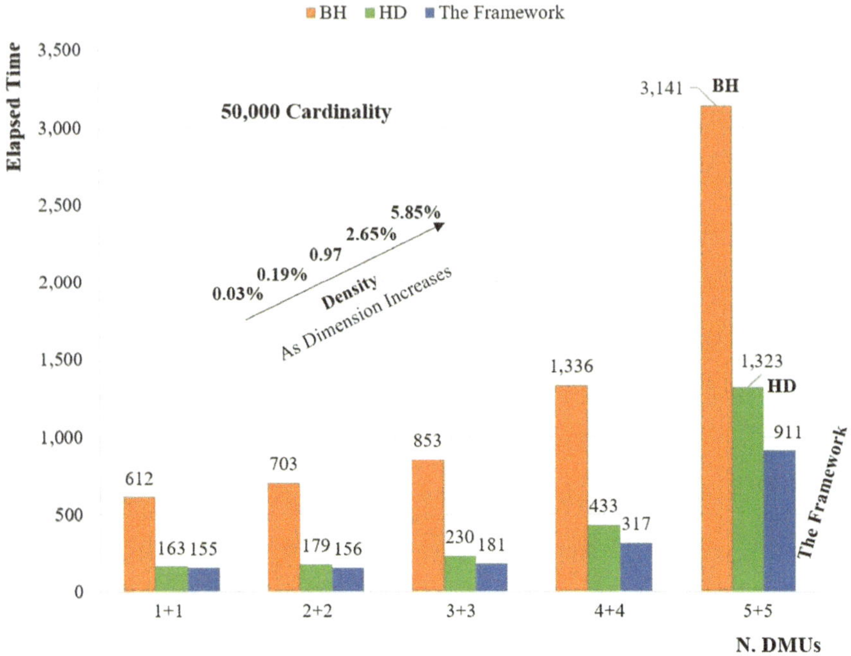

Fig. 1.11 The total elapsed times when dimension is changed from 1 + 1 to 5 + 5, using CPU i7

In this case, the Framework decreases the total elapsed time of BH by 55%, where density is 46%, using a single computer with CPU i7.

1.6.4 The Performance of HD

It is clear that as density in each block increases, HD becomes weaker in comparison with BH. For example, as shown in Table 1.7, when the dimension is 9 + 9 and the cardinality is 25,000, HD builds the hull by 19,328 DMUs to find the efficient DMUs, whereas the Framework only uses 10,714 DMUs to build the hull and finds the efficient DMUs (BH uses 10,694 DMUs). Similarly, this can be seen when cardinality is 50,000 and dimension is 10 + 10. This is due to the fact that HD uses the maximum number of DMUs to build the hull in comparison with the Framework and BH. As discussed, the Framework can be run by selecting a subsample of size $\lfloor\sqrt{n}\rfloor$ and still be as powerful as when a subsample of size $\min\left(\lfloor\sqrt{(m+s)n}\rfloor, \lfloor n/2\rfloor\right)$ is selected; however, this is not true for HD.

When dimensions are 9 + 9 and the cardinality is 25,000, the size of each block in the first iteration of HD is 671. Therefore, HD divides the 25,000 DMUs into 37 distinct blocks (each includes 671 DMUs) and a block of 173 DMUs. On average, the

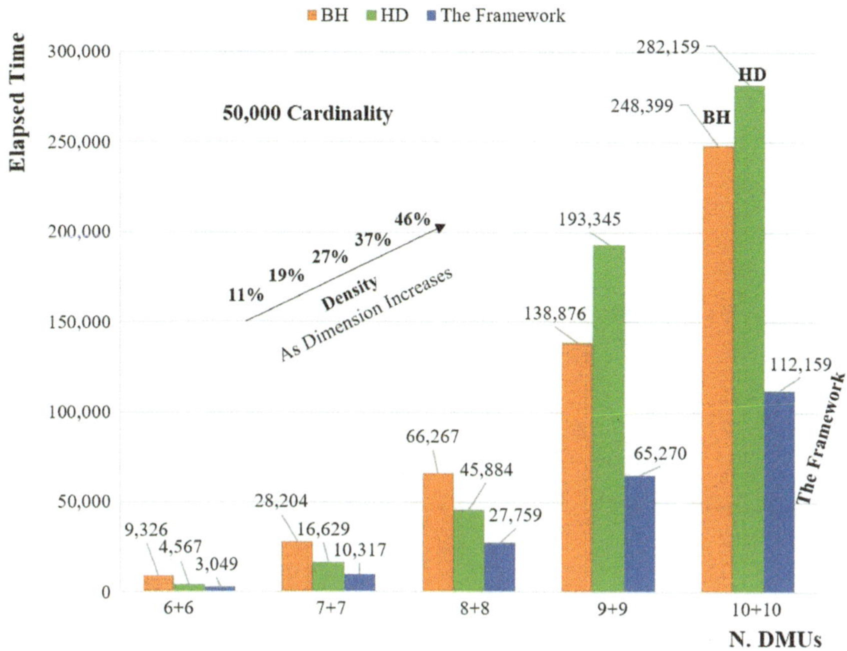

Fig. 1.12 The total elapsed times when dimension is changed from 6 + 6 to 10 + 10, using CPU i7

number of best-practice DMUs in each block is 562 DMUs, and the DMUs in the last block are also introduced as the best-practice DMUs. As a result, there are 20,955 DMUs introduced as best-practice DMUs. Since 20,955/25,000 is about 0.84, HD increases the size of blocks to 1,006 (1.5×671) and divides the best-practice DMUs into 20 blocks (each contains 1,006 best-practice DMUs) and a block of size 835. Now, the average number of best-practice DMUs in each block is 928, and in the last block there are 768 best-practice DMUs. Therefore, there are 19,328 best-practice DMUs in the set of 20,955 DMUs. Since 19,328/20,955 is about 0.92, HD stops re-blocking the best-practice DMUs and aggregates all the found best-practice DMUs into one block; that is, HD builds the hull with 19,328 DMUs. In this situation, if we select the parameter beta and gamma as $\beta = 1.5$ and $\gamma = 0.95$, we can decrease the elapsed times of HD by 20%, that is, the total elapsed time of HD in this case decreases to 19,665 s (vs. 24,466 for BH), as HD builds the hull by 11,024 DMUs (after 8 times re-blocking best practices). As discussed, the Framework finishes the task in 9,152.77 s only (in comparison with 19,665 s for HD and 24,466 for BH). We note that increasing the value of the parameter γ may substantially increase the elapsed times of HD in comparison with that of BH.

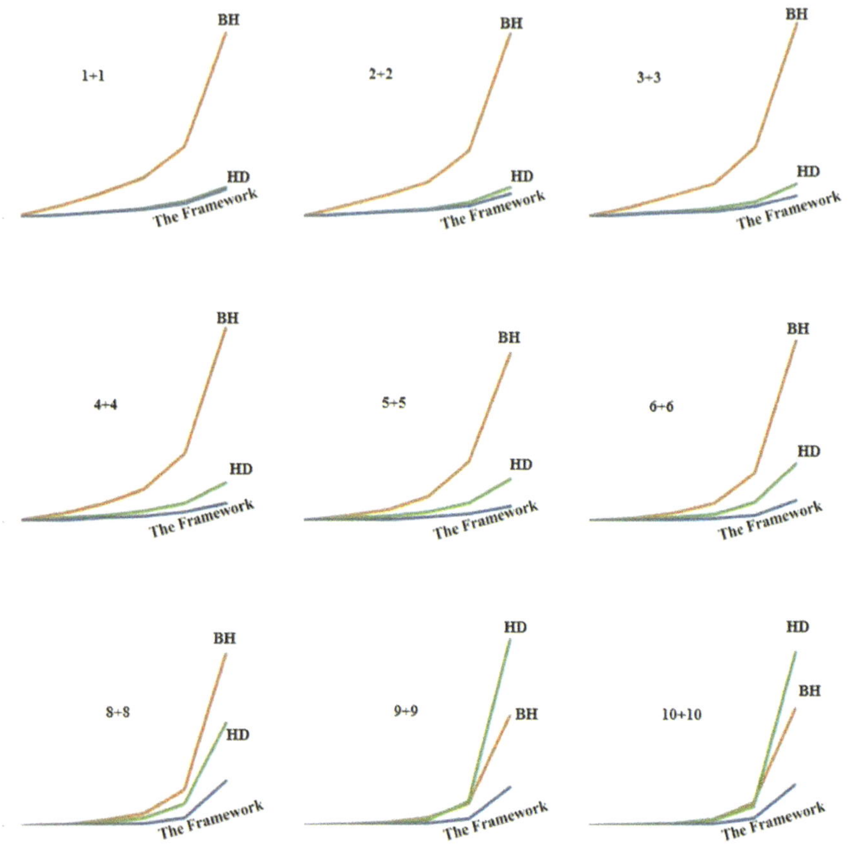

Fig. 1.13 The total elapsed times when cardinality is changed from 1,000 to 50,000, using CPU i7

1.6.5 Outcomes Using CPU I9

We now use a single computer with CPU i9 to illustrate the differences in elapsed time of BH and the Framework. For the first time in the literature, we consider the cardinalities 200,000, 500,000 and 1,000,000 when dimensions are $1 + 1$, $2 + 2$ and $5 + 5$.

Table 1.8 represents the elapsed times of BH and the Framework, using CPU i9, when cardinality is changed up to 1,000,000 DMUs. The last two columns of Table 1.8 illustrate how much the BH's elapsed times are decreased when the Framework is used.

As shown in Table 1.6, using CPU i7, the Framework decreases the elapsed time of BH by 80%, when the cardinality is 10,000 and dimension is $1 + 1$. Almost the same result is concluded when CPU i9 is used. In other words, 80.6% of the BH's elapsed time is decreased using the Framework and CPU i9. In contrast, when cardinality is 50,000 (and dimension $1 + 1$), using CPU i7, the Framework decreases

Table 1.8 Comparing the elapsed times of BH and the framework, using CPU i9

Dim.	Card.	Eff.	BH			The framework			Elapsed time decrease	
			S.1	Total	R#DB	S.1	Total	R#DB	S.1 (%)	Total (%)
1 + 1	10,000	6	62	67	6	8	13	6	87.1	80.6
	50,000	13	456	478	9	35	56	9	92.3	88.3
	100,000	25	1,384	1427	8	59	59	8	95.7	95.9
	200,000	29	5,217	5,303	9	109	196	9	97.9	96.3
	500,000	59	30,867	31,101	10	256	475	14	99.2	98.5
	1,000,000	122	124,422	124,855	11	528	966	19	99.6	99.2
2 + 2	10,000	64	66	70	58	9	14	61	86.4	80.0
	50,000	97	556	578	82	31	54	95	94.4	90.7
	100,000	128	1,752	1,797	107	56	101	109	96.8	94.4
	200,000	180	6,592	6,692	130	104	201	140	98.4	97.0
	500,000	226	39,802	40,032	127	245	476	141	99.4	98.8
	1,000,000	335	159,023	159,482	136	504	969	139	99.7	99.4
5 + 5	10,000	1,293	158.66	169.60	1,289	10.23	20.81	1,292	93.6	87.7
	50,000	2,877	2,164	2,343	2,861	65	240	2,875	97.0	89.8
	100,000	4,248	7,925	8,630	4,196	178	882	4244	97.8	89.8
	200,000	5,502	26,515	28,781	5,396	431	2,726	5481	98.4	90.5
	500,000	8,931	159,977	174,308	8,666	1,792	16,722	8891	98.9	90.4
	1,000,000	11,061	529,267	572,393	10,499	3,963	50,399	10,894	99.3	91.2

the elapsed time of BH by 75% versus 88.3% when CPU i9 is used. As a result, when cardinality (and dimension) is large, the elapsed time is also affected substantially using different CPUs.

When the cardinality is 1,000,000 and the dimension is 1 + 1, the Framework decreases the elapsed time of BH (using CPU i9) by 99.2%. In this case, BH uses 34.56 h (almost one and half a day) to find 11 efficient DMUs. In contrast, the Framework calculates the efficiency scores of all DMUs in 8 min. Similarly, when the cardinality is 1,000,000 and the dimension is 5 + 5, BH uses almost 7 days (one week) to calculate the efficiency scores of DMUs, whereas the Framework finishes the task in half a day only. In this case, the Framework decreases the elapsed time of BH in Stage 1 by 99%. Now, it is clear how fast the Framework can finish the task when two computers are used in parallel.

1.7 A Real Dataset

We now use the real dataset of Khezrimotlagh et al. (2019), including 30,099 DMUs with 2 + 4 dimensions to compare HD, BH and the Framework. They selected the size of the subsample using $\lfloor \sqrt{n} \rfloor$, that is, 173. Here, we select the size of the subsample (blocks) using $\min(\lfloor \sqrt{(m+s)n} \rfloor, \lfloor n/2 \rfloor)$, which is 425. We also select $\beta = 1.5$ and $\gamma = 0.90$ for both HD and the Framework.

There are 2,450 VRS efficient DMUs in this dataset. BH uses only 84 DMUs to build the hull and measures the efficiency of DMUs with a linear programming problem of size 85 (84 lambdas and 1 alpha). However, HD uses 2,545 DMUs to build the hull. In contrast, the Framework builds the hull with 457 DMUs. Here the density for BH is 0.28%, whereas it is 8.46% for HD and 1.52% for the Framework. The actual density is 8.14%.

Tables 1.9 and 1.10 illustrate the elapsed times of the Framework versus HD and BH, using CPU i7 and CPU i9, respectively.

Table 1.9 Comparing the elapsed times, using CPU i7

Outcome	BH	HD	The framework
Elapsed time in Stage 1	402.78	104.71	62.45
Elapsed time in total	448.98	367.25	122.79
R#DB	84	2545	457

Table 1.10 Comparing the elapsed times, using CPU i9

Outcome	BH	HD	The framework
Elapsed time in Stage 1	340.57	34.66	22.60
Elapsed time in total	354.34	103.48	40.51
R#DB	84	2545	457

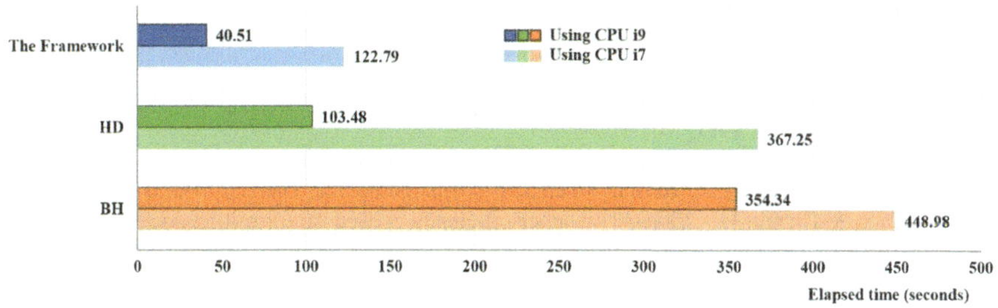

Fig. 1.14 Comparing the elapsed times using a single computer with different CPUs

Although BH only uses 84 DMUs to build the hull; BH finds these 84 DMUs in 402.78 s, while HD and the Framework complete the total task in 367.25 and 122.79 s, respectively. In other words, BH can only solve one problem at a time in Stage 1. In contrast, the Framework using a single computer with CPU i7 decreases the running time of BH by 71.10% in this example. Both stages of the Framework and HD are completed before BH finds the 84 efficient DMUs to build the hull.

Now, if we use the computer with CPU i9, the Framework decreases the elapsed time of BH by 93.36% in the first stage. HD also decreases the elapsed time of BH by 89.82% in Stage 1. In the total elapsed time, HD decreases the elapsed time of BH by 70.8%, whereas the Framework decreases the elapsed time of BH by 88.57%.

Note that, although HD uses parallel processing, it builds the hull with 2,545 DMUs, using more DMUs than the number of available efficient DMUs in the dataset to build the hull.

Nevertheless, as shown in Fig. 1.14, if we use a computer with CPU i9 instead of CPU i7, the elapsed times of HD sharply decrease from 367.25 to 103.48, that is, 72% decreasing the elapsed time of HD just by using a more powerful CPU. Now, if we use two computers in parallel, the disadvantage of BH can be completely transparent in comparison with both HD and the Framework.

1.8 Conclusion

We provided several examples to clarify the steps of applying the proposed framework by Khezrimotlagh et al. (2019), called "the Framework" in this study. The examples clearly show the superiority of the Framework to the existing methodologies when dealing with large-scale DEA problems. The elapsed times of the existing methodologies are decreased up to 100% when the Framework is used, using a single computer. The Framework is easily applied without computational complexity. It uses all power of a used computer and it is the framework of choice to handle large-scale DEA.

References

Ali, A. I. (1993). Streamlined computation for data envelopment analysis. *European Journal of Operational Research, 64*(1), 61–67.

Banker, R. D., Charnes, A., & Cooper, W. W. (1984). Some models for estimating technical and scale inefficiencies in data envelopment analysis. *Management Science, 30*(9), 1078–1092.

Banker, R. D., & Chang, H. (2006). The super-efficiency procedure for outlier identification, not for ranking efficient units. *European Journal of Operational Research, 175*(2), 1311–1320.

Barr, R. S., & Durchholz, M. L. (1997). Parallel and hierarchical decomposition approaches for solving large-scale data envelopment analysis models. *Annals of Operations Research, 73*, 339–372.

Charnes, A., Cooper, W. W., & Rhodes, E. (1978). Measuring the inefficiency of decision making units. *European Journal of Operational Research, 2*(6), 429–444.

Chen, W. C., & Cho, W. J. (2009). A procedure for large-scale DEA computations. *Computers & Operations Research, 36*(6), 1813–1824.

Chen, W. C., & Lai, S. Y. (2017). Determining radial efficiency with a large data set by solving small-size linear programs. *Annals of Operations Research, 250*(1), 147–166.

Dulá, J. H., & Helgason, R. V. (1996). A new procedure for identifying the frame of the convex hull of a finite collection of points in multidimensional space. *European Journal of Operational Research, 92*(2), 352–367.

Dulá, J. H., & Thrall, R. M. (2001). A computational framework for accelerating DEA. *Journal of Productivity Analysis, 16*(1), 63–78.

Dulá, J. H. (2008). A computational study of DEA with massive data sets. *Computers & Operations Research, 35*(4), 1191–1203.

Dulá, J. H. (2011). A method for data envelopment analysis. *INFORMS Journal on Computing, 23*(2), 284–296.

Dulá, J. H., & López, F. J. (2013). DEA with streaming data. *Omega, 41*(1), 41–47.

Khezrimotlagh, D., Zhu, J., Cook, W., & Toloo, M. (2019). Data Envelopment Analysis and Big Data. *European Journal of Operational Research, 274*(3), 1047–1054.

Zhu, Q., Wu, J., & Song, M. (2018). Efficiency evaluation based on data envelopment analysis in the big data context. *Computers & Operations Research, 98*, 291–300.

Chapter 2
Data Envelopment Analysis (DEA): Algorithms, Computations, and Geometry

José H. Dulá

Abstract Data Envelopment Analysis (DEA) has matured but remains vibrant and relevant, in part, because its algorithms, computational experience, and geometry have a broad impact within and beyond the field. Algorithmic, computational, and geometric results in DEA allow us to solve larger problems faster; they also contribute to various other fields including computational geometry, statistics, and machine learning. This chapter reviews these topics from a historical viewpoint, as they currently stand, and as to how they will evolve in the future.

2.1 Introduction

DEA is an operationalization of the problem of measuring, comparing, and making recommendations about the relative efficiency of a transformation process of many, functionally similar, entities or "Decision Making Units", using only measurements in a common list of inputs and outputs. DEA is computationally intensive since it requires that each entity be classified as either efficient or inefficient and this entails the application of a specialized procedure. Depending on the procedure, additional information which can be interpreted in economic terms may be obtained about either type of unit. Although DEA is predicated on specific economic premises, the basic classification problem of the entities is a geometrical problem involving polyhedral sets. DEA began to be widely studied, understood, and applied when linear programming (LP) was introduced for the classification and analyses in 1978 (Charnes et al. 1978). LP remains the main computational tool for DEA although it is not exclusive. The framework for performing DEA that evolved around the original LP formulations remains standard, to this day.

DEA matured as a research topic a while back, perhaps as early as the late 90s. By then most of the interpretations of the LP formulations in terms of the economic assumptions behind DEA had been worked out. DEA's inherent geometric nature

J. H. Dulá (✉)
University of Alabama, Tuscaloosa, USA
e-mail: jhdula@cba.ua.edu

V. Charles et al. (eds.), *Data Science and Productivity Analytics*,
International Series in Operations Research & Management Science 290,
https://doi.org/10.1007/978-3-030-43384-0_2

was well understood, due, in large part, to LP itself. It was in the 90s that the community finally realized that full facial decomposition of the polyhedral hull in DEA is unrealistic. Finally, the important computational issues associated with the use of LP to classify individual entities had been fully developed, as well as computational enhancements that resulted in substantial time reductions had been exploited. Most of DEA's idiosyncrasies such as weak efficiency had been resolved.

After maturity, contributions to DEA algorithms, computations, and geometry have been more specialized. For example, geometrical contributions focused on subcategories of efficient DMUs and new faster algorithms, such as Dulá's `BuildHull` (Dulá 2011) that deviated from the standard approach were proposed, albeit, still based on LP. During this time there have been explorations to alternatives to the simplex method for solving LPs; namely, interior point methods (IPMs).

Let's present some preliminary conventions and notation; more will follow. We refer to the entities being compared in a DEA study as "Decision Making Units" or DMUs. We imagine we are comparing n such units each having the same inputs and outputs: m_1 of the former and m_2 of the latter. We refer to the space where the problem is being solve as $\Re^m$ where $m = m_1 + m_2$. These inputs and outputs are quantities that have been measured and reported as nonnegative magnitudes. We assume the reader is familiar with the concept of *efficient frontier*, *production possibility set*, how DMUs are classified as *efficient, extreme efficient, weak efficient*, and *inefficient*. Two of DEA's seminal works: one by Charnes, Cooper, and Rhodes, from 1978 and the other by Banker, Charnes, and Cooper from 1984 will be abbreviated as "CCR" and "BCC".

2.2 Notation and Modeling Issues

DEA data define a point set with cardinality n, the number of DMUs in the study. Each data point holds two types of measurements: those for the m_1 inputs and the other for the m_2 outputs in the same transformation process in which all the DMUs are involved. The vectors x_j and y_j collect these measurements for DMU j and they look like this:

$$x^j = \begin{bmatrix} x_1^j \\ \vdots \\ x_{m_1}^j \end{bmatrix}, \qquad y^j = \begin{bmatrix} y_1^j \\ \vdots \\ y_{m_2}^j \end{bmatrix}; \; j = 1, \ldots, n.$$

It makes economic sense to assume that these measurements cannot be negative although in terms of geometry this is not important. We organize the data in the following way:

$$\mathcal{A} = \{a^1, \ldots, a^n\}, \quad \text{where,} \quad a^j = \begin{bmatrix} -x^j \\ y^j \end{bmatrix}.$$

The point set $\mathcal{A}$ is transformed into an $m_1 + m_2$ by n matrix, A, by making the points a^j its columns. Negating the input vectors simplifies notation.

Any polyhedral set, bounded or unbounded, can be expressed "internally" (Rockafellar 1970) as a combination of a convex hull and a cone. We refer to such sets as simply *hulls*. The convex hull of a set of points, $\mathcal{P} = \{p^1, \ldots, p^K\}$ is a bounded polyhedron, a *polytope*. The conical hull of a set of directions, $\mathcal{V} = \{v^1, \ldots, v^L\}$ is a cone. These two sets are in $\Re^m$ and the shapes can be combined:

$$\mathcal{Q}(\mathcal{P}, \mathcal{V}) = \left\{ z | z = \sum_{k=1}^{K} p^k \lambda_k + \sum_{\ell=1}^{L} v^\ell \mu_\ell, \ \sum_k \lambda_k = 1, \lambda_k \geq 0, \mu_\ell \geq 0; \ \forall k, \ell \right\}.$$

When $\mathcal{P}$ is empty, $\mathcal{Q}$ is a cone and when $\mathcal{V}$ is empty, the polyhedral set is bounded. The vectors in $\mathcal{V}$ define the polyhedral set's *recession cone*. The same polyhedral set can be described "externally" as the intersection of half spaces. Going from internal to external representation, and vice-versa, is called *facial decomposition*. Facial decomposition is numerically complex and will, for large enough problems, eventually overwhelm any computer.

The production possibility set is a fundamental geometric object in DEA. It is a polyhedral set defined internally by the data in $\mathcal{A}$ and a recession cone expressed by the $m_1 + m_2$ vectors in $\Re^m : e^1, \ldots, e^m$ where e^i is $\Re^m$'s ith unit vector.

There are four standard production possibility sets in DEA, "Constant", "Variable", "Increasing", and "Decreasing" Returns to Scale, depending on the economic assumption about the transformation process. These are:

$$\mathcal{P}^{CRS} = \left\{ z | z = \sum_{j=1}^{n} a^j \lambda_j + \sum_{i=1}^{m} e^i \mu_i; \ \mu_i, \lambda_j \geq 0; \ \forall i, j \right\}; \tag{2.1}$$

$$\mathcal{P}^{VRS} = \left\{ z | z = \sum_{j=1}^{n} a^j \lambda_j + \sum_{i=1}^{m} e^i \mu_i; \ \sum_j \lambda_j = 1; \mu_i, \lambda_j \geq 0; \ \forall i, j \right\}; \tag{2.2}$$

$$\mathcal{P}^{IRS} = \left\{ z | z = \sum_{j=1}^{n} a^j \lambda_j + \sum_{i=1}^{m} e^i \mu_i; \ \sum_j \lambda_j \geq 1; \mu_i, \lambda_j \geq 0; \ \forall i, j \right\}; \tag{2.3}$$

$$\mathcal{P}^{DRS} = \left\{ z | z = \sum_{j=1}^{n} a^j \lambda_j + \sum_{i=1}^{m} e^i \mu_i; \ \sum_j \lambda_j \leq 1; \mu_i, \lambda_j \geq 0; \ \forall i, j \right\}; \tag{2.4}$$

where a^j are the vectors composed of the input and output subvectors that define the j-th entity's data point in $\Re^m$. $\mathcal{P}^{CRS}$ is a cone whereas $\mathcal{P}^{VRS}$, $\mathcal{P}^{IRS}$, $\mathcal{P}^{DRS}$ are polyhedrons with, possibly, many extreme points. The representations for $\mathcal{P}^{IRS}$ and $\mathcal{P}^{DRS}$ deviate

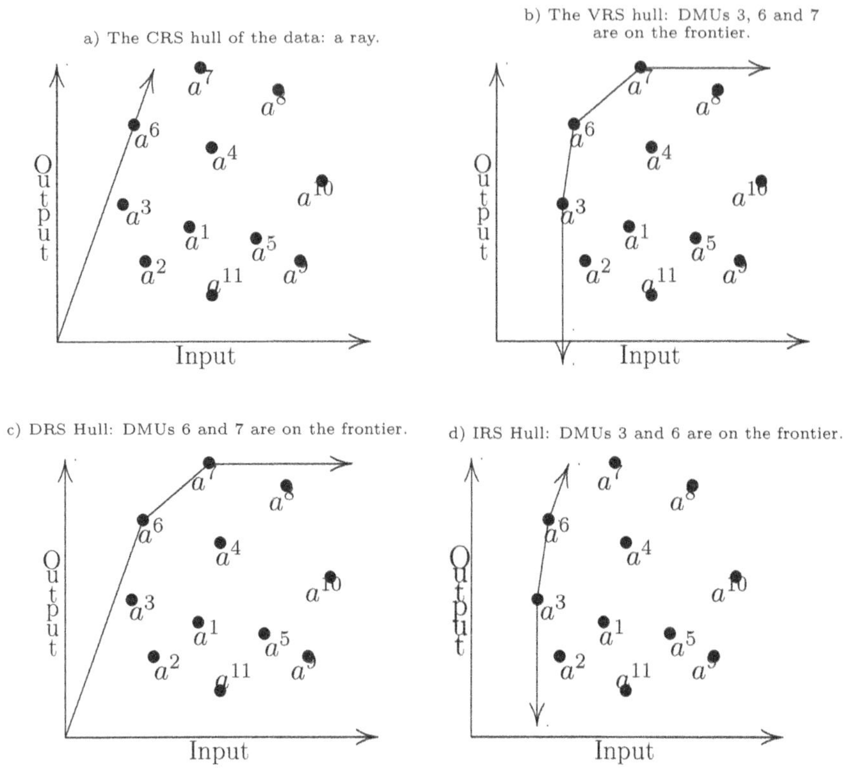

Fig. 2.1 The four standard DEA production possibility sets generated by a two-dimensional point set

from our convention about expressing a polyhedral set in that the condition on the sum of the λ's is not 1. These inequalities hide additional operations on the data: in the case of $\mathcal{P}^{IRS}$, the data points themselves are allowed to act both as generators of a convex hull and as directions of recession which amounts to the convex hull and an infinite expansion away from the origin; in the case of $\mathcal{P}^{DRS}$ this occurs in the opposite direction, i.e., the production possibility set is the convex hull and its contraction, which ends up being equivalent to making the origin an implied element of the data. All four production possibility sets are unbounded and full-dimensional since the m unit recession directions define a complete orthant of $\Re^m$. Figure 2.1 depicts an example of these four polyhedral sets in 2D generated by the same eleven DMUs each with a single input and output.

A production possibility set in DEA collects points corresponding to all theoretically admissible production combinations of inputs and outputs as represented by the data and subject to economic assumptions such as convexity and "free-disposability". The DEA data are a finite subset of a production possibility set. A production possibility set, as with any polyhedral set, has an interior and a boundary. DEA further classifies a production possibility set's boundary into its "efficient frontier" and the rest. The efficient frontier is composed of all points which cannot be dominated in the sense that there is no point in the production possibility set which has more of any output or less of any input. The standard DEA analysis requires locating the points in a production possibility set with respect to one of these three regions. The question about location may be asked about any point in $\Re^m$; not just the DEA data. For data points not on the efficient frontier, DEA may be tasked with answering additional questions such as "how inefficient" is it and which DMUs are its "benchmark" in its efforts to attain efficiency. Only points on the efficient frontier are efficient—these include some of the DMUs—all the rest are inefficient although it may be difficult to distinguish between efficient and inefficient points on the boundary. A point on the efficient frontier may be extreme in which case it corresponds to an "extreme efficient" DMU. Note that the extreme points of the production possibility set correspond to actual DMUs in the data and the set of extreme points (or rays in the CRS case) are minimally sufficient to describe the entire production possibility set. This set of extreme points is referred to as the production possibility set's *frame*. Each of the four standard production possibility sets has potentially a different frame for given data although they are closely related (Dulá and Thrall 2001).

The principal task in DEA is to classify the DMUs as efficient or inefficient by locating their data points in a specified production possibility set. Additionally, DEA answers economic questions about inefficient DMUs. The question of location can be asked about any interior or exterior point with respect to a polyhedral set, internally or externally defined. A DEA study answers other questions. For example, an inefficiency "score" is calculated which provides information about how separated a point is from the boundary along a ray that "pierces" the boundary of the production possibility set. The place at which a judiciously chosen ray makes contact with the boundary can be used as a benchmark for an inefficient DMU providing a prescription for how to attain efficiency. Such a point is on the face of the hull and can be expressed as a combination of extreme points (and, possibly, unit directions). These extreme points, which correspond to extreme efficient DMUs, become a *reference set* for the inefficient point being scored. This reference set can be seen as attainable ideals for inefficient DMUs. This translation along a ray can also be applied—and interpreted—in the case of exterior points. Benchmarks and reference sets depend on which ray is used to direct a point to the boundary of the production possibility set. If the ray is restricted to lie on the m_1 input dimensions and in a direction where the input coordinates are reduced, the benchmark and reference set information reflect an input orientation to the analysis where the prescription for efficiency involves reducing the inputs; analogously, if the ray is in the space of outputs and the direction is towards greater output values. All these questions have a geometrical interpretation based on polyhedral hulls, projections, translations along rays, etc.

It's important to note the role of weak efficiency in DEA. A DMU is weakly efficient if its data point is on the boundary of a production possibility set but necessarily involves a unit direction in its description. These points can be dominated by other points on the boundary, therefore, they are inefficient. Weak efficiency has confounded DEA since CCR where the issue is acknowledged but only sufficient conditions to identify it are provided. The problem was finally properly addressed theoretically and computationally in BCC by introducing the concept of the "non-Archimedean" constant. Weak efficiency remains a problem in DEA since much of the literature simply avoids the issue by ignoring it risking misclassifying weak efficient DMUs as efficient.

2.3 DEA and LP

The seminal article by Charnes, Cooper, and Rhodes (CCR) from 1978 generalized and formulated Farrell's DEA under constant returns to scale assumption as an LP. This is when the long, intimate, and, sometimes, problematic relation between DEA and LP began. In the first LP formulation, the variables are weights associated with the input and output values in what became known as the *multiplier* form. There is one constraint for every DMU plus an additional constraint moved down from the denominator in the objective function of the original fractional program. Since the denominator is where the inputs are combined using the weights, the LP formulation is referred to as "input" oriented. An analogous "output" orientation LP formulation emerges when this operation is applied to the numerator. CCR assumed that scaling a DMU's inputs and outputs by a nonnegative constant did not alter its efficiency relative to the rest of the DMUs. This is what is known as a *constant* returns to scale (CRS) transformation environment to economists. CCR's first formulation, like Farrell's, assumed CRS.

Every LP has a dual. The dual of the multiplier DEA LP is known as the *envelopment* LP. In this formulation, there is a constraint for each input and output in the model and there is a variable for each DMU. Other LP formulations quickly followed to reflect different returns to scale most prominent among them the variable returns to scale (VRS) of Banker Charnes and Cooper (BCC) in 1984. VRS assumes that convex combinations of DMUs (along with free-disposability) produce other viable elements of the production possibility set.

Problems with DEA's LP formulations emerged almost immediately. CCR observes that the solution to the LP formulation they proposed provides necessary but insufficient conditions for classifying a DMU as efficient. The condition is that the optimal objective function value is 1; however there is a class of inefficient DMUs that can also attain this score; namely, weak efficient DMUs. The data point for a weak efficient DMU lies on the boundary of the production possibility set but is dominated by other points in terms of either fewer inputs or more outputs. Necessary and sufficient conditions for efficiency require that all alternate optimal basic feasible solutions be identified; then, if at least one has a positive slack, the DMU being scored is weak, otherwise, it is efficient. This, of course, is impractical.

A practical work-around for the problem of efficiency identification appears in BCC and relies on a "non-Archimedean" constant introduced to the formulation; simply stated, the new formulation involved the slacks in the LP's objective function multiplied by a tiny factor ϵ. This forces weak efficient DMUs to reveal themselves because their score can get at most within ϵ's of 1. The role of the non-Archimedean constant in DEA LPs was debated for many years with evidence showing that the choice for the value for ϵ could affect classifications of the DMUs.

CCR's assumption about constant returns to scale makes the CRS production possibility set a cone. BCC's VRS assumption means the production possibility set contains extreme points which are not the origin; the production possibility set is no longer a cone. This makes the production possibility set a polyhedral hull with potential for multiple extreme points while receding in the positive unit directions (recall we have negated the input data). Soon after, formulations for increasing and decreasing returns to scale (IRS & DRS) transformation assumptions were introduced. The shape of the production possibility set for these models differs in that they recede in additional directions (IRS) or include the origin (DRS).

Solving LPs for DEA—and in any other circumstance—requires a decision about which algorithm to use: the simplex or one of the many versions of an interior point method (IPM). The first efficient IPM algorithm for LP was introduced by Karmarkar (1984). It became a viable alternative to Dantzig's simplex algorithm which had been around since the 40s. The two algorithms solve LPs in different ways: the simplex follows edges connecting extreme points that improve the objective function while IPMs try to follow a smooth internal path. If the optimal solution to an LP is unique, it occurs at an extreme point of the feasible region, and both algorithms find it. In the presence of alternate (multiple) optima, however, the algorithms arrive at different optimal solutions. Alternate optima occur when an entire face of the feasible region with a dimension greater than zero (edges and higher) is optimal. If there are alternate optima, the simplex will arrive at one of the extreme point optimal solutions. Traditional IPMs, in particular "primal-dual, path-following" methods (see Wright 1997; Caron et al. 2002), find an "analytic center" optimal solution which will be in the interior of the optimal face. This is consequential for DEA.

Whenever the point being scored in DEA is on a face with fewer than $m - 1$ dimension (or when an interior DMU projects onto one) the optimal solution of a multiplier LP will have multiple optima (hence degeneracy in the envelopment form) which means there is not a unique optimal supporting hyperplane to the production possibility set for the DMU being scored. This is exaggerated to the limit when the point being scored (or a projection—rather rare) is extreme to the production possibility set. This presents challenges to the economic interpretation and implementation of a solution since the supporting hyperplane provides information about the interactions, tradeoffs, and rates of substitution among the inputs and outputs; if this is not unique or predictable, the interpretations and explanations become ambiguous since they depend on which hyperplane the simplex will select. More importantly, if an extreme point is in the support set of a hyperplane where one of its m coefficients is zero, there is a complete loss of information about that dimension's economic role; such hyperplanes are common in practice (they are related to anchor points discussed below). IPMs provide reproducible solutions in the interior of faces and the production possibility set's hyperplanes derived from the solution will not have zero coefficients when scoring points on the efficient frontier.

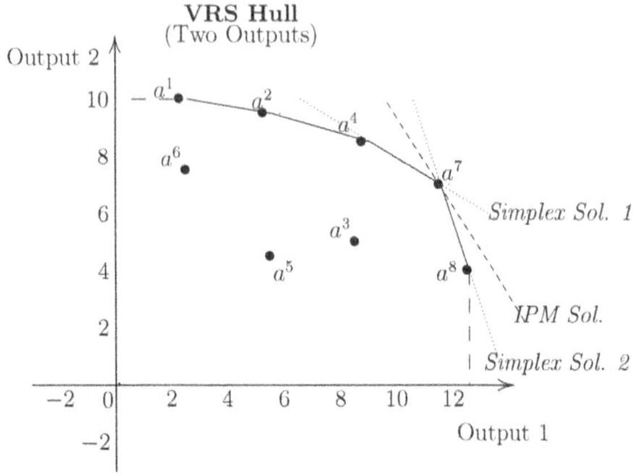

Fig. 2.2 Supporting hyperplanes at Point a^7: two are from optimal basic feasible solutions obtained using the simplex algorithm. The third supporting hyperplane is the IPM solution

IPMs offer other advantages such as proven polynomiality and immunity to degeneracy. Although the simplex has not been shown to be polynomial in its complexity, it performs well compared to IPMs in many different settings, including DEA (Dulá 2008). This is due to the fact that DEA LPs are dense.

Some commercial IPM implementations will proceed by default to identify an optimal extreme point solution—sometimes referred to as a "crossover" solution—after the analytical center has been found. This is due to complementary slackness advantages such as standard sensitivity analyses. This default feature needs to be deactivated if the interior analytical center solution is desired.

Figure 2.2 illustrates how the simplex algorithm and an IPM generate supporting hyperplanes for the production possibility set by solving multiplier LPs. The figure depicts a two-dimensional VRS production possibility set in a DEA problem with only two outputs. Let's direct our attention to Point a^7 in Fig. 2.2. The two dotted lines: one through a^4 & a^7 and the other through a^7 & a^8 labeled "*Simplex Sol. 1*" and "*Simplex Sol. 2*" are supporting hyperplanes generated by the two optimal basic feasible solutions to a multiplier LP (or by looking at shadow prices in the envelopment LP solution) formulated to score a^7 generated by the simplex algorithm. It cannot be predicted which of the two solutions the simplex will generate. The dashed line making contact with the production possibility set only at Point a^7 labeled "*IPM Sol.*" is generated by solving a multiplier LP using an IPM such as the one in Bougnol et al. (2012). Notice that an LP formulated to score a^1 (or a^8) could generate a horizontal (vertical) supporting hyperplane if solved using the simplex algorithm. Such a hyperplane would have a zero as one of its coefficients. This will not happen when using an IPM.

2.4 Computational Geometry

There is a large intersection between DEA and computational geometry. Much of computational geometry deals with the geometry, algorithms, and computations of polyhedral sets generated by linear operations on point sets or (dualy) by the intersection of halfspaces in multidimensional space. Once a DEA problem is reduced to the data and returns to scale assumption is made, it becomes a special case in computational geometry.

At some point, the DEA problem is described by its data and a working assumption about the returns to scale. Before and after there are economic considerations but those do not interest us. The data define a point set in $\Re^m$ with cardinality n. Each of the standard production possibility sets is a polyhedral set. This set is generated by special linear combinations of the points. In DEA the production possibility set is the union of a complex polyhedral hull and directions of recession where by "complex" we mean polyhedral sets with possibly multiple extreme points, as opposed to just cones. For example, the VRS production possibility set is the union of the convex hull of the data with the m unit directions of recession $e^1, \ldots, e^m$ in $\Re^m$. This defines an unbounded polyhedral set usually with many extreme points. The recession cone over which the VRS production possibility set is unbounded reflects the "free-disposability" economic assumption about the transformation where any point with more of any of the inputs or fewer of the outputs than any convex combination of the data is considered viable. The CRS production possibility set is a cone because under this assumption, all production possibilities can be scaled up or down any positive amount. The other two standard returns to scale assumptions, DRS and IRS, are in a sense hybridizations of the VRS and CRS models where only upward (IRS) or downward (DRS) scaling is possible. DEA has had to deal with other shapes. For example modeling a situation where increasing outputs indefinitely is not possible due to congestion effects, the corresponding direction of recession direction is truncated. The resultant production possibility set recedes in a smaller (lower dimensional) cone.

The process of classifying DMUs as efficient or inefficient in DEA is exactly the "frame" problem in computational geometry sometimes ambiguously referred to as the "convex hull" problem. Finding the frame of a polyhedral hull of a finite point set consists of identifying all its extreme elements: points and rays. One way to do this is by using LP. Data points that are extreme in the production possibility set correspond to the extreme efficient DMUs. In the case of CRS, these are points on extreme rays that are not on a combination of unit directions. Interior points are inefficient. Boundary points may be truly efficient or just "weak" efficient depending on whether they can be expressed using a strictly positive combination of one or more of the unit directions of recession; these can be detected by the presence of positive slacks in an optimal solution in many of the standard LP formulations. Unfortunately, this is just a sufficient condition.

Notice how the frame problem is defined on a polyhedral set generated by combining points in a finite point set. Rockafellar (1970) refers to this as their *internal* representation. The extreme points are a subset of the data, so an algorithm needs only search over the data. This makes such a procedure tractable. A polyhedral set can also be generated by taking the intersection of halfspaces. In this case, the halfspaces are bounded by hyperplanes in $\Re^m$ and these support the polyhedral set at its faces although the only ones needed are those that support at facets assuming the polyhedral set has full dimension. This is an *external* representation. Such a polyhedral set has its own extreme points and directions. These extreme elements will provide an internal representation of the same polyhedral set. This is a duality relation. Finding the extreme points and directions of an externally generated polyhedral set is a different proposition than the frame problem. Going from one representation to another is referred to as *facial decomposition* and it is what Farrell proposed as a solution approach to DEA. Facial decomposition in either direction is intrinsically combinatorial and, although the size of problems which can be handled in a reasonable time has steadily increased due to algorithmic and computational advances, the point when a problem becomes intractable arrives quickly. This applies to computational geometry in general, and therefore, to DEA as well.

2.5 History: 1957–1978

Farrell (1957) raised the first computational issues in DEA in 1957. Farrell's work is remarkably clear in laying the foundations of modern DEA. His model applies to the constant returns to scale transformation assumption with a single output which reduces to a variable returns equivalent with just inputs. Farrell recognized that the efficient frontier in DEA is the union of facets each defined by the convex hull of a special subset of the data; namely the extreme points. Unfortunately, Farrell's contribution occurred before linear programming was well known. His computational approach to DEA was to identify facets of the frontier directly. This was done by calculating hyperplanes using groups of m (affinely independent) data points and testing if they supported the production possibility set. Calculating the efficiency score of a DMU becomes a simple algebraic operation after that. This is a direct facial decomposition of a point set and is, of course, exponential in its complexity. Farrell recognized the method "... is practicable only for a very limited range of values of n and m" in his day; an assessment which has not changed much in 60 years. He did foresee, however, that his procedure could be accelerated if the data corresponding to inefficient DMUs were removed before starting the search for facets. A version of this would later be developed and referred to as "Reduced Basis Entry" (RBE) by Ali (1993).

The next important contribution to DEA was by Charnes, Cooper, and Rhodes in 1978 (CCR). It is here where the connection between DEA and linear programming (LP) was made. DEA is not an "LP-based procedure" as we are told in countless introductions to articles about DEA since, as seen above, DEA analysis can be performed without LP. The basic algorithm implied in CCR for carrying out a DEA analysis referred to as the "standard" approach, requires solving one LP for each

DMU. The optimal solution to any such m by n (roughly) LP provides an optimal solution the objective value of which (the *score*) can be used to classify the DMU as efficient or inefficient. If the latter, the solution then provides information used for benchmarking and recommending a path to efficiency. LP is the prevalent tool for processing DEA. CCR observed that there was an issue with classifying weak efficient DMUs using their LP formulation; more about that later.

LP made DEA practical and scalable. The size of the DEA problems which can be solved grows as computing hardware and software evolve. The former refers to the machinery and has been occurring in leaps and bounds. The latter refers almost exclusively to LP algorithmic improvements which result in faster solvers which, in these days, means implementations of the variations of the simplex algorithm and interior point methods (IPMs); both of which have also been advancing at great speed.

2.6 History: 1978–2000

CCR produced the marriage between DEA and LP but was restricted to the CRS transformation assumption. The next major work in DEA, in what was still a straight sequence, was by Banker, Charnes, and Cooper in 1984 (BCC). This paper introduced the axiomatic premises for the VRS transformation assumption where only convex hull interpolations, as opposed to unlimited positive scaling, are required for inclusion in the production possibility set. This variation applies in many situations and is now more widely used than CRS. The discussion in BCC suggested the other returns to scale assumptions; namely IRS and DRS. It turned out that the LP formulation for VRS is a minor modification of the CRS LP; it requires an additional unrestricted variable in the multiplier form which translates to a single convexity-type constraint in the dual envelopment form. BCC discussed geometrical aspects of DEA in a clear way including how the optimal solution to the multiplier LP defines a supporting hyperplane for the production possibility set and how this hyperplane would make contact over a facet. An important algorithmic and computational contribution of BCC was formally addressing the complications from processing weak efficient DMUs. The problem was resolved theoretically by introducing the use of a "non-Archimedean" infinitesimal in the LP formulation. This solution turned out to be impractical and somewhat unreliable in actual practice.

Applications of DEA appeared early and are interesting from a computational point of view. Indeed, E. Rhodes' 1978 dissertation discusses a DEA application in education. The work in Rhodes' dissertation was later used in Charnes et al. (1981) where they report on the effects of government experimental programs in public schools by applying DEA. The article compares two groups of DMUs, the largest of which had 49 units, using a total of eight inputs plus outputs. Perhaps the first DEA application appearing in a mainstream academic journal was by Bessent and Bessent in 1980. This work was also about education. There is scarce information about how the DEA LPs were solved in these early works. One concrete indication comes from the discussion in Bessent et al. (1982) also about DEA in education, where mention is made of a specially designed program in Fortran (see Bessent and Kennington 1980). I. Ali is acknowledged as having participated as a Ph.D. student. One imagines it must have been a difficult task to process even moderately sized DEA problems in those early days of computers, programming languages, and LP.

Five important computational and algorithmic developments affected DEA in the following years until 2000.

1. The Additive Model. DEA's "additive" model generates a different LP formulation than the previous "radial" oriented CRS, VRS, IRS, and DRS formulations. The difference is that this model's envelopment LP objective function maximizes the sum of input and output slacks. This formulation first appeared in 1982 (Charnes et al. 1982). A feature of this model is that its optimal solution provides necessary and sufficient conditions to classify DMUs. This has immediate implications for DEA computations in that it obviates the need for involving the non-Archimedean infinitesimal. Another advantage is that there is no input or output orientation to the LPs. A drawback is that, unlike the traditional radial oriented models, inefficient DMUs are thrown onto the boundary of the efficient frontier at the farthest point according to the L1 norm. This makes the projections of little economic value.
2. Preprocessing. There are ways to classify DMUs in advance as either efficient or inefficient without having to solve an LP. Farrell knew in 1957 that some inefficient DMUs were easily identified and probably had in mind testing for "domination". A DMU is dominated if there exists another with fewer inputs and more outputs. Finding these involve simple comparisons. One of the first comprehensive studies about this preprocessing scheme is by Sueyoshi and Chang in 1989, where they note that this simple approach can identify a large number of inefficient DMUs. Another quick classification idea applies to efficient DMUs in the case of VRS when one of the input or output components is the smallest or largest in the data. This was used by Ali in 1993 and requires easy sorting operations. Another computational saving device to avoid LPs is to translate/rotate hyperplanes towards/supporting the efficient frontier and locate new boundary contact points. These procedures require calculating and sorting inner products. This is discussed by Dulá and López in 2009.
3. Reduced Basis Entry (RBE) and Early Identification of Efficients (EIE). These two "enhancements" formalized by I. Ali in 1993 are a consequence of the same principle: the variables associated with an optimal basic feasible solution for any DMU in an envelopment DEA LP must be boundary points. Such points almost always correspond to extreme efficient DMUs (non-extreme DMUs on the boundary are rare). This is a direct corollary of the geometrical fact that only boundary points can be in the support set of a supporting hyperplane for a convex set. Farrell knew about this in 1957. RBE is a modification applied to the standard DEA algorithm as it iterates over each of the DMUs by solving an LP, which consists of excluding (i.e., removing) the data of units previously classified as inefficient. The effect is that the number of variables of the envelopment LPs

is progressively reduced as the iterations progress. As the LPs become smaller, the time to solve them is reduced. This may not have much of an impact at first but as the number of scored DMUs accumulates, the procedure makes faster progress. This effect can be dramatic when the proportion of inefficient DMUs is high, as is commonly the case. Dulá (2008) reports the effect of RBE to be as much as halving computation times. RBE is easily implemented and should be part of every DEA procedure. The principle behind RBE provides the theoretical foundation for the "Hierarchical Decomposition" (HD) algorithm for DEA of Barr and Durchholz (1997) discussed below. EIE is essentially the converse of RBE; it saves computations by obviating an LP for any efficient DMU previously identified as part of an optimal basic feasible solution of an envelopment LP. This idea has not been implemented or tested directly much in part because DEA data typically have a low proportion of efficient DMUs added to the fact that it is frequently the case that a few of the DMUs tend to have a disproportionate presence in LP bases; i.e., they re-appear frequently.

4. Deleted Domain (DD). DD is the simple idea of excluding from the coefficient matrix the data of the DMU being scored in an envelopment LP. It was originally proposed by Andersen and Petersen in 1993 as a way of distinguishing among efficient DMUs by providing individual scores different from 1. Geometrically the solution to a DD LP is a hyperplane that supports the truncated production possibility set and separates it from the excluded point. This method is interesting as a computational device in that it brought two issues in DEA to the forefront: infeasibility and degeneracy of the envelopment LP. Until DD, envelopment LPs were always feasible; after DD it was noticed that they could be infeasible (Dulá and Hickman 1997). This had an impact on later algorithms for DEA which meant they had to deal with this contingency directly. It had been noticed as far back as 1994 (Ali 1994) that DEA envelopment LPs were prone to degeneracy. The reason is simple: the data of one of the variables in the standard DEA LP appears on the right-hand side. Geometrically this can be interpreted as resulting from the multiple optimal supporting hyperplanes at a point on a face of the production possibility set with fewer than m dimensions, the extreme case of which occurring when this point is extreme. This sort of "induced" degeneracy is not present under DD.
5. Hierarchical Decomposition (HD). HD proposed by Barr and Durchholz in 1997 processes DEA using LP in two phases. The first phase identifies inefficient DMUs using uniformly sized blocks that partition a DEA data set. Each partition is processed as a separate DEA problem, and since what is inefficient in a partition will also be inefficient to the entire data, inefficient DMUs can be identified using smaller LPs. Notice that the size of the envelopment LPs depend directly on the block sizes which means they could have many fewer than n variables. After the data in the blocks are processed, the unclassified DMUs are grouped to form new blocks and the process is repeated to identify additional inefficient DMUs. This goes on until a decision is made to place the remaining unclassified DMUs into a single block to be processed to extract what will be the final and complete list of efficient DMUs. In a second phase, the efficient DMUs are used in an LP to

properly score the inefficient DMUs. Notice that such an LP will be small if the density of efficient DMUs is low. HD demonstrated time reductions of one order of magnitude compared to standard approaches with RBE enhancements. HD requires user-defined parameters such as initial block size, size of subsequent aggregated blocks, and "depth" (number of times new blocks will be created). HD is particularly well suited for a parallel implementation since blocks can be processed independently. HD was at the forefront of applying the two-phase approach for DEA where the efficient DMUs (or a small superset that contains them) are isolated and collected in the first phase and the rest of the DMUs are scored in the second phase using only the efficient DMUS and correspondingly smaller LPs. There is a distinct possibility in HD that more than n LPs will be solved in the first phase. HD is important also because it was one of the first algorithmic and computational contributions specifically designed for large scale DEA.

A final note about what is known about how general LP relates to DEA. The LPs that are solved in many of the procedures developed for DEA are usually similar from one instance to the next. In many cases, the difference is only on the right-hand side and perhaps a column in the coefficient matrix of an envelopment formulation. This suggests that LP re-optimization techniques such as "hot starts" where information about a previous optimal solution is used to solve the next LP can be used. Indeed, hot starts implementations in commercial LP solvers have been shown to have a substantial impact on reducing computation times when applied to DEA. Dulá (2008) tested this with CPLEX and found 80% reductions in solution times when hot starts procedures are applied.

In the area of DEA computations what followed next of major importance were more specialized procedures to process DMUs and for specific types of extreme efficient DMUS known as "anchor points".

2.7 History: 2000–Present

The next big push in algorithmic and computational developments in DEA was directed at making gains in speed with an eye on the clear trends towards larger data. There was also an interest in specialized aspects of the geometry of the production possibility set specifically with regards to classifications of extreme points accompanied by algorithms to identify them.

1. Frame-Based Procedures BuildHull (BH). A DEA production possibility set is a polyhedral hull, the extreme elements of which (the frame) are a subset of the data and are the minimum data needed to describe it. A procedure based on efficiently and directly identifying the frame of a convex hull was proposed by Dulá and Helgason in 1996. This procedure was adapted to DEA in Dulá (2011). The procedure has two phases: the first phase identifies the elements of the frame; the second phase scores the rest of the DMUs using this frame. The number of

variables in the LPs used to score the DMUs in the second phase is the cardinality of the frame (plus one). Frame algorithms apply the following properties: (i) an element of a hull of any subset of the data is an element of the full hull; (ii) if the hull of a subset of the data is a strict subset of the full hull, then there exists at least one frame element outside the hull of the subset; and (iii) the support set of a supporting hyperplane for the full hull of the data contains at least one extreme point (frame element) (Dulá and López 2012). Frame-based procedures initialize with any subset of the frame including a single element, which is easy to find. At each iteration, a test point is identified as either internal or external to the current "partial" frame; internal points are discarded and a new unclassified test point is used. This continues until an external test point to the partial hull is found. The existence of an external point assures the existence of a new frame element (from property *ii* above). Each test requires an LP the variables of which are the elements of the partial hull which, by construction, are frame elements of the full hull. A new frame element will be found by translating the separating hyperplane obtained from the solution to the LP that was used to establish that the test point was external. Translating hyperplanes involves inner products. This way, the partial hull grows one frame element at the time. A depiction of the algorithm's progress can be seen in Fig. 2.1 of Dulá (2011). Note that BH relies heavily on the calculation of inner products. Algorithm BH is fully deterministic, conclusive, requires no user-defined parameters, and solves exactly $n - 1$ LPs in the first phase. These LPs grow one variable every time a new frame element is identified and will not exceed the number of frame elements in the full hull (plus one).

2. Reference-Searching (RS) Procedures. A series of papers by W. C. Chen and other coauthors (Chen and Cho 2009; Chen and Lai 2017) present variations on a procedure for DEA based on accreting/swapping variables in an envelopment LP until the LP contains the optimal reference set for the DMU being scored. The idea is driven by the fact mentioned earlier that, when scoring an inefficient DMU, m of the basic variables of an optimal solution of the envelopment LP correspond to efficient DMUs; they are the, so-called, *reference set*. (Efficient DMUs also generate a reference set but it has a different interpretation and is not as interesting). Every inefficient DMU has a reference set and knowing it would mean it would be possible to score the DMU using an LP with only m variables (plus maybe a few other variables, depending on the model, e.g., the VRS case). This would represent the smallest possible LP needed to score a particular DMU. Of course, it is not possible to know in advance a DMU's reference set. RS procedures formulate an initial LP with a relatively small number of—preferably—efficient DMUs in its data; this number can be as small as m. The reduced size envelopment LP is solved not expecting that the score will be the true one for that DMU; although it could happen. Whether it is or not can be checked by looking at how many dual constraints in the full problem are violated; a simple calculation involving inner products. It could be that none are violated in which case the problem is solved! This is, of course, unlikely especially if the number of DMUs used in this first LP formulation is relatively small (but

$\geq m$). The idea then is to judiciously select a specified number of new data points (DMUs) to be added to the next envelopment LP. Which ones to include is at the crux of the problem and several ideas have been proposed including assessing the likelihood of belonging to a reference set using trigonometric arguments. Another idea proposed and tested is to use the magnitude of dual infeasibilities. The LP grows if no data points are removed, which they may have to be if the number of variables is to be kept capped. Eventually, it is hoped that the data of the LP will contain the DMU's reference set in which case its optimal solution will be the same as the full LP. The background geometry is that assuming only efficient DMUs make it into the LP when the reference set is not yet part of the LP's data, the solution defines a hyperplane that cuts through the production possibility set, i.e., not supporting it. Therefore, there are efficient points of the production possibility set on the "outside" of the hyperplane with different separations which can be measured using inner products. By selecting those which are more separated to be included in the next LP and, if required, removing points also applying a separation criterion but for DMUs in the inside part of the hyperplane, the next LP will be different. This procedure is repeated until the LP contains the DMU's reference set. Experiments show that an LP with the DMU's actual reference set frequently occurs relatively quickly with surprisingly small LPs although this is not guaranteed. RS procedures are designed for when there is a cap on the number of variables in the envelopment LP or when a few select DMUs need to be scored. RS procedures do not yet compete with procedures specifically designed for speed to process all the DMUs.

3. Single Decomposition (SD) Scheme. Khezrimotlagh et al. in 2019 proposed a DEA procedure operating on a single subset of the data as its premise. As with HD and BH, SD has two phases: in the first phase the boundary points of the full production possibility set are identified and this is used in a second phase to score the rest of the DMU's. The procedure starts with a subset of the data with p points. SD proceeds to identify the subset's boundary points using p LPs with $p + 1$ variables. The boundary points found will contain the subset's frame of which we will denote its cardinality by $|f(p)|$. Next, SD identifies all points outside the subset's VRS hull requiring $n - p$ LPs with $|f(p)| + 1$ variables. The first phase concludes with the identification of the boundary points of the full production possibility set which requires the union of the frame of the original subset and the exterior points. This involves solving $n - (p - |f(p)|)$ LPs with $|f(p)| + (n - p) + 1$ variables. Notice that the total number of LPs is $2n - (p - |f(p)|) > n$ sized between $|f(p)| + 1$ and $|f(p)| + (n - p) + 1$. The next phase scores the DMUs using only the boundary points of the full production possibility set just like in HD and BH. The initial subset's cardinality in SD is a user-defined parameter which the authors recommend at $\sqrt{n}$. The authors provide ideas for constructing the initial subset based on heuristics with the goal of maximizing the presence of boundary points from the full hull. The composition of the initial subset in terms of how many of the full hull's boundary points it contains is unpredictable. It is possible that the first phase of SD will be concluded sooner without having to check for external points if the starting subset contains the

frame of the full production possibility set. This can occur only if the cardinality of the full frame is less than or equal to p but, even then, it appears unlikely even after applying the heuristics provided. The authors of SD claim their procedure can be two to three times faster than their implementation of BH using real and synthetic data. The comparison, however, applies an implementation of BH where additional (dual) LPs (presumably using the data points of the unclassified DMUs) are solved to identify new extreme points of the full production possibility set.

4. Anchor Points (AP). All DEA hulls (CRS, VRS, IRS, and DRS) are unbounded due entirely in some cases (e.g., VRS) to the "free-disposability" axiom (Banker et al. 1984). In the VRS production possibility set, the efficient frontier is limited to the union of bounded faces. An extreme point on an unbounded face of a VRS production possibility set is called an Anchor Point (AP). APs correspond to efficient DMUs since they are extreme but they define geometrically where the efficient frontier abruptly ends. Researchers have attributed interesting economic properties to APs (see Allen and Thanassoulis 2004; Edvardsen et al. 2008). The problem of defining and identifying APs is intensely geometric and provides theoretical and computational challenges which have been tackled in several works since the early 2000s. Allen and Thanassoulis in 2004 presented an algorithm to detect APs for the case of a single input and multiple outputs. Edvardsen et al. in 2008 provide an algorithm that was able to detect a subset of these points. Bougnol and Dulá in 2009 designed a procedure to identify APs in a general VRS production possibility set with multiple inputs and outputs. The procedure reduces to a DEA analysis in each of m simple projections of the data. (A simple projection is a DEA hull of the data where one of the dimensions is omitted). Mostafaee and Soleimani-damaneh in 2014 provided a different procedure to identify APs where, by looking at the neighborhood of an extreme point, they can detect if it lies on an unbounded face. Krivonozhko et al. in 2015 defined a special subset of APs they termed as "terminal" and provided a procedure to detect them. Terminal APs are APs on an unbounded edge of the production possibility set (APs may actually be "tucked" away from unbounded edges).

APs, like non-extreme efficient DMUs and weak efficient DMUs, can easily be artificially generated; it's a matter of combining extreme points and unit directions. An interesting observation about APs is their preponderance in large problems from real applications. In some cases, 100% of the extreme points of a VRS hull are APs. This has been independently verified (e.g., Bougnol and Dulá 2009; Krivonozhko et al. 2015). The reasons why non-extreme efficient or weak efficient units are rare in real data can be explained using numerical and precision arguments, e.g., any small perturbation or imprecision of any of a data point's component values is enough to move it away from the boundary making it either interior or extreme. In other words, the probability that a point is located exactly on the face of a production possibility set without it being extreme is low. The preponderance of APs in real data remains a mystery.

2.8 Future of DEA

Some trends in DEA are easy to predict: DEA applications are becoming more varied and widespread, new ones appearing almost daily, many applied to economic activities in the developing world. Interest in DEA algorithms, computations, and geometry will continue. It is hard to predict who will be the last researcher to produce the fastest algorithm to perform a DEA analysis, but almost certainly, the algorithm that will achieve this will rely on new geometrical insights. The day will come when DEA will be a standard tool in everyone's analytical toolbox—in the same way as simple linear regression—and Excel will bundle a dedicated solver for it!

A more immediate and specific prediction about DEA algorithms and computations is that the knowledge acquired so far in designing and testing algorithms will be used in other areas.

1. DEA and the Frame Problem. As mentioned before, the frame problem—a venerable problem in computational geometry—consists of finding the extreme points of a polyhedral hull defined internally by constrained linear combinations of the elements of a point set. In computational geometry the problem is usually limited to convex hulls. DEA finds the frame of different, more varied, hulls; the convex hull being just a special case. In this respect, DEA and computational geometry speak the same language. An algorithm to find the efficient DMUs that works well for the VRS hull in DEA is easily adapted to the frame of the convex hull and would interest the computational geometry community (and vice-versa, see, e.g., Dulá and López 2012).
   The frame problem is of interest beyond DEA and computational geometry. In statistics, there is an interest in using "point depth" as a way to partially order multivariate data (Tukey 1974; Barnett 1976). The idea is to count how many convex hull layers need to be "peeled" away before a specified point is on the boundary. The last points reached are, in some way, at a center. The simple median of a collection of scalar values is at a maximal point depth. This operation is used in non-parametric analyses. Convex hull peeling appears in astronomy (Babu and McDermott 2002).
   Another application of the frame problem is *Archetypal Analysis* (Cutler and Breiman 1994). In Archetypal Analysis points are treated as mixtures (or blends, or combinations) of single archetypes which may or may not be data elements. If they are part of the data, then finding them is a frame problem. In some applications, the archetypes may naturally emerge from the data. For example, in a face recognition application, human faces can be constructed by combining a limited number of facial features such as special types of noses, shapes of eyes, mouths, eyebrows, etc. (see Lee and Seung 1999). Other applications are in text mining and hyperspectral imaging (see Salmani 2018; Ang and Gillis 2019).
   An area where the frame problem looms large is Nonnegative Matrix Factorization (NMF). Let $W_{m\times n}$ be a nonnegative matrix. NMF factors W into the product of two nonnegative matrices $U_{m\times r}$ and $V_{r\times n}$; thus: $W = UV$. Many times, the best that can be done is to find U, V to approximate W: $W \approx UV$. This is what

occurs when r, the "rank" of the factorization, is capped at some value $r \ll m$ as a way to capture information in data using fewer dimensions; in what's known as *dimensionality reduction*, a topic of intensive research in statistics and machine learning. The approximation can be made tight by minimizing $\|W - UV\|$ using a specific norm and subject to additional constraints.
There are situations when U, V can be found such that $W = UV$. When this happens, W is said to be "separable" (Donoho and Stodden 2003). New applications where separability of W plays an important role is a topic of current research (Gillis 2014). This is a familiar territory for DEA. Collect the DEA data for the m inputs and outputs for n DMUs into a matrix W. Let there be r frame elements to the production possibility set and collect their data points into a matrix U along with a full complement of the m unit directions appended as columns to account for the recession directions for the VRS case; this gives U the dimension $m \times (r + m)$. Every DEA data point can be expressed as a combination of the frame elements plus a combination of unit directions of recessions. The $r + m$ coefficients needed to express each of the n data points can be collected into an $(r + m) \times n$ matrix V. With this, we have an exact nonnegative factorization of W using U and V. DEA's data are, therefore separable, and indeed the extreme DMUs represent the "archetypes" of the data (also referred to as "basis elements"). There is much that can be applied in these emerging areas of machine learning and statistics from what we have learned in DEA; especially with respect to algorithms, computations, and geometry.
2. DEA and the L1 norm. Encounters between DEA and the $L1$ norm have happened at least twice. The first is major and consequential: the advent of the additive LP formulation early in DEA's history. This formulation's importance for DEA theory and computations was discussed earlier. Charnes et al. in 1985 state that the optimal solution to the additive LP defines a projection that maximizes the $L1$ distance to the efficient frontier. Another encounter occurred when Briec in 1998 observed that the projection of an interior point onto the boundary of a DEA efficient frontier using the $L1$ can be found by solving m LPs. Briec's result is important because nothing until then—and since—has addressed so conclusively the issue of a true projection onto the boundary of the production possibility set. All analyses in DEA involving standard LP formulations provide solutions which locate points on the boundary onto which a data point being scored may *arrive* but these are not true projections in the sense that they minimize a distance based on a proper norm. Notice that finding the projection on the efficient frontier of an interior point using the familiar Euclidean norm would be of great value to DEA but, unfortunately, this is not linear or convex and remains impractical.
3. DEA and the Recommender Problem (RP). RPs are a collection of machine learning methods used to predict what an individual/entity will purchase/consume based on historical data (Hill et al. 1995). A specific problem in the *collaborative filtering* type of RP uses data from similar "peer" consumers to make a prediction. The data being handled are usually magnitude measurements such as ratings or counts, e.g., number of clicks, etc. There is no doubt this is an important problem for the retail and entertainment industry and the intensity with which it is studied

serves as evidence. A connection between DEA and the RP comes from the latter's critical directive of having to identify a community of "peer" users for the prediction. DEA is familiar with a similar concept; namely, the reference set of a DMU being scored. Any DMU has a reference set determined by the optimal solution to the LP used to score it. This reference set has been used, traditionally, as a collection of special DMUs on the efficient frontier that can be combined to define a "virtual" DMU which can be used as a benchmark for the DMU being scored. Geometrically, the reference set is composed of the extreme points on a supporting hyperplane of the production possibility set on a facet where the DMU being scored lands. Depending on the LP used to score the DMU, the benchmark represents an empirical aspiration for an inefficient DMU and provides specific instructions as to how inputs and/or outputs should be decreased and/or increased to attain efficiency. The benchmark and the reference set share the same facet of the hull so they have quite a bit in common (e.g., the same supporting hyperplane) and could be treated as a geometrically coherent group, i.e., a set of geometric peers. Moreover, the extreme point peers are archetypes in the sense above. The idea of using reference sets to define peers for predictions for collaborative filtering RPs needs to be tested against current practices for finding peers by calculating k nearest neighbors using simple metrics.

2.9 Conclusions

DEA has had a distinguished career as an analytical tool derived to analyze efficiency and productivity, and this promises to continue. The methods used to perform DEA rely on sophisticated algorithms and are computationally demanding and this is because the underlying objects behind the applications and interpretations have a complicated geometry. Understanding, and contributing to, the algorithmic, computational, and geometric aspects of DEA requires having a tight grasp on LP, duality, and computational geometry. Without this knowledge, major breakthroughs are not possible. This knowledge is also the currency in other areas; some established and some new and emerging. Knowledge acquired in DEA can be used for contributions in areas such as computational geometry and machine learning where researchers are just now beginning to understand—and apply—the fundamental importance of the basic operation of extracting extreme points from polyhedral hulls of point sets.

In this chapter, we have discussed the history of DEA algorithms, computations, and geometry. This is a history of how these three aspects come together in DEA for a steady stream of contributions. The chapter is also about how algorithm design can be used in other new and exciting areas in analytics; especially machine learning. DEA is ready to be unfastened from its economics/efficiency/productivity moorings and be generalized into a machine learning, big data, analytics tool.

References

Ali, I. (1993). Streamlined computation for data envelopment analysis. *European Journal of Operational Research, 64*, 61–7.

Ali, I. (1994). Computational aspects of data envelopment analysis. In A. Charnes, W. W. Cooper, A. Y. Lewin, & L. M. Seiford (Eds.), *DEA: Theory, methodology and applications*. Boston: Kluwer Academic Publishers.

Allen, R., & Thanassoulis, E. (2004). Improving envelopment in data envelopment analysis. *European Journal of Operational Research, 154*(4), 363–379.

Andersen, P., & Petersen, N. C. (1993). A procedure for ranking efficient efficient units in data envelopment analysis. *Management Science, 39*, 1261–1264.

Ang, M. S., & Gillis, N. (2019). *Algorithms and comparisons of nonnegative matrix factorizations with volume regularization for hyperspectral unmixing*. arXiv:1903.04362.

Babu, G. J., & McDermott, J. P. (2002). Statistical methodology for massive datasets and model selection. In *Astronomical data analysis II* (Vol. 4847, pp. 228–238). International Society for Optics and Photonics.

Banker, R. D., Charnes, A., & Cooper, W. W. (1984). Some models for estimating technological and scale inefficiencies in data envelopment analysis. *Management Science, 30*, 1078–1092.

Barnett, V. (1976). The ordering of multivariate data. *Journal of the Royal Statistical Society: Series A*, 318–344.

Barr, R. S., & Durchholz, M. L. (1997). Parallel and hierarchical decomposition approaches for solving large scale data envelopment analysis models. *Annals of Operations Research, 73*, 339–372.

Bessent, A., & Bessent, W. (1980). Determining the comparative efficiency of schools through data envelopment analysis. *Education Administration Quarterly, 16*, 57–75.

Bessent, A., Bessent, W., Kennington, J., & Reagan, B. (1982). An application of mathematical programming to assess productivity in the houston independent school district. *Management Science, 28*, 1355–1475.

Bessent, A., & Kennington, J. (1980). *A primal simplex code for computing the efficiency of decision making units*, version 2.0. Technical report, Center for Cybernetics Studies, University of Texas at Austin.

Bougnol, M.-L., & Dulá, J. H. (2009). Anchor points in DEA. *European Journal of Operational Research, 192*, 668–676.

Bougnol, M. L., Dulá, J. H., & Rouse, P. (2012). Interior point methods in DEA to determine non-zero multiplier weights. *Computers & Operations Research, 39*, 698–708.

Briec, W. (1998). Holder distance function and measurement of technical efficiency. *Journal of Productivity Analysis, 11*, 111–131.

Caron, R. J., Greenberg, H. J., & Holder, A. G. (2002). Analytic centers and repelling inequalities. *European Journal of Operations Research, 143*, 268–290.

Charnes, A., Cooper, W. W., Golany, B., Seiford, L., & Stutz, J. (1985). Foundations of data envelopment analysis for Pareto-Koopmans efficient empirical production functions. *Journal of Econometrics, 30*, 91–107.

Charnes, A., Cooper, W. W., & Rhodes, E. (1978). Measuring the efficiency of decision making units. *European Journal of Operational Research, 2*, 429–444.

Charnes, A., Cooper, W. W., & Rhodes, E. (1981). Evaluating program and managerial-efficiency: An application of data envelopment analysis to program follow through. *Management Science, 27*, 607–730.

Charnes, A., Cooper, W. W., Seiford, L., & Stutz, J. (1982). A multiplicative model for efficiency analysis. *Socio-Economic Planning Sciences, 16*, 223–224.

Chen, W.-C., & Cho, W.-J. (2009). A procedure for large-scale DEA computations. *Computers & Operations Research, 36*, 1813–1824.

Chen, W.-C., & Lai, S.-Y. (2017). Determining radial efficiency with a large data set by solving small-size linear programs. *Annals of Operations Research, 250*, 147–166.

Cutler, A., & Breiman, L. (1994). Archetypal analysis. *Technometrics*, *36*, 338–347.
Donoho, D., & Stodden, V. (2003). When does non-negative matrix factorization give a correct decomposition into parts? In *Advances in neural information processing systems* (Vol. 16).
Dulá, J. H. (2008). A computational study of DEA with massive data sets. *Computers & Operations Research*, *35*, 1191–1203.
Dulá, J. H. (2011). An algorithm for data envelopment analysis. *INFORMS Journal on Computing*, *23*, 284–296.
Dulá, J. H., & Hickman, B. L. (1997). Effects of excluding the column being scored from the DEA envelopment LP technology matrix. *Journal of the Operational Research Society*, *48*, 1001–1012.
Dulá, J. H., & Helgason, R. V. (1996). new procedure for identifying the frame of the convex hull of a finite collection of points in multidimensional space. *European Journal of Operational Research*, *92*, 352–367.
Dulá, J. H., & López, F. J. (2009). Preprocessing DEA. *Computers & Operations Research*, *36*, 1204–1220.
Dulá, J. H., & López, F. J. (2012). Competing output-sensitive frame algorithms. *Computational Geometry: Theory and Applications*, *45*, 186–197.
Dulá, J. H., & Thrall, R. M. (2001). A computational framework for accelerating DEA. *Journal of Productivity Analysis*, *16*, 63–78.
Edvardsen, D. F., Forsund, F. R., & Kittelsen, S. A. C. (2008). Far out or alone in the crowd: A taxonomy of peers in DEA. *Journal of Productivity Analysis*, *29*, 201–210.
Farrell, M. J. (1957). The measurement of productive efficiency. *Journal of the Royal Statistical Society. Series A*, *120*, 253–290.
Gillis, N. (2014). *The why and how of nonnegative matrix factorization*. arXiv:1401.5226.
Hill, W., Stead, L., Rosenstein, M., & Furnas, G. (1995). Recommending and evaluating choices in a virtual community of use. In *Proceedings of the SIGCHI conference on human factors in computing systems* (pp. 194–201). Denver, CO.
Karmarkar, N. (1984). A new polynomial-time algorithm for linear programming. *Combinatorica*, *4*, 373–395.
Khezrimotlagh, D., Zhu, J., Cook, W., & Toloo, M. (2019). Data envelopment analysis and big data. *European Journal of Operational Research*, *274*, 1047–1054.
Krivonozhko, V. E., Forsund, F. R., & Lychev, A. V. (2015). Terminal units in DEA: Definition and determination. *Journal of Productivity Analysis*, *43*, 151–164.
Lee, D. D., & Seung, H. S. (1999). Learning the parts of objects by non-negative matrix factorization. *Nature*, *401*, 788–791.
Mostafaee, A., & Soleimani-damaneh, M. (2014). Identifying the anchor points in DEA using sensitivity analysis in linear programming. *European Journal of Operational Research*, *237*, 383–388.
Rockafellar, R. T. (1970). *Convex analysis*. Princeton, NJ: Princeton University Press.
Salmani, G. (2018). *Rotating supporting hyperplanes and snug circumscribing simplexes*. Ph.D. dissertation, Virginia Commonwealth University, Richmond, VA.
Sueyoshi, T., & Chang, Y. L. (1989). Efficient algorithm for additive and multiplicative models in data envelopment analysis. *Operations Research Letters*, *8*, 205–213.
Tukey, J. (1974). Mathematics and the picturing of data. In *Proceedings of the international congress of mathematicians* (pp. 523–531). Vancouver, BC.
Wright, S. J. (1997). *Primal-dual interior-point methods*. Philadelphia, PA: SIAM.

Chapter 3
An Introduction to Data Science and Its Applications

Alex Rabasa and Ciara Heavin

3.1 Introduction

Data science has become a fundamental discipline, both in the field of basic research and in the resolution of applied problems, where statistics and computer science intersect. Thus, from the perspective of the data itself, machine learning, operation research, methods and algorithms, and data mining techniques are aligned to address new challenges characterised by the complexity, volume and heterogeneous nature of data. Researchers affirm that data science is not just another discipline, but it responds better to the concept of a new scientific paradigm (Hey et al. 2009).

Consultants, vendors and the media continue to propagate technology trends and fads. This can lead to confusion when terms that have different meanings are used indiscriminately. It is common for the terms big data, data science and data mining to be used interchangeably; however, they are not the same. Without going into the technical details, it is important to clarify that data science, apart from a paradigm, is a discipline in which many areas of expertise converge (statistics, computing, decision theory and artificial intelligence, among others) and that puts data as a central element of any hypothesis or modelling. Provost and Fawcett (2013) define data science as "a set of fundamental principles that support and guide the principled extraction of information and knowledge from data" (p. 51). Big data, a very fashionable term, is understood as an integral process from collection, storage, analysis to the final representation of large volumes of data using advanced technologies (Chen et al. 2012), while data mining aims to extract valuable knowledge from data. Data mining provides techniques for the data analysis phase. From a broader perspective,

A. Rabasa (✉)
Center of Operations Research, University Miguel Hernandez, Elche, Alicante, Spain
e-mail: a.rabasa@umh.es

C. Heavin
Business Information Systems, University College Cork, Cork, Ireland

V. Charles et al. (eds.), *Data Science and Productivity Analytics*,
International Series in Operations Research & Management Science 290,
https://doi.org/10.1007/978-3-030-43384-0_3

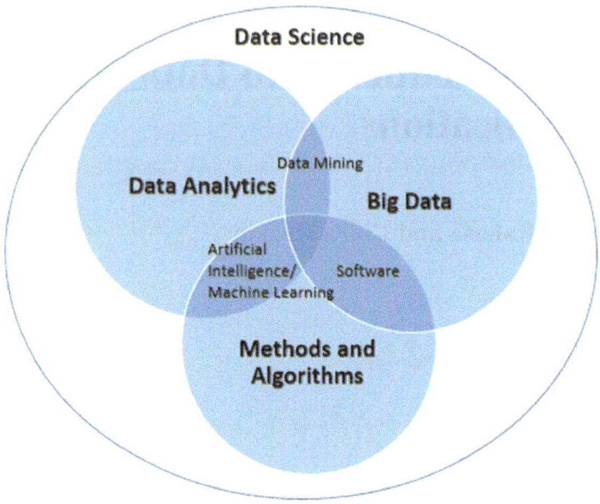

Fig. 3.1 Positioning data science, big data and data mining

data science is not restricted to big data contexts, nor does it focus exclusively on formalising and testing analytical techniques handled by data mining. Figure 3.1 illustrates the relationship between data science, big data, and data mining.

From an algorithmic perspective, much research is focused on the design of new predictive models that incorporate increasingly precise heuristics; the creation of new classifiers dynamically adaptable to data stream contexts; and the design of new optimisation algorithms. Data stream research is oriented to design and test new scalable and secure cloud architectures, as well as the design and application of parallel and ubiquitous computing methodologies. From a practical viewpoint, data science is required to tackle problems across a range of application areas, including healthcare, public services, financial services, retail, telecommunications, and manufacturing. In each of these application areas, the objective is to be able to extract the largest (and best) knowledge from the available data, facilitating the decision-making process with the aim of improving decision outcomes. To achieve this, a data scientist requires both domain knowledge and a broad set of quantitative skills (Waller and Fawcett 2013).

Data science and decision-making are inextricably linked, as data science enables an in-depth analysis of data to support data-driven decision-making (Provost and Fawcett 2013). "The ultimate goal of data science is improving decision making, as this generally is of paramount interest to business" (Provost and Fawcett 2013, p. 53). With this in mind, this chapter explores data science and its applications, specifically focusing on classification rules to enable better decision-making. Section 3.2 considers different types of decision support systems (DSS) and highlights how rule systems are practically implemented at the core of DSS. Section 3.3 explores data and model-driven approaches driving organisational decision-making. The principles of classification methods are formally described in Sect. 3.4, where the fundamentals of decision trees and associated metrics are described, as well as generation and reduction rule systems procedures. Examples of classification rules applications are outlined in Sect. 3.5. Here we describe four case studies explaining the input data sets, the data science analysis tasks, the classification model generation and the strategic decisions based on such rule systems. Section 3.6 leverages the cases by characterising these four DSS and elucidating the role of the different classification algorithms in defining the rules set for each of the decision scenarios. Finally, this chapter concludes by outlining some opportunities for future research on decision-making based on classification rules systems.

In summary, in this chapter the authors intend to provide a review of data science techniques in the field of predictive classification models. This review is accompanied by specific cases of application in areas as different as e-commerce, tourism, public health and the analysis of social network data. The examples presented show how the application of data science techniques contributes to improving the efficiency of decision systems, regardless of their application framework.

3.2 Characterising Decision Support Systems (DSS)

The term decision support system (DSS) was first used by Gorry and Scott-Morton (1971). They advocated that supporting information systems for semi-structured and unstructured decisions should be characterised as a DSS. Further, Power (1997) defined DSS as "an interactive computer-based system or subsystem intended to help decision makers use communications technologies, data, documents, knowledge, and/or models to identify and solve problems, complete decision process tasks, and make decisions. Decision support system is a general term for any computer application that enhances a person or group's ability to make decisions. In general, decision support systems are a class of computerised information system that supports decision-making activities". A DSS is a type or category of computerised information systems that supports decision-making activities in organisations (Power and Heavin 2017). The term DSS is quite and may be used to refer to different systems or types of technologies that support the decision maker.

According to Sprague and Watson (1979), at its core, a DSS involves a database, a model, a user interface and a decision maker. A database "consists of the database and the software system for managing it" (Sprague and Watson 1979, p. 63). A model may rely on internal organisational data and/or external depending on whether the model is strategic, tactical or operational (Sprague and Watson 1979). Further, Sprague and Watson (1979, p. 64) purported that "there are no huge corporate models built from scratch, but small tentative models tested, integrated, revised, and tested again". A range of modelling approaches may be leveraged in DSS, from a well-defined knowledge domain, for example accounting where definitional relationships and formulas are used to calculate the consequences of actions, to representational and optimisation models (Alter 1977). The latter are leveraged where the data is more complex and the decisions are semi-structured or unstructured in nature. These approaches do not offer definitive answers rather they predict possible outcomes and make suggestions based on the data available (Alter 1977). More recently, the

Table 3.1 Five types of decision support system (adapted from Power and Heavin (2017))

Decision support system	DSS description
Communications-driven DSS	Offers cooperation and collaboration capabilities to two or more decision makers to enhance group decision-making. DeSanctis and Gallupe (1987, p. 589) referred to group decision support systems (GDSS) as "sophisticated rule-based systems that enable a group to pursue highly structured and novel decision paths"
Data-driven DSS	Leverages large volumes of organised, accessible data for analysis, visualisation and reporting. Early versions of data-driven decision support systems were called data-oriented DSS (Alter 1980) or retrieval-only DSS (Bonczek et al. 1981). Some data-driven DSSs use real-time data to assist in operational performance monitoring
Document-driven DSS	Provides document retrieval and analysis capabilities. Document-driven DSSs are also classified as a type of knowledge management system
Knowledge-driven DSS	Stores knowledge that may be used by decision makers to provide domain-specific knowledge to a variety of specific problems and tasks
Model-driven DSS	Leverages different quantitative models to offer decision alternatives based on well-defined constraints and business rules. Early versions of model-driven DSSs were called model-oriented DSS (Alter 1980) and computationally oriented DSS (Bonczek et al. 1981)

elements of a DSS have been extended to include DSS architecture, network and analytical capabilities (Power and Heavin 2017).

Power (2001, 2008) proposed five generic DSS types. These are identified and defined based upon the dominant technology component. The types of DSS include: (a) communications-driven; (b) data-driven; (c) document-driven; (d) knowledge-driven; and (e) model-driven (Power 2001, 2008) (see Table 3.1).

The different types of DSS may be used to provide decision makers with the right data at the right time to serve a particular decision need in a specific knowledge domain. Existing research explains that the differentiator between the two approaches is the key architectural component underpinning the DSS, that is, a model versus a data set (Power and Sharda 2007). Model-driven and data-driven DSS are the primary types of DSS that one might consider developing using a simple spreadsheet package (Power and Sharda 2007). Spreadsheet capabilities are appropriate for building a DSS using one or more simple models. A developer implements the model and then adds form objects and other tools to support a decision maker in "what if?" and sensitivity analysis (Power and Sharda 2007). A data-driven DSS can also be developed using spreadsheet technology. For example, the developer downloads a large data set to the DSS application from a database, a website or a flat file, for example csv. Pivot tables and other complementary data visualisation techniques are used in helping a decision maker to summarise, manipulate and understand the decision relevant data

(Power and Sharda 2007). Arguably, both approaches need to access and process data sets or a "database" to support decision-making by managers and other stakeholders (Sprague and Watson 1979; Power and Heavin 2017).

The insurance industry was among the earliest to use computerised DSS capabilities to support their structured and semi-structured decision-making needs (Alter 1980). Data-driven models are used by insurance organisations to improve fraud detection and other criminal activity, as well as cross sell and upsell to existing customers (Exastax 2017). In terms of model-driven decision support, a modelling approach can be used to determine customer segmentation and predict the potential value of the customer to the insurance provider (Verhoef and Donker 2001), and to determine business forecasts based on assets and insurance liability (Rusov and Misita 2016).

With the increased focus on developing digital tools in the health domain, there are many examples of DSS in healthcare. Given the nature and complexity of the area, decision support needs vary across healthcare services and knowledge workers. For example, administrative decision makers need data-driven DSS for performance monitoring, reporting and improved cost control (Murdoch and Detsky 2013). These decision makers could also benefit from more model-driven DSS to improve the scheduling and rationalisation of scarce resources (Murdoch and Detsky 2013), that is, staff scheduling and bed allocation. Clinicians are keen to embrace predictive modelling approaches such as decision trees to better diagnose headache disorders (Potter et al. 2018), to better assess patients for communicable disease (Hardy et al. 2017), and to better predict patient outcomes (Hunt et al. 1998). The health domain continues to generate an abundance of data; subsequently, the need for data-driven decision support is prevalent (Murdoch and Detsky 2013). The next section briefly considers data science in the context of DSS and decision-making.

3.3 From Data Science to Decision-Making

3.3.1 Model-Driven and Data-Driven Approaches

Data science advocates principles, processes, and techniques for understanding the phenomena through automated data analysis (Provost and Fawcett 2013). The goal of this approach is to utilise sophisticated analysis techniques to improve decision-making. Leveraging Alter's (1977) classification of decision support systems (DSS) and Provost and Fawcett's (2013) approach to data science and its relationship to data-driven decision-making in the organisation, Fig. 3.2 illustrates "the simple dichotomy" (Alter 1977) between model-driven and data-driven approaches. It is important to acknowledge that these two approaches are highly complex and nuanced, with some commonality.

Model-driven approaches attempt to capture knowledge and derive decisions through defining explicit representations and rules through the use of simulation

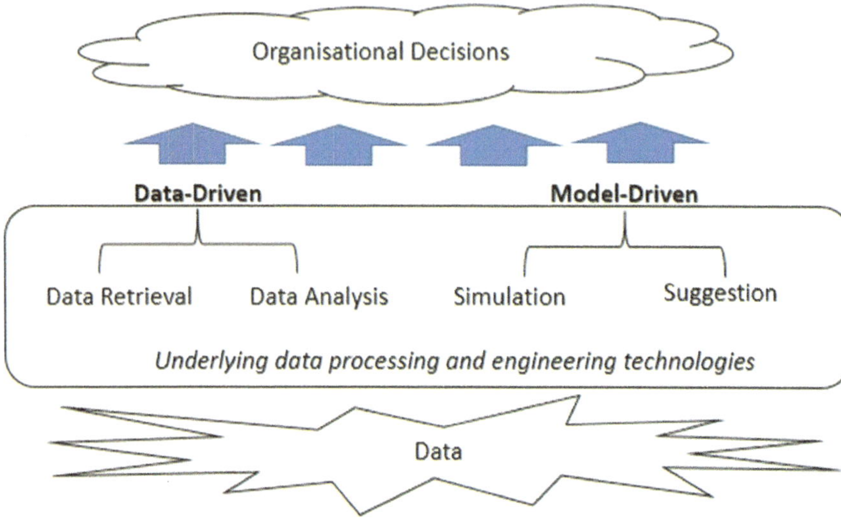

Fig. 3.2 Model-driven versus data-driven approach. Adapted from Alter (1977) and Provost and Fawcett (2013)

and optimisation technologies (Power and Sharda 2007). Model-driven approaches are used to support and assist in formulating alternatives, analysing a range of possibilities, and interpreting and identifying suitable options (Power and Heavin 2017). A data-driven approach involves identifying the right answer based on interrogating, retrieving, and analysing a large data set(s). The availability of inexpensive data storage, fast processors, and advancements in neural net algorithms and other data-centric techniques has made it possible to derive significant value from data (Ashri 2018). According to Power and Heavin (2018) modern decision support may have a well-defined data-driven subsystem and a model-driven subsystem. A modern DSS may be characterised to include: "(1) Broad domain of applications with diverse functionality, (2) Faster access to data stored in very large data sets, (3) Faster deployment, (4) Faster response, (5) Integrated DSS with TPS, multiple decision support subsystems, (6) Lower cost per user, (7) Multi-user and collaborative interaction, (8) Real-time data and real-time DSS use, (9) Ubiquitous, (10) User friendly and a better user experience, and (11) Visualisation" (Power and Heavin 2018, p. 18). These modern DSSs are designed to support data-based decision-making in organisations (Power and Heavin 2018). Analytic capabilities are important in data-driven and model-driven DSS and analysis with quantitative and statistical tools (Power and Heavin 2018), so the next section considers two types: (1) descriptive and (2) predictive.

3.3.2 Descriptive Versus Predictive Models

It is assumed that descriptive models try to explain the existing data up to a given moment in time. They use techniques capable of finding and modelling changes of tendency and patterns. Descriptive models do not contemplate the existence of a specific objective variable. Predictive models try to infer possible future values of specific target variables, from their previous values and their direct or indirect correlations with other variables (explanatory) that intervene in the problem.

Within the descriptive scope, the most frequently used analytical tasks are:

- Segmentation or grouping (Clustering): consists of generating groups of items according to criteria of similarity between members of the same group. The reference algorithm is the K-means (MacQueen 1967) that has been improved from different perspectives as CLARA (including random seeds) or PAM (implementing faster procedures). The clustering algorithms have been applied successfully for many years in very different scopes, such as detection of abnormal values (He et al. 2003), or classifying customers on risk of abandonment (Bloemer et al. 2003). Although the techniques of segmentation or grouping can deal with discrete variables, the most frequent thing is to find them working on variables of a continuous nature. Data scientists are making great efforts to develop new clustering techniques adapted to data stream contexts (Yin et al. 2018).
- Association consists of finding recurring patterns, that is, groups of values that tend to occur simultaneously, within a sample. The Apriori (Agrawal and Srikant 1996) algorithm is the main reference in association. It has also been studied and updated from different approaches, such as the case of AprioriTid (Adamo 2001) with a parallel approach. As a reference algorithm, Apriori has been widely used for a long time in very different contexts. Association techniques do not consider any particular objective variable and deal only with discrete variables.

Within the predictive scope, analytical methods are framed in these two types of tasks:

- Classification consists of modelling an objective variable of discrete nature (not continuous). Among the most used classification methods is the classification trees, where ID3 is the main reference algorithm (Quinlan 1986). As with the reference algorithms cited above, ID3 has been modified with a series of variants such as C4.5 (Quinlan 1993) that is able to handle numerical antecedent variables. Supervised learning in quest (SLIQ) (Mehta et al. 1996) is another classifier that manages numerical and categorical antecedents by maintaining an ordered list of continuous attributes and another ordered list of classes (discrete attributes). Different classification algorithms have also been applied successfully in areas as diverse as medicine (Lashari et al. 2018), transportation, telecommunication (Tsami et al. 2018) and finance (Pérez-Martín et al. 2018).
- Regression: consists of modelling a target variable of a continuous nature (whole numbers or decimals and dates, mainly). The methods to solve regression tasks range from the algorithms that generate systems of regression equations CART

(Breiman et al. 1984) or CHAID (Kass 1980), to temporary series analysis systems such as the ARIMA models (Box and Jenkins 1973). Predictive methods, independent of their details of operation and implementation, associate possible values to a certain objective variable together with a margin of precision. The accuracy of predictive models is one of its most important aspects. Apart from calculating their corresponding average accuracies, it is also important to identify (and quantify) which cases in the sample are predicted with greater precision and which with less.

Rule sets are an integral part of a DSS. They are used to model and predict a target variable based on a number of included input variables (Berner and La Lande 2016). Rule sets are typically context-specific; they are defined according to the environment in which the DSS is embedded (Bonczek et al. 1981). Rule sets are defined using selected classification models, such as ID3, CART and CHAID algorithms (Kumar et al. 2011). The resulting model then uses algorithmic processes to address the information needs of the decision maker (Sprague and Watson 1993). For example, forecasting models guide decision makers by analysing the underlying relationships between different variables based on a bounded set of parameters (Power and Sharda 2007). The principles on which classification is based, and the specific details of its operation, are explained in detail in the next section.

3.4 Principles of Classification Methods

3.4.1 The Type of Attributes

The choice of a specific analytical method is absolutely dependent on the type of data that must be handled in each case.

The input data sets to a classification process consist of:

- A single objective or dependent variable (also called consequent, in the rules that are going to be generated). On classification problems, the dependent variable is discrete (non-numerical) and its prediction or modelling is the objective of classification problems.
- Explanatory variables (also called antecedents in the rules that will be generated) may be discrete or continuous. The continuous explanatory variables will be someway discretised.

Discretisation is a process by which a numeric variable is segmented into a finite number of labels (García et al. 2013). Discretisation is sometimes carried out internally by the tree-generating algorithms themselves. Other times, a previous discretisation is used based on the frequency distributions of the variable, even resorting to criteria dictated by the experts in the context of the problem.

3.4.2 The First Decision Tree Algorithms

Decision trees are one of the most widely used predictive tools in data mining. Depending on the nature of the target variable, these trees can be: classification trees (objective variable of nominal or discrete type) and regression trees (objective variable of a numerical or continuous type). There are some trees capable of carrying out both classification and regression tasks. CART (Breiman et al. 1984) or M5 (Yang et al. 2017) are examples of algorithms that generate this type of classification/regression trees.

Figure 3.3 shows a WEKA application window with graphic output corresponding to a classification tree J48 (Java implementation of the classification tree C4.5), capable of processing discrete and numerical information in its input attributes. Each branch on the tree can be interpreted from the root (upper node) to each leaf (terminal nodes) as a classification rule where the different attributes are instantiated to possible values. Each leaf node is accompanied by the total number of occurrences corresponding to that rule (or its percentage of occurrence over the total number of input tuples).

There are multiple implementations for both classification trees and regression trees. Such implementations differ, fundamentally, in the pruning heuristics they use, in the expanding criteria they implement and in the different forms of tree paths. However, the biggest difference from the analyst's point of view is the nature of the variables (in this case the internal, non-terminal nodes of the branches). Thus, there are trees such as ID3 (Quinlan 1979), the main reference model in this scope which work only with nominal variables (both in the internal nodes and in the leaves). Others, such as C4.5 (Quinlan 2006) are also capable of handling numerical variables in their internal nodes. In fact, C4.5 is an evolution of the original ID3 where the authors incorporated a series of improvements. The most important of these improvements, from a functional perspective, was this ability to handle numerical attributes.

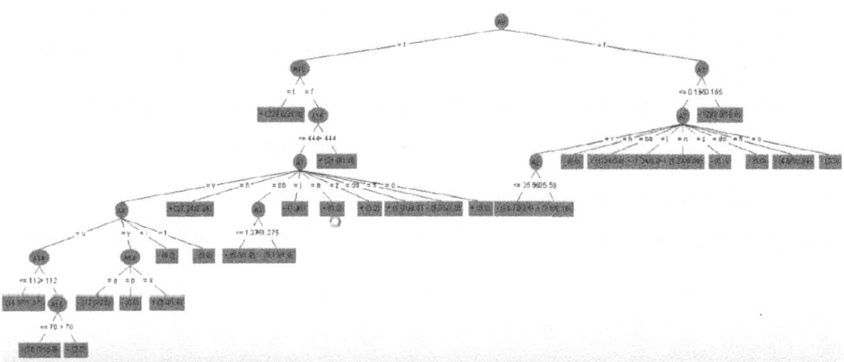

Fig. 3.3 J48 decision tree, WEKA

Other tree-generating algorithms are designed to solve both classification and regression problems. The reference model in this sense is CART (Breiman et al. 1984) from which numerous implementations and variants have been developed that are better suited to certain contexts of application.

Focusing on classification problems, the depth path of each branch of a tree from the root node to one of its leaves produces a classification rule. There are methods of generating rules which do not need to physically build the tree in memory. These methods are based on the classical technique of frequency counting where rules are formed from occurrences in the data set that exceed a certain minimum support. That is, they occur with a certain probability within the sample. This procedure is operational. It is described in more detail in Sect. 3.4.4.

3.4.3 Measuring Accuracy (Confusion Matrices)

Accuracy is the main measure of the overall quality of a predictive model. It basically consists on the proportion between correctly classified instances over the total number of classifications. Accuracy takes values between 0 and 1, with 1 being the maximum, when the total of the sample was correctly classified from the training set. However, since not all the values of the objective variable are predicted with the same precision, it is absolutely essential to measure how the precision is distributed over each of the possible values of the consequent.

The importance of identifying how prediction errors are distributed is highlighted with a simple example: medical tests to diagnose a certain disease, with a binary consequent. The results of the test of the disease can be either: "+" (have the disease) or "–" (not have it). In this case, the false positives are the errors made predicting that the result of the test would be "+" (and actually took value "–"), while the false negatives are the errors made predicting that the test was going to provide "–" as a result. The first mistakes tell the patient that the test for this disease has been positive (when in fact the patient is healthy). In this sense, the second errors are worse, because they are telling the patient that the test has been negative (and in fact, he/she is sick).

In this matrix (Table 3.2), 300 positive cases were correctly classified and 200 negative cases were correctly classified. However, 40 cases that were negative were diagnosed as positive, and 10 cases that were positive were classified as negative. These cases were incorrectly classified.

The medium accuracy is: (300 + 200)/(300 + 200 + 10 + 40) = 0.91 (91%).

Table 3.2 Confusion matrix, binary class example

It is ↓ and is classified as →	+	–
+	300	10
–	40	200

The confusion matrices are also applicable to non-binary consequents, that is, they are capable of taking n values, becoming square $n \times n$ matrices. In general, the total correctly classified instances are obtained by adding the positive diagonal of the confusion matrix. The average accuracy of the model is the quotient between success and the sum of all the values of the matrix.

3.4.4 Generating and Reducing Rule Systems

The problem of high dimensions in a data set (a high number of attributes) implies the need to eliminate those that are not very influential on the target variable. Thus, it is essential to use feature selection techniques which show the attributes (and combinations of them) that have a greater correlation with a certain target variable and distinguish them from those that have a practically random relationship with the variable to be modelled. The extraction of characteristics is especially important in problems where, in addition to many attributes, the data samples are not too large, that is, "wide" problems (Hall and Xue 2014). The extraction of relevant characteristics can be carried out applying different methodologies such as support vector machine (SVM) (Chapelle et al. 2002) and techniques such as principal component analysis (PCA) (Peres-Neto et al. 2005). This type of algorithm pursues an iterative approach to improve a certain metric of significance (Lu et al. 2014). Focusing on the business world, the automatic extraction of features has been critically relieved in different situations such as in customer relationship management (CRMs) of airline companies (Ismail and Husnayati 2013), in prediction of bankruptcy in financial systems (Tsai 2009), and in the prediction of stock breakage (Tsai and Hsiao 2010).

Irrespective of the mechanisms of feature selection, tree generation algorithms also have mechanisms to choose the attributes and their corresponding value ranges on which to build the tree under maximum information gain criteria in each node. The elimination of potential branches that do not pass during the generation of the predictive model is known as pre-pruning.

Some branches or rules can be deleted after their creation. This may occur for a number of reasons: for example, when the rule does not have significant support on the sample, or it incurs overlapping or contradictions with other rules of the same rule set, or it offers low confidence. The elimination of branches or rules after their generation is known as post-pruning task.

Sometimes, the reduction of the final rule system can be carried out with mixed pre-pruning and post-pruning techniques. An example of this that already operates on the rules without the need of generating the tree itself is RBS algorithm (Almiñana et al. 2012), which is based on the concepts of support and confidence of the rules.

RBS formally defines an antecedent as a group of instanced variables, without considering the class attribute:

$$A = \langle v_{s_1}, \ldots v_{s_p} \rangle,$$

with $v_{s^{\cdot}} \in X$ and $P \leq m - 1$.

The consequent is defined as follows:

$$C = \langle v_m \rangle.$$

This way, a rule, $r_k^{\overline{AC}} : \overline{A} \to \overline{C}$ with $k = 1, \ldots, n$ is a specific tuple of the original data set D.

where:

$\overline{A}$ denotes a specific tuple of antecedent values instancing antecedent A
$\overline{C}$ denotes a specific tuple of consequent values instancing consequent C

A rule set $RS_{A \to C}$, formed by n rules, is defined as follows:

$$RS_{A \to C} = \left\{ r_1^{\overline{AC}}, \ldots, r_n^{\overline{AC}} \right\},$$

Next, Fig. 3.4 shows an example of rule set. The first rule must be understood as follows:

The antecedent: (aloja = Hotels, motive = Leisure, A16 = Yes, ccaa = Canarias) occurs 8.52% on the sample. When this antecedent occurs, the target variable GastoTotalD is [17.1.67e+03] with 87.86% probability.

After that, RBS associates all the rules to a specific region depending on two-rule metrics:

Supp (%)	aloja	motivo	A16	ccaa	Conf (%)	GastoTotalD
8.52	Hotels	Leisure	Yes	Canarias	87.86	[17,1.67e+03)
8.70	Hotels	Leisure	No	Cataluña	80.51	[17,1.67e+03)
4.44	Over-The-Counter Accommodation	Leisure	No	Comunitat Valenciana	86.66	[17,1.67e+03)
3.89	Hotels	Leisure	Yes	Illes Balears	94.61	[17,1.67e+03)
3.64	Over-The-Counter Accommodation	Leisure	No	Cataluña	87.69	[17,1.67e+03)
3.45	Hotels	Leisure	No	Illes Balears	92.13	[17,1.67e+03)
3.81	Hotels	Leisure	No	Canarias	81.29	[17,1.67e+03)
3.62	Over-The-Counter Accommodation	Leisure	No	Andalucia	85.44	[17,1.67e+03)
3.09	Hotels	Leisure	No	Andalucia	86.11	[17,1.67e+03)
2.75	Over-The-Counter Accommodation	Leisure	No	Illes Balears	93.58	[17,1.67e+03)
2.74	Over-The-Counter Accommodation	Others	No	Comunitat Valenciana	93.42	[17,1.67e+03)
2.33	Hotels	Leisure	No	Comunitat Valenciana	92.41	[17,1.67e+03)
2.45	Hotels	Leisure	Yes	Cataluña	87.68	[17,1.67e+03)
2.13	Over-The-Counter Accommodation	Others	No	Cataluña	85.71	[17,1.67e+03)

Fig. 3.4 Output screen, rule set provided by RBS algorithm

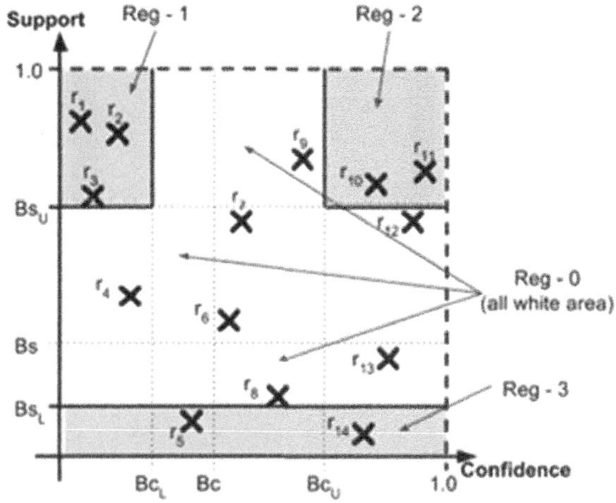

Fig. 3.5 Rules on their significant regions

- the antecedent support (the proportion of tuples in the data set containing the antecedent) $\overline{A}$
- the rule confidence (the proportion of tuples with antecedent, $\overline{A}$ that are also containing as consequent) $\overline{C}$

All rules are placed, according to coordinates: (Fig. 3.5) in its corresponding significance region, depending on its values: antecedent support (axis Y), and rule confidence (axis X). Only rules on regions 1 (discriminant rules), 2 (positive rules) and 3 (very infrequent cases) are able to provide significant information. So, rules on region 0 are deleted from the final rule set. Rabasa (2009) provides greater detail on the RBS methodology.

The high number of attributes in a data set usually forces the selection of those values that have a greater correlation with the variable to be predicted. This can be done with ad hoc techniques of feature selection before generating the predictive models (SVM, PCA, …). Besides, rules that provide poor information with low support or confidence are frequently deleted by pre-pruning or post-pruning procedures.

The classification rule sets to be incorporated as a predictive engine in the DSS come from algorithms that generate classification rules (sometimes from trees of type ID3, C4.5, other times from incremental iterative methods of construction of the rules). The classification trees in each branch represent a rule that indicates a series of conditions to be fulfilled from the root node to each of the leaves.

These rule sets, despite the pre-pruning of one type or another in the process in which they were generated, usually require a new filter that selects the most reliable ones to be transferred to the core of a DSS or an expert system. Figure 3.6 illustrates the generation of classification rules.

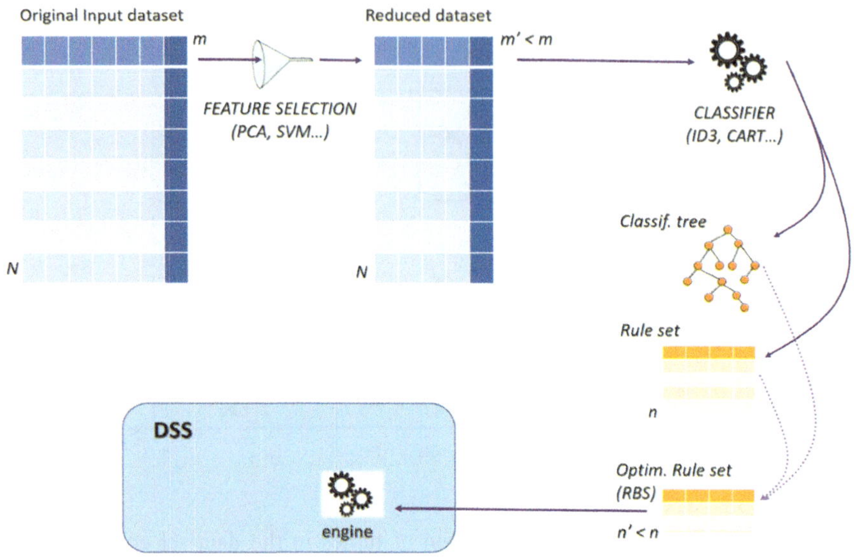

Fig. 3.6 From the data set to the optimised rule system

3.5 Real Applications of Classification Methods

This section presents four cases where the classification methods outlined above are applied. These cases have been chosen because they are diverse, not only in terms of application context, and input data structures, but also in terms of the general data management and forecasting precision ratios employed.

In order to facilitate understanding and comparison, each case is divided into four sections: the problem, the data, the data science techniques and the classification model and decisions based on rules.

3.5.1 Predicting Customer Behaviour on Vehicle Reservations (Rent-a-Car Company)

- The problem:

An international rent-a-car company manages on-line reservations, but according to market-specific policies they do not charge a fee for making the reservation. However, this causes large economic losses in the company due to the opportunity costs derived from trying to satisfy the demands of customers who, despite having made the reservation, did not pick up their vehicle. This amounted to more than 14% over the sample and they were named as "not-shown" by the company.

The other two possible types of reservation were: cancellations (they did not generate benefits, but neither generated losses) and effective (those that resulted in a service).

The company is keen to discover the patterns that lead to this type of unsuccessful booking ("not-shown"), searching among the database of online reservations.

- The data:

There is a database of online reservations corresponding to the last 3 years of customer bookings. With a total of 734,352 reservations, each of which had a maximum of 17 fields, the nationality, the advance with which the reservation was made, the details of the type of vehicle and extra services, and office where the vehicle is to be collected. The target variable (variable to be classified) is "state of the reservation" which can be: ok, cancel or not-shown.

- The data science techniques and the classification model:

First, pre-processing tasks were carried out over the original data set. Some outliers were detected and deleted. Some variables were discretised and techniques for automatic feature selection were applied to reduce the number of relevant attributes from 17 to 8.

A classification tree of the ID3 family was used with subsequent rules filtering and ordering by using RBS method, focusing on not-shown patterns.

- Decisions based on rules:

The rules revealed the reservation patterns that most likely resulted in customers "not-shown". For example, young customers reserving very cheap cars for some specific destinations, with no extra services (no GPS, no extra insurance, etc.), and reserving only 2 or 3 days in advance, had very high probability (up to 80%) of "not-shown". Implementing these rules (patterns) on a business intelligence engine, the company could anticipate, in real time, those reservations which according to the profile of the same had a high probability (higher than a certain threshold that the company was adjusting) of not showing up or cancelling. As a result, the company implemented the decision to require these reservations types to pay a deposit as part of the vehicle reservation process.

3.5.2 *Extracting Spending Patterns on Tourism (Tourism Valencian Agency)*

- The problem:

Due to the diversification of the tourist offering in Mediterranean countries, in certain areas the profiles of foreign visitors are changing rapidly. Although sun and beach tourism is still the main attraction, other destinations and activities now motivate cultural, oenological or even work-related trips. Each of these typologies is characterised by particular spending patterns. The public authorities responsible for tourism (one of the main industries of the Valencian Community) highlighted the need to predict the spending thresholds of tourists visiting south-east Spain. This was achieved by collecting tourist data via thousands of surveys that reflected tourist travel patterns in the region capturing their consumption of transport, accommodation and/or leisure services.

- The data:

The original data set had a total number of 22,742 records with 285 variables. So, several target variables were selected, including expenses on leisure, transport, accommodation or total expense. So, a concrete data set was created for each expense.

- The data science techniques and the classification model:

In this case, as the data were collected from thousands of surveys, a thorough pre-processing stage was necessary where, among other actions, records with inconsistent (or outliers) values were eliminated, values of certain attributes were recoded and some numerical values were assigned to concepts previously responded in an aggregate manner. The expenses by type were discretised by using segmentation criteria provided by K-means cluster algorithm (see Fig. 3.7).

For each of the target expenditure variables, CART algorithms were designed and both classification trees and regression models were obtained that calculated the expenditure with error margins below 15%.

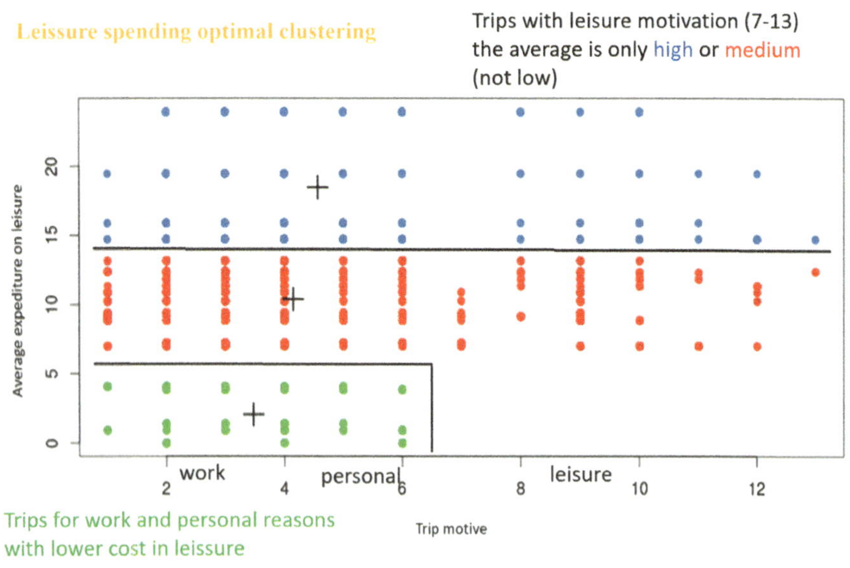

Fig. 3.7 Expenditure on leisure discretisation from K means cluster segmentation

- Decisions based on rules:

Based on the detected travels and spending patterns, the tourist board was able to better organise their promotional campaigns at international fairs, that is, offer discounts on trips hired at the destination and offer targeted promotions encouraging certain overnight stays focusing on certain destinations. They were able to design these promotional products from a reliable prediction of how the expense would be developed according to the traveller circumstances and travel typologies identified.

3.5.3 Avoiding Unnecessary Pre-surgery Tests (Healthcare)

- The problem:

In the Valencian community (Spain) public health service, before a patient enters the operating room to undergo a major surgery (where anaesthesia is required), he/she is always subjected to three tests: a complete blood test, an electrocardiogram and chest X-ray plate. Public health administrators encounter the problem of organising and conducting expensive and/or invasive pre-surgery tests on patients. Although these tests are mandatory and standard as ordered by anaesthesiology services, it is critical to schedule the anaesthesiology process during the surgery. In fact, often some of these tests do not provide relevant information.

In certain scenarios, medical experts question the necessity of conducting such pre-operative tests, such as an X-ray of a patient who has never had a pneumopathy. The health services could avoid unnecessary expense from additional extraneous medical assessments that must be informed by specialist consultants, such as cardiologists, pulmonologists and radiologists. These additional assessments also slow down patient's time to operation.

This is a critical decision: The elimination of a pre-operative test can only be taken in cases with sufficient statistical support and with 100% confidence indicating that there is no need for a specific patient assessment (patterns located in the upper right quadrant of the RBS system).

- The data:

A data set of more than 200 attributes (some multi-labelled) and less than 1,000 records was a very difficult challenge to address. Given the high number of attributes (compared to the small number of instances), it was necessary to divide the input sample by pathologies and type of surgery, so that in each analysis it was possible to dispense with all the attributes that corresponded to different pathologies. The data were collected by a multi-form application (Fig. 3.8).

- The data science techniques and the classification model:

The original sample was divided by type of surgery, and all blank columns were deleted. After the automatic selection of most relevant attributes, each subsample

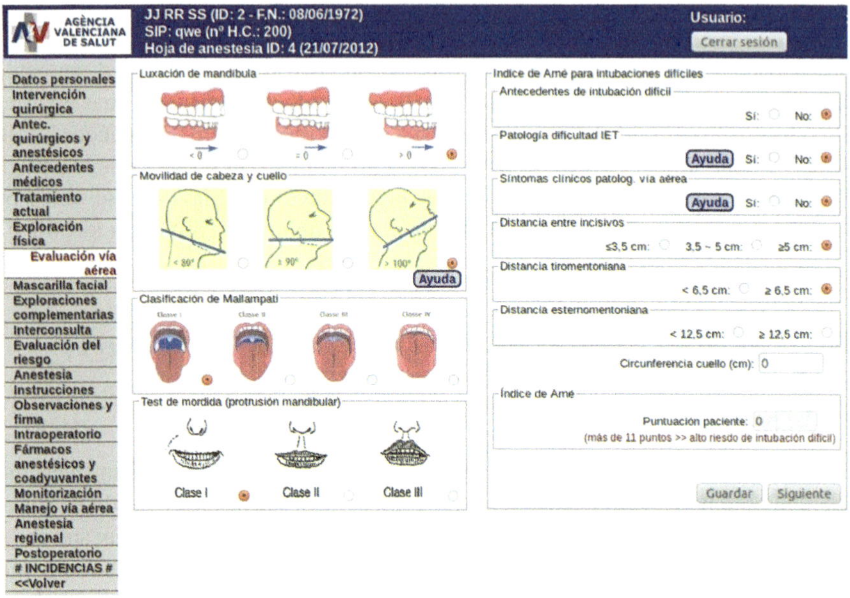

Fig. 3.8 Input data form number 7 (of 20) for patient evaluation of the respiratory tract

was submitted to association tasks (Apriori algorithm). Subsequently, classification algorithms (type ID3) were run for each subsample providing the corresponding rule sets that were filtered by operation_status = OK.

- Decisions based on rules:

No rules were obtained with 100% confidence and enough support to suggest the elimination of any medical tests. As a result of this study, hospital managers were able to demonstrate empirically, based on the available data set, that no pre-surgery test could be safely eliminated.

3.5.4 Classifying Violent/Radicalism on Twitter (Security Surveillance)

- The problem:

Many users freely express their opinions on different events through a variety of social media channels. For example, some of the messages that are posted on Twitter reveal potentially dangerous radicalisation movements. Academic and police authorities in Europe are addressing the problem of early detection of radicalisation, based on the available messages (tweets).

- The data:

The original input data was extracted from Twitter databases through the developers API. The data set is related to the London Bridge and Charlie Hebdo terrorist attacks. Almost half a million tweets (records) were collected. They consisted of 15 attributes (for Charlie Hebdo sample) and 17 (for London Bridge sample).

- The data science techniques and the classification model:

This case is characterised by a hybrid classification: by content (by accumulation of words previously catalogued as "radical" by expert criminologists) and by context (from environmental parameters of the tweet and the account, previously identified as very significant in contents of violent speeches) (Esteve et al. 2018). For example, a tweet containing the words: "kill", "bomb", "blood" (or some other derivatives) is automatically classified as radical by its content. If it also responds to certain patterns (such as not having geolocation active, exhausting the maximum space of characters, including URLs and being close in date to certain events) the tweet is also classified as radical by context (see Fig. 3.9).

The *RandomForestClassifier* model, provided by Python, was parameterised in this case study, for tweets classification.

- Decisions based on rules:

At the end of the tweet classification process, the DSS provides a list of Twitter users ordered by degree of potential radicalisation, in a given context and period of time. From this, national security authorities and social services can carry out preventive actions either by monitoring and controlling more interconnected users with the potentially radical individual, or by directly infiltrating their home/family environment.

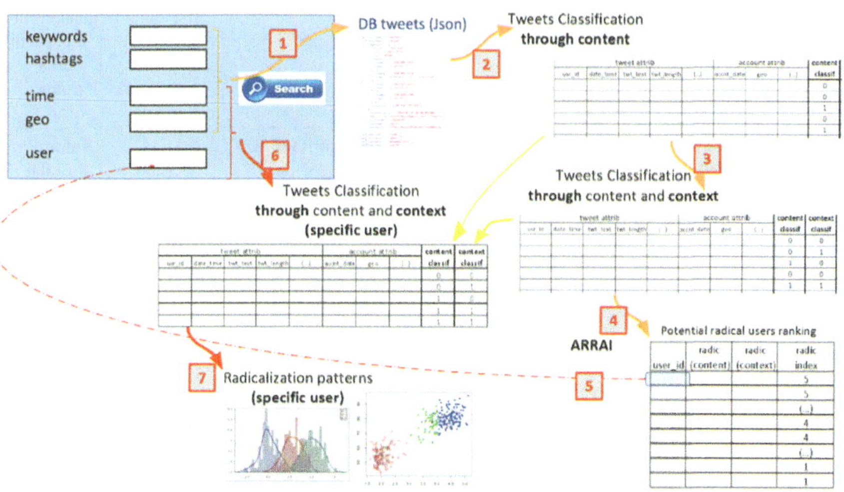

Fig. 3.9 General process for hybrid tweets radicalisation

As illustrated in the four cases presented above, classification rule systems derived from small data sets may form the core of a computerised DSS in diverse fields of application.

3.6 From Data Science to DSS in Four Scenarios

This chapter moves beyond building models based on spreadsheets (Power and Sharda 2007), to leverage data sets or "databases" to build classification rules for key decision makers in four diverse settings. The cases presented in Table 3.3 pursue a data-centric approach, and the model is built based on four diverse historic data sets or "snapshots" of data by case.

As highlighted in Table 3.3, in each case a relatively small data set (in big data terms) was used to build and train the respective models. In the four scenarios, the

Table 3.3 Synthesis of the four cases

Scenario/Decision maker	Data set description	Decision problem	Classification	Decision rules	Decision
Car rental/Management	734,352 Customer booking records with 17 variables	Identify likelihood of a customer "no-show"	Preprocessing, ID3, RBS method	Detected customers most likely to "no show"	Specific customers required to pay booking deposit to secure vehicle
Tourism/Valencian Agency of Tourism	22,742 customer expenditure records with 285 variables	Predict tourist spending threshold in specific geographic region	Preprocessing, K means cluster algorithm	Detected customer spending patterns and behaviours	Targeted promotions and events for specific tourist groups
Surgical—Public Hospital/Hospital managers	>1,000 patient records with 200 attributes	Identify unnecessary patient tests in surgical scenario	Apriori algorithm, ID3	No decision rules obtained with confidence	Inform hospital managers that no test may be ruled out based on available data
Terrorism/Security Authorities	Almost 500,000 Twitter records with 15 and 17 attributes	Identify radicals to prevent terrorist activity	Hybrid classification—by content and context, RandomForestClassifier model	Detected potential radical individuals in an ordered list	Increased monitoring of suspect individuals

data was pre-processed and analysed using a range of classification techniques. In three of the four scenarios, there was a positive outcome based on the data analysis and rules, where decision makers were presented with a rule set that they could act on. In the car reservation scenario, patterns were identified to support managers to better predict the customer base that were most likely not to show up ("not shown") for their vehicle rental. Young customers making low value bookings for specific destinations, with no extra services were identified as the most likely "no-shows". Subsequently, management implemented the decision to levy a booking deposit on these customer types. Considering the tourism scenario, the spending patterns of tourists were analysed providing new insights into tourist travel movements and prospective spending patterns. Based on these insights, the Valencian tourist board took the decision to design and implement new promotional activities and events targeted at specific groups of tourists. The analysis of almost half a million tweets allowed us to devise a rule set to identify potentially threatening radicalised individuals. From this, national security surveillance has been implemented to monitor this situation in an attempt to prevent future terrorist events.

The healthcare scenario was exceptional, and with the data set available we were unable to determine individual patients where a pre-operative medical assessment could be ruled out. This finding was reported to hospital managers. Noteworthy, the patient assessment scenario is highly complex with over 200 attributes and the potential negative outcome (risk) from not conducting the correct patient tests is high. In the case of a classification model where the rule set is intended to replicate the decision-making of a subject-matter expert, uncertainty can be reduced by using expert input (Uusitalo et al. 2015) and model testing using simulation data (Power and Sharda 2007). This surgical scenario is highly risky and complex. The model requires a larger data set to further train and refine the model. For the foreseeable future, it is highly likely that this type of decision scenario will require expert input and the DSS will be leveraged primarily as support. The four cases presented in Table 3.3 are not automated decision-making systems and (Power and Sharda 2007), uncertainty in any of these cases may be used by the decision maker to support the decision based on the information presented by the DSS.

3.7 Opportunities for Future Research

The objective of this chapter was to leverage data science approaches, specifically focusing on building classification rules to improve data-based decision-making in organisations. To achieve this, we analysed static data sets to produce classification models that are also static. Implemented as part of a DSS, this approach provided the four diverse organisations with key insights that allowed them to leverage a data-based approach to their decision-making.

One of the biggest difficulties to overcome is the management of business key performance indicators (KPIs) that change over time, both in terms of how they are defined and also in terms of their relative weight on the target variables should be predicted. The challenge is twofold. First, we need to design dynamic predictive models capable of generating answers in real time (or at times assumable by the context of the problem) without renouncing the precision of static models. Second, moving from data snapshots to a real-time view of data provides key decision makers with a more up-to-date view of the business. Thus, it requires decision makers to "match the velocity of streaming data with high-velocity, data-based decision making and agile execution to successfully compete" (Power and Heavin 2018, p. 48). Decision makers must engage in new ways to make greater use of data by understanding the importance of analysing and using data in decision making to improve the overall functioning and success of an organisation (Power and Heavin 2018).

The cases presented here primarily focus on the data set and building the classification rules. From a DSS perspective, there is an opportunity to undertake a user-centric approach to design, evaluate, and refine the DSS user interface in conjunction with key decision makers. For these DSS to be used, it is essential that the data needs of the decision makers are well understood and that these needs are reflected in the DSS design. Also, these systems must be designed to be flexible enough to be modified and updated over time to adapt to changing business KPIs.

There is a growing need to consider the development of increasingly automatic and efficient data pre-processing and data integration techniques. To better deal with the adaptation of static predictive models to dynamic predictive models that can be integrated into DSS, both technological and algorithmic improvements are required. From the point of view of computational efficiency, as the volume of data to be handled is becoming more critical, parallel computing techniques are being successfully incorporated and they are providing very good results. These techniques are presented as optimal approaches in real application contexts where it is likely that the input data will be fragmented. Technological advances in the collection of massive data sets and real-time monitoring systems are increasingly frequent, particularly in areas such as medicine/health, eBusiness, and manufacturing. New decision-making processes will demand additional methodological changes, thus emphasising the need for increasingly agile and powerful technology.

This article highlights an important opportunity to promote further collaboration between the data science and the decision support communities. This study highlights new opportunities to undertake further research to develop the heuristics that lead to increased accuracy (or at least maintain it) in response to contexts in real time. The challenge that the data stream contexts pose to both communities highlights a real opportunity to design new joint strategies of treatment and analysis of changing data, from which decision-making in public and private organisations can be supported. In dynamic environments, as new situations (new data) lead to the emergence of new patterns, it is common for rule systems to grow exponentially over time. Therefore, it becomes important to design efficient algorithms capable of dynamic weighing of rule precedence. This will enable us to reduce and re-order the predictive rule systems. So, rule systems may be adapted at each moment depending on changing input data.

Moreover, the growing trend of incorporating new data sources, often open data sets, is leading us to problems with a very high number of variables with increasing complexity. It is important to have more powerful and adaptable feature selection mechanisms that will enable us to automatically choose those that have a greater correlation with the target variable. Considering only the really relevant variables, the predictive models generated are more precise, more efficient in their computation times and, above all, more easily integrated in the DSS engine. Beyond the precision of the model, future research would also involve understanding the value derived from implementing these data-based decisions (based on the model), the decision quality and the efficiency and effectiveness of decision processes.

References

Adamo, J. M. (2001). *Data Mining for association rules and sequential patterns. Sequential and parallel algorithms*. Springer.

Agrawal, R., & Srikant, R. (1996). Fast algorithms for mining association rules. In *Proceedings of the 20th VLDB Conference,* Santiago, Chile.

Almiñana, M., Escudero, L. F., Pérez-Martín, A., Rabasa, A., & Santamaría, L. (2012). A classification rule reduction algorithm based on significance domains. *TOP, 22,* 397–418.

Alter, S. L. (1977). A taxonomy of decision support systems. *Sloan Management Review, 19*(1), 39–56.

Alter, S. L. (1980). *Decision support systems: Current practice and continuing challenge*. Reading, MA: Addison-Wesley.

Ashri, R. (2018). Building AI software: Data-driven vs. model-driven AI and why we need an AI-specific software development paradigm. https://hackernoon.com/building-ai-software-data-driven-vs-model-driven-ai-and-why-we-need-an-ai-specific-software-640f74aaf78f.

Berner, E. S., & La Lande, T. J. (2016). Overview of clinical decision support systems. *Clinical Decision Support Systems*, 1–17.

Bloemer, J. M., Brijs, T., Vanhoof, K., & Swinnen, G. (2003). Comparing complete and partial classification for identifying customers at risk. *Research in Marketing, 604,* 1–15.

Bonczek, R. H., Holsapple, C. W., & Whinston, A. B. (1981). *Foundations of decision support systems*. New York: Academic Press.

Box, G. E. P., & Jenkins, G. M. (1973). Some comments on a paper by Chatfield and Prothero and on a review by Kendall. *Journal of the Royal Statistical Society. Series A (General), 136*(3), 337–352.

Breiman, L., Friedman, J., Stone, C. J., & Olshen, R. A. (1984). *Classification and regression trees. The Wadsworth and Brooks-Cole statistics-probability Series*. Taylor & Francis.

Chapelle, O., Vapnik, V., & Bousquet, O. (2002). *Choosing multiple parameters for support vector machines machine learning, 46,* 131.

Chen, H., Chiang, R. H. L., & Storey, V. C. (2012). Business intelligence and analytics: From Big Data to big impact. *MIS Quarterly, 36*(4), 1165–1188.

Desanctis, G., & Gallupe, R. B. (1987). A foundation for the study of group decision support systems. *Management Science, 33*(5), 589–609.

Esteve, M., Miró, F., & Rabasa, A. (2018). Classification of tweets with a mixed method based on pragmatic content and meta-information. *International Journal of Design & Nature and Ecodynamics, 13*(1), 60–70.

Exastax. (2017). Top 7 Big Data Use Cases in Insurance Industry. Retrieved December 31, 2018 from https://www.exastax.com/big-data/top-7-big-data-use-cases-in-insurance-industry/.

García, S., Luengo, J., Sáez, J. A., López, V., & Herrera, F. (2013). A survey of discretization techniques: Taxonomy and empirical analysis in supervised learning. *IEEE Transactions on Knowledge and Data Engineering, 25*(4), 734–750.

Gorry, G. A. and Scott-Morton M. A. (1971). A Framework for Management Information Systems, Sloan Management Review, October, pp 55–70.

Hall, P., & Xue, J. H. (2014). On selecting interacting features from high-dimensional data. *Computational Statistics & Data Analysis, 71*, 694–708. https://doi.org/10.1016/j.csda.2012.10.010.

Hardy, V., O'Connor, Y., Heavin, C., Mastellos, N., Tran, T., O'Donoghue, J., et al. (2017). The added value of a mobile application of Community Case Management on under-5 referral, re-consultation and hospitalization rates in two districts in Northern Malawi: Study protocol for a pragmatic stepped wedge cluster-randomized controlled trial. *Trials, 18*, 475. https://doi.org/10.1186/s13063-017-2213-z.

He, Z., He, Z., Xu, X., & Deng, S. (2003). Discovering cluster-based local outliers. *Pattern Recognition Letters, 24*, 1641–1650.

Hey, T., Tansley, S., & Tolle, K. (2009). The fourth paradigm: Data-intensive scientific discovery. Ed. Microsoft Research.

Hunt, D. L., Haynes, R. B., Hanna, S. E., & Smith, K. (1998). Effects of computer-based clinical decision support systems on physician performance and patient outcomes: A systematic review. *JAMA, 280*(15), 1339–1346. https://doi.org/10.1001/jama.280.15.1339.

Ismail, N. A., & Hussin H. (2013). E-CRM features in the context of airlines e-ticket purchasing: A conceptual framework. In *5th International Conference on Information and Communication Technology for the Muslim World (Ict4m)*.

Kass, G. V. (1980). An exploratory technique for investigating large quantities of categorical data. *Applied Statistics, 29*(2), 119–127.

Kumar, D. S., Sathyadevi, G., & Sivanesh, S. (2011). Decision support system for medical diagnosis using Data Mining. *International Journal of Computer Science Issues, 8*(3), 147–153.

Lashari, S. A., Ibrahim, R., Senan, N., & Taujuddin, N. S. (2018). Application of Data Mining techniques for medical data classification: A review. In *Proceedings of the MATEC Web of Conferences* (Vol. 150, p. 06003).

Lu, Z. C., Qin, Z., Zhang, & Fang, J. (2014). A fast feature selection approach based on rough set boundary regions. *Pattern Recognition Letters, 36*, 81–88. https://doi.org/10.1016/j.patrec.2013.09.012.

MacQueen, J. B. (1967). Some methods for classification and analysis of multivariate observations. In: *Proceedings of 5th Berkeley Symposium on Mathematical Statistics and Probability* (pp. 281–297). University of California Press.

Mehta, M., Agrawal, R., Rissanen, J. (1996). SLIQ: A fast scalable classifier for Data Mining. In: P. Apers, M. Bouzeghoub, G. Gardarin (Eds), *Advances in database technology—EDBT 1996* (Vol. 1057). Lecture notes in computer science. Springer.

Murdoch, T. B., & Detsky, A. S. (2013). The inevitable application of Big Data to health care. *JAMA, 309*(13), 1351–1352.

Peres-Neto, P. R., Jackson, D. A., & Somers, K. M. (2005). How many principal components? Stopping rules for determining the number of non-trivial axes revisited. *Computational Statistics & Data Analysis, 49*(4), 974–997.

Pérez-Martín, A., Pérez-Torregrosa, A., & Vaca, M. (2018). Big data techniques to measure credit banking risk in home equity loans. *Journal of Business Research, 89*, 448–454.

Potter, R., Probyn, K., Bernstein, C., Pincus, T., Underwood, M., & Matharu, M. (2018). Diagnostic and classification tools for chronic headache disorders: A systematic review. *Cephalalgia*. https://doi.org/10.1177/0333102418806864.

Power, D. J. (1997). What is DSS? The Online Executive Journal for Data-Intensive Decision Support, *1*(3).

Power, D. J. (2001). Supporting decision makers: An expanded framework. In A. Harriger (Ed.), *E-Proceedings 2001 Informing Science Conference* (pp. 431e–436e).

Power, D. J. (2002). *Decision support systems: Concepts and resources for managers*. Westport, CT: Greenwood/Quorum.

Power, D. J., & Sharda, R. (2007). Model-driven decision support systems: Concepts and research directions. *Decision Support Systems, 43*(3), 1044–1061.

Power, D. J. (2008). Decision support systems concept. In F. Adam, P. Humphreys (Eds.), *Encyclopedia of decision making and decision support technologies* (pp. 232–235). IGI-Global.

Power, D. J., & Heavin, C. (2017). *Decision support, analytics, and business intelligence* (3rd ed.). New York, NY: Business Expert Press.

Power, D., & Heavin, C. (2018). *Data-based decision making and digital transformation*. New York, NY: Business Expert Press.

Provost, F., & Fawcett, T. (2013). Data Science and its relationship to Big Data and data-driven decision making. *Big Data, 1*(1), 51–59. https://doi.org/10.1089/big.2013.1508.

Quinlan, J. R. (1986). *Machine Learning, 1*, 81. https://doi.org/10.1007/BF00116251.

Quinlan, J. R. (1993). C4.5: *Programs for machine learning. Series in machine learning*. USA: Morgan Kaufmann Publishers.

Rusov, J., & Mishita, M. (2016). Model of decision support system used for assessment of insurance risk. *Journal of Applied Engineering Science, 14*(1), 13–20. https://doi.org/10.5937/jaes14-8845.

Shim, J. P., Warkentin, M., Courtney, J. F., Power, D. J., Sharda, R., & Carlsson, C. (2002). Past, present, and future of decision support technology, decision support systems 33, 111–126.

Simon, H. (1960). *The new science of management decision*. New-York: Harper and Row.

Sprague, R. H., & Watson, M. J. (1979). Bit by Bit: Toward decision support systems. *California Management Review, 22*(1), 60–67.

Sprague, R., & Watson, H. (1993). *Decision Support Systems: Putting Theory into Practice*. Englewood Cliffs, New Jersey: Prentice Hall

Tsai, C. F. (2009). Feature selection in bankruptcy prediction. *Knowledge-Based Systems, 22*(2), 120–127.

Tsai, C. F., & Hsiao, Y. C. (2010). Combining multiple feature selection methods for stock prediction: Union, intersection, and multi-intersection approaches. *Decision Support Systems, 50*(1), 258–269.

Tsami, M., Adamos, G., Nathanail, E., Budilovich, E., Yatskiv, I., & Magginas, V. (2018). A decision tree approach for achieving high customer satisfaction at urban interchanges. *Transport and Telecommunication, 19*(3), 194–202.

Uusitalo, L., et al. (2015). An overview of methods to evaluate uncertainty of deterministic models in decision support. *Environmental Modelling and Software, 63*, 24–31.

Verhoef, P. C., & Donkers, B. (2001). Predicting customer potential value an application in the insurance industry. *Decision Support Systems, 32*, 189–199.

Waller, M., & Fawcett, S. (2013). Data Science, predictive analytics, and Big Data: A revolution that will transform supply chain design and management. *Journal of Business Logistics, 34*(2), 77–84.

Yang, L., Liu, S., Tsoka, S., & Papageorgiou, L. G. (2017). A regression tree approach using mathematical programming. *Expert Systems with Applications, 78*(15), 347–357.

Yin, Ch., Xia, L., Zhang, S., & Wang, J. (2018). Improved clustering algorithm based on high-speed network data stream. *Soft Computing, 22*(13), 4185–4195.

Chapter 4
Identification of Congestion in DEA

Mahmood Mehdiloo, Biresh K. Sahoo, and Joe Zhu

Abstract Productivity is a common descriptive measure for characterizing the resource-utilization performance of a production unit, or decision making unit (DMU). The challenge of improving productivity is closely related to a particular form of congestion, which reflects waste (overuse) of input resources at the production unit level. Specifically, the productivity of a production unit can be improved not only by reducing some of its inputs but also simultaneously by increasing some of its outputs, when such input congestion is present. There is thus a need first for identifying the presence of congestion, and then for developing congestion-treatment strategies to enhance productivity by reducing the input wastes and the output shortfalls associated with such congestion. Data envelopment analysis (DEA) has been considered a very effective method in evaluating input congestion. Because the assumption of strong input disposability precludes congestion, it should not be incorporated into the axiomatic modeling of the true technology involving congestion. Given this fact, we first develop a production technology in this contribution by imposing no input disposability assumption. Then we define both weak and strong forms of congestion based on this technology. Although our definitions are made initially for the output-efficient DMUs, they are well extended in the sequel for the output-inefficient DMUs. We also propose in this contribution a method for identifying congestion. The essential tool for devising this method is the technique of finding a maximal element of a non-negative polyhedral set. To our knowledge, our method is the only

M. Mehdiloo
Department of Mathematics and Applications,
University of Mohaghegh Ardabili, Ardabil, Iran
e-mail: m.mehdiloozad@gmail.com; m.mehdiloo@uma.ac.ir

B. K. Sahoo (✉)
Xavier Institute of Management, Xavier University Bhubaneswar,
Bhubaneswar 751013, India
e-mail: biresh@ximb.ac.in

J. Zhu
Foisie Business School, Worcester Polytechnic Institute,
Worcester, MA 01609, USA
e-mail: jzhu@wpi.edu

V. Charles et al. (eds.), *Data Science and Productivity Analytics*,
International Series in Operations Research & Management Science 290,
https://doi.org/10.1007/978-3-030-43384-0_4

reliable method for precisely detecting both weak and strong forms of congestion. This method is computationally more efficient than the other congestion-identification methods developed in the literature. This is due to the fact that, unlike the others, our method involves solving a single linear program. Unlike the other methods, the proposed method also deals effectively with the presence of negative data, and with the occurrence of multiple projections for the output-inefficient DMUs. Based on our theoretical results, three computational algorithms are developed for testing the congestion of any finite-size sample of observed DMUs. The superiority of these algorithms over the other congestion-identification methods is demonstrated using four numerical examples, one of which is newly introduced in this contribution.

Keywords Production technology · Congestion · Data envelopment analysis · Negative data · Multiple projections · Maximal element

4.1 Introduction

Increasing competitive pressure today forces production units, or decision making units (DMUs) in the data envelopment analysis (DEA) terminology, to improve their performance by efficient utilization and allocation of their input resources. The objectives of efficient resource utilization by a production unit are (1) producing as much outputs as possible from specific quantities of inputs, and (2) producing specific quantities of outputs using as little inputs as possible (Ray, 2004). In line with these objectives, a descriptive measure for characterizing the resource-utilization performance of a DMU, called *productivity*, is defined as the ratio of output to input. In the rare case of a single-input and single-output production technology, this ratio can be easily computed for all DMUs. In contrast, in the more likely situations involving multiple inputs and multiple outputs, the inputs and outputs of each DMU have to be aggregated in some economically sensible fashion, so that its productivity remains the ratio of two scalars, the aggregate output and the aggregate input (Fried et al., 2008).

The challenge of improving productivity is closely related to the concept of efficiency. By definition, the productivity of a production unit is enhanced by reducing its inputs and/or increasing its outputs. It is also known that, a DMU is (technically) full-efficient in a production technology if, no feasible DMU results from a reduction in some of its inputs and/or an increase in some of its outputs without worsening the remaining inputs and outputs. Therefore, the inefficiency of a DMU is a sufficient condition for the possibility of improving its productivity.[1] Established examples of severely inefficient DMUs are those who waste (overuse) their inputs in the sense of

[1]The stated condition is not necessary because all efficient DMUs are not necessarily the most productive ones. This means that the productivity of an efficient DMU may be further improved by moving toward the most productive efficient DMUs.

exhibiting *input congestion*. From the *presence* point of view,[2] congestion means that improvement in inputs can be associated with improvement in outputs. This concept must be defined at the output-efficient boundary of production technology so that its associated output shortfalls can be distinguished from the output shortfalls arising from the technical output inefficiency (if any). Taking this into account, an output-efficient DMU in a production technology is called (weakly) congested if, some of its outputs can be increased simultaneously by reducing some of its inputs without worsening the remaining inputs and outputs, and the resulting DMU is feasible (Mehdiloozad et al., 2018; Tone & Sahoo, 2004).[3]

From the definitions of productivity and congestion, it follows that the productivity of a congested DMU can be improved not only by reducing some of its inputs but also simultaneously by increasing some of its outputs. This shows the need first for identifying congested DMUs in any finite-size sample, and then for developing congestion-treatment strategies to enhance their productivities by reducing the input wastes and the output shortfalls associated with their congestion levels. The non-parametric method of DEA has been established to be effective in evaluating the input congestion. Ever since this method was introduced by Charnes et al. (1978), the evaluation of congestion has been widely studied within its framework. The concept of congestion was first introduced by Färe and Svensson (1980) and was later given an operationally implementable form by Färe and Grosskopf (1983) and Färe et al. (1985). Subsequently, Cooper et al. (1996) developed an alternative approach to evaluating congestion that has seen various extensions and applications.[4]

The assumption of disposability of inputs is central to the modeling of an unknown production technology involving congestion. The most common assumption of strong input disposability (SID) in production theory states that, if a DMU produces a vector of outputs from a vector of inputs, then the same vector of outputs can be produced from the same or a larger (component-wise) vector of inputs (Färe et al., 1985; Ray, 2004). Any production technology incorporating the SID assumption, e.g, the standard variable returns-to-scale (VRS) technology of Banker et al. (1984), precludes congestion and is therefore congestion-free (Färe & Grosskopf, 1983). This means that this assumption should not be incorporated into the modeling process so that the presence of congestion is allowed. An approach to fulfilling this condition is to assume no disposability for inputs.[5] Following this approach, the so-called

---

[2]Note that the concept of congestion can be investigated either from the presence or occurrence perspectives. For more description, see Sect. 4.3.1 of this chapter.

[3]Throughout this chapter, congestion is meant in the sense of Cooper et al. (1996). This note is made here because the economic concept of congestion is differently interpreted in the literature. Such interpretations are briefly stated in Mehdiloozad et al. (2018).

[4]The DEA literature on investigating and applying congestion has been extensive to date. The important part of this literature can be found in Cooper et al. (2011), Khodabakhshi et al. (2014), Mehdiloozad et al. (2018), and Zare Haghighi et al. (2014).

[5]Another approach, suggested by Färe and Grosskopf (1983), Färe and Svensson (1980) and Färe et al. (1985), for not allowing the strong disposability of inputs is to weaken it by replacing it with the assumption of weak input disposability. For correct characterization of technologies with weakly disposable inputs, the reader may refer to Mehdiloozad and Podinovski (2018).

congestion technology is axiomatically characterized in Mehdiloozad et al. (2018), Tone and Sahoo (2004) and Wei and Yan (2004) as a non-parametric estimation of the unknown technology. With respect to this technology, Tone and Sahoo (2004) present the above-mentioned definition of "weak congestion", and further introduce a special form of weak congestion as "strong congestion". Mehdiloozad et al. (2018) later redefine the concept of strong congestion in order to make its definition compatible with the presence of negative data.[6] Precisely, they call a weakly congested DMU as strongly congested if a reduction (not necessarily proportional) of all its inputs can be associated with an increase in all its outputs.

As we mentioned earlier, congestion is a frontier concept that is well-defined only for DMUs on the output-efficient frontier of the congestion technology. The commonly accepted approach in DEA for extending the definition of a frontier concept to non-frontier DMUs is to define that concept at the projection of such a DMU on the frontier of the technology. Following this approach, the congestion of an inefficient DMU is defined at its projection on the output-efficient frontier of the congestion technology. More explicitly, an output-inefficient DMU is said to be weakly (resp., strongly) congested if its projection is weakly (resp., strongly) congested. Clearly, this definition is mathematically well-defined if the projection set (i.e., the set of all possible projections) is singleton. However, as rightly argued by Sueyoshi and Sekitani (2009), the well-definedness is not guaranteed when multiple projections occur and one of them is chosen arbitrarily. This is because different projections of an output-inefficient DMU may produce different results in terms of its congestion status and its associated inefficiency.

Tone and Sahoo (2004) are the first who develop an approach to identifying the presence of weak and strong congestion. Their approach is based on two assumptions: (a) input–output data are positive, (b) the projection of any output-inefficient DMU is unique. To relax the first assumption, Khoveyni et al. (2013) propose a two-stage slack-based procedure to identify both weak and strong congestions in the presence of non-negative data. Subsequently, Khoveyni et al. (2017) modify this procedure to deal with negative data. As shown in Sect. 4.6 of this chapter, this modified procedure fails to accurately distinguish between the weakly, but not strongly, DMUs and the strongly congested ones. Concerning the second assumption, though Khoveyni et al. (2013) claim that their proposed procedure addresses the issue of occurring multiple projections, they provide no evidence for their claim. Surprisingly, the next study by Khoveyni et al. (2017) also remains silent on this issue.

From the above paragraph, it follows that the three methods suggested in Khoveyni et al. (2013, 2017) and Tone and Sahoo (2004) cannot deal with the occurrence of multiple projections and the presence of negative data. By the proper use of the technique developed in Mehdiloozad et al. (2018) for finding a maximal element of a non-negative convex set, Mehdiloozad et al. (2018) propose a reliable method for precisely detecting both weak and strong forms of congestion. Their proposed method is computationally more efficient than the other congestion-identification

[6]For the DEA literature in dealing with negative data, the interested reader may refer to Mehdiloozad et al. (2018) and the references therein.

methods in the literature. This is due to the fact that, unlike the others, their method involves solving a single linear program (LP). Unlike the other methods, their proposed method also deals effectively with the presence of negative data, and the occurrence of multiple projections for output-inefficient DMUs. To the best of our knowledge, the method of Mehdiloozad et al. (2018) is the only method available so far in the literature to deal with both the occurrence of multiple projections and the presence of negative data. Therefore, this chapter is devoted to thoroughly analyzing this method.

The reminder of this chapter unfolds as follows. In Sect. 4.2 we introduce notation and provide some preliminaries that we need in the following sections of the current chapter. Applying the results of Mehdiloozad et al. (2018), we also develop an original LP to find a maximal element of a non-negative polyhedral set defined by a finite set of linear equalities and inequalities. We devote Sect. 4.3 to investigating the congestion of output-efficient DMUs. First, we give a general definition of congestion. Second, we define and characterize the congestion technology. Third, we define both weak and strong forms of congestion. Fourth, we formulate an output-oriented model based on the congestion technology. Finally, we propose a single-stage LP to identify both weak and strong forms of congestion. In Sect. 4.4 we extend the concept of congestion for faces of the congestion technology, any of which is proved to be a bounded polyhedral set. Precisely, we call such a face as weakly (resp., strongly) congested if all DMUs on its relative interior are weakly (resp., strongly) congested. By linking congestion to the concept of global reference set, we characterize *minimal face* of an output-inefficient DMU (i.e., the smallest face containing all projections of the DMU) and its relative interior as well. We also develop an LP to find max-projection of such a DMU on the relative interior of its minimal face. Then, we call an output-inefficient DMU weakly (resp., strongly) congested if its corresponding minimal face is weakly (resp., strongly) congested. Based on the fact that all DMUs on the relative interior of a face of the congestion technology have the same status of congestion, we evaluate the congestion status of an output-inefficient DMU at its max-projection. In Sect. 4.5 we develop three computational algorithms for testing the weak and strong congestions of any finite-size sample of observed DMUs. In Sect. 4.6 we demonstrate the superiority of our algorithms over the other congestion-identification approaches using four numerical examples, one of which is newly introduced in this chapter. In Sect. 4.7 we present our concluding remarks. The proofs of all mathematical results are given in Sect. 4.8.

4.2 Preliminaries

4.2.1 Notation

Let $\mathbb{R}^d$ denote the d-dimensional Euclidean space, and let $\mathbb{R}^d_+$ (resp., $\mathbb{R}^d_{++}$) denote its non-negative (resp., positive) orthant. We denote sets by uppercase calligraphic letters, vectors by boldface lowercase letters, and matrices by boldface uppercase letters. We denote the relative interior and the relative boundary of a non-empty set $\mathcal{S}$ in $\mathbb{R}^d$ by $ri\,(\mathcal{S})$ and $rb\,(\mathcal{S})$, respectively.[7]

By convention, all vectors are column vectors. Vectors $\mathbf{0}_d$ and $\mathbf{1}_d$ are the d-dimensional vectors, all components of which are equal to 0 and 1, respectively. For simplicity, notation $(\mathbf{a}; \mathbf{b})$ is used to show the vertical vector obtained by adding vector $\mathbf{b} \in \mathbb{R}^{d'}$ below vector $\mathbf{a} \in \mathbb{R}^d$. The transpose of a vector or matrix is indicated by superscript T. By the support of a non-negative vector $\mathbf{a} \in \mathbb{R}^d_+$, we mean the index set $\sigma\,(\mathbf{a}) = \{i \in 1, \ldots, d \colon a_i > 0\}$. For vectors $\mathbf{a}, \mathbf{b} \in \mathbb{R}^d$, the inequality $\mathbf{a} \geq \mathbf{b}$ (resp., $\mathbf{a} > \mathbf{b}$) means that $a_i \geq b_i$ (resp., $a_i > b_i$), for all $i = 1, \ldots, d$. Vector $\mathbf{a}$ weakly dominates vector $\mathbf{b}$ (denoted $\mathbf{a} \gneqq \mathbf{b}$) if $\mathbf{a} \geq \mathbf{b}$ and $\mathbf{a} \neq \mathbf{b}$. Furthermore, vector $\mathbf{a}$ strongly dominates vector $\mathbf{b}$ if $\mathbf{a} > \mathbf{b}$. Matrix $\mathbf{0}_{d \times d'}$ is the $d \times d'$ matrix, all components of which are equal to 0. Matrix $\mathbf{I}_d$ is the d-dimensional identity matrix, and vector $\mathbf{e}_i$ $(i = 1, \ldots, d)$ is the ith column of this matrix, i.e., $\mathbf{e}_i = \big(0, \ldots, \underset{i\text{th}}{\underline{1}}, \ldots, 0\big)^T \in \mathbb{R}^d_+$.

Throughout this chapter, we represent a DMU by the vector $(\mathbf{x}; \mathbf{y})$, where $\mathbf{x} = (x_1, \ldots, x_m)^T$ is the vector of its inputs and $\mathbf{y} = (y_1, \ldots, y_s)^T$ is the vector of its outputs. We assume that there are n observed DMUs, denoted $(\mathbf{x}_j; \mathbf{y}_j)$ $(j \in \mathcal{J} = \{1, \ldots, n\})$, and that $\mathbf{X}$ and $\mathbf{Y}$ are, respectively, the corresponding $m \times n$ input and $s \times n$ output data matrices. The DMU under evaluation that is not necessarily observed is denoted by $(\mathbf{x}_o; \mathbf{y}_o)$. We assume that all observed DMUs and the DMU under evaluation have at least one non-zero input and one non-zero output, i.e., $\mathbf{x}_j \neq \mathbf{0}_m$ and $\mathbf{y}_j \neq \mathbf{0}_s$, for all $j \in \mathcal{J} \cup \{o\}$.

4.2.2 The VRS Model

A production technology $\mathcal{T}$ that transforms input vectors $\mathbf{x} \in \mathbb{R}^m_+$ into output vectors $\mathbf{y} \in \mathbb{R}^m_+$ is interpreted as follows.

$$\mathcal{T} = \left\{(\mathbf{x}; \mathbf{y}) \in \mathbb{R}^{m+s} : \mathbf{x} \in \mathbb{R}^m_+ \text{ can produce } \mathbf{y} \in \mathbb{R}^m_+\right\}.$$

Throughout this chapter, we assume that the inputs and outputs are non-negative, i.e., $\mathcal{T} \subset \mathbb{R}^{m+s}_+$. The only exception is Sect. 4.5.3 where the inputs and outputs are allowed to be unrestricted in sign. Because in empirical applications the technology is usually unknown, it should be approximated based on observed data and a set of assumed properties of production technology, stated as *production axioms* (Banker et al., 1984; Färe et al., 1985; Shephard 1974). The following are the main axioms that we use in this chapter:

Axiom IO (*Inclusion of Observations*) $(\mathbf{x}_j; \mathbf{y}_j) \in \mathcal{T}$, for all $j \in \mathcal{J}$.
Axiom SID (*Strong Input Disposability*) If $(\mathbf{x}; \mathbf{y}) \in \mathcal{T}$, then $(\tilde{\mathbf{x}}; \mathbf{y}) \in \mathcal{T}$ for all $\tilde{\mathbf{x}} \geq \mathbf{x}$.
Axiom SOD (*Strong Output Disposability*) If $(\mathbf{x}; \mathbf{y}) \in \mathcal{T}$, then $(\mathbf{x}; \tilde{\mathbf{y}}) \in \mathcal{T}$ for all $\mathbf{0}_s \leq \tilde{\mathbf{y}} \leq \mathbf{y}$.

---

[7]The *relative interior* (resp., *relative boundary*) of a non-empty set $\mathcal{S}$ in $\mathbb{R}^d$ is meant by its interior (resp., boundary) relative to the affine hull of $\mathcal{S}$ (Tuy, 1998).

Axiom CT (*Convexity of Technology*) $\mathcal{T}$ is a convex set.

An approach to approximating an unknown technology is to consider the intersection of all technologies that satisfy the production axioms describing its properties. If it is further verified that such intersection itself satisfies all considered axioms, then it is the smallest technology that satisfies those axioms. This means that the approximated technology includes only those DMUs that are needed to satisfy the assumed axioms, and does not include any arbitrary DMUs (Mehdiloozad & Podinovski, 2018). The described approach was originally pioneered in DEA by Banker et al. (1984) and is known as the *minimum extrapolation principle*. Using this approach, Banker et al. (1984) define the standard VRS technology as follows:

Definition 1 Technology $\mathcal{T}_{\text{VRS}}$ is the intersection of all technologies $\mathcal{T} \subset \mathbb{R}_+^{m+s}$ that satisfy Axioms IO, SID, SOD and CT.

It is straightforward to show that technology $\mathcal{T}_{\text{VRS}}$ itself satisfies all Axioms IO, SID, SOD and CT. Therefore, it is the smallest technology that satisfies the stated axioms. It can be proved that the explicit statement of the VRS technology is as follows:

$$\mathcal{T}_{\text{VRS}} = \left\{ (\mathbf{x};\mathbf{y}) \in \mathbb{R}_+^{m+s} : \mathbf{X}\boldsymbol{\lambda} \le \mathbf{x},\ \mathbf{Y}\boldsymbol{\lambda} \ge \mathbf{y},\ \mathbf{1}_n^T\boldsymbol{\lambda} = 1,\ \boldsymbol{\lambda} \ge \mathbf{0}_n \right\}. \tag{4.1}$$

Based on this statement of technology $\mathcal{T}_{\text{VRS}}$, the output-oriented VRS model can be written in the following form:

$$\begin{aligned}
&\max \quad \varphi + \varepsilon\left(\mathbf{1}_m^T\mathbf{p} + \mathbf{1}_s^T\mathbf{q}\right) \\
&\text{subject to} \\
&\mathbf{X}\boldsymbol{\lambda} \le \mathbf{x}_o + \mathbf{p}, \\
&\mathbf{Y}\boldsymbol{\lambda} \ge \varphi\mathbf{y}_o + \mathbf{q}, \\
&\mathbf{1}_n^T\boldsymbol{\lambda} = 1, \\
&\boldsymbol{\lambda} \ge \mathbf{0}_n,\ \mathbf{p} \ge \mathbf{0}_m,\ \mathbf{q} \ge \mathbf{0}_s,\ \varphi \text{ sign free,}
\end{aligned} \tag{4.2}$$

where $\varepsilon > 0$ is a small (theoretically non-Archimedean infinitesimal) value epsilon.[8]

Let $(\varphi^*, \boldsymbol{\lambda}^*, \mathbf{p}^*, \mathbf{q}^*)$ be an optimal solution to program (4.2). Then the output radial efficiency of DMU $(\mathbf{x}_o;\mathbf{y}_o)$ in technology $\mathcal{T}_{\text{VRS}}$ is defined as the inverse of φ^*.

4.2.3 *Efficiency*

The concept of *efficiency* can be explained in different ways. To our development, we define below this concept in terms of (a) the reduction of the inputs while the outputs are kept unchanged (input efficiency), (b) the expansion of the outputs while

[8]For details on the use of epsilon, the reader may refer to Podinovski and Bouzdine-Chameeva (2017).

the inputs are kept unchanged (output efficiency), and (c) the simultaneous reduction of the inputs and the expansion of the outputs (full efficiency). The given definitions are applicable to any technology $\mathcal{T}$.

Definition 2 A DMU $(\mathbf{x}; \mathbf{y}) \in \mathcal{T}$ is

- *input-efficient* if there does not exist a DMU $(\mathbf{x}'; \mathbf{y}) \in \mathcal{T}$ such that $-\mathbf{x}' \gneqq -\mathbf{x}$.
- *output-efficient* if there does not exist a DMU $(\mathbf{x}; \mathbf{y}') \in \mathcal{T}$ such that $\mathbf{y}' \gneqq \mathbf{y}$.
- *full-efficient* if there does not exist a DMU $(\mathbf{x}'; \mathbf{y}') \in \mathcal{T}$ such that $(-\mathbf{x}'; \mathbf{y}') \gneqq (-\mathbf{x}; \mathbf{y})$.

It is clear that the input efficiency is weaker than the full efficiency, in the sense that the latter implies the former but the converse is not generally true. A similar statement holds between the output efficiency and the full efficiency. We use the notation $\partial_{\mathrm{O}}\mathcal{T}$ and $\partial_{\mathrm{F}}\mathcal{T}$ to denote, respectively, the output-efficient and full-efficient frontiers of technology $\mathcal{T}$.

4.2.4 Finding a Maximal Element of a Non-negative Polyhedral Set

In this section we deal with finding a maximal element of a non-negative polyhedral set. From Mehdiloozad et al. (2018), a convex, and therefore polyhedral, set in $\mathbb{R}^d$ is *non-negative* if it is a subset of $\mathbb{R}^d_+$. An element of a non-negative convex set is *maximal* if it has the maximum number of positive components. For any arbitrary maximal element $\mathbf{z}^{\max}$ of a non-negative convex set $\mathcal{Z}$, the following equality is true:

$$\sigma\left(\mathbf{z}^{\max}\right) = \bigcup_{\mathbf{z}\in\mathcal{Z}} \sigma\left(\mathbf{z}\right). \tag{4.3}$$

This statement shows that all maximal elements of $\mathcal{Z}$ have the same support, the union of the supports of all elements of $\mathcal{Z}$.

Let $\mathbf{A}$ and $\mathbf{B}$ be two matrices of the respective dimensions $h \times k$ and $h \times l$, and let $\mathbf{c}$ be a h-dimensional vector. We consider the non-empty set $\mathcal{P}$ given by

$$\mathcal{P} = \left\{\mathbf{u} \in \mathbb{R}^k : \mathbf{A}\mathbf{u} + \mathbf{B}\mathbf{v} \lesseqqgtr \mathbf{c},\ \mathbf{u} \geq \mathbf{0}_k,\ \mathbf{v} \geq \mathbf{0}_l\right\}, \tag{4.4}$$

where the symbol $\lesseqqgtr$ means that each equation can be an equality or inequality. By the *projection lemma*,[9] it follows that $\mathcal{P}$ is a polyhedral set. In particular, $\mathcal{P}$ is a non-negative polyhedral set because $\mathcal{P} \subset \mathbb{R}^k_+$.

[9]The projection lemma follows that the projection $\mathcal{A}(Q) = \left\{\mathbf{a} \in \mathbb{R}^d : \exists \mathbf{b} \in \mathbb{R}^{d'} \text{ such that } (\mathbf{a}; \mathbf{b}) \in Q\right\}$ of a polyhedral set $Q \subset \mathbb{R}^{d+d'}$ is still a polyhedral set. See Corollary 2.4 in Bertsimas and Tsitsiklis (1997).

More recently, Mehdiloozad et al. (2018) proposed a general convex program for finding a maximal element of a non-negative convex set. As a special case of their program, an LP is developed below to identify a maximal element of $\mathcal{P}$.

Theorem 1 *Let* $(\mathbf{s}^*, \mathbf{t}^*, \mathbf{v}^*, w^*)$ *be an optimal solution to the following LP*[10]*:*

$$\begin{aligned} &\max \quad \mathbf{1}_k^T \mathbf{t} \\ &\text{subject to} \\ &\mathbf{A}(\mathbf{s}+\mathbf{t}) + \mathbf{B}\mathbf{v} \lesseqqgtr \mathbf{c}w, \\ &\mathbf{s} \geq \mathbf{0}_k,\ \mathbf{1}_k \geq \mathbf{t} \geq \mathbf{0}_k,\ \mathbf{v} \geq \mathbf{0}_l,\ w \geq 1. \end{aligned} \tag{4.5}$$

Then $\frac{1}{w^*}(\mathbf{s}^* + \mathbf{t}^*)$ *is a maximal element of* $\mathcal{P}$.

The proofs of Theorem 1 and the other statements are given in the appendix of this chapter.

4.3 Congestion of Output-Efficient DMUs

4.3.1 General Definition of Input Congestion

Throughout this chapter we follow Cooper et al. (1996) who defined the congestion faced by a DMU as follows.

Definition 3 An output-efficient DMU $(\mathbf{x}; \mathbf{y})$ in technology $\mathcal{T}$ is *congested* if, some of its outputs can be increased simultaneously by reducing some of its inputs without worsening its remaining inputs and outputs, and the resulting DMU is in technology $\mathcal{T}$.[11]

Concerning Definition 3, two remarks are worth noting. First, though the concept of congestion is defined above from the *presence* point of view, it can be alternatively defined and evaluated from *occurrence* perspective. If the decision-maker wants to decrease overall inputs of the system (i.e., the set of all observed DMUs), (s)he needs to identify those DMUs who are actually experiencing congestion. This is due to the fact that, even if a congested DMU is output-efficient, reducing its inputs not only is in line with the decision-maker's main objective, but can also be associated

[10]By assumption, the set $\mathcal{P}$ is non-empty. Let $\bar{\mathbf{u}} \in \mathcal{P}$. Then there exists a vector $\bar{\mathbf{v}} \in \mathbb{R}^l_+$, such that the vector $(\bar{\mathbf{u}}; \bar{\mathbf{v}})$ satisfies (4.4). Let us define $s'_g = \max\{0, \bar{u}_g - 1\}$ and $t'_g = \min\{1, \bar{u}_g\}$, for all $g = 1, \ldots, k$, and $\mathbf{v}' = \bar{\mathbf{v}}$, $w' = 1$. Then $(\mathbf{s}', \mathbf{t}', \mathbf{v}', w')$ is a feasible solution of program (4.5). Because the objective value of program (4.5) is upper bounded by $k+1$, it has a finite optimal solution.

[11]Though the definition of congestion has been limited to output-efficient DMUs, it will be extended to output-inefficient DMUs in Sect. 4.4 of this chapter.

with increasing its outputs. This is precisely the same presence viewpoint on which Definition 3 is based.

There are, however, situations where the decision-maker wants to identify those DMUs who are likely to suffer from congestion if their inputs are increased. This means that the increment of some inputs of such a DMU are likely to result in the reduction of some of its outputs. An example of such situations is when the decision-maker is interested in increasing the overall input of the system. This is the occurrence viewpoint.[12]

The second remark regarding Definition 3 is that, because all interior units of a technology are always strongly dominated (in the sense of Definition 5) by another unit in technology $\mathcal{T}$, the concept of congestion is considered only for the boundary units. In addition, out of the boundary units, only output-efficient DMUs must be considered for the evaluation of congestion because their outputs cannot be improved unless their inputs are changed.

4.3.2 The Congestion Technology

If a production technology contains at least one congested DMU, then it is said to be congested; otherwise, it is called congestion-free. The so-called congestion technology is the widely used technology in the DEA literature as the estimation of an unknown congested technology. This technology was first applied for the evaluation of congestion by Tone and Sahoo (2004) and Wei and Yan (2004). It was axiomatically defined by Wei and Yan (2004), and was subsequently more analyzed by Mehdiloozad et al. (2018).

To justify the need for developing the congestion technology, we show that the VRS technology cannot be considered as a correct estimation for a true congested technology, because it precludes congestion by incorporating Axiom SID.

Theorem 2 *If technology $\mathcal{T}$ satisfies Axiom SID, then it is congestion-free.*

From Theorem 2, the incorporation of Axiom SID into the modeling precess must be avoided. The congestion technology defined below fulfills this condition by incorporating no disposability assumption for inputs.

Definition 4 Technology $\mathcal{T}_{\text{CONG}}$ is the intersection of all technologies $\mathcal{T} \subset \mathbb{R}_+^{m+s}$ that satisfy Axioms IO, SOD and CT.

It is straightforward to prove that technology $\mathcal{T}_{\text{CONG}}$ itself satisfies all Axioms IO, SOD and CT. Therefore, it is the smallest technology that satisfies the stated axioms. The following theorem gives a constructive statement of this technology.

[12] While the sets of DMUs who are all currently suffering from congestion and who are likely to suffer from congestion may have overlap, they may not necessarily be equal. For example, consider DMU F in Fig. 4.1 that is not currently congested, but is likely to face it if its input is increased.

Theorem 3 *Technology $\mathcal{T}_{\text{CONG}}$ is stated as follows*[13]*:*

$$\mathcal{T}_{\text{CONG}} = \left\{(\mathbf{x}; \mathbf{y}) \in \mathbb{R}_{+}^{m+s} : \mathbf{X}\boldsymbol{\lambda} = \mathbf{x},\ \mathbf{Y}\boldsymbol{\lambda} \geq \mathbf{y},\ \mathbf{1}_n^T \boldsymbol{\lambda} = 1,\ \boldsymbol{\lambda} \geq \mathbf{0}_n\right\}. \tag{4.6}$$

The next result further characterizes the structure of technology $\mathcal{T}_{\text{CONG}}$.

Theorem 4 *Technology $\mathcal{T}_{\text{CONG}}$ is a bounded polyhedral set, or a polytope, which implies that all its faces*[14] *are polytopes.*

Note that, unlike technology $\mathcal{T}_{\text{VRS}}$, technology $\mathcal{T}_{\text{CONG}}$ requires the input constraints to hold as equality in (4.6) and, therefore, disallows the inputs to be freely disposed of. Consequently, as proved in Theorem 4, technology $\mathcal{T}_{\text{CONG}}$ is a polytope, unlike technology $\mathcal{T}_{\text{VRS}}$ that is an unbounded polyhedral set.[15]

By definition, technology $\mathcal{T}_{\text{VRS}}$ satisfies all the axioms that define technology $\mathcal{T}_{\text{CONG}}$. Because technology $\mathcal{T}_{\text{CONG}}$ is the smallest technology that satisfies these axioms, we have the following corollary.

Corollary 1 *The following embedding is true: $\mathcal{T}_{\text{CONG}} \subset \mathcal{T}_{\text{VRS}}$.*

For a graphical illustration of the above findings, consider the following example which was originally studied in Tone and Sahoo (2004).

Example 1 Consider the seven observed DMUs A, B, C, D, E, F, and G, whose input–output data set has been shown in Table 4.1. Figure 4.1 shows technologies $\mathcal{T}_{\text{CONG}}$ and $\mathcal{T}_{\text{VRS}}$ generated by these DMUs. Technology $\mathcal{T}_{\text{CONG}}$ is the polytope bounded by the broken line $A'ABCDEFGG'$. However, technology $\mathcal{T}_{\text{VRS}}$ is the unbounded polyhedral set constructed from horizontally expanding technology $\mathcal{T}_{\text{CONG}}$ along the input axis. This expansion is, indeed, the result of incorporating Axiom SID by technology $\mathcal{T}_{\text{VRS}}$.

The output-efficient frontier of technology $\mathcal{T}_{\text{VRS}}$ consists of the broken line $ABCDD'$. It is clear that no output-efficient DMU in technology $\mathcal{T}_{\text{VRS}}$ can increase its output by reducing its input. Therefore, technology $\mathcal{T}_{\text{VRS}}$ is congestion-free (see Theorem 2). However, technology $\mathcal{T}_{\text{CONG}}$ with the broken line $ABCDEFG$ being its output-efficient frontier allows for the presence of congestion. This is because any DMU on the broken segment EFG (except E) can increase its output by decreasing its input, and therefore exhibits congestion.

[13]The axiom of *weak input disposability* (WID) states that any DMU remains feasible if its inputs are simultaneously scaled up in the same proportion. For clarification, we emphasize that technology $\mathcal{T}_{\text{CONG}}$ does not satisfy Axiom WID. The interested readers may refer to Mehdiloozad and Podinovski (2018) for details on correct axiomatic characterization of technologies that exhibit weak disposability of inputs.

[14]A convex subset $\mathcal{F}$ of a convex set $C \subseteq \mathbb{R}^d$ is a *face* of C if, for every $\mathbf{x}, \mathbf{y} \in C$ and every $\lambda \in (0, 1)$ such that $\lambda\mathbf{x} + (1 - \lambda)\,\mathbf{y} \in C$, we have $\mathbf{x}, \mathbf{y} \in \mathcal{F}$ (Tuy, 1998). Of course, the empty set and C itself are faces of C.

[15]Because the polyhedral technology $\mathcal{T}_{\text{VRS}}$ satisfies Axioms SID and SOD, the vectors $(\mathbf{e}_i; \mathbf{0}_s) \in \mathbb{R}_{+}^{m+s}$, $i = 1, \ldots, m$, and the vectors $(\mathbf{0}_m; \mathbf{e}_r) \in \mathbb{R}_{+}^{m+s}$, $r = 1, \ldots, s$, are the recession directions of $\mathcal{T}_{\text{VRS}}$. This implies that technology $\mathcal{T}_{\text{VRS}}$ is unbounded.

Table 4.1 The data set in Example 1

DMU	A	B	C	D	E	F	G
x	2	3	4	6	7	8	9
y	1	3	4	5	5	4	2

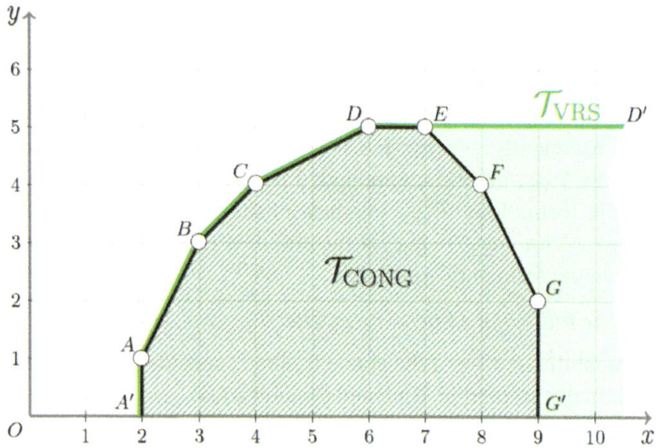

Fig. 4.1 The VRS and congestion technologies in Example 1

Turning back to Fig. 4.1, it is clear that, the full-efficient frontiers of technologies $\mathcal{T}_{\mathrm{VRS}}$ and $\mathcal{T}_{\mathrm{CONG}}$ are equal to the broken line $ABCD$. The following result shows that the full-efficient frontier of technology $\mathcal{T}_{\mathrm{VRS}}$ coincides generally with that of technology $\mathcal{T}_{\mathrm{CONG}}$.

Theorem 5 *The following equality is true:* $\partial_{\mathrm{F}}\mathcal{T}_{\mathrm{VRS}} = \partial_{\mathrm{F}}\mathcal{T}_{\mathrm{CONG}}$.

4.3.3 Weak and Strong Congestions

Tone and Sahoo (2004) refer to the congestion introduced by Definition 3 as *weak* congestion. Under the positivity assumption of input–output data, they further call a particular form of weak congestion as *strong* congestion. By their definition, an output-efficient DMU is strongly congested if a proportional reduction in all its inputs can be associated with a proportional increase in all its outputs. With a slight modification, Mehdiloozad et al. (2018) make the definition of strong congestion compatible with the presence of negative data. Specifically, they redefine the strong congestion in terms of changes in inputs and outputs that are not necessarily proportional. Using the concept of dominance, their definitions are stated as follows.

Table 4.2 The data set in Example 2

DMU	A	B	C	D	E	F	G	H
x	1	1	3	6	7	8	8	7
y_1	5	4	6	6	4	4	5	6
y_2	4	5	6	6	6	5	4	4

Source Mehdiloozad et al. (2018)

Definition 5 A DMU $(\mathbf{x}; \mathbf{y}) \in \mathcal{T}_{\text{CONG}}$ is

- *weakly congested* if there exists a DMU $(\mathbf{x}'; \mathbf{y}') \in \mathcal{T}_{\text{CONG}}$ such that $-\mathbf{x}' \gneqq -\mathbf{x}$ and $\mathbf{y}' \gneqq \mathbf{y}$.
- *strongly congested* if exists a DMU $(\mathbf{x}'; \mathbf{y}') \in \mathcal{T}_{\text{CONG}}$ such that $(-\mathbf{x}'; \mathbf{y}') > (-\mathbf{x}; \mathbf{y})$.

The above definition of weak congestion, which is equivalent to Definition 3, is the same one considered by Tone and Sahoo (2004). However, unlike the original definition of strong congestion, our modified definition implies that a (not-necessarily proportional) reduction in all inputs is associated with an increase in all outputs. Consequently, with the positivity assumption of input–output data, the strong congestion in the sense of Tone and Sahoo (2004) implies it in the sense of Definition 6, but not vice versa.

Note that the strong dominance is a special case of the weak dominance in the sense that the former implies the latter, but the converse is not generally true. Therefore, by definition, the strong congestion implies the weak congestion, but not vice versa.[16] To visualize the difference between the weak and the strong forms of congestion, we present a numerical example.

Example 2 Consider the eight observed DMUs A, B, C, D, E, F, G, and H defined in Table 4.2, where each consumes one input x to produce two outputs y_1 and y_2. Figure 4.2 shows technology $\mathcal{T}_{\text{CONG}}$ generated by these DMUs. Clearly, the output-efficient frontier of this technology consists of the triangle ABC, the segment CD and the pentagon $DEFGH$. This shows that all the observed DMUs are output-efficient and, therefore, satisfy the first condition of the evaluation of congestion.

First, we consider DMU G. From Fig. 4.2, it is observed that $x_D < x_G$ and $\mathbf{y}_D > \mathbf{y}_G$. This means that DMU G is strongly, and therefore weakly, congested. Let us now consider DMU H. Because $x_D < x_H$, $y_{1D} = y_{1E}$ and $y_{2D} > y_{2H}$, DMU H is weakly, but not strongly, congested.

[16] In the special case of a single-input and single-output production technology (Fig. 4.1), there is no distinction between the weak and strong forms of congestion.

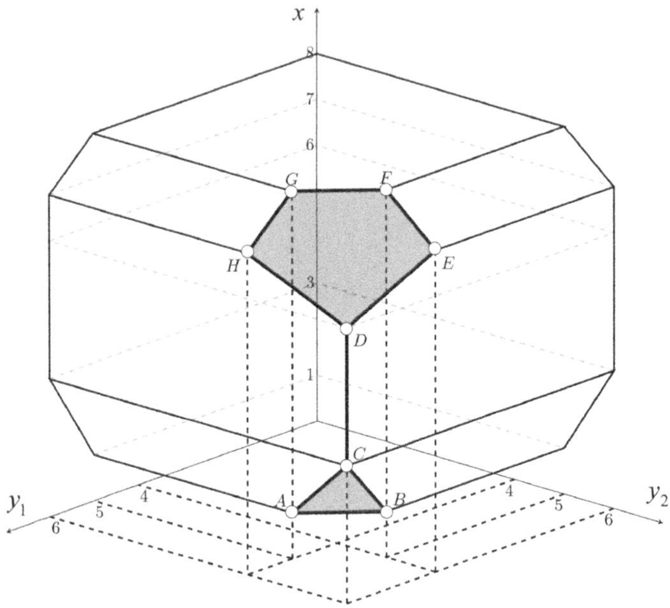

Fig. 4.2 Technology $\mathcal{T}_{\text{CONG}}$ in Example 2

4.3.4 The Congestion Model

By Definition 5, the first condition for a DMU to be considered for the evaluation of congestion is that it is output-efficient in technology $\mathcal{T}_{\text{CONG}}$. In this subsection we develop a model for testing this condition.

Let $\varepsilon > 0$ be a small (theoretically infinitesimal) value epsilon. Based on the statement of technology $\mathcal{T}_{\text{CONG}}$ obtained in Theorem 3, the output-oriented congestion model is developed as follows by Tone and Sahoo (2004):

$$\begin{aligned} &\max \quad \varphi + \varepsilon\left(\mathbf{1}_s^T \mathbf{q}\right) \\ &\text{subject to} \\ &\mathbf{X}\boldsymbol{\lambda} = \mathbf{x}_o, \\ &\mathbf{Y}\boldsymbol{\lambda} \geq \varphi \mathbf{y}_o + \mathbf{q}, \\ &\mathbf{1}_n^T \boldsymbol{\lambda} = 1, \\ &\boldsymbol{\lambda} \geq \mathbf{0}_n,\ \mathbf{q} \geq \mathbf{0}_s,\ \varphi \text{ sign free.} \end{aligned} \tag{4.7}$$

Using an optimal solution of program (4.7), the next result (without proof) provides a necessary and sufficient condition for the output efficiency of DMU $(\mathbf{x}_o; \mathbf{y}_o)$.

Theorem 6 *Let* $(\varphi^*, \boldsymbol{\lambda}^*, \mathbf{q}^*)$ *be an optimal solution of program* (4.7). *Then*
(i) $(\mathbf{x}_o; \varphi^* \mathbf{y}_o + \mathbf{q}^*) \in \partial_{\text{O}} \mathcal{T}_{\text{CONG}}$.

(ii) $(\mathbf{x}_o; \mathbf{y}_o) \in \partial_{\mathrm{O}}\mathcal{T}_{\mathrm{CONG}}$ *if and only if,* $\varphi^* = 1$ *and* $\mathbf{q}^* = \mathbf{0}_s$.

Theorem 6 shows that model (4.7) determines whether DMU $(\mathbf{x}_o; \mathbf{y}_o)$ is output-efficient in technology $\mathcal{T}_{\mathrm{CONG}}$; and if the answer is negative, the model projects that DMU onto the output-efficient frontier $\partial_{\mathrm{O}}\mathcal{T}_{\mathrm{CONG}}$. Based on part (i) of this theorem, an output-efficient projection of the assessed DMU, denoted $\left(\mathbf{x}_o^{\mathrm{P}}; \mathbf{y}_o^{\mathrm{P}}\right)$, can be identified as follows:

$$\left(\mathbf{x}_o^{\mathrm{P}}; \mathbf{y}_o^{\mathrm{P}}\right) = \left(\mathbf{x}_o; \varphi^* \mathbf{y}_o + \mathbf{q}^*\right). \tag{4.8}$$

Because of the non-radial nature of model (4.7), the projection $\left(\mathbf{x}_o^{\mathrm{P}}; \mathbf{y}_o^{\mathrm{P}}\right)$ is not generally unique. Specifically, if $\mathbf{q}^* \neq \mathbf{0}_s$, then multiple projections may be determined by optimal solutions of program (4.7). Sueyoshi and Sekitani (2009) argue that the occurrence of multiple projections for output-inefficient DMUs is an important issue that affects the evaluation of congestion, and thus needs to be addressed. In Sect. 4.4, we deal effectively with this issue.

4.3.5 The Congestion-Identification Model

To recognize the presence of weak and strong congestions, we first prove that the problem under consideration turns into finding a maximal element of a non-negative polyhedral set. Based on this finding and employing Theorem 1, we then develop a single-stage LP as the congestion-identification model.

Corresponding to DMU $(\mathbf{x}_o; \mathbf{y}_o)$, we consider the set $\mathcal{S}_o \subset \mathbb{R}_+^{m+s}$ given by

$$\mathcal{S}_o = \left\{(\boldsymbol{\alpha}; \boldsymbol{\beta}) \in \mathbb{R}_+^{m+s} : \ (\mathbf{x}_o - \boldsymbol{\alpha}; \mathbf{y}_o + \boldsymbol{\beta}) \in \mathcal{T}_{\mathrm{CONG}}\right\}.$$

Clearly, $\mathbf{0}_{m+s} \in \mathcal{S}_o$, implying $\mathcal{S}_o \neq \emptyset$. By Theorem 3, the set $\mathcal{S}_o$ is stated as follows:

$$\mathcal{S}_o = \left\{(\boldsymbol{\alpha}; \boldsymbol{\beta}) \in \mathbb{R}^{m+s} : \mathbf{X}\boldsymbol{\lambda} = \mathbf{x}_o - \boldsymbol{\alpha},\ \mathbf{Y}\boldsymbol{\lambda} \geq \mathbf{y}_o + \boldsymbol{\beta},\ \mathbf{1}_n^T\boldsymbol{\lambda} = 1,\ \boldsymbol{\lambda} \geq \mathbf{0}_n,\ \boldsymbol{\alpha} \geq \mathbf{0}_m,\ \boldsymbol{\beta} \geq \mathbf{0}_s\right\}. \tag{4.9}$$

By the projection lemma, it follows that $\mathcal{S}_o$ is a non-negative polyhedral set in $\mathbb{R}_+^{m+s}$. Using the maximal elements of $\mathcal{S}_o$, the next theorem provides some necessary and sufficient conditions for the full efficiency, the input efficiency, and the congestion status of DMU $(\mathbf{x}_o; \mathbf{y}_o)$. To brief the conditions, we present the following lemma.

Lemma 1 *Let* $(\mathbf{x}_o; \mathbf{y}_o) \in \partial_{\mathrm{O}}\mathcal{T}_{\mathrm{CONG}}$*, and let* $(\boldsymbol{\alpha}; \boldsymbol{\beta}) \in \mathcal{S}_o$*. If* $\boldsymbol{\beta} \neq \mathbf{0}_s$*, then* $\boldsymbol{\alpha} \neq \mathbf{0}_m$.

Theorem 7 *Let* $(\mathbf{x}_o; \mathbf{y}_o) \in \partial_{\mathrm{O}}\mathcal{T}_{\mathrm{CONG}}$*, and let* $\left(\boldsymbol{\alpha}_o^{\max}; \boldsymbol{\beta}_o^{\max}\right)$ *be a maximal element of* $\mathcal{S}_o$*. Then, DMU* $(\mathbf{x}_o; \mathbf{y}_o)$ *is*

(i) full-efficient in technology $\mathcal{T}_{\mathrm{CONG}}$ *if and only if* $\boldsymbol{\alpha}_o^{\max} = \mathbf{0}_m$.

(ii) input-inefficient, but not congested, in technology $\mathcal{T}_{\text{CONG}}$ if and only if, $\boldsymbol{\alpha}_o^{\max} \neq \mathbf{0}_m$ *and* $\boldsymbol{\beta}_o^{\max} = \mathbf{0}_s$.

(iii) weakly congested if and only if $\boldsymbol{\beta}_o^{\max} \neq \mathbf{0}_s$.

(iv) strongly congested if and only if $\left(\boldsymbol{\alpha}_o^{\max}; \boldsymbol{\beta}_o^{\max}\right) > \mathbf{0}_{m+s}$.

Part (i) of Theorem 7, being equivalent to Theorem 3.2 in Mehdiloozad et al. (2018), states that the trivialness of $\mathcal{S}_o$ is both necessary and sufficient for DMU $(\mathbf{x}_o; \mathbf{y}_o)$ to be full-efficient in technology $\mathcal{T}_{\text{CONG}}$, and therefore in technology $\mathcal{T}_{\text{VRS}}$ by Theorem 5. By parts (ii) and (iii) of Theorem 7, it follows that the input efficiency of an output-efficient DMU in technology $\mathcal{T}_{\text{CONG}}$ is a necessary but not sufficient condition for the presence of weak congestion.

By parts (iii) and (iv) of Theorem 7, any maximal element of $\mathcal{S}_o$ can determine the presence of weak and strong congestions at DMU $(\mathbf{x}_o; \mathbf{y}_o)$. To obtain such an element, we propose the following LP by employing Theorem 1:

$$\begin{aligned} &\max \quad \mathbf{1}_m^T \mathbf{t}^- + \mathbf{1}_s^T \mathbf{t}^+ \\ &\text{subject to} \\ &\mathbf{X}\boldsymbol{\delta} + \mathbf{s}^- + \mathbf{t}^- = \mathbf{x}_o w, \\ &\mathbf{Y}\boldsymbol{\delta} - \mathbf{s}^+ - \mathbf{t}^+ \geq \mathbf{y}_o w, \\ &\mathbf{1}_n^T \boldsymbol{\delta} = w, \\ &\boldsymbol{\delta} \geq \mathbf{0}_n, \ \mathbf{s}^- \geq \mathbf{0}_m, \ \mathbf{s}^+ \geq \mathbf{0}_s, \\ &\mathbf{1}_m \geq \mathbf{t}^- \geq \mathbf{0}_m, \ \mathbf{1}_s \geq \mathbf{t}^+ \geq \mathbf{0}_s, \ w \geq 1. \end{aligned} \tag{4.10}$$

Let $\left(\boldsymbol{\delta}^*, \mathbf{s}^{-*}, \mathbf{s}^{+*}, \mathbf{t}^{-*}, \mathbf{t}^{+*}, w^*\right)$ be an optimal solution to program (4.10). Then, by Theorem 1, we derive a maximal element of $\mathcal{S}_o$ as follows:

$$\left(\boldsymbol{\alpha}_o^{\max}; \boldsymbol{\beta}_o^{\max}\right) = \frac{1}{1 + w^*}\left(\mathbf{s}^{-*} + \mathbf{t}^{-*}; \mathbf{s}^{+*} + \mathbf{t}^{+*}\right). \tag{4.11}$$

The advantages of using model (4.10) for the identification of weak and strong congestions are outlined as follows:

- In contrast to all the existing congestion-identification methods, the proposed model precisely distinguishes between the weak and the strong congestion statuses of DMUs.
- Besides the correct identification of weak and strong congestion statuses of a DMU, our model is able to test its full efficiency and input efficiency statuses.
- In contrast to all the existing congestion-identification methods, the proposed model deals effectively with input–output data sets containing negative values.
- In contrast to all the existing congestion-identification methods, the proposed model determines all improvable inputs and outputs, i.e., all reducible inputs and all increasable outputs.

- The proposed model is a single-stage LP. Therefore, applying this model to identifying the weak and the strong congestion statuses is computationally more efficient than the existing congestion-identification methods, where each requires solving separate LPs to accomplish the task.

4.4 Congestion of Output-Inefficient DMUs

4.4.1 Congestion of Faces of Technology $\mathcal{T}_{\text{CONG}}$

By Theorem 4, each face of technology $\mathcal{T}_{\text{CONG}}$ is a polytope. In this section we show that all relative interior points of such a face have the same status of congestion.

Let $\mathcal{F}$ be a face of technology $\mathcal{T}_{\text{CONG}}$, and let $\left(\mathbf{x}^{\text{ri}}; \mathbf{y}^{\text{ri}}\right) \in ri\,(\mathcal{F})$ be arbitrarily taken. Further, assume that $\mathcal{I}_{\text{ri}}$ and $\mathcal{O}_{\text{ri}}$ are, respectively, the index sets of all improvable inputs and outputs of DMU $\left(\mathbf{x}^{\text{ri}}; \mathbf{y}^{\text{ri}}\right)$, viz.,

$$\mathcal{I}_{\text{ri}} = \bigcup_{(\boldsymbol{\alpha};\boldsymbol{\beta})\in\mathcal{S}_{\text{ri}}} \sigma\,(\boldsymbol{\alpha}), \qquad \mathcal{O}_{\text{ri}} = \bigcup_{(\boldsymbol{\alpha};\boldsymbol{\beta})\in\mathcal{S}_{\text{ri}}} \sigma\,(\boldsymbol{\beta}), \tag{4.12}$$

where $\mathcal{S}_{\text{ri}} \subset \mathbb{R}^{m+s}_{+}$ denotes the set defined in (4.9) for $\left(\mathbf{x}^{\text{ri}}; \mathbf{y}^{\text{ri}}\right)$.

Considering the fact that the set $\mathcal{S}_{\text{ri}}$ is a non-negative convex polyhedral set, denote $\left(\boldsymbol{\alpha}^{\max}_{\text{ri}}; \boldsymbol{\beta}^{\max}_{\text{ri}}\right)$ a maximal element of $\mathcal{S}_{\text{ri}}$. Then the following equalities are obtained by (4.3):

$$\mathcal{I}_{\text{ri}} = \sigma\left(\boldsymbol{\alpha}^{\max}_{\text{ri}}\right), \qquad \mathcal{O}_{\text{ri}} = \sigma\left(\boldsymbol{\beta}^{\max}_{\text{ri}}\right). \tag{4.13}$$

Because $\mathcal{F}$ is a polytope, DMU $\left(\mathbf{x}^{\text{ri}}; \mathbf{y}^{\text{ri}}\right)$ can be expressed as a convex combination of all observed DMUs spanning $\mathcal{F}$. This means that there exists a subset of observed DMUs, namely $\left\{\left(\mathbf{x}_{j_1}; \mathbf{y}_{j_1}\right), \ldots, \left(\mathbf{x}_{j_l}; \mathbf{y}_{j_l}\right)\right\}$, and a positive weight vector $\boldsymbol{\rho} \in \mathbb{R}^{l}_{++}$ such that

$$\left(\mathbf{x}^{\text{ri}}; \mathbf{y}^{\text{ri}}\right) = \sum_{k=1}^{l} \rho_k \left(\mathbf{x}_{j_k}; \mathbf{y}_{j_k}\right), \quad \sum_{k=1}^{l} \rho_k = 1. \tag{4.14}$$

For each $k = 1, \ldots, l$, let $\left(\boldsymbol{\delta}^*_k, \mathbf{s}^{-*}_k, \mathbf{t}^{-*}_k, \mathbf{s}^{+*}_k, \mathbf{t}^{+*}_k, w^*_k\right)$ denote the optimal solution obtained from the evaluation of DMU $\left(\mathbf{x}_{j_k}; \mathbf{y}_{j_k}\right)$ by model (4.10). Then, a maximal element of $\mathcal{S}_{j_k}$ is obtained by (4.11) as follows:

$$\left(\boldsymbol{\alpha}^{\max}_k; \boldsymbol{\beta}^{\max}_k\right) = \frac{1}{1+w^*_k}\left(\mathbf{s}^{-*}_k + \mathbf{t}^{-*}_k; \mathbf{s}^{+*}_k + \mathbf{t}^{+*}_k\right). \tag{4.15}$$

To state the next theorem, we define

$$\mathcal{I}_{\mathcal{F}} = \bigcup_{k=1}^{l} \sigma\left(\boldsymbol{\alpha}_k^{\max}\right), \qquad \mathcal{O}_{\mathcal{F}} = \bigcup_{k=1}^{l} \sigma\left(\boldsymbol{\beta}_k^{\max}\right). \tag{4.16}$$

Theorem 8 *Let $\mathcal{F}$ be a face of technology $\mathcal{T}_{\text{CONG}}$. Then $\mathcal{I}_{\mathcal{F}} = \mathcal{I}_{\text{ri}}$ and $\mathcal{O}_{\mathcal{F}} = \mathcal{O}_{\text{ri}}$, for all $\left(\mathbf{x}^{\text{ri}}; \mathbf{y}^{\text{ri}}\right) \in ri\ (\mathcal{F})$.*

Theorem 8 shows that, all relative interior points of face $\mathcal{F}$ must exhibit the same form of congestion, weak or strong. By this finding, we propose the following definition.

Definition 6 A face $\mathcal{F}$ of technology $\mathcal{T}_{\text{CONG}}$ is *weakly* (resp., *strongly*) congested if all DMUs on $ri\ (\mathcal{F})$ are weakly (resp., strongly) congested.

By Theorem 8, if any arbitrary DMU on the relative interior of face $\mathcal{F}$ is congested, then all the DMUs on its relative interior are also congested. Therefore, as a consequence of Theorem 8 and Definition 6, we derive the following corollary.

Corollary 2 *A face $\mathcal{F}$ of technology $\mathcal{T}_{\text{CONG}}$ is weakly (resp., strongly) congested if and only if an arbitrary DMU on $ri\ (\mathcal{F})$ is weakly (resp., strongly) congested.*

Corollary 2 shows that testing the congestion of any DMU on the relative interior of face $\mathcal{F}$ is enough for identifying the congestion of this face. The congestion of such a DMU can be successfully identified by solving the congestion-identification model (4.5). Nonetheless, the identification task can be accomplished alternatively using the following equalities, which follow from (4.13) and (4.16) by Theorem 8:

$$\sigma\left(\boldsymbol{\alpha}_{\text{ri}}^{\max}\right) = \bigcup_{k=1}^{l} \sigma\left(\boldsymbol{\alpha}_k^{\max}\right), \qquad \sigma\left(\boldsymbol{\beta}_{\text{ri}}^{\max}\right) = \bigcup_{k=1}^{l} \sigma\left(\boldsymbol{\beta}_k^{\max}\right). \tag{4.17}$$

This is stated precisely as the following corollary.

Corollary 3 *Let $\mathcal{F}$ be a face of technology $\mathcal{T}_{\text{CONG}}$ that is spanned by the observed DMUs $\left(\mathbf{x}_{j_1}; \mathbf{y}_{j_1}\right), \ldots, \left(\mathbf{x}_{j_l}; \mathbf{y}_{j_l}\right)$, and let $\left(\mathbf{x}^{\text{ri}}; \mathbf{y}^{\text{ri}}\right) \in ri\ (\mathcal{F})$. Then DMU $\left(\mathbf{x}^{\text{ri}}; \mathbf{y}^{\text{ri}}\right)$, and therefore face $\mathcal{F}$, is*

(i) weakly congested if and only if $\bigcup_{k=1}^{l} \sigma\left(\boldsymbol{\alpha}_k^{\max}\right) \neq \emptyset$.

(ii) strongly congested if and only if, $\bigcup_{k=1}^{l} \sigma\left(\boldsymbol{\alpha}_k^{\max}\right) = \{1, \ldots, m\}$ and $\bigcup_{k=1}^{l} \sigma\left(\boldsymbol{\beta}_k^{\max}\right) = \{1, \ldots, s\}$.

Part (i) of the above corollary states that, a face of technology $\mathcal{T}_{\text{CONG}}$ is weakly congested if and only if at least one of the its spanning DMUs is weakly congested. Concerning part (ii), however, note that the presence of strong congestion at one of the spanning DMUs is a sufficient, but not necessary, condition for the presence of strong congestion at the corresponding face.

4.4.2 The Minimal Face of an Output-Inefficient DMU

Let DMU $(\mathbf{x}_o; \mathbf{y}_o)$ be output-inefficient in technology $\mathcal{T}_{\text{CONG}}$, and let Π_o denote its projection set, i.e., the set of all output-efficient projections obtained by (4.8). Further, assume that $\mathcal{F}_o^{\min}$ is the intersection of all faces of $\mathcal{T}_{\text{CONG}}$ that contain Π_o. Then $\mathcal{F}_o^{\min}$ is non-empty because $\mathcal{T}_{\text{CONG}}$ is a face of itself and contains Π_o. Moreover, from convex analysis, $\mathcal{F}_o^{\min}$ is a face of $\mathcal{T}_{\text{CONG}}$ and is therefore the smallest face containing Π_o. Consequently, it is referred to as the *minimal face* of DMU $(\mathbf{x}_o; \mathbf{y}_o)$.

In this subsection, we are concerned about finding an element of the intersection $\Pi_o \cap ri\left(\mathcal{F}_o^{\min}\right)$, which is always non-empty.[17] We start with the definition of (unique) *global reference set* (GRS)[18] of DMU $(\mathbf{x}_o; \mathbf{y}_o)$:

$$\mathcal{G}_o = \left\{\left(\mathbf{x}_j; \mathbf{y}_j\right) : \lambda_j > 0 \text{ in some optimal solution of program (4.7)}\right\}.$$

By Theorem 4, the minimal face $\mathcal{F}_o^{\min}$ is a polytope. The following theorem further characterizes the structure of this face by showing that it coincides with the convex hull of the GRS.

Theorem 9 *The following equality is true:*

$$\mathcal{F}_o^{\min} = \operatorname{conv}\left(\mathcal{G}_o\right). \tag{4.18}$$

Denote $\mathcal{J}_o$ the index set of all reference DMUs in $\mathcal{G}_o$. Then, by Theorem 9, the minimal face $\mathcal{F}_o^{\min}$ is stated as follows:

$$\mathcal{F}_o^{\min} = \left\{(\mathbf{x}; \mathbf{y}) \in \mathbb{R}_+^{m+s} : (\mathbf{x}; \mathbf{y}) = \sum_{j \in \mathcal{J}_o} \lambda_j \left(\mathbf{x}_j; \mathbf{y}_j\right), \sum_{j \in \mathcal{J}_o} \lambda_j = 1, \lambda_j \geq 0, \forall j \in \mathcal{J}_o\right\}. \tag{4.19}$$

From convex analysis, the relative interior of the convex hull of a finite collection of points is characterized by their strict convex combinations. Therefore, the next corollary is immediate from (4.19).

Corollary 4 *The relative interior of $\mathcal{F}_o^{\min}$ is characterized as follows:*

$$ri\left(\mathcal{F}_o^{\min}\right) = \left\{(\mathbf{x}; \mathbf{y}) \in \mathbb{R}_+^{m+s} : (\mathbf{x}; \mathbf{y}) = \sum_{j \in \mathcal{J}_o} \lambda_j \left(\mathbf{x}_j; \mathbf{y}_j\right), \sum_{j \in \mathcal{J}_o} \lambda_j = 1, \lambda_j > 0, \forall j \in \mathcal{J}_o\right\}. \tag{4.20}$$

---

[17]By contradiction, assume that $\Pi_o \cap ri\left(\mathcal{F}_o^{\min}\right) = \emptyset$ or, equivalently, that $\Pi_o \subseteq rb\left(\mathcal{F}_o^{\min}\right)$. Then, there is a face of $\mathcal{F}_o^{\min}$ of minimum dimension, namely $\mathcal{F}'$, for which $\Pi_o \subseteq \mathcal{F}' \subseteq rb\left(\mathcal{F}_o^{\min}\right) \subsetneqq \mathcal{F}_o^{\min}$. From convex analysis, $\mathcal{F}'$ is a face of $\mathcal{T}_{\text{CONG}}$ (see Tuy 1998, p. 24). Because the dimension of $\mathcal{F}'$ is less than that of $\mathcal{F}_o^{\min}$, we have a contradiction. Therefore, $\Pi_o \cap ri\left(\mathcal{F}_o^{\min}\right) \neq \emptyset$.

[18]For detailed information on the concept of GRS, the reader may refer to Mehdiloozad (2017), Mehdiloozad et al. (2015) and Mehdiloozad and Sahoo (2016).

Corollary 4 shows that a relative interior point of $\mathcal{F}_o^{\min}$ can be found by identifying the set $\mathcal{G}_o$. To make such identification, we assume Λ_o to be the set of all intensity vectors that are associated with all optimal solutions of program (4.7). Given an optimal solution $(\varphi^*, \mathbf{q}^*, \boldsymbol{\lambda}^*)$ to program (4.7), the set Λ_o is stated as follows:

$$\Lambda_o = \left\{ \boldsymbol{\lambda} \in \mathbb{R}_+^n : \begin{bmatrix} \mathbf{X} \\ \mathbf{Y} \\ \mathbf{1}_n^T \\ \mathbf{0}_n^T \end{bmatrix} \boldsymbol{\lambda} + \begin{bmatrix} \mathbf{0}_{m\times s} \\ -\mathbf{I}_s \\ \mathbf{0}_s^T \\ \mathbf{1}_s^T \end{bmatrix} \mathbf{q} = \begin{bmatrix} \mathbf{x}_o \\ \varphi^* \mathbf{y}_o \\ 1 \\ \mathbf{1}_s^T \mathbf{q}^* \end{bmatrix}, \ \boldsymbol{\lambda} \geq \mathbf{0}_n, \ \mathbf{q} \geq \mathbf{0}_s \right\}. \quad (4.21)$$

By the projection lemma, Λ_o is a non-negative polyhedral set in $\mathbb{R}_+^n$. The following theorem links the maximal elements of this set to the problem of identifying $\mathcal{G}_o$.

Theorem 10 *Let $\boldsymbol{\lambda}_o^{\max}$ be a maximal element of Λ_o. Then the following equality is true:*

$$\mathcal{J}_o = \sigma\left(\boldsymbol{\lambda}_o^{\max}\right). \quad (4.22)$$

From Corollary 4 and Theorem 10, it is clear that a relative interior point of $\mathcal{F}_o^{\min}$ can be obtained through the vector $\boldsymbol{\lambda}_o^{\max}$. To find this vector, we formulate the following LP based on Theorem 1:

$$\begin{aligned} & \max \ \mathbf{1}_n^T \boldsymbol{\lambda}^2 \\ & \text{subject to} \\ & \begin{bmatrix} \mathbf{X} \\ \mathbf{Y} \\ \mathbf{1}_n^T \\ \mathbf{0}_n^T \end{bmatrix} \left(\boldsymbol{\lambda}^1 + \boldsymbol{\lambda}^2\right) + \begin{bmatrix} \mathbf{0}_{m\times s} \\ -\mathbf{I}_s \\ \mathbf{0}_s^T \\ \mathbf{1}_s^T \end{bmatrix} \mathbf{q} = \begin{bmatrix} \mathbf{x}_o \\ \varphi^* \mathbf{y}_o \\ 1 \\ \mathbf{1}_s^T \mathbf{q}^* \end{bmatrix} z, \\ & \boldsymbol{\lambda}^1 \geq \mathbf{0}_n, \ \mathbf{1}_n \geq \boldsymbol{\lambda}^2 \geq \mathbf{0}_n, \ \mathbf{q} \geq \mathbf{0}_s, \ z \geq 1. \end{aligned} \quad (4.23)$$

Let $\left(\boldsymbol{\lambda}^{1*}, \boldsymbol{\lambda}^{2*}, \mathbf{q}^*, z^*\right)$ be an optimal solution to program (4.23). Then we have

$$\boldsymbol{\lambda}_o^{\max} = \frac{1}{1+z^*}\left(\boldsymbol{\lambda}^{1*} + \boldsymbol{\lambda}^{2*}\right). \quad (4.24)$$

To conclude the results of this subsection, we present the following interesting corollary.

Corollary 5 *Let $\boldsymbol{\lambda}_o^{\max}$ be a maximal element of Λ_o. Then*

$$\left(\mathbf{X}\boldsymbol{\lambda}_o^{\max}; \mathbf{Y}\boldsymbol{\lambda}_o^{\max}\right) \in \Pi_o \cap ri\left(\mathcal{F}_o^{\min}\right). \quad (4.25)$$

4.4.3 A Precise Definition of Congestion for Output-Inefficient DMUs

For any output-inefficient DMU in technology $\mathcal{T}_{\text{CONG}}$, the concept of congestion is conventionally defined at its projection on the output-efficient frontier of this technology. More explicitly, an output-inefficient DMU is defined to be weakly (resp., strongly) congested if its projection is weakly (resp., strongly) congested.[19]

Throughout this subsection, we assume that DMU $(\mathbf{x}_o; \mathbf{y}_o)$ is output-inefficient. As in Tone and Sahoo (2004), if the projection $(\mathbf{x}_o^{\text{P}}; \mathbf{y}_o^{\text{P}})$ as defined in (4.8) is used for evaluating the congestion of DMU $(\mathbf{x}_o; \mathbf{y}_o)$, there arise two cases: $\mathbf{q}^* = \mathbf{0}_s$ or $\mathbf{q}^* \neq \mathbf{0}_s$. In the case of $\mathbf{q}^* = \mathbf{0}_s$, the projection is unique and, therefore, the definition of congestion is well-defined. Consequently, the congestion of DMU $(\mathbf{x}_o; \mathbf{y}_o)$ in this case is straightforwardly evaluated at $(\mathbf{x}_o^{\text{P}}; \mathbf{y}_o^{\text{P}})$. However, as noted in Sect. 4.3.4, the difficulty may arise when $\mathbf{q}^* \neq \mathbf{0}_s$. This is because in this case the non-radial nature of model (4.7) may generate multiple projections that, in turn, may yield conflicting results in terms of the congestion and efficiency statuses of DMU $(\mathbf{x}_o; \mathbf{y}_o)$. For example, for a DMU that its projection set includes both congested and non-congested units, no presence of congestion is identified when the DMU is arbitrarily projected via model (4.7) onto a non-congested DMU.

From the above argument, the question arises as to how the concept of congestion can be defined for any output-inefficient DMU so that the offered definition is mathematically well-defined. We address this question by proposing the following original definition.

Definition 7 Let $(\mathbf{x}_o; \mathbf{y}_o) \notin \partial_{\text{O}}\mathcal{T}_{\text{CONG}}$. Then DMU $(\mathbf{x}_o; \mathbf{y}_o)$ is weakly (resp., strongly) congested if the corresponding minimal face $\mathcal{F}_o^{\min}$ is weakly (resp., strongly) congested.

It is worth noting that the conventional definition for the congestion of DMU $(\mathbf{x}_o; \mathbf{y}_o)$ in the case $\mathbf{q}^* = \mathbf{0}_s$ is a special case of Definition 7. This is due to the fact that, in this case we have $ri\left(\mathcal{F}_o^{\min}\right) = \mathcal{F}_o^{\min} = \{(\mathbf{x}_o; \mathbf{y}_o)\}$.

By Corollary 2, the congestion status of any face of technology $\mathcal{T}_{\text{CONG}}$ is determined by that of an arbitrary DMU on its relative interior. By Definition 7, it follows that the congestion of DMU $(\mathbf{x}_o; \mathbf{y}_o)$ can be identified by any DMU on $ri\left(\mathcal{F}_o^{\min}\right)$, namely the max-projection of DMU $(\mathbf{x}_o; \mathbf{y}_o)$ defined by

$$\left(\mathbf{x}_o^{\max}; \mathbf{y}_o^{\max}\right) = \left(\mathbf{X}\boldsymbol{\lambda}_o^{\max}; \mathbf{Y}\boldsymbol{\lambda}_o^{\max}\right), \tag{4.26}$$

which is on $\Pi_o \cap ri\left(\mathcal{F}_o^{\min}\right)$ (see Corollary 5).

Consequently, the congestion of DMU $(\mathbf{x}_o; \mathbf{y}_o)$ is identified by carrying out the following stages:

---

[19]The output shortfalls of an output-inefficient congested DMU is made up of two components. The first component represents the output shortfalls associated with the distance from the DMU itself to its projection, i.e., the DMU's technical output inefficiency. The second component represents, however, the output shortfalls arising from the congestion of the DMU's projection.

Stage 1: Find the max-projection $\left(\mathbf{x}_o^{\max}; \mathbf{y}_o^{\max}\right)$ by solving program (4.23).
Stage 2: Replace $(\mathbf{x}_o; \mathbf{y}_o)$ with $\left(\mathbf{x}_o^{\max}; \mathbf{y}_o^{\max}\right)$ and solve program (4.10). Then recognize the congestion status by applying Theorem 7 to the vector $\left(\boldsymbol{\alpha}_o^{\max}; \boldsymbol{\beta}_o^{\max}\right)$ obtained from (4.11).

4.5 Three Congestion-Identification Algorithms

4.5.1 Algorithm 1: Incorporating the Non-negativity Assumption

Under the non-negativity assumption of input–output data, we propose Algorithm 1 for evaluating the congestion of all the observed DMUs.

Algorithm 1

---

for $o \leftarrow 1$ **to** n, **do**

STEP O.1: EVALUATE DMU $(\mathbf{x}_o; \mathbf{y}_o)$ BY MODEL (4.7) TO FIND THE OPTIMAL SOLUTION $(\varphi_o^*, \boldsymbol{\lambda}_o^*, \mathbf{q}_o^*)$.

if $\mathbf{q}_o^{+*} = \mathbf{0}_s$, **then**
find the projection $\left(\mathbf{x}_o^{\mathrm{P}}; \mathbf{y}_o^{\mathrm{P}}\right)$ by (4.8),
$(\mathbf{x}_o; \mathbf{y}_o) \leftarrow \left(\mathbf{x}_o^{\mathrm{P}}; \mathbf{y}_o^{\mathrm{P}}\right)$,
go to Step o.2.
else
solve model (4.23) and find the max-projection $\left(\mathbf{x}_o^{\max}; \mathbf{y}_o^{\max}\right)$ by (4.26),
$(\mathbf{x}_o; \mathbf{y}_o) \leftarrow \left(\mathbf{x}_o^{\max}; \mathbf{y}_o^{\max}\right)$,
go to Step o.2.
end if

STEP O.2: SOLVE MODEL (4.10) AND OBTAIN $\left(\boldsymbol{\alpha}_o^{\max}; \boldsymbol{\beta}_o^{\max}\right)$ BY (4.11).
if $\left(\boldsymbol{\alpha}_o^{\max}; \boldsymbol{\beta}_o^{\max}\right) > \mathbf{0}_{m+s}$, **then**
DMU $(\mathbf{x}_o; \mathbf{y}_o)$ is strongly congested.
else if $\boldsymbol{\beta}_o^{\max} \neq \mathbf{0}_s$, **then**
DMU $(\mathbf{x}_o; \mathbf{y}_o)$ is weakly, but not strongly, congested.
else [DMU $(\mathbf{x}_o; \mathbf{y}_o)$ is not congested]
if $\boldsymbol{\alpha}_o^{\max} \neq \mathbf{0}_s$, **then**
DMU $(\mathbf{x}_o; \mathbf{y}_o)$ is output-efficient and input-inefficient, but is not congested.
else
DMU $(\mathbf{x}_o; \mathbf{y}_o)$ is full-efficient.
end if
end if

end for

4.5.2 Algorithm 2: Enhancing Computational Efficiency

In this section we develop Algorithm 2 for evaluating the congestion of all the observed DMUs which all face non-negative input–output data. This algorithm can be computationally more efficient than Algorithm 1 because the number of LPs required to be solved is less. This is due to the fact that, for any output-inefficient DMU $(\mathbf{x}_o; \mathbf{y}_o)$ with non-zero optimal output slacks, Algorithm 2 recognizes its congestion status by looking at those of its reference DMUs, instead of evaluating its max-projection $\left(\mathbf{x}_o^{\max}; \mathbf{y}_o^{\max}\right)$ by model (4.23).

4.5.3 Algorithm 3: Removing the Non-negativity Assumption

To this point, the input-output data are assumed to be non-negative. This section first develops an output-oriented additive model to deal with the presence of negative data in the measurement of efficiency. Using this model, it then designs an algorithm for evaluating congestion of DMUs with inputs and outputs unrestricted in sign. The new algorithm is a modification of Algorithm 1.

With the assumption that inputs and outputs are unrestricted in sign (i.e., $\mathcal{T} \subset \mathbb{R}^{m+s}$), we define technology $\tilde{\mathcal{T}}_{\text{CONG}}$ to be the intersection of all technologies $\mathcal{T} \subset \mathbb{R}^{m+s}$ that satisfy the axioms considered in the definition of technology $\mathcal{T}_{\text{CONG}}$ with the slight modification that the disposal of outputs is not limited by the condition of their non-negativity. Then, $\tilde{\mathcal{T}}_{\text{CONG}}$ is the smallest technology that satisfies all Axioms IO, SID, modified SOD and CT and is explicitly stated as follows:

$$\tilde{\mathcal{T}}_{\text{CONG}} = \left\{(\mathbf{x}; \mathbf{y}) \in \mathbb{R}^{m+s} : \mathbf{X}\boldsymbol{\lambda} = \mathbf{x},\ \mathbf{Y}\boldsymbol{\lambda} \geq \mathbf{y},\ \mathbf{1}_n^T\boldsymbol{\lambda} = 1,\ \boldsymbol{\lambda} \geq \mathbf{0}_n\right\}. \tag{4.27}$$

Based on this technology, the output-oriented additive model is presented in the following form (Wei and Yan, 2009):

$$\begin{aligned} &\max \quad \mathbf{1}_s^T\mathbf{q} \\ &\text{subject to} \\ &\mathbf{X}\boldsymbol{\lambda} = \mathbf{x}_o, \\ &\mathbf{Y}\boldsymbol{\lambda} \geq \mathbf{y}_o + \mathbf{q}, \\ &\mathbf{1}_n^T\boldsymbol{\lambda} = 1, \\ &\boldsymbol{\lambda} \geq \mathbf{0}_n,\ \mathbf{q} \geq \mathbf{0}_s. \end{aligned} \tag{4.28}$$

Let $(\boldsymbol{\lambda}^*, \mathbf{q}^*)$ be an optimal solution to problem (4.28). It is clear that $\mathbf{1}_s^T\mathbf{q}^* \geq 0$. DMU $(\mathbf{x}_o; \mathbf{y}_o)$ is output-efficient in $\tilde{\mathcal{T}}_{\text{CONG}}$ if and only if, $\mathbf{1}_s^T\mathbf{q}^* = 0$ or, equivalently, $\mathbf{q}^* = \mathbf{0}_s$. Model (4.28) identifies the output-efficient projection $\left(\tilde{\mathbf{x}}_o^{\text{P}}; \tilde{\mathbf{y}}_o^{\text{P}}\right)$ of the assessed DMU as follows:

$$\left(\tilde{\mathbf{x}}_o^{\text{P}}; \tilde{\mathbf{y}}_o^{\text{P}}\right) = \left(\mathbf{x}_o; \mathbf{y}_o + \mathbf{q}^*\right). \tag{4.29}$$

Algorithm 2

STEP 1: DECOMPOSING THE SET OF ALL OBSERVED DMUS INTO THE DISJOINT CATEGORIES C_1 and C_2.

Initialize $C_1 = C_2 = \emptyset$.
for $o = 1 \to n$ **do**
  Evaluate DMU $(\mathbf{x}_o; \mathbf{y}_o)$ by model (4.7) and obtain the optimal solution $\left(\varphi_o^*, \boldsymbol{\lambda}_o^*, \mathbf{q}_o^*\right)$.
  if $\mathbf{q}_o^{+*} = \mathbf{0}_s$, **then**
    $C_1 \leftarrow C_1 \cup \{(\mathbf{x}_o; \mathbf{y}_o)\}$.
  else
    $C_2 \leftarrow C_2 \cup \{(\mathbf{x}_o; \mathbf{y}_o)\}$.
  end if
end for
return C_1 and C_2.
go to Step 2.

STEP 2: EVALUATING DMUS IN C_1.

for $(\mathbf{x}_o; \mathbf{y}_o) \in C_1$ **do**
  find the projection $\left(\mathbf{x}_o^{\mathrm{P}}; \mathbf{y}_o^{\mathrm{P}}\right)$ by (4.8),
  $(\mathbf{x}_o; \mathbf{y}_o) \leftarrow \left(\mathbf{x}_o^{\mathrm{P}}; \mathbf{y}_o^{\mathrm{P}}\right)$,
  solve model (4.10) and obtain the vector $\left(\boldsymbol{\alpha}_o^{\max}; \boldsymbol{\beta}_o^{\max}\right)$ by (4.11).
  if $\left(\boldsymbol{\alpha}_o^{\max}; \boldsymbol{\beta}_o^{\max}\right) > \mathbf{0}_{m+s}$, **then**
    DMU $(\mathbf{x}_o; \mathbf{y}_o)$ is strongly congested.
  else if $\boldsymbol{\beta}_o^{\max} \neq \mathbf{0}_s$, **then**
    DMU $(\mathbf{x}_o; \mathbf{y}_o)$ is weakly, but not strongly, congested.
  else [DMU $(\mathbf{x}_o; \mathbf{y}_o)$ is not congested]
    if $\boldsymbol{\alpha}_o^{\max} \neq \mathbf{0}_s$, **then**
      DMU $(\mathbf{x}_o; \mathbf{y}_o)$ is output-efficient and input-inefficient, but is not congested.
    else
      DMU $(\mathbf{x}_o; \mathbf{y}_o)$ is full-efficient.
    end if
  end if
end for
go to Step 3.

STEP 3: EVALUATING DMUS IN C_2.

for $(\mathbf{x}_o; \mathbf{y}_o) \in C_3$ **do**
  Solve model (4.23) and obtain $\boldsymbol{\lambda}_o^{\max}$ by (4.24).
  Set $\left(\boldsymbol{\alpha}_o^{\mathrm{GRS}}; \boldsymbol{\beta}_o^{\mathrm{GRS}}\right) = \sum_{j \in \sigma(\boldsymbol{\lambda}_o^{\max})} \lambda_{jo}^{\max} \left(\boldsymbol{\alpha}_j^{\max}; \boldsymbol{\beta}_j^{\max}\right)$.
  if $\left(\boldsymbol{\alpha}_o^{\mathrm{GRS}}; \boldsymbol{\beta}_o^{\mathrm{GRS}}\right) > \mathbf{0}_{m+s}$, **then**
    DMU $(\mathbf{x}_o; \mathbf{y}_o)$ is strongly congested.
  else if $\boldsymbol{\beta}_o^{\mathrm{GRS}} \neq \mathbf{0}_s$, **then**
    DMU $(\mathbf{x}_o; \mathbf{y}_o)$ is weakly, but not strongly, congested.
  else [DMU $(\mathbf{x}_o; \mathbf{y}_o)$ is not congested]
    if $\boldsymbol{\alpha}_o^{\mathrm{GRS}} \neq \mathbf{0}_s$, **then**
      DMU $(\mathbf{x}_o; \mathbf{y}_o)$ is output-efficient and input-inefficient, but is not congested.
    else
      DMU $(\mathbf{x}_o; \mathbf{y}_o)$ is full-efficient.
    end if
  end if
end for

Using the above results, we develop Algorithm 3 for the evaluation of congestion in the presence of negative data. We note that the computational efficiency of this algorithm can be improved in the same way that the computational efficiency of Algorithm 1 was enhanced by Algorithm 2.

Algorithm 3

---

for $o \leftarrow 1$ **to** n, **do**

STEP O.1: EVALUATE DMU $(\mathbf{x}_o; \mathbf{y}_o)$ BY MODEL (4.28) TO OBTAIN THE OPTIMAL SOLUTION $\left(\boldsymbol{\lambda}_o^*, \mathbf{q}_o^*\right)$.

if $\mathbf{q}_o^{+*} = \mathbf{0}_s$, **then**

  go to Step o.2.

else

  solve model (4.23) and find the max-projection $\left(\mathbf{x}_o^{\max}; \mathbf{y}_o^{\max}\right)$ by (4.26),

  $(\mathbf{x}_o; \mathbf{y}_o) \leftarrow \left(\mathbf{x}_o^{\max}; \mathbf{y}_o^{\max}\right)$,

  go to Step o.2.

end if

STEP O.2: SOLVE MODEL (4.10) AND OBTAIN $\left(\boldsymbol{\alpha}_o^{\max}; \boldsymbol{\beta}_o^{\max}\right)$ BY (4.11).

if $\left(\boldsymbol{\alpha}_o^{\max}; \boldsymbol{\beta}_o^{\max}\right) > \mathbf{0}_{m+s}$, **then**

  DMU $(\mathbf{x}_o; \mathbf{y}_o)$ is strongly congested.

else if $\boldsymbol{\beta}_o^{\max} \neq \mathbf{0}_s$, **then**

  DMU $(\mathbf{x}_o; \mathbf{y}_o)$ is weakly, but not strongly, congested.

else [DMU $(\mathbf{x}_o; \mathbf{y}_o)$ is not congested]

  if $\boldsymbol{\alpha}_o^{\max} \neq \mathbf{0}_s$, **then**

    DMU $(\mathbf{x}_o; \mathbf{y}_o)$ is output-efficient and input-inefficient, but is not congested.

  else

    DMU $(\mathbf{x}_o; \mathbf{y}_o)$ is full-efficient.

  end if

end if

end for

4.6 Numerical Examples

In this section we present four numerical examples to demonstrate the superiority of our proposed method over the existing methods for detecting weak and strong congestion statuses of DMUs. For brevity, throughout this section we write a DMU is weakly congested to mean that it is weakly, but not strongly, congested.

Example 3 illustrates how Algorithm 1 accurately distinguishes the DMUs that are weakly congested from the ones that are strongly congested. Example 4 further demonstrates how Algorithm 1 copes effectively with the occurrence of multiple projections. Additionally, it shows that the application of Algorithm 2 results in reducing computational effort. Example 5 shows that Algorithm 3 can deal with negative data. This example also argues that the procedure of Khoveyni et al. (2017) is not at all reliable in correctly distinguishing the weakly congested DMUs from the strongly congested ones. More precisely, Example 5 reveals that their procedure may erroneously declare a DMU to be weakly congested when it is actually strongly congested. Example 6 shows that Algorithm 4 by Noura et al. (2010) also cannot correctly recognize the presence of weak congestion based on the optimal solution of the VRS model.

Algorithm 4 The algorithm developed by Noura et al. (2010) based on model (4.2)

STEP 1: EVALUATING ALL OBSERVED DMUS BY MODEL (4.2).

Initialize $\mathcal{E} = \emptyset$.
for $o = 1 \to n$ **do**
  Evaluate DMU $(\mathbf{x}_o; \mathbf{y}_o)$ by model (4.2) and obtain the optimal solution $\left(\varphi_o^*, \boldsymbol{\lambda}_o^*, \mathbf{p}_o^*, \mathbf{q}_o^*\right)$.
  if $\varphi_o^* = 1$, **then**
    $\mathcal{E} \leftarrow \mathcal{E} \cup \{(\mathbf{x}_o; \mathbf{y}_o)\}$.
  end if
end for
return $\mathcal{E}$
go to Step 2.

STEP 2: IDENTIFYING THE PRESENCE OF CONGESTION.

Set $x_i^{\max} = \left\{x_{ij} : \left(\mathbf{x}_j; \mathbf{y}_j\right) \in \mathcal{E}\right\}$, for all $i = 1, \ldots, m$.
for $o = 1 \to n$ **do**
  if (c1) $x_{io} > x_i^{\max}$ for some $i \in \{1, \ldots, m\}$ **and** (c2) $\varphi_o^* > 1$ and/or $q_{ro}^* > 0$ for some $r \in \{1, \ldots, s\}$, **then**
    DMU $(\mathbf{x}_o; \mathbf{y}_o)$ is weakly congested.
  else
    DMU $(\mathbf{x}_o; \mathbf{y}_o)$ is not congested.
  end if
end for

Example 3 Consider the four observed DMUs A, B, C, and D shown in Table 4.3, each of which consumes two inputs x_1 and x_2 to produce two outputs y_1 and y_2. For each of these DMUs, the execution of Step o.1 of Algorithm 1 demonstrated its output efficiency with respect to technology $\mathcal{T}_{\text{CONG}}$. Therefore, each of them was then evaluated by model (4.10) at Step o.2 of the applied algorithm. Table 4.4 presents the obtained results.

From our results, DMUs A and B are fully efficient in technology $\mathcal{T}_{\text{CONG}}$. However, DMUs C and D exhibit weak and strong congestions, respectively. Indeed, taking the output efficiency of these two units into account, DMU C is weakly congested because it can increase its second output by decreasing its second input, without changing its first input and first output. Additionally, DMU D is strongly congested because both its outputs can be increased by decreasing both of its inputs.

Example 4 Consider the seven observed DMUs A, B, C, D, E, F, and G, whose input–output data are shown in Table 4.5. Each DMU uses two inputs x_1 and x_2 to produce four outputs y_r, $r = 1, \ldots, 4$. We applied Algorithm 1 to these DMUs in

Table 4.3 The data set in Example 3

DMU	x_1	x_2	y_1	y_2
A	1	1	1	1
B	2	2	2	2
C	2	3	2	1
D	3	3	1	1

Source Tone and Sahoo (2004)

Table 4.4 The results of applying Algorithm 1 in Example 3

DMU	φ^*	q_1^*	q_2^*	$\alpha_1^{\max}$	$\alpha_1^{\max}$	$\beta_1^{\max}$	$\beta_1^{\max}$	Status
A	1	0	0	0	0	0	0	Full-efficient
B	1	0	0	0	0	0	0	Full-efficient
C	1	0	0	0	1	0	1	Weakly congested
D	1	0	0	1	1	1	1	Strongly congested

order to evaluate their congestion statuses. We found only DMU A to be output-inefficient in technology $\mathcal{T}_{\text{CONG}}$.

Because DMU A has non-zero optimal output slack vector (see Table 4.6), we evaluated it by model (4.23) to find its GRS and thereby its max-projection $\left(\mathbf{x}_A^{\max}; \mathbf{y}_A^{\max}\right)$ (see Table 4.7). Then, before proceeding to Step A.2, we first replaced DMU $(\mathbf{x}_A; \mathbf{y}_A)$ with its max-projection.

Table 4.8 presents the results that we obtained for each of the seven observed DMUs by model (4.10) in its corresponding Step o.2 of the applied algorithm. As can be observed, DMUs A, C, and D are weakly congested and the remaining DMUs are all full-efficient.

To prove the superiority of our proposed method over those by Tone and Sahoo (2004) and Khoveyni et al. (2017) in dealing with multiple projections, we consider DMU A for our illustration. This DMU has been recognized by Algorithm 1 to be weakly congested (see Table 4.8). However, no presence of congestion would be found at it if it was evaluated by either of the methods by Tone and Sahoo (2004) or Khoveyni et al. (2017). This is because both methods arbitrarily project DMU A onto DMU B, which is a full-efficient DMU.[20]

Now, we turn to illustrate how Algorithm 2 evaluates the congestion of an output-inefficient DMU without employing model (4.10). Specifically, we show that the congestion status of such a DMU can be recognized using those of its reference DMUs. For this illustration, we again consider DMU A that its GRS consists of DMUs B, C, and D with the respective weights of 0.333, 0.333 and 0.333 (Table 4.7). For these reference DMUs, Table 4.8 shows that $\left(\boldsymbol{\alpha}_B^{\max}, \boldsymbol{\beta}_B^{\max}\right) = \mathbf{0}_6$, $\left(\boldsymbol{\alpha}_C^{\max}, \boldsymbol{\beta}_C^{\max}\right) =$

[20]If DMU A is evaluated by model (4.7), the optimal value of φ is 1. This means that the second stage of model (4.7) is equal to model (4.28). Therefore, both models (4.7) and (4.28) identify DMU B as the output-efficient projection of DMU A (see Table 4.6).

Table 4.5 The data set in Example 4

DMU	x_1	x_2	y_1	y_2	y_3	y_4
A	2	2	2	2	2	2
B	2	2	2	3	2	2
C	2	2	2	2	3	2
D	2	2	2	2	2	3
E	1	1	2	2.5	3	2
F	1	3	2	2	2	4
G	2	1	2	2.5	2.25	3

Source Sueyoshi and Sekitani (2009)

Table 4.6 The results obtained by model (4.7) for DMU A

DMU	φ^*	$\mathbf{1}_s^T \mathbf{q}^*$	x_1^{P}	x_2^{P}	y_1^{P}	y_2^{P}	y_3^{P}	y_4^{P}
A	1	1	2	2	2	3	2	2

Table 4.7 The GRS and the max-projection of DMU A

DMU	GRS			Max-projection					
	λ_B	λ_C	λ_D	$x_1^{\max}$	$x_2^{\max}$	$y_1^{\max}$	$y_2^{\max}$	$y_3^{\max}$	$y_4^{\max}$
A	0.333	0.333	0.333	2	2	2	2.333	2.333	2.333

Table 4.8 The results obtained by model (4.10) in Example 4

DMU	$\alpha_1^{\max}$	$\alpha_2^{\max}$	$\beta_1^{\max}$	$\beta_2^{\max}$	$\beta_3^{\max}$	$\beta_4^{\max}$	Status
A	0.75	0.25	0	0.167	0.167	0.167	Weakly congested
B	0	0	0	0	0	0	Full-efficient
C	1	1	0	0.5	0	0	Weakly congested
D	0.333	0.238	0	0.238	0.238	0.238	Weakly congested
E	0	0	0	0	0	0	Full-efficient
F	0	0	0	0	0	0	Full-efficient
G	0	0	0	0	0	0	Full-efficient

$(1, 1, 0, 0.5, 0, 0)^T$ and $\left(\boldsymbol{\alpha}_D^{\max}, \boldsymbol{\beta}_D^{\max}\right) = (0.333, 0.238, 0, 0.238, 0.238, 0.238)^T$. According to Step 3 of Algorithm 2, we have

$$\left(\boldsymbol{\alpha}_A^{\mathrm{GRS}}; \boldsymbol{\beta}_A^{\mathrm{GRS}}\right)^T = 0.333 \times (1, 1, 0, 0.5, 0, 0)^T + 0.333 \times (0.333, 0.238, 0, 0.238, 0.238, 0.238)^T,$$

which implies that DMU A is weakly congested because $\boldsymbol{\beta}_A^{\mathrm{GRS}} \neq \mathbf{0}_4$ and $\beta_{1A}^{\mathrm{GRS}} = 0$. This is the same result presented in Table 4.8 using the optimal solution of program (4.10).

Table 4.9 The data set in Example 5

DMU	$D1$	$D2$	$D3$	$D4$	$D5$	$D6$	$D7$	$D8$	$D9$	$D10$	$D11$	$D12$	$D13$	$D14$	$D15$
x_1	−1	−3	0	−2	−2	2	4	−2	−2	4	3	−2	3	2	3
x_2	−3	−1	−2	0	2	−2	−2	4	2	4	−2	3	3	1	3
y	−1	−1	1	1	1	1	0	0	1	0	0.5	0.5	0.5	−2	−3

Source Khoveyni et al. (2017)

Table 4.10 The results obtained by model (4.28) for DMUs $D14$ and $D15$

DMU	$\mathbf{1}_s^T \mathbf{q}^*$	$\bar{x}_1^{\mathrm{P}}$	$\bar{x}_2^{\mathrm{P}}$	$\bar{y}^{\mathrm{P}}$
$D14$	2.75	2	1	0.75
$D15$	3.5	3	3	0.5

Example 5 Table 4.9 displays the data set of fifteen observed DMUs $D1, \ldots, D15$, each of which uses two inputs x_1 and x_2 to produce one output y. We applied Algorithm 3 to this data set because it contains negative values. Following Step o.1 of this algorithm, we first evaluated all the DMUs by model (4.28), whereby we found DMUs $D14$ and $D15$ output-inefficient and the remaining ones output-efficient in technology $\tilde{\mathcal{T}}_{\mathrm{CONG}}$. For each of these two output-inefficient DMUs, Table 4.10 shows the optimal objective value of program (4.28) and its corresponding projection.

From Table 4.10, it is observed that the optimal output slack vector $\mathbf{q}^*$ is non-zero for DMUs $D14$ and $D15$. It means that multiple projections may occur for each of these DMUs. In view of this, we solved program (4.23) for each of them and thereby identified their corresponding GRSs and max-projections, which are shown in Table 4.11.

Table 4.12 presents the results obtained from the evaluation of the output-efficient DMUs $D1, \ldots, D13$ themselves and the max-projections of the output-inefficient DMUs $D14$ and $D15$ by model (4.23). The four DMUs $D1$, $D2$, $D3$, and $D4$ are full-efficient. Despite being output-efficient, DMUs $D5$, $D6$, and $D9$ are input-inefficient, but are not congested. However, all the remaining DMUs are strongly congested.

Note that the presence of congestion at DMUs $D14$ and $D15$ can be directly identified. In fact, besdies being a reference DMU for each of these DMUs (see Table 4.11), DMU $D13$ is strongly congested (see Table 4.12). By part (ii) of Corollary 3, it follows that DMUs $D14$ and $D15$ are also strongly congested. Note also that the approach of Khoveyni et al. (2017) is unable to recognize the true congestion status for the DMUs $D7$, $D8$, $D11$, and $D12$. Indeed, as shown in Table 4.11, these DMUs are strongly congested. However, as Khoveyni et al. (2017) have reported in Table 4.5 of their paper, their approach erroneously declares these DMUs to be weakly congested. This demonstrates that Algorithm 3 scores over their approach in precisely distinguishing the DMUs that are weakly congested from the ones that are strongly congested.

Table 4.11 The GRSs and the max-projections of DMUs $D14$ and $D15$

DMU	GRS				Max-projection		
	$\lambda_5^{\max}$	$\lambda_6^{\max}$	$\lambda_9^{\max}$	$\lambda_{13}^{\max}$	$x_1^{\max}$	$x_2^{\max}$	$y^{\max}$
$D14$	0.062	0.375	0.062	0.5	2	1	0.75
$D15$				1	3	3	0.5

Table 4.12 The congestion and efficiency status of DMUs $D1, \ldots, D15$ in Example 5

DMU	$D1$	$D2$	$D3$	$D4$	$D5$	$D6$	$D9$	$D7$	$D8$	$D10$	$D11$	$D12$	$D13$	$D14$	$D15$
$\alpha_1^{\max}$	0	0	0	0	0	2	0	4.333	0.333	4	3.167	0.167	3	2	3
$\alpha_2^{\max}$	0	0	0	0	2	0	2	0.333	4.333	6	0.167	3.167	5	3	5
$\beta^{\max}$	0	0	0	0	0	0	0	0.333	0.333	1	0.167	0.167	0.5	0.25	0.5
Status	Full-efficient				Non-congested			Strongly congested							

Table 4.13 The data set in Example 6

DMU	A	B	C	D	E	F
x	1	4	6	6	5	5.5
y_1	6	6	4	6	6	5.5
y_2	5	5	5	3	4	4

Table 4.14 The results of applying Algorithm 4 in Example 6

DMU	A	B	C	D	E	F
φ^*	1	1	1	1	1	1.09
p^*	0	3	5	5	4	4.5
q_1^*	0	0	2	0	0	0
q_2^*	0	0	0	2	1	0.64
Status	Non-congested					

Table 4.15 The results obtained by model (4.5) in Example 6

DMU	A	B	C	D	E	F
$\alpha^{\max}$	0	1	2.5	4	4	4.5
$\beta_1^{\max}$	0	0	1	0	0	0.5
$\beta_2^{\max}$	0	0	0	1	1	0.5

Table 4.16 The efficiency and congestion status of DMUs $A, \ldots, F$ in Example 6

DMU	A	B	C	D	E	F
Status	Full-efficient	Non-congested	Weakly congested			Strongly congested

Example 6 Consider the six observed DMUs A, B, C, D, E, and F whose input-output data are presented in Table 4.13, wherein x is input and y_1 and y_2 are outputs. Table 4.14 shows the optimal output improvement factor φ^* and the corresponding input and output slacks $\left(p^*, q_1^*, q_2^*\right)$ obtained from evaluating each of these six DMUs by model (4.2). From the results, $\mathcal{E} = \{A, B, C, D, E\}$, implying $x^{\max} = 6$. Because none of the observed DMUs satisfies the condition (c1) stated in Step 2 of Algorithm 4, none of them is congested.

By pairwise comparison of the input–output vectors of these six DMUs, it can be found that DMU B weakly dominates DMUs C, D, and E, and strongly dominates DMU F. Consequently, DMUs C, D, and E are weakly congested, and DMU F is strongly congested. These findings reveal that Algorithm 4 of Noura et al. (2010) cannot recognize the presence of congestion.

Tables 4.15 and 4.16 presented the results obtained by applying Algorithm 1 to the given data set. Because all the six DMUs were identified to be output-efficient in $\mathcal{T}_{\text{CONG}}$, they were evaluated by model (4.5). For each DMU, Table 4.15 shows its corresponding vector $\left(\alpha_o^{\max}; \boldsymbol{\beta}_o^{\max}\right)$. Table 4.15 also presents the efficiency and congestion status of each of these six DMUs.

4.7 Concluding Remarks

In this chapter we are concerned with the precise identification of both weak and strong forms of congestion of production units. As an essential tool for our development, we formulate an LP to find a maximal element of a non-negative polyhedral set. This LP is derived from the convex program proposed by Mehdiloozad et al. (2018) for finding a maximal element of a non-negative convex set. Then we argue that the assumption of strong input disposability embedded in the standard VRS technology is inappropriate for modeling a true technology involving congestion. The most common approach in the DEA literature to such modeling is to make no assumption at all about the disposability of inputs. Using this approach, we axiomatically develop technology $\mathcal{T}_{\text{CONG}}$ to estimate the true technology involving congestion.

We define both weak and strong forms of congestion with respect to technology $\mathcal{T}_{\text{CONG}}$. While our definition of weak congestion is equivalent to the original one suggested by Tone and Sahoo (2004), we redefine the concept of strong congestion in order to make its evaluation possible in technologies that their data contain negative values. Using the LP proposed for finding a maximal element of a non-negative polyhedral set, we develop a single-stage congestion-identification model. This model determines the index sets of all improvable inputs and outputs of DMUs and thereby, recognizes the presence of weak and strong congestions. Furthermore, it evaluates the full efficiency and input efficiency statuses of DMUs. The use of our proposed model is computationally more efficient than the two-stage approaches by Tone and Sahoo (2004), Khoveyni et al. (2013), and Khoveyni et al. (2017). Moreover, unlike the approach of Khoveyni et al. (2017), our congestion-identification model deals effectively with the negative data.

From Tone and Sahoo (2004), an output-inefficient DMU is weakly (resp., strongly) congested if its projection on the output-efficient boundary of technology $\mathcal{T}_{\text{CONG}}$ is weakly (resp., strongly) congested. Sueyoshi and Sekitani (2009) argue that this definition may not be mathematically well-defined when multiple projections are present. To deal effectively with this issue, we prove that all DMUs on the relative interior of any face of technology $\mathcal{T}_{\text{CONG}}$ have the same status of congestion. Based on this finding, we say that a face of technology $\mathcal{T}_{\text{CONG}}$ is weakly (resp., strongly) congested if all DMUs on its relative interior are weakly (resp., strongly) congested. We then extend the conventional definition of the congestion of an output-inefficient DMU. Specifically, we say that an output-inefficient DMU is weakly (resp., strongly) congested if its corresponding minimal face is weakly (resp., strongly) congested. Our new definition is always well-defined, and reduces to the conventional one if a unique projection is present.

Note that the congestion of any face of technology $\mathcal{T}_{\text{CONG}}$ can be determined by any arbitrary DMU on its relative interior. Therefore, the presence of congestion at any DMU can be identified by finding a projection on the relative interior of its corresponding minimal face. To find such a projection, we show that the minimal face is a polytope which is spanned by the GRS of the DMU under evaluation. As a consequence of this, we characterize the relative interior of the minimal face as the strict convex hull of the GRS. Then, using the LP proposed for finding a maximal element of a non-negative polyhedral set, we develop a model that finds the max-projection of the DMU under evaluation as a strict convex combination of its reference DMUs.

As an inserting finding, we also demonstrate that the congestion status of any output-inefficient DMU faced with multiple projections can be determined by using those of its reference DMUs. This finding eliminates the need for evaluating the max-projection of the DMU under evaluation via our congestion-identification model. Therefore, apart from its theoretical attractiveness, our finding helps reduce the computational effort.

Based on our proposed results, we develop three computational algorithms for identifying the congestion statuses of all DMUs in any finite-size sample. Then, we present three numerical examples to illustrate the superiority of our algorithms over the existing approaches by Tone and Sahoo (2004), Khoveyni et al. (2013), and Khoveyni et al. (2017). Finally, by developing an original counterexample, we show that the VRS model-based approach of Noura et al. (2010) cannot even identify the presence of weak congestion, whereas our proposed method correctly recognizes the presence of both weak and strong congestions.

4.8 Appendix: Proofs

Proof of Theorem 1 Let $(\mathbf{u}^*, \mathbf{s}^*, \mathbf{t}^*, w^*)$ be an optimal solution to program (4.5). By dividing both sides of the linear constraints of program (4.5) by w^* and assuming $\mathbf{u}^{\max} = \frac{1}{w^*}(\mathbf{s}^* + \mathbf{t}^*)$, we have $\mathbf{u}^{\max} \in \mathcal{P}$. If $\mathbf{u}^{\max} > \mathbf{0}_k$, there is nothing to prove.

Otherwise, we assume without loss of generality that $\sigma(\mathbf{t}^*) = \{1, \ldots, e\}$ $(e < k)$. Because the linear constraints of program (4.5) are homogeneous, it is straightforward to show that $t_j^* = 1$ for all $j = 1, \ldots, e$. Therefore, $e = \mathbf{1}_k^T \mathbf{t}^*$.

By contradiction, we assume that $\mathbf{u}^{\max}$ is not a maximal element of $\mathcal{P}$. Then, by (4.3), there is a vector $\tilde{\mathbf{u}} \in \mathcal{P}$ that takes positive values in some zero components of $\mathbf{u}^{\max}$. This means that $\sum_{j=e+1}^{k} \tilde{u}_j > 0$. Let us define

$$\tilde{s}_j = \max\left\{0, s_j^* + t_j^* + \tilde{u}_j - 1\right\} \ \forall j, \ \tilde{t}_j = \min\left\{1, s_j^* + t_j^* + \tilde{u}_j\right\} \ \forall j, \ \tilde{\mathbf{v}} = \mathbf{v}^*, \ \tilde{w} = w^* + 1.$$

Then, $\left(\tilde{\mathbf{s}}, \tilde{\mathbf{t}}, \tilde{\mathbf{v}}, \tilde{w}\right)$ is a feasible solution to program (4.5) that its corresponding objective value is greater than e. This contradicts the optimality of $(\mathbf{u}^*, \mathbf{s}^*, \mathbf{t}^*, w^*)$. Therefore, $\mathbf{u}^{\max}$ is a maximal element of $\mathcal{P}$. □

Proof of Theorem 2 By contradiction, assume that technology $\mathcal{T}$ satisfying Axiom SID is congested. This means that there is an output-efficient DMU $(\bar{\mathbf{x}}; \bar{\mathbf{y}}) \in \mathcal{T}$ that is weakly dominated by some $\left(\hat{\mathbf{x}}; \hat{\mathbf{y}}\right) \in \mathcal{T}$. Because technology $\mathcal{T}$ satisfies Axiom SID, we have $\left(\bar{\mathbf{x}}; \hat{\mathbf{y}}\right) \in \mathcal{T}$, which contradicts the output efficiency of $(\bar{\mathbf{x}}; \bar{\mathbf{y}})$. Therefore, technology $\mathcal{T}$ is congestion-free. □

Lemma 2 *Denote $\mathcal{V}$ the set on the right-hand side of* (4.6). *Then $\mathcal{V}$ is a polyhedral set, which implies that it satisfies Axioms CT. Furthermore, $\mathcal{V}$ satisfies Axioms IO and SOD.*

Proof of Lemma 2 From the projection lemma, it follows that $\mathcal{V}$ is a polyhedral set and, consequently, satisfies Axiom CT. Clearly, $\mathcal{V}$ also satisfies Axiom IO.

Let $(\mathbf{x}; \mathbf{y}) \in \mathcal{V}$. Then $(\mathbf{x}; \mathbf{y})$ satisfies (4.6) with some vector $\bar{\boldsymbol{\lambda}}$. To prove that $\mathcal{V}$ satisfies Axiom SOD, let $\mathbf{0}_s \leq \hat{\mathbf{y}} \leq \mathbf{y}$. Then $\left(\mathbf{x}; \hat{\mathbf{y}}\right)$ satisfies (4.6) with the same vector $\bar{\boldsymbol{\lambda}}$. Therefore, $\left(\mathbf{x}; \hat{\mathbf{y}}\right) \in \mathcal{V}$, and $\mathcal{V}$ satisfies Axiom SOD. □

Proof of Theorem 3 Let $\mathcal{V}$ be the set on the right-hand side of (4.6). By Lemma 2, the set $\mathcal{V}$ satisfies Axioms IO, CT, and SOD. Because $\mathcal{T}_{\text{CONG}}$ is the smallest technology that satisfies Axioms IO, CT, and SOD, we have $\mathcal{T}_{\text{CONG}} \subseteq \mathcal{V}$.

Conversely, let $(\mathbf{x}; \mathbf{y}) \in \mathcal{V}$. Then $(\mathbf{x}; \mathbf{y})$ satisfies (4.6) with some vector $\bar{\boldsymbol{\lambda}}$. Because technology $\mathcal{T}_{\text{CONG}}$ satisfies Axiom IO and CT, we have $\left(\mathbf{X}\bar{\boldsymbol{\lambda}}; \mathbf{Y}\bar{\boldsymbol{\lambda}}\right) \in \mathcal{T}_{\text{CONG}}$. Because this technology also satisfies Axiom SOD, it follows that $(\mathbf{x}; \mathbf{y}) \in \mathcal{T}_{\text{CONG}}$. Therefore, $\mathcal{V} \subseteq \mathcal{T}_{\text{CONG}}$. □

Proof of Theorem 4 The fact that $\mathcal{T}_{\text{CONG}}$ is a polyhedral set follows from the projection lemma. To prove that $\mathcal{T}_{\text{CONG}}$ is also bounded, let $(\mathbf{x}; \mathbf{y}) \in \mathcal{T}_{\text{CONG}}$. Then $(\mathbf{x}; \mathbf{y})$ satisfies (4.6) with some vector $\bar{\boldsymbol{\lambda}}$. By the normalizing equality $\mathbf{1}_n^T \bar{\boldsymbol{\lambda}} = 1$, it follows from the input and output constraints that $x_i \leq \max_{j \in \mathcal{J}} \left\{x_{ij}\right\}$ for all $i = 1, \ldots, m$, and $y_r \leq \max_{j \in \mathcal{J}} \left\{y_{rj}\right\}$ for all $r = 1, \ldots, s$. This implies that technology $\mathcal{T}_{\text{CONG}}$ has no recession direction and is, therefore, bounded. Because any face of a polyhedral set is itself a polyhedral set, if follows that all faces of $\mathcal{T}_{\text{CONG}}$ are polytopes. □

Proof of Theorem 6 By Definition 2 and Corollary 1, it follows that $\partial_{\mathrm{F}}\mathcal{T}_{\mathrm{VRS}} \subseteq \partial_{\mathrm{F}}\mathcal{T}_{\mathrm{CONG}}$. Conversely, let $(\hat{\mathbf{x}}; \hat{\mathbf{y}}) \in \partial_{\mathrm{F}}\mathcal{T}_{\mathrm{CONG}}$. By Corollary 1, $(\hat{\mathbf{x}}; \hat{\mathbf{y}})$ is in $\mathcal{T}_{\mathrm{VRS}}$. By contradiction, let there exist a DMU $(\mathbf{x}'; \mathbf{y}') \in \mathcal{T}_{\mathrm{VRS}}$ such that $(-\mathbf{x}'; \mathbf{y}') \gneq (-\hat{\mathbf{x}}; \hat{\mathbf{y}})$. Without loss of generality, assume also that $(\mathbf{x}'; \mathbf{y}') \in \partial_{\mathrm{F}}\mathcal{T}_{\mathrm{VRS}}$. Denote $\mathcal{F}'$ as the smallest face of $\mathcal{T}_{\mathrm{VRS}}$ containing $(\mathbf{x}'; \mathbf{y}')$, i.e., the intersection of the collection of faces which contain $(\mathbf{x}'; \mathbf{y}')$. Then $(\mathbf{x}'; \mathbf{y}') \in ri\,(\mathcal{F}')$ (see Proposition 1.19 in Tuy (1998)). Moreover, $\mathcal{F}'$ is a strong face of $\mathcal{T}_{\mathrm{VRS}}$ and is therefore a polytope (see Theorem 2 in Davtalab Olyaie et al. (2014)). Therefore, $(\mathbf{x}'; \mathbf{y}')$ can be expressed as a convex combination of extreme points of the face $\mathcal{F}'$. Because extreme points of face $\mathcal{F}'$ are extreme in $\mathcal{T}_{\mathrm{VRS}}$ itself (see Rockafellar (1970), page 163), $(\mathbf{x}'; \mathbf{y}')$ is a convex combination of extreme observed DMUs. By the convexity of technology $\mathcal{T}_{\mathrm{CONG}}$, it follows that $(\mathbf{x}'; \mathbf{y}') \in \mathcal{T}_{\mathrm{CONG}}$. This contradicts the full efficiency of $(\hat{\mathbf{x}}; \hat{\mathbf{y}})$ in technology $\mathcal{T}_{\mathrm{CONG}}$. Therefore, $(\hat{\mathbf{x}}; \hat{\mathbf{y}}) \in \partial_{\mathrm{F}}\mathcal{T}_{\mathrm{VRS}}$, implying $\partial_{\mathrm{F}}\mathcal{T}_{\mathrm{CONG}} \subseteq \partial_{\mathrm{F}}\mathcal{T}_{\mathrm{VRS}}$. □

Proof of Lemma 1 By contradiction, assume that $\boldsymbol{\alpha} = \mathbf{0}_m$. Then we have $(\mathbf{x}_o; \mathbf{y}_o + \boldsymbol{\beta}) \in \mathcal{T}_{\mathrm{CONG}}$. Because $\boldsymbol{\beta} \neq \mathbf{0}_s$, it follows that DMU $(\mathbf{x}_o; \mathbf{y}_o)$ is output-inefficient, which is a contradiction. Therefore $\boldsymbol{\alpha} \neq \mathbf{0}_m$. □

Proof of Theorem 7 *Part (i)* Let DMU $(\mathbf{x}_o; \mathbf{y}_o)$ be full-efficient in technology $\mathcal{T}_{\mathrm{CONG}}$. Then we have $\mathcal{S}_o = \{\mathbf{0}_{m+s}\}$, because otherwise the full efficiency of DMU $(\mathbf{x}_o; \mathbf{y}_o)$ is contradicted. Therefore, $\boldsymbol{\alpha}_o^{\max} = \mathbf{0}_m$. Conversely, assume by contradiction that DMU $(\mathbf{x}_o; \mathbf{y}_o)$ is not full-efficient. By definition, there is a DMU $(\mathbf{x}'; \mathbf{y}') \in \mathcal{T}$, such that $(-\mathbf{x}'; \mathbf{y}') \gneq (-\mathbf{x}; \mathbf{y})$. Then the non-zero vector $(\boldsymbol{\alpha}'; \boldsymbol{\beta}') = (\mathbf{x}_o - \mathbf{x}'; \mathbf{y}' - \mathbf{y}_o)$ satisfies all conditions in (4.9) with some $\boldsymbol{\lambda}'$. Taking into account (4.3), it follows that $(\boldsymbol{\alpha}_o^{\max}; \boldsymbol{\beta}_o^{\max}) \neq \mathbf{0}_{m+s}$. If $\boldsymbol{\beta}_o^{\max} = \mathbf{0}_s$, then it is clear that $\boldsymbol{\alpha}_o^{\max} \neq \mathbf{0}_m$. Otherwise, Lemma 1 implies that $\boldsymbol{\alpha}_o^{\max} \neq \mathbf{0}_m$. In both cases, the assumption is contradicted. Therefore, DMU $(\mathbf{x}_o; \mathbf{y}_o;)$ is full-efficient.

Part (ii) Let DMU $(\mathbf{x}_o; \mathbf{y}_o)$ be input-inefficient in technology $\mathcal{T}_{\mathrm{CONG}}$. Then, by part (i) of the theorem, it follows that $\boldsymbol{\alpha}_o^{\max} \neq \mathbf{0}_m$. Conversely, let $\boldsymbol{\alpha}_o^{\max} \neq \mathbf{0}_m$. Because $(\mathbf{x}_o - \boldsymbol{\alpha}_o^{\max}; \mathbf{y}_o + \boldsymbol{\beta}_o^{\max}) \in \mathcal{T}_{\mathrm{CONG}}$ we have $(\mathbf{x}_o - \boldsymbol{\alpha}_o^{\max}; \mathbf{y}_o) \in \mathcal{T}_{\mathrm{CONG}}$ Therefore, DMU $(\mathbf{x}_o; \mathbf{y}_o)$ is input-inefficient.

Part (iii) Let DMU $(\mathbf{x}_o; \mathbf{y}_o)$ be weakly congested. By definition, there exists a DMU $(\mathbf{x}'; \mathbf{y}')$ in technology $\mathcal{T}_{\mathrm{CONG}}$, such that $-\mathbf{x}' \gneq -\mathbf{x}$ and $\mathbf{y}' \gneq \mathbf{y}$. The vector $(\boldsymbol{\alpha}'; \boldsymbol{\beta}') = (\mathbf{x}_o - \mathbf{x}'; \mathbf{y}' - \mathbf{y}_o)$ satisfies all conditions in (4.9) with some $\boldsymbol{\lambda}'$. Because $\boldsymbol{\beta}' \neq \mathbf{0}_s$, (4.3) follows that $\boldsymbol{\beta}_o^{\max} \neq \mathbf{0}_s$. Conversely, let $\boldsymbol{\beta}_o^{\max} \neq \mathbf{0}_s$. Then Lemma 1 implies that $\boldsymbol{\alpha}_o^{\max} \neq \mathbf{0}_m$. We also have $(\mathbf{x}_o - \boldsymbol{\alpha}_o^{\max}; \mathbf{y}_o + \boldsymbol{\beta}_o^{\max}) \in \mathcal{T}_{\mathrm{CONG}}$. Therefore, DMU $(\mathbf{x}_o; \mathbf{y}_o)$ is weakly congested.

Part (iv) Let DMU $(\mathbf{x}_o; \mathbf{y}_o)$ be strongly congested. By definition, there exists a DMU $(\mathbf{x}'; \mathbf{y}')$ in technology $\mathcal{T}_{\mathrm{CONG}}$, such that $(-\mathbf{x}'; \mathbf{y}') > (-\mathbf{x}; \mathbf{y})$. The vector $(\boldsymbol{\alpha}'; \boldsymbol{\beta}') = (\mathbf{x}_o - \mathbf{x}'; \mathbf{y}' - \mathbf{y}_o)$ satisfies all conditions in (4.9) with some $\boldsymbol{\lambda}'$. Because $(\boldsymbol{\alpha}'; \boldsymbol{\beta}') > \mathbf{0}_{m+s}$, (4.3) follows that $(\boldsymbol{\alpha}_o^{\max}; \boldsymbol{\beta}_o^{\max}) > \mathbf{0}_{m+s}$. Conversely, let $(\boldsymbol{\alpha}_o^{\max}; \boldsymbol{\beta}_o^{\max}) > \mathbf{0}_{m+s}$. Because $(\mathbf{x}_o - \boldsymbol{\alpha}_o^{\max}; \mathbf{y}_o + \boldsymbol{\beta}_o^{\max}) \in \mathcal{T}_{\mathrm{CONG}}$, DMU $(\mathbf{x}_o; \mathbf{y}_o)$ is strongly congested. □

Proof of Theorem 8 Assume that $(\mathbf{x}^{\mathrm{ri}}; \mathbf{y}^{\mathrm{ri}})$ and $\boldsymbol{\rho} \in \mathbb{R}^l_{++}$ are as in (4.14). For each $k = 1, \ldots, l$, let $(\delta_k^*, \mathbf{s}_k^{-*}, \mathbf{t}_k^{-*}, \mathbf{s}_k^{+*}, \mathbf{t}_k^{+*}, w_k^*)$ be the optimal solution obtained from the evaluation of DMU $(\mathbf{x}_{j_k}; \mathbf{y}_{j_k})$ by model (4.10). Then, from the constraints of program (4.10) at optimality, we have the following equations:

$$\mathbf{X}\left(\frac{1}{1+w_k^*}\boldsymbol{\delta}_k^*\right)+\boldsymbol{\alpha}_k^{\max}=\mathbf{x}_{j_k},\ \mathbf{Y}\left(\frac{1}{1+w_k^*}\boldsymbol{\delta}_k^*\right)-\boldsymbol{\beta}_k^{\max}\geq\mathbf{y}_{j_k},\ \mathbf{1}_n^T\left(\frac{1}{1+w_k^*}\boldsymbol{\delta}_k^*\right)=1,\ k=1,\ldots,l, \tag{4.30}$$

where the vector $(\boldsymbol{\alpha}_k^{\max}; \boldsymbol{\beta}_k^{\max})$ is as defined in (4.15). Multiplying both sides of the kth input, output, and normalizing equations in (4.30) by ρ_k, and then separately summing up the resulting input, output and normalizing equations over k lead to the following equations:

$$\mathbf{X}\boldsymbol{\lambda}' = \mathbf{x}^{\mathrm{ri}} - \boldsymbol{\alpha}', \quad \mathbf{Y}\boldsymbol{\lambda}' \geq \mathbf{y}^{\mathrm{ri}} + \boldsymbol{\beta}', \quad \mathbf{1}_n^T\boldsymbol{\lambda}' = 1, \tag{4.31}$$

where the vectors $\boldsymbol{\lambda}'$, $\boldsymbol{\alpha}'$, and $\boldsymbol{\beta}'$ are as follows:

$$\boldsymbol{\lambda}' = \sum_{k=1}^{l}\frac{\rho_k}{1+w_k}\boldsymbol{\delta}_k^*, \quad \boldsymbol{\alpha}' = \sum_{k=1}^{l}\rho_k\boldsymbol{\alpha}_k^{\max}, \quad \boldsymbol{\beta}' = \sum_{k=1}^{l}\rho_k\boldsymbol{\beta}_k^{\max}. \tag{4.32}$$

The equations obtained in (4.31) show that $(\mathbf{x}^{\mathrm{ri}}; \mathbf{y}^{\mathrm{ri}})$ satisfies all conditions in (4.10) with $(\boldsymbol{\lambda}', \boldsymbol{\alpha}', \boldsymbol{\beta}')$. This means that $(\boldsymbol{\alpha}', \boldsymbol{\beta}') \in \mathcal{S}_{\mathrm{ri}}$. Because $\boldsymbol{\rho} \in \mathbb{R}_{++}^l$, it is clear from (4.16) and (4.32) that $\mathcal{I}_{\mathcal{F}} = \sigma(\boldsymbol{\alpha}')$ and $\mathcal{O}_{\mathcal{F}} = \sigma(\boldsymbol{\beta}')$. Consequently, taking into account (4.13), it follows that $\mathcal{I}_{\mathcal{F}} \subseteq \mathcal{I}_{\mathrm{ri}}$ and $\mathcal{O}_{\mathcal{F}} \subseteq \mathcal{O}_{\mathrm{ri}}$.

To prove that $\mathcal{I}_{\mathrm{ri}} \subseteq \mathcal{I}_{\mathcal{F}}$, let $\hat{i} \in \mathcal{I}_{\mathrm{ri}}$. Then, taking into account (4.12), it follows that some $(\hat{\boldsymbol{\alpha}}; \hat{\boldsymbol{\beta}}) \in \mathcal{S}_{\mathrm{ri}}$ exists such that $\hat{\alpha}_{\hat{i}} > 0$. This means that $(\mathbf{x}^{\mathrm{ri}} - \hat{\boldsymbol{\alpha}}; \mathbf{y}^{\mathrm{ri}} + \hat{\boldsymbol{\beta}}) \in \mathcal{T}_{\mathrm{CONG}}$ such that $x_{\hat{i}}^{\mathrm{ri}} - \hat{\alpha}_{\hat{i}} < x_{\hat{i}}^{\mathrm{ri}}$. From (4.14), we have $x_{\hat{i}}^{\mathrm{ri}} = \sum_{k=1}^{l}\rho_k x_{\hat{i}j_k}$. Because $\sum_{k=1}^{l}\rho_k = 1$, it follows that $x_{\hat{i}}^{\mathrm{ri}} - \hat{\alpha}_{\hat{i}} < x_{\hat{i}j_{\hat{k}}}$ for some $\hat{k} \in \{1,\ldots,l\}$. Because $(\boldsymbol{\alpha}_{\hat{k}}^{\max}; \boldsymbol{\beta}_{\hat{k}}^{\max})$ is a maximal element of $\mathcal{S}_{j_{\hat{k}}}$, we therefore have $\hat{i} \in \sigma(\boldsymbol{\alpha}_{\hat{k}}^{\max})$, so $\hat{i} \in \mathcal{I}_{\mathcal{F}}$. Therefore, $\mathcal{I}_{\mathrm{ri}} \subseteq \mathcal{I}_{\mathcal{F}}$. Similarly, it can be proved that $\mathcal{O}_{\mathrm{ri}} \subseteq \mathcal{O}_{\mathcal{F}}$. □

Proof of Theorem 9 Because $\mathcal{F}_o^{\min}$ is a face of the polyhedral set $\mathcal{T}_{\mathrm{CONG}}$, there is a supporting hyperplane of $\mathcal{T}_{\mathrm{CONG}}$, namely $\mathcal{H}^{\min}$, such that $\mathcal{F}_o^{\min} = \mathcal{H}^{\min} \cap \mathcal{T}_{\mathrm{CONG}}$.[21] Because $\mathcal{F}_o^{\min}$ contains Π_o, the hyperplane $\mathcal{H}^{\min}$ is binding at all projections in Π_o, and therefore passes through all the reference DMUs in $\mathcal{G}_o$. By the convexity of $\mathcal{H}^{\min}$, it follows that $\mathrm{conv}(\mathcal{G}_o) \subseteq \mathcal{H}^{\min}$. Therefore, $\mathrm{conv}(\mathcal{G}_o) \subseteq \mathcal{F}_o^{\min}$.

As shown in Footnote 17, $\Pi_o \cap ri(\mathcal{F}_o^{\min}) \neq \emptyset$. This implies that all the observed DMUs on $\mathcal{F}_o^{\min}$, and therefore all the vertices of $\mathcal{F}_o^{\min}$, belong to $\mathcal{G}_o$. By Theorem 4, $\mathcal{F}_o^{\min}$ is a polytope. Therefore, $\mathcal{F}_o^{\min} \subseteq \mathrm{conv}(\mathcal{G}_o)$. □

[21] If $C \subseteq \mathbb{R}^d$ is a convex set and $\mathcal{H}$ is a supporting hyperplane of C, then the intersection $C \cap \mathcal{H}$ is called an *exposed face* of C. The two notions of face and exposed face coincide for polyhedral sets, but this may not be the case for convex sets. Precisely, every exposed face of C is a face of C, but the converse is not generally true (Rockafellar, 1970).

Proof of Theorem 10 Assume that $(\varphi^*, \mathbf{q}^*, \boldsymbol{\lambda}^*)$ is that optimal solution of program (4.7) by which Λ_o has been stated as in (4.21). Let $\boldsymbol{\lambda}_o^{\max}$ be a maximal element of Λ_o, and let $\boldsymbol{\lambda}_o^{\max}$ satisfy (4.21) with some $\mathbf{q}_o^{\max}$. Then $\left(\varphi^*, \boldsymbol{\lambda}_o^{\max}, \mathbf{q}_o^{\max}\right)$ is an optimal solution of program (4.7). This proves that $\sigma\left(\boldsymbol{\lambda}_o^{\max}\right) \subseteq \mathcal{J}_o$.

Conversely, let $\hat{j} \in \mathcal{J}_o$. Then there is exists an optimal solution $\left(\varphi', \boldsymbol{\lambda}', \mathbf{q}'\right)$ to program (4.7) such that $\hat{j} \in \sigma\left(\boldsymbol{\lambda}'\right)$. Because $\varphi' = \varphi^*$ and $\mathbf{1}_s^T\mathbf{q}' = \mathbf{1}_s^T\mathbf{q}^*$, it follows that $\boldsymbol{\lambda}' \in \Lambda_o$. By (4.3), we have $\sigma\left(\boldsymbol{\lambda}'\right) \subseteq \sigma\left(\boldsymbol{\lambda}_o^{\max}\right)$, and so $\hat{j} \in \sigma\left(\boldsymbol{\lambda}_o^{\max}\right)$. Therefore, $\mathcal{J}_o \subseteq \sigma\left(\boldsymbol{\lambda}_o^{\max}\right)$. □

References

Banker, R. D., Charnes, A., & Cooper, W. W. (1984). Some models for estimating technical and scale efficiencies in data envelopment analysis. *Management Science*, *30*(9), 1078–1092.

Bertsimas, D., & Tsitsiklis, J. N. (1997). *Introduction to linear optimization*. Massachusetts: Athena Scientific.

Charnes, A., Cooper, W. W., & Rhodes, E. (1978). Measuring the efficiency of decision making units. *European Journal of Operational Research*, *2*(6), 429–441.

Cooper, W. W., Deng, H., Seiford, L. M., & Zhu, J. (2011). Congestion: Its identification and management with DEA. In: W. W. Cooper, L. M. Seiford, & J. Zhu (Eds.), *Handbook on data envelopment analysis*, pp. 173–193. Boston: Springer.

Cooper, W. W., Thompson, R. G., & Thrall, R. M. (1996). Chapter 1 Introduction: Extensions and new developments in DEA. *Annals of Operations Research*, *66*(1), 1–45.

Davtalab Olyaie, M., Roshdi, I., Jahanshahloo, G., & Asgharian, M. (2014). Characterizing and finding full dimensional efficient facets in DEA: A variable returns to scale specification. *Journal of the Operational Research Society*, *65*(9), 1453–1464.

Färe, R., & Grosskopf, S. (1983). Measuring congestion in production. *Zeitschrift für Nationalökonomie*, *43*(3), 257–271.

Färe, R., Grosskopf, S., & Lovell, C. A. K. (1985). *The measurement of efficiency of production*. Boston: Kluwer Academic Publishers.

Färe, R., & Svensson, L. (1980). Congestion of production factors. *Econometrica*, *48*(7), 1745–1753.

Khodabakhshi, M., Hosseinzadeh Lotfi, F., & Aryavash, K. (2014). Review of input congestion estimating methods in DEA. *Journal of Applied Mathematics*. https://doi.org/10.1155/2014/963791.

Khoveyni, M., Eslami, R., Khodabakhshi, M., Jahanshahloo, G. R., & Hosseinzadeh Lotfi, F. (2013). Recognizing strong and weak congestion slack based in data envelopment analysis. *Computers & Industrial Engineering*, *64*(2), 731–738.

Khoveyni, M., Eslami, R., & Yang, G. (2017). Negative data in DEA: Recognizing congestion and specifying the least and the most congested decision making units. *Computers & Operations Research*, *79*, 39–48.

Fried, H. O., Lovell, C. A. K., & Schmidt, S. S. (2008). Efficiency and productivity. In H. O. Fried, C. A. K. Lovell, & S. S. Schmidt (Eds.), *The measurement of productive efficiency and productivity growth* (pp. 3–91). New York: Oxford University Press.

Mehdiloozad, M. (2017). Identifying the global reference set in DEA: A mixed 0–1 LP formulation with an equivalent LP relaxation. *Operational Research*, *17*(1), 205–211.

Mehdiloozad, M., Mirdehghan, S. M., Sahoo, B. K., & Roshdi, I. (2015). The identification of the global reference set in data envelopment analysis. *European Journal of Operational Research*, *214*(3), 679–688.

Mehdiloozad, M., & Podinovski, V. V. (2018). Nonparametric production technologies with weakly disposable inputs. *European Journal of Operational Research, 266*(1), 247–258.
Mehdiloozad, M., & Sahoo, B. K. (2016). Identifying the global reference set in DEA: An application to the determination of returns to scale. In S.-N. Huang, H.-S. Lee, & J. Zhu (Eds.), *Handbook of operations analytics using data envelopment analysis* (pp. 299–330). US, Boston: Springer.
Mehdiloozad, M., Tone, K., Askarpour, R., & Ahmadi, M. B. (2018). Finding a maximal element of a non-negative convex set through its characteristic cone: An application to finding a strictly complementary solution. *Computational & Applied Mathematics, 37*(1), 53–80.
Mehdiloozad, M., Zhu, J., & Sahoo, B. K. (2018). Identification of congestion in data envelopment analysis under the occurrence of multiple projections: A reliable method capable of dealing with negative data. *European Journal of Operational Research, 265*(2), 644–654.
Noura, A. A., Hosseinzadeh Lotfi, F., Jahanshahloo, G. R., Rashidi, S. F., & Parker, B. R. (2010). A new method for measuring congestion in data envelopment analysis. *Socio-Economic Planning Sciences, 44*(4), 240–246.
Podinovski, V. V., & Bouzdine-Chameeva, T. (2017). Solving DEA models in a single optimization stage: Can the non-Archimedean infinitesimal be replaced by a small finite epsilon? *European Journal of Operational Research, 257*(2), 412–419.
Ray, S. C. (2004). *Data envelopment analysis: Theory and techniques for economics and operations research*. Cambridge: Cambridge University Press.
Rockafellar, R. T. (1970). *Convex analysis*. Princeton: Princeton University Press.
Shephard, R. W. (1974). Semi-homogeneous production functions and scaling of production. In W. Eichhorn, R. Henn, O. Opitz, & R. W. Shephard (Eds.), *Production theory* (pp. 253–285). New York: Springer-Verlag.
Sueyoshi, T., & Sekitani, K. (2009). DEA congestion and returns to scale under an occurrence of multiple optimal projections. *European Journal of Operational Research, 194*(2), 592–607.
Tone, K., & Sahoo, B. K. (2004). Degree of scale economies and congestion: A unified DEA approach. *European Journal of Operational Research, 158*(3), 755–772.
Tuy, H. (1998). *Convex analysis and convex optimization*. Boston: Kluwer Academic Publisher.
Wei, Q. L., & Yan, H. (2004). Congestion and returns to scale in data envelopment analysis. *European Journal of Operational Research, 153*(3), 641–660.
Wei, Q. L., & Yan, H. (2009). Weak congestion in output additive data envelopment analysis. *Socio-Economic Planning Sciences , 43*(1), 40–54.
Zare Haghighi, H., Khodabakhshi, M., & Jahanshahloo, G. R. (2014). Review of the methods for evaluating congestion in DEA and computing output losses due to congestion. *International Journal of Industrial Mathematics, 6*, 1–17.

Chapter 5
Data Envelopment Analysis and Non-parametric Analysis

Gabriel Villa and Sebastián Lozano

Abstract This chapter gives an introduction to Data Envelopment Analysis (DEA), presenting an overview of the basic concepts and models used. Emphasis is made on the non-parametric derivation of the Production Possibility Set (PPS), on the multiplicity of DEA models and on how to handle different types of situations, namely, undesirable outputs, ratio variables, multi-period data, negative data non-discretionary variables, and integer variables.

Keywords Axioms · Non-parametric approach · Production possibility set · Efficient frontier · Efficiency score · Orientation · Metric · Productivity change · Efficiency change · Technical change

5.1 Introduction

Extracting the relevant information that underlies the available data and using it to maximize the benefits to the organization is a key feature of data science. Among the most important information that can be derived from operational data are efficiency and productivity indicators. The most useful tool for that purpose is Data Envelopment Analysis (DEA), which is a data-driven, non-parametric technique that is able to effectively evaluate such indicators under multiple scenarios. This chapter is intended as a broad introduction to the DEA methodology. The object of study is a set of homogeneous entities (generally termed Decision-Making Units, DMU) that carry out an input–output transformation process. Inputs are assumed to be the-smaller-the-better variables while output variables are assumed to be the-larger-the-better. Each DMU is therefore characterized by an input–output vector whose components are the amounts of the different inputs consumed and of the different outputs produced.

G. Villa (✉) · S. Lozano
Department of Industrial Management, University of Seville, Seville, Spain
e-mail: gvilla@us.es

V. Charles et al. (eds.), *Data Science and Productivity Analytics*, International Series in Operations Research & Management Science 290,
https://doi.org/10.1007/978-3-030-43384-0_5

The basis of the DEA methodology is the derivation of the PPS from the set of input–output vectors corresponding to the observed DMU. This is done in a non-parametric way that does not make any assumption on the input–output transformation function. Thus, it does not require nor make use of any knowledge on how such transformation takes place or the physical mechanisms or processes behind it.

The PPS contains all operating points (i.e., input–output vectors) that are deemed feasible, based on the information provided by the observed DMUs. Once the PPS is known, many different questions can be posed. The most basic one is that given a DMU, are there feasible operating points that operate more efficiently, i.e., consuming less (or equal) input and producing more (or equal) output? If the answer is yes, then the DMU is inefficient and its input–output vector is dominated by other input–output vectors of the PPS. Changing the operating point from the observed DMU to any operating point that dominates (i.e., using that operating point as target) means input reductions and/or output increases and, in any case, an efficiency improvement.

The non-dominated subset of the PPS forms the efficient frontier. These operating points are efficient in the Pareto sense, i.e., no variable can be improved (meaning reduced if it is an input and increased if it is an output) without worsening any other variable. Moreover, those operating points represent the best practices and cannot be improved in the same way inefficient operating points can. Also, of all the multiple operating points that dominate an inefficient DMU only those that belong to the efficient frontier are rational targets. When information on the inputs and outputs unit prices is available then it may be profitable for a DMU to choose a target that does not dominate the DMU but, in that case also, only efficient operating points can be considered as targets.

Another question frequently asked is how to determine the level of efficiency of each DMU. Efficiency scores are generally non-negative and normalized so that a score of one means efficiency, and the lower the score the more inefficient the DMU is. There are different ways to compute the efficiency score but, as a general rule, the efficiency score is related to the distance from the DMU to the efficient frontier. This distance generally depends on the metric used or on the direction along which the projection is performed, information which needs to be provided together with the numerical score. It is also related to the efficient target selected (which in turn depends on the metric used or on the direction along which the projection is performed) as the efficiency score represents the distance between the DMU and its target.

Note that although in order to intuitively understand DEA, it is better to think of the input–output process as a transformation or production process so that inputs are resources and outputs are products; actually the framework is more general. Thus, for example, DEA can be used to benchmark different product models considering as inputs those attributes that are the-smaller-the-better and as outputs those that are the-larger-the-better. No actual transformation of the inputs into outputs takes place in that situation. Another example of this is the use of DEA to assess the performance of nations at the Olympics. The outputs in that case are the number of gold, silver, and bronze medals won by a country while the inputs are the population and the Gross Domestic Product (GDP) of the country. Cook et al. (2014) presents an insightful discussion on some basic decisions that have to be made prior to choosing a model, such as the purpose of the performance measurement exercise, the orientation of the model and the selection of the inputs and outputs.

The chapter unfolds as follows. In the next section, we deal with the first step of the DEA methodology, which is how to derive the PPS from the observations using some basic axioms. In Sect. 5.3, different DEA models are presented, classifying them according to their orientation (or lack of) and their metric. The rest of the chapter is dedicated to showing how DEA can handle different situations that can occur when applying the methodology. In particular, we will discuss undesirable outputs, ratio variables, multi-period data, negative data non-discretionary variables, and integer variables. In the last section, some additional topics not covered in this introductory chapter are commented and limitations of the methodology are discussed.

5.2 Inferring the PPS

Let n be the number of DMUs and $\left(X_j, Y_j\right)$, $j = 1, \ldots, n$ the corresponding input–output vectors. Let $X_j = \left(x_{1j}, \ldots, x_{mj}\right) \in \mathbb{R}_+^m$ and $Y_j = \left(y_{1j}, \ldots, y_{pj}\right) \in \mathbb{R}_+^p$. The Production Possibility Set (PPS), a.k.a. the DEA technology, is the set $T = \left\{(X, Y) \in \mathbb{R}_+^{m+p} : Y \text{ can be produced from } X\right\}$. To infer the PPS the following axioms can be considered:

A1. Envelopment: $\left(X_j, Y_j\right) \in T,\ j = 1, \ldots, n$

A2. Free disposability of inputs and outputs:

$$(X, Y) \in T, \hat{X} \geq X, \hat{Y} \leq Y \Rightarrow \left(\hat{X}, \hat{Y}\right) \in T.$$

A3. Convexity:

$$(X, Y), \left(\hat{X}, \hat{Y}\right) \in T \Rightarrow \left(\lambda X + (1-\lambda)\hat{X}, \lambda Y + (1-\lambda)\hat{Y}\right) \in T, \forall \lambda \in [0, 1].$$

A4. Scalability: $(X, Y) \in T \Rightarrow (\lambda X, \lambda Y) \in T, \forall \lambda \geq 0$

These basic axioms mean that the observed DMUs are feasible (A1), that it is always possible to waste resources and to fall short of output (A2), that any mixture of two feasible operating points defines a feasible operating point (A3) and that scaling up or down any operating point is feasible (A4).

Depending on which of these four axioms are assumed, and applying the Minimum Extrapolation Principle, different PPS results. Thus, the Free Disposal Hull (FDH) technology corresponds to assuming just A1 + A2.

$$T^{FDH} = \left\{ (X, Y) : \sum_{j=1}^{n} \lambda_j X_j \le X, \sum_{j=1}^{n} \lambda_j Y_j \ge Y, \sum_{j=1}^{n} \lambda_j = 1, \lambda_j \in \{0, 1\} \forall j \right\}. \tag{5.1}$$

The Variable Returns to Scale (VRS) technology corresponds to assuming A1 + A2 + A3

$$T^{VRS} = \left\{ (X, Y) : \sum_{j=1}^{n} \lambda_j X_j \le X, \sum_{j=1}^{n} \lambda_j Y_j \ge Y, \sum_{j=1}^{n} \lambda_j = 1, \lambda_j \ge 0 \forall j \right\}. \tag{5.2}$$

And the Constant Returns to Scale (CRS) technology corresponds to assuming all four axioms A1 + A2 + A3 + A4

$$T^{CRS} = \left\{ (X, Y) : \sum_{j=1}^{n} \lambda_j X_j \le X, \sum_{j=1}^{n} \lambda_j Y_j \ge Y, \lambda_j \ge 0 \forall j \right\}. \tag{5.3}$$

Note that $T^{FDH} \subset T^{VRS} \subset T^{CRS}$, i.e., the FDH technology is more conservative in what it assumes as feasible (based on the observed DMUs) than the VRS technology and this, in turn, is more conservative than the CRS technology. The latter is bolder in determining what is feasible based, containing efficient operating points of any size, including the origin $(0_m, 0_p) \in T^{CRS}$.

Although less used than the above three technologies that are two technologies that are intermediate between the VRS and the CRS technologies and correspond to substituting the unlimited scalability axiom (A4) by one of these other two

A5. Downward scalability: $(X, Y) \in T \quad \Rightarrow \quad (\lambda X, \lambda Y) \in T, \ \forall 0 \le \lambda \le 1$

A6. Upward scalability: $(X, Y) \in T \quad \Rightarrow \quad (\lambda X, \lambda Y) \in T, \ \forall \lambda \ge 1$

Note that A5 + A6 $\Leftrightarrow$ A4. However, A1 + A2 + A3 + A5 leads to the Non-Increasing Returns to Scale (NIRS) technology

$$T^{NIRS} = \left\{ (X, Y) : \sum_{j=1}^{n} \lambda_j X_j \le X, \sum_{j=1}^{n} \lambda_j Y_j \ge Y, \sum_{j=1}^{n} \lambda_j \le 1, \lambda_j \ge 0 \forall j \right\}. \tag{5.4}$$

while A1 + A2 + A3 + A6 leads to the Non-Decreasing Returns to Scale (NDRS) technology

$$T^{NDRS} = \left\{ (X, Y) : \sum_{j=1}^{n} \lambda_j X_j \le X, \sum_{j=1}^{n} \lambda_j Y_j \ge Y, \sum_{j=1}^{n} \lambda_j \ge 1, \lambda_j \ge 0 \forall j \right\}. \tag{5.5}$$

It can be easily seen that

$$T^{VRS} \subset T^{NIRS} \subset T^{CRS}$$
$$T^{VRS} \subset T^{NDRS} \subset T^{CRS}$$
$$T^{NIRS} \cup T^{NDRS} = T^{CRS}$$
$$T^{NIRS} \cap T^{NDRS} = T^{VRS}.$$

This all can be seen in Figs. 5.1 and 5.2, which show all these technologies for the case of a single input and a single output.

Given any of the above DEA technologies T, the weak efficient frontier of T is defined as $\partial^W(T) = \left\{(X, Y) \in T : \hat{Y} > Y, \hat{X} < X \Rightarrow \left(\hat{X}, \hat{Y}\right) \notin T\right\}$. The true efficient frontier, a.k.a. strong efficient frontier, is, however, a subset of $\partial^W(T)$ and corresponds to $\partial^S(T) = \left\{(X, Y) \in T : \hat{Y} \geq Y, \hat{X} \leq X, \left(\hat{X}, \hat{Y}\right) \neq (X, Y) \Rightarrow \left(\hat{X}, \hat{Y}\right) \notin T\right\}$.

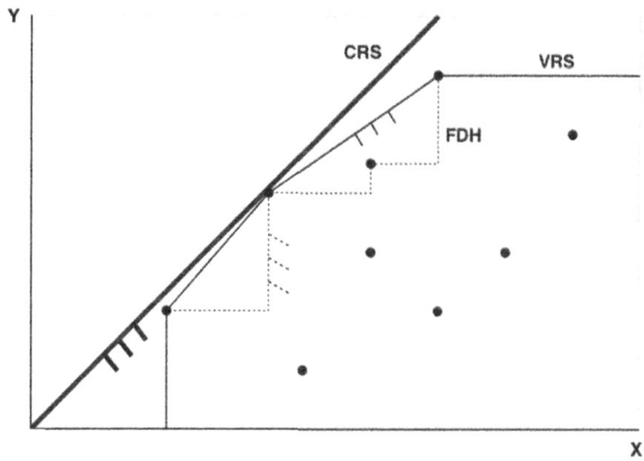

Fig. 5.1 CRS, VRS, and FDH technologies for one input one output case

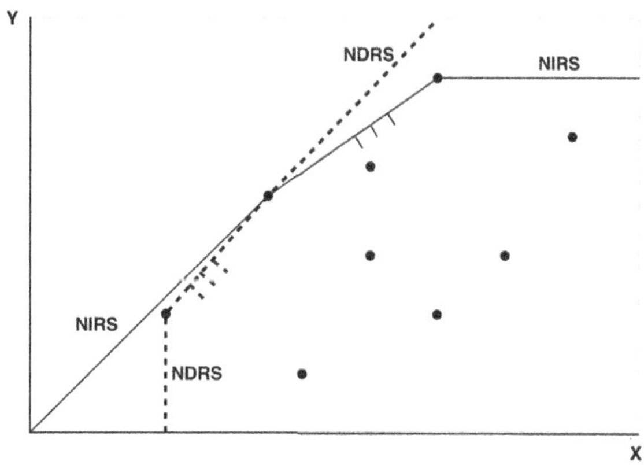

Fig. 5.2 NRIS and NDRS technologies for one input one output case

The importance of this first step of inferring the PPS cannot be overstated as the PPS is the basis of all DEA models. The PPS defines the playing field of what is feasible. All DEA models have to limit themselves to search within the assumed PPS. A relevant question is how to choose between these different DEA technologies. To simplify matters, the non-convex FDH technology is considered when the convexity assumption is not reasonable, for example, because the resources or the outputs are not divisible. Otherwise, convexity is a reasonable assumption generally used. As regards scalability, CRS is a strong assumption which leads to an enlarged PPS, more ambitious targets, and lower efficiency scores. CRS is usually assumed when scale size effects do not exist or the DMUs operate in a perfect competitive market. Otherwise, VRS is a safer and more flexible option as it allows that some parts of the efficient frontier can locally exhibit CRS or Increasing Returns to Scale (IRS) or Decreasing Returns to Scale (DRS). The former region, i.e., the CRS part of the VRS efficient frontier, is called the Most Productive Scale Size (MPSS) and corresponds to the intersection between the CRS and the VRS efficient frontiers $\partial^S(T^{VRS}) \cap \partial^S(T^{CRS})$ as shown in Fig. 5.1.

As regards the intermediate NIRS and NDRS technologies, they are useful for determining the local Returns to Scale (RTS) of a DMU. Thus, if a DMU (X_0, Y_0) is projected onto the VRS, the CRS, and the NIRS technologies, giving three different targets $(X_0^{VRS}, Y_0^{VRS}) \in \partial^S(T^{VRS})$, $(X_0^{CRS}, Y_0^{CRS}) \in \partial^S(T^{CRS})$, and $(X_0^{NIRS}, Y_0^{NIRS}) \in \partial^S(T^{NIRS})$, then the following three cases are possible:

$(X_0^{CRS}, Y_0^{CRS}) = (X_0^{NIRS}, Y_0^{NIRS}) = (X_0^{VRS}, Y_0^{VRS})$: DMU 0 exhibits CRS locally.
$(X_0^{CRS}, Y_0^{CRS}) \neq (X_0^{NIRS}, Y_0^{NIRS}) \neq (X_0^{VRS}, Y_0^{VRS})$: DMU 0 exhibits DRS locally.
$(X_0^{CRS}, Y_0^{CRS}) = (X_0^{NIRS}, Y_0^{NIRS}) \neq (X_0^{VRS}, Y_0^{VRS})$: DMU 0 exhibits IRS locally.

This way of projecting a DMU onto more than one efficient frontier also allows determining its scale efficiency. Thus, the scale efficiency of DMU (X_0, Y_0) measures the distance between its corresponding CRS and VRS targets (X_0^{CRS}, Y_0^{CRS}) and (X_0^{VRS}, Y_0^{VRS}), and represents the separation between the CRS and VRS efficient frontiers $\partial^S(T^{CRS})$ and $\partial^S(T^{VRS})$ measured at those points. Scale efficiency and RTS are related. Thus, when a DMU exhibits IRS it means that its scale efficiency would increase, if the DMU increased its size. The opposite occurs if it exhibits DRS, in which case the scale efficiency would increase if the DMU reduced its size. A DMU exhibiting CRS has a scale efficiency of one, which means that it does not need to increase or decrease its size as it cannot increase its scale efficiency.

5.3 Formulating DEA Models

In this section, we will present several DEA models, differentiating them based on some features. In particular, we will classify DEA models according to their oriented/non-oriented character and the radial/non-radial metric considered.

5.3.1 *Defining the Aims of the Model*

As we said above, once the PPS derived from the observed DMUs is available, one can use it for different purposes. The two most common aims of a DEA model are (1) efficiency assessment, i.e., classifying the DMUs into efficient and inefficient and computing an efficiency score, and (2) target setting, i.e., determining an efficient target with corresponding input and output improvements. These two aims can concur or only one of them may be of interest. Thus, for example, when the emphasis is on target setting then instead of projecting onto a distant efficient target (as it is generally done when assessing efficiency), it is preferable to compute closest targets (e.g., Aparicio et al. 2007; Aparicio 2016) or to use a multi-objective optimization approach in which the preferences of the Decision-Maker (DM) can be taken into account. And when the emphasis is on measuring efficiency then the efficient target computed is secondary and depends on the metric used and the orientation considered. It can also occur, especially in additive DEA models, that the efficiency assessment model has alternative optima and hence the target is not uniquely determined; something which is not utterly important when target setting is not the main goal.

Another usual aim of DEA models is ranking the DMUs. Ranking the inefficient DMUs is, in principle, easy since their corresponding efficiency scores can be used for that purpose. However, that does not work for ranking the efficient DMUs, which may require other DEA methods, e.g., super-efficiency approaches (e.g., Tone 2002).

The DEA methodology can also be used for other purposes, like, for example, estimating cost, revenue or profit efficiency (e.g., Zofio et al. 2013), centralized resource reallocation (e.g., Lozano and Villa 2004), production planning (Lozano 2014), estimating merger gains (e.g., Lozano 2010), estimating the benefits of cooperation between organizations (e.g., Lozano 2012, 2013a), selecting the best partner for a joint venture (e.g., Lozano 2013b), etc. What all these seemingly different DEA approaches have in common is that all of them compute operating points that lie on the efficient frontier of the PPS derived from the observed DMUs. That is the essence of the DEA methodology. And the fact that this is done using a non-parametric approach is its main strength. It also helps that the optimization models used are generally linear programs and therefore easy to solve using any off-the-shelf optimization software.

Summarizing this point, the DEA methodology is very flexible and can be used for different purposes. Hence, before building or choosing a DEA model the analyst must bear in mind the aim of the study. In what follows, and since this chapter is just an introduction to DEA, we will focus on the most basic aim of DEA, which is efficiency assessment. But the reader should be aware that this is only a fraction of what DEA can be used for.

Finally, an important feature that is commonly demanded of a DEA model, whatever the aim of the study, is unit invariance, i.e., that the efficiency scores, the ranking derived from them, the targets computed, etc., which do not change if the unit of measurement of any of the variables is changed. Thus, if one of the inputs considered is energy, then the results should not depend on whether that variable is measured in kJ, kcal, kWh, BTU or any other equivalent unit of measurement of energy.

5.3.2 Selecting the Orientation and the Metric

As it was said above, in order to detect whether a DMU is efficient or not the DEA methodology looks if there are any feasible operating points which can produce as much or more and consume as much or less. When the DMU is inefficient then there are many such points so it is rational to demand that the input and output improvements be maximized. This can be done in an input or output orientation, i.e., giving priority to improving the inputs over the outputs or the opposite, or in a non-oriented manner, i.e., improving inputs and outputs simultaneously. The term non-oriented is used here in the sense just mentioned and it does not mean that the DEA model may not use a specific projection direction or that the relative importance of the inputs and output improvements may not be weighted. In other words, we have three possibilities: a DEA model can be input-oriented, output-oriented or, else, it is non-oriented (Fig. 5.3).

Apart from the oriented or non-oriented character there is another dimension that can be used to differentiate the existing DEA models and it is the metric used to measure the input and output improvements. Thus, additive DEA models (Charnes et al. 1985) use the rectangular (a.k.a. Manhattan) distance l_1 which basically sums

Fig. 5.3 Input, output, and non-oriented projections

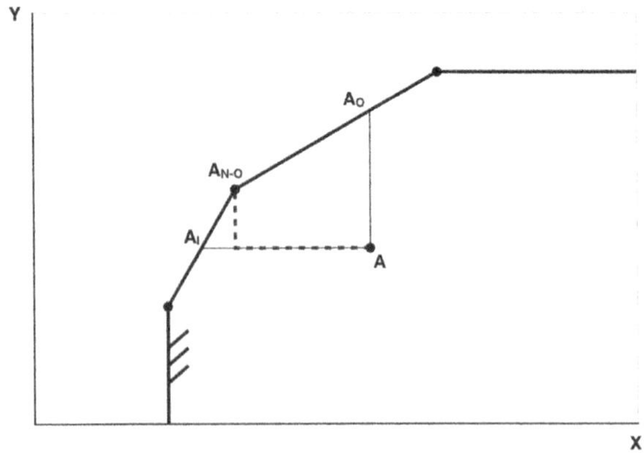

Table 5.1 Some examples of the existing DEA models

	Radial	Non-radial
Oriented	CCR-I/BCC-I CCR-O, BCC-O	Russell non-radial
Non-oriented	Hyperbolic	BAM, DDF, SBM

the input and output slacks, preferably after normalizing them to make them dimensionless. One example is the bounded additive measure of efficiency (BAM) model (Cooper et al. 2011).

Another commonly used metric is the so-called radial metric and its characteristic is that it contracts the inputs and/or expands the outputs using a uniform multiplication factor. That means that all the inputs are reduced equi-proportionally and all outputs are also increased equi-proportionally. The main DEA models in this category are the CCR (Charnes et al. 1978) and BCC (Banker et al. 1984) DEA models. The hyperbolic graph efficiency measure (Färe et al. 1985) can also be considered within this radial metric category.

Other metrics often used are the Russell non-radial technical efficiency measure (Färe et al. 1985), the directional distance function (DDF, Chambers et al. 1996), and the slacks-based measure of efficiency (SBM) (Tone 2001) a.k.a. enhanced Russell graph measure (ERGM, Pastor et al. 1999). In order to simplify, we can group all non-radial DEA approaches together so that we basically have two types of DEA models: radial and non-radial. Table 5.1 shows some examples of existing DEA models belonging to each of the four combinations of orientation and metric.

These models will be presented in the following sections but let us for the moment formulate the constraints that are common to all those DEA models. Those constraints correspond to establishing that the target input and output variables are computed as linear combinations of the observed DMUs and that the corresponding multipliers must comply with the corresponding DEA technology considered. These constraints appear in all the DEA models below. In other words, the DEA models presented in the next sections have some common and some specific constraints. The common constraints are

$$\hat{x}_i \geq \sum_{j=1}^{n} \lambda_j x_{ij} \ \forall i \quad \hat{y}_k \leq \sum_{j=1}^{n} \lambda_j y_{kj} \ \forall k. \tag{5.6}$$

$$(\lambda_1, \lambda_2, .., \lambda_n) \in \Lambda^T. \tag{5.7}$$

$$\Lambda^{CRS} = \left\{(\lambda_1, \lambda_2, .., \lambda_n) : \lambda_j \geq 0 \, \forall j\right\}. \tag{5.8}$$

$$\Lambda^{VRS} = \left\{(\lambda_1, \lambda_2, .., \lambda_n) : \lambda_j \geq 0 \, \forall j, \sum_{j=1}^{n} \lambda_j = 1\right\}. \tag{5.9}$$

$$\Lambda^{NIRS} = \left\{ (\lambda_1, \lambda_2, .., \lambda_n) : \lambda_j \geq 0 \forall j, \sum_{j=1}^{n} \lambda_j \leq 1 \right\}. \tag{5.10}$$

$$\Lambda^{FDH} = \left\{ (\lambda_1, \lambda_2, .., \lambda_n) : \lambda_j \in \{0, 1\} \forall j, \sum_{j=1}^{n} \lambda_j = 1 \right\}. \tag{5.11}$$

These constraints guarantee that the computed target $\left(\hat{X}, \hat{Y}\right)$, where $\hat{X} = \left(\hat{x}_1, \ldots, \hat{x}_m\right)$ and $\hat{Y} = \left(\hat{y}_1, \ldots, \hat{y}_p\right)$, belong to the corresponding PPS.

5.3.3 *Radial Oriented DEA Models*

The CCR-I model (Charnes et al. 1978) is a two-phase procedure whose main aim is carrying out a uniform (i.e., equi-proportional) reduction of the inputs. Thus, the first phase (a.k.a radial phase) solves

$$\begin{aligned}
&Min \quad \theta \\
&s.t. \\
&\sum_{j=1}^{n} \lambda_j X_j \leq \theta X_0 \\
&\sum_{j=1}^{n} \lambda_j Y_j \geq Y_0 \\
&\lambda_j \geq 0 \forall j \ \theta \ free.
\end{aligned} \tag{5.12}$$

However, the above model can project the unit 0 onto a weak efficient point. That occurs if $(\theta^* X_0, Y_0)$ admits additional improvements while remaining in T^{CRS}. Therefore, the second phase (a.k.a. rectangular phase) incorporates these improvements by slack variables that represent the input excess (S^-) and the output shortfalls (S^+):

$$\begin{aligned}
&Max \quad \mathbf{1}S^- + \mathbf{1}S^+ \\
&s.t. \\
&\hat{X} = \sum_{j=1}^{n} \lambda_j X_j = \theta^* X_0 - S^- \\
&\hat{Y} = \sum_{j=1}^{n} \lambda_j Y_j = Y_0 + S^+
\end{aligned}$$

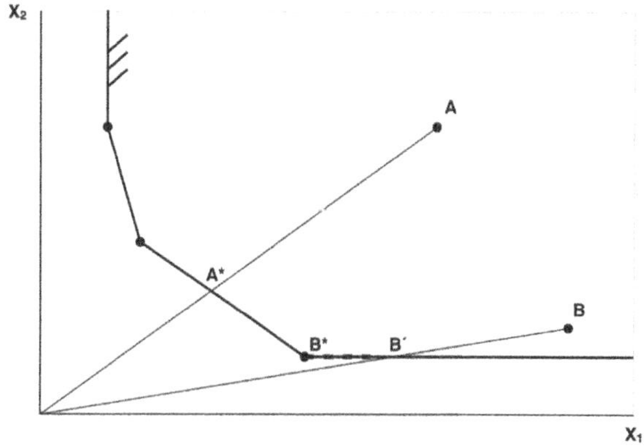

Fig. 5.4 Radial input orientation in two-input–one constant output CRS case

$$\lambda_j \geq 0\, \forall j \quad S^-, S^+ \geq 0. \tag{5.13}$$

where $\mathbf{1}S^- + \mathbf{1}S^+ = \sum_{i=1}^{m} s_i^- + \sum_{k=1}^{p} s_k^+$. The aim of the rectangular phase is to find a solution that maximizes the possible improvements of inputs and outputs after the input contraction θ^* of the radial phase.

In Fig. 5.4, we show the solutions computed by models (5.12) and (5.13) in two inputs (X_1, X_2)—one constant output CRS problem. Model (5.12) contracts the units A and B radially reaching the points A* and B′, respectively. The rectangular phase (5.13) finds an additional reduction of input X_1 for B′ obtaining B* as the definitive reference point for unit B.

When the problem exhibits VRS, the above models have to be modified by adding the convexity constraint (5.9) in both phases. The resultant model was proposed in Banker et al. (1984) and it is known as BCC-I model.

The CCR-O model is an output-oriented model, and analogously to CCR-I, it is solved in two phases. The radial phase seeks to maximize the outputs radially by multiplying them by a factor γ:

$$\begin{aligned} &Max \quad \gamma \\ &s.t. \\ &\sum_{j=1}^{n} \lambda_j X_j \leq X_0 \\ &\sum_{j=1}^{n} \lambda_j Y_j \geq \gamma Y_0 \\ &\lambda_j \geq 0\, \forall j \quad \gamma \ free. \end{aligned} \tag{5.14}$$

Once the solution of (5.14) is obtained, the rectangular phase is solved in order to find additional improvements of inputs and outputs:

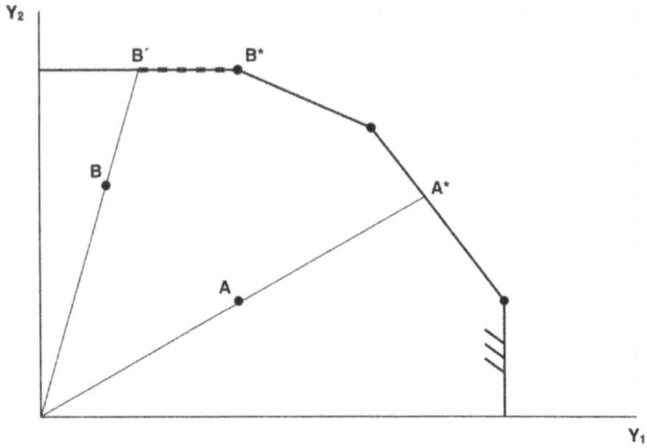

Fig. 5.5 Radial output orientation in one constant input–two-output CRS case

$$\begin{aligned}
&Max \quad \mathbf{1}S^{-} + \mathbf{1}S^{+} \\
&s.t. \\
&\hat{X} = \sum_{j=1}^{n} \lambda_j X_j = X_0 - S^{-} \\
&\hat{Y} = \sum_{j=1}^{n} \lambda_j Y_j = \gamma^{*} Y_0 + S^{+} \\
&\lambda_j \geq 0 \forall j \quad S^{-}, S^{+} \geq 0.
\end{aligned} \tag{5.15}$$

Similarly, in Fig. 5.5, a one constant input–two-output (Y1, Y2) CRS problem is illustrated. Note that the rectangular phase (5.15) obtains an additional improvement in output Y_1 shown in the figure as the segment $B' - B^{*}$.

The BCC-O model is applied when VRS is assumed and the formulation is obtained by the addition of the constraint (5.9) in the radial and rectangular phases (5.14) and (5.15), respectively. In the CRS case, the input radial measure of efficiency equals to the reciprocal of the output radial measure of efficiency, i.e., $\theta = 1/\gamma$. This does not hold, however, in the VRS case.

5.3.4 *Radial Non-oriented DEA Models*

In this section, we are interested in the case in which a uniform improvement in both inputs and outputs is sought. The hyperbolic graph efficiency measure proposed in Färe et al. (1985) makes inputs decrease and outputs increased concurrently and by the same factor ω:

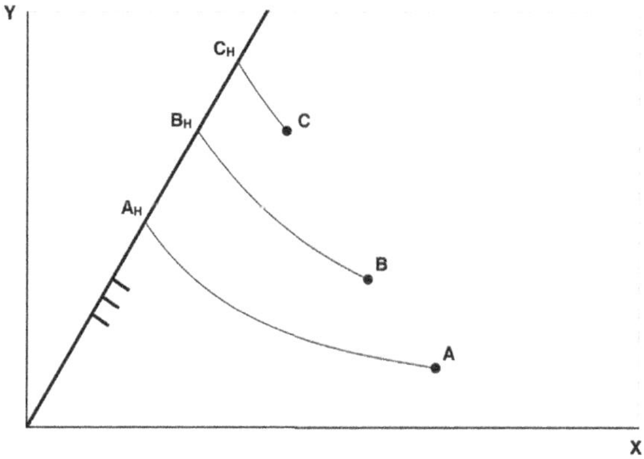

Fig. 5.6 Hyperbolic DEA projection

$$
\begin{aligned}
& Min \quad \omega \\
& s.t. \\
& \sum_{j=1}^{n} \lambda_j X_j \le \omega X_0 \\
& \sum_{j=1}^{n} \lambda_j Y_j \ge \frac{1}{\omega} Y_0 \\
& \lambda_j \ge 0 \forall j \ \omega \ free.
\end{aligned}
\tag{5.16}
$$

Figure 5.6 shows a single-input–single-output CRS case in which the hyperbolic model projects units A, B, and C onto A_H, B_H, and C_H which belong to the efficient frontier. The curve line represents the hyperbolic function defined by the restrictions of (5.16).

Although in principle this optimization model is non-linear, it can be transformed into a linear program considering the following variables:

$$
\begin{aligned}
& \lambda' = \omega \cdot \lambda \\
& \tau = \omega^2.
\end{aligned}
\tag{5.17}
$$

Thus (5.16) can be rewritten as

$$
\begin{aligned}
& Min \quad \tau \\
& s.t. \\
& \sum_{j=1}^{n} \lambda'_j X_j \le \tau X_0
\end{aligned}
$$

$$\sum_{j=1}^{n} \lambda'_j Y_j \geq Y_0$$
$$\lambda'_j \geq 0 \, \forall j \;\; \tau \geq 0. \tag{5.18}$$

The target, in this case, can be computed as

$$\begin{aligned} \hat{X} &= \sum_{j=1}^{n} \lambda_j^* X_j = \tfrac{1}{\omega^*} \sum_{j=1}^{n} \lambda_j'^* X_j \left(\leq \tfrac{1}{\omega^*} \tau^* X_0 = \omega^* X_0\right) \\ \hat{Y} &= \sum_{j=1}^{n} \lambda_j^* Y_j = \tfrac{1}{\omega^*} \sum_{j=1}^{n} \lambda_j'^* Y_j \left(\geq \tfrac{1}{\omega^*} Y_0\right). \end{aligned} \tag{5.19}$$

In the VRS case, the hyperbolic graph efficiency measure is formulated as model (5.17) plus the convexity constraint (5.9). In this case, a linearization of the model is not possible although an efficient iterative approach using linear programs can be used (see Färe et al. 2016).

5.3.5 Non-radial Oriented DEA Models

The so-called Russell input measure of technical efficiency uses a different reduction factor θ_i for each input (Färe et al. 1985). The corresponding model minimizes the average of those input reduction factors. Actually, the procedure also involves two phases. Phase I is

$$Min \;\; \frac{1}{m} \sum_{i=1}^{m} \theta_i$$
$$s.t.$$
$$\sum_{j=1}^{n} \lambda_j x_{ij} \leq \theta_i x_{i0} \quad \forall i$$
$$\sum_{j=1}^{n} \lambda_j y_{kj} \geq y_{k0} \quad \forall k$$
$$\lambda_j \geq 0 \, \forall j \quad 0 \leq \theta_i \leq 1 \, \forall i. \tag{5.20}$$

Once model (5.20) is solved, and in order to guarantee an efficient target, a Phase II that exhausts the possible output slacks is run:

$$Max \;\; \sum_{k=1}^{p} s_k^+$$
$$s.t.$$

$$\hat{x}_i = \sum_{j=1}^{n} \lambda_j x_{ij} = \theta_i^* x_{i0} \quad \forall i$$

$$\hat{y}'_k = \sum_{j=1}^{n} \lambda_j y_{kj} = y_{k0} + s_k^+ \quad \forall k$$

$$\lambda_j \geq 0 \, \forall j \quad s_k^+ \geq 0 \, \forall k. \tag{5.21}$$

Note that, unlike the Phase II of the CCR-I and BCC-I models, (5.21) only seeks additional improvements in outputs, since the optimal solution of (5.20) does not allow further input reductions.

Analogously, the Russell output measure of technical efficiency also involves two phases:

$$Max \quad \frac{1}{p} \sum_{k=1}^{p} \gamma_k$$

$$s.t.$$

$$\sum_{j=1}^{n} \lambda_j x_{ij} \leq x_{i0} \quad \forall i$$

$$\sum_{j=1}^{n} \lambda_j y_{kj} = \gamma_k y_{k0} \quad \forall k$$

$$\lambda_j \geq 0 \, \forall j \quad \gamma_k \geq 1 \, \forall k. \tag{5.22}$$

and

$$Max \quad \sum_{i=1}^{m} s_i^-$$

$$s.t.$$

$$\hat{x}_i = \sum_{j=1}^{n} \lambda_j x_{ij} = x_{i0} - s_i^- \quad \forall i$$

$$\hat{y}_k = \sum_{j=1}^{n} \lambda_j y_{kj} = \gamma_k^* y_{k0} \quad \forall k$$

$$\lambda_j \geq 0 \, \forall j \quad s_i^- \geq 0 \, \forall i. \tag{5.23}$$

5.3.6 Non-radial Non-oriented DEA Models

There are several non-radial, non-oriented DEA models that maximize the sum of the normalized input and output slacks. One such model is the bounded additive measure (BAM), proposed by Cooper et al. (2011), that normalizes the input and output slacks of DMU 0 using the corresponding lower and upper ranges L_0 and U_0, respectively, computed as follows:

$$\begin{aligned} L_0 &= X_0 - \underline{X} \\ U_0 &= \overline{Y} - Y_0. \end{aligned} \tag{5.24}$$

where

$$\begin{aligned} \underline{X} &= \min_{j=1,\ldots,n} X_j \\ \overline{Y} &= \max_{j=1,\ldots,n} Y_j. \end{aligned} . \tag{5.25}$$

The model used to evaluate the BAM is thus

$$\begin{aligned} & Max \quad \phi = \frac{1}{m+p}\left(\mathbf{1}\frac{S^-}{L} + \mathbf{1}\frac{S^+}{U}\right) \\ & s.t. \\ & \hat{X} = \sum_{j=1}^{n} \lambda_j X_j = X_0 - S^- \\ & \hat{Y} = \sum_{j=1}^{n} \lambda_j Y_j = Y_0 + S^+ \\ & \sum_{j=1}^{n} \lambda_j = 1 \\ & \lambda_j \geq 0\,\forall j \quad S^-, S^+ \geq 0. \end{aligned} \tag{5.26}$$

Once this model is solved, BAM is computed as $1-\phi^*$. Note that the above model assumes VRS and can be modified for the CRS technology adding the following constraints to model (5.26):

$$\begin{aligned} & \sum_{j=1}^{n} \lambda_j X_j \geq \underline{X} \\ & \sum_{j=1}^{n} \lambda_j Y_j \leq \overline{Y}. \end{aligned} \tag{5.27}$$

Another well-known non-radial, non-oriented DEA model is the directional distance function (DDF, Chambers et al. 1996). Given a technology T, and a non-zero directional vector $(g^x, g^y) \in (\Re_+^m \times \Re_+^p)$, the DDF $D_T(X_0, Y_0, g^x, g^y)$ is defined as

$$
\begin{aligned}
D_T(X_0, Y_0, g^x, g^y) = Max \quad & \beta \\
s.t. \quad & \\
& \hat{X} = \sum_{j=1}^{n} \lambda_j X_j \le X_0 - \beta g^x \\
& \hat{Y} = \sum_{j=1}^{n} \lambda_j Y_j \ge Y_0 + \beta g^y \\
& (\lambda_1, \lambda_2, .., \lambda_n) \in \Lambda^T \quad \beta \; free.
\end{aligned} \tag{5.28}
$$

where Λ^T can be, for example, (5.8), (5.9), (5.10), or (5.11). From the first and second constraints of model (5.28) it can be easily seen that $D_T(X_0, Y_0, g^x, g^y)$ reduces the inputs and increases the outputs simultaneously, using (g^x, g^y) as projection direction.

Figure 5.7 shows the projection obtained when model (5.28) is applied to unit A in a single-input–single-output VRS case. The resulting efficient reference point is labeled as A_{DDF}. Note that the components g^x and g^y of vector g are also shown.

Another non-radial, non-oriented DEA model is the Russell graph measure of technical efficiency (Färe et al. 1985), which can be formulated as

$$
\begin{aligned}
Min \quad & \frac{1}{m+p} \left(\sum_{i=1}^{m} \theta_i + \sum_{k=1}^{p} \frac{1}{\gamma_k} \right) \\
s.t. \quad &
\end{aligned}
$$

Fig. 5.7 DDF projection

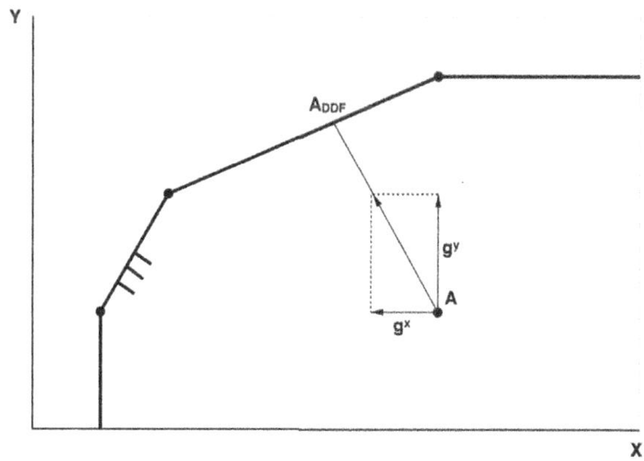

$$\hat{x}_i = \sum_{j=1}^{n} \lambda_j x_{ij} = \theta_i x_{i0} \quad \forall i$$
$$\hat{y}_k = \sum_{j=1}^{n} \lambda_j y_{kj} = \gamma_k y_{k0} \quad \forall k$$
$$\lambda_j \geq 0 \, \forall j \quad 0 \leq \theta_i \leq 1 \, \forall i \quad \gamma_k \geq 1 \, \forall k. \tag{5.29}$$

This model can be used under VRS by just adding the convexity constraint (5.9). Pastor et al. (1999) pointed out two drawbacks when model (5.29) is applied: it is a non-linear problem, and the efficiency is measured as a weighted average of arithmetic and harmonic means, what makes it not readily understood. To overcome those drawbacks, they proposed the enhanced Russell graph measure of efficiency (ERGM):

$$Min \quad \left(\frac{\frac{1}{m} \sum_{i=1}^{m} \theta_i}{\frac{1}{p} \sum_{k=1}^{p} \gamma_k} \right)$$
$$s.t.$$
$$\hat{x}_i = \sum_{j=1}^{n} \lambda_j x_{ij} = \theta_i x_{i0} \quad \forall i$$
$$\hat{y}_k = \sum_{j=1}^{n} \lambda_j y_{kj} = \gamma_k y_{k0} \quad \forall k$$
$$\lambda_j \geq 0 \, \forall j \quad \theta_i \leq 1 \, \forall i \quad \gamma_k \geq 1 \, \forall k. \tag{5.30}$$

As before, the version for VRS is obtained by adding the convexity constraint (5.9). Note also that carrying out the following change of variables:

$$\theta_i = \frac{x_{i0} - s_i^-}{x_{i0}} = 1 - \frac{s_i^-}{x_{i0}} \quad \forall i$$
$$\gamma_k = \frac{y_{k0} + s_k^+}{y_{k0}} = 1 + \frac{s_k^+}{y_{k0}} \quad \forall k. \tag{5.31}$$

model (5.30) can be equivalently formulated as

$$Min \quad \left(\frac{1 - \frac{1}{m} \sum_{i=1}^{m} \frac{s_i^-}{x_{io}}}{1 + \frac{1}{p} \sum_{k=1}^{p} \frac{s_k^+}{y_{ko}}} \right)$$
$$s.t.$$

$$\hat{x}_i = \sum_{j=1}^{n} \lambda_j x_{ij} = x_{i0} - s_i^- \quad \forall i$$
$$\hat{y}_k = \sum_{j=1}^{n} \lambda_j y_{kj} = y_{k0} + s_k^+ \quad \forall k$$
$$\lambda_j \geq 0 \, \forall j \quad s_i^- \geq 0 \, \forall i \quad s_k^+ \geq 0 \, \forall k. \tag{5.32}$$

This model is also known as the slacks-based measure of efficiency (SBM, Tone 2001), and can be linearized introducing a new positive variable t that equates the denominator of the objective function of (5.32) and defining the new variables

$$\hat{S}^- = tS^-$$
$$\hat{S}^+ = tS^+$$
$$\hat{\lambda} = t\lambda. \tag{5.33}$$

Thus, (5.32) can be equivalently formulated as

$$Min \quad t - \frac{1}{m}\sum_{i=1}^{m} \frac{\hat{s}_i^-}{x_{io}}$$
$$s.t.$$
$$t + \frac{1}{p}\sum_{k=1}^{p} \frac{\hat{s}_k^+}{y_{ko}} = 1$$
$$\sum_{j=1}^{n} \hat{\lambda}_j x_{ij} = t x_{i0} - \hat{s}_i^- \quad \forall i$$
$$\sum_{j=1}^{n} \hat{\lambda}_j y_{kj} = t y_{k0} + \hat{s}_k^+ \quad \forall k$$
$$\hat{\lambda}_j \geq 0 \, \forall j \quad \hat{s}_i^- \geq 0 \, \forall i \quad \hat{s}_k^+ \geq 0 \, \forall k \quad t > 0. \tag{5.34}$$

For VRS, it is necessary to add the following restriction:

$$\sum_{j=1}^{n} \hat{\lambda}_j = t. \tag{5.35}$$

Which corresponds to the convexity constraints (5.9) expressed with the new variables $\hat{\lambda} = t\lambda$. In any case, the targets can be computed as

$$\hat{x}_i = \sum_{j=1}^{n} \lambda_j^* x_{ij} = \frac{1}{t^*}\sum_{j=1}^{n} \hat{\lambda}_j^* x_{ij} = x_{i0} - \frac{1}{t^*}\hat{s}_i^{-*} \quad \forall i$$

$$\hat{y}_k = \sum_{j=1}^{n} \lambda_j^* y_{kj} = \frac{1}{t^*} \sum_{j=1}^{n} \hat{\lambda}_j^* y_{kj} = y_{k0} + \frac{1}{t^*} \hat{s}_k^{+*} \quad \forall k. \tag{5.36}$$

5.4 Handling Negative Data

Most DEA models assume that inputs and outputs are non-negative. However, this assumption fails in some cases, such as, for instance, when the variable of interest corresponds to a magnitude that may take negative values (net profit of a company) or when the variable is measured as a difference from one period to another (growth of the number of clients).

Pastor (1996) proposed a simple approach to this problem through some data transformation. The idea was to add a large positive value to the negative variables, in order to turn them into positive. However, this is only possible when the DEA model is translation invariant. Seiford and Zhu (2002) treated undesirable inputs and outputs multiplying them by minus 1 and then adding an arbitrary large number to let all negative data become positive, as proposed in Pastor (1996).

Silva Portela et al. (2004) defined the input and output ranges of possible improvements for unit 0 as

$$\begin{aligned} R_0^- &= X_0 - \min_{j=1,\ldots,n} X_j \\ R_0^+ &= \max_{j=1,\ldots,n} Y_j - Y_0. \end{aligned} \tag{5.37}$$

and proposed the range directional model (RDM)

$$\begin{aligned} & Max \quad \beta \\ & s.t. \\ & \sum_{j=1}^{n} \lambda_j X_j \leq X_0 - \beta R_0^- \\ & \sum_{j=1}^{n} \lambda_j Y_j \geq Y_0 + \beta R_0^+ \\ & \sum_{j=1}^{n} \lambda_j = 1 \\ & \lambda_j \geq 0 \quad \forall j. \end{aligned} \tag{5.38}$$

Silva Portela et al. (2004) showed that, under VRS assumption, the RDM is translation and units invariant and hence can be used when there are negative data.

Note that the direction taken by RDM is biased toward the variables with largest ranges of potential improvement.

Sharp et al. (2006) proposed a modified slacks-based measure (MSBM) that solves problems with negative inputs and outputs using the ranges R_0^- and R_0^+ (5.37):

$$\begin{aligned}
& Min \quad \left(\frac{1 - \sum_{i=1}^{m} \frac{u_i s_i^-}{R_{i0}^-}}{1 + \sum_{k=1}^{p} \frac{\upsilon_k s_k^+}{R_{k0}^+}} \right) \\
& s.t. \\
& \sum_{j=1}^{n} \lambda_j x_{ij} \leq x_{i0} - s_i^- \quad \forall i \\
& \sum_{j=1}^{n} \lambda_j y_{kj} \geq y_{k0} + s_k^+ \quad \forall k \\
& \sum_{j=1}^{n} \lambda_j = 1 \\
& \lambda_j \geq 0 \, \forall j \quad s_i^- \geq 0 \, \forall i \quad s_k^+ \geq 0 \, \forall k.
\end{aligned} \tag{5.39}$$

where, in the case that any input or output range is zero, the corresponding term would be eliminated from the objective function. Note that this model considers normalized input and output weights, i.e., $\sum_{i=1}^{m} u_i = 1$ and $\sum_{k=1}^{p} \upsilon_k = 1$. Carrying out the change of variables (5.33), this model can be linearized as

$$\begin{aligned}
& Min \quad t - \sum_{i=1}^{m} \frac{u_i s_i^-}{R_{i0}^-} \\
& s.t. \\
& t + \sum_{k=1}^{p} \frac{\upsilon_k s_k^+}{R_{k0}^+} = 1 \\
& \sum_{j=1}^{n} \hat{\lambda}_j x_{ij} = t x_{i0} - \hat{s}_i^- \quad \forall i \\
& \sum_{j=1}^{n} \hat{\lambda}_j y_{kj} = t y_{k0} + \hat{s}_k^+ \quad \forall k \\
& \hat{\lambda}_j \geq 0 \, \forall j \quad \hat{s}_i^- \geq 0 \, \forall i \quad \hat{s}_k^+ \geq 0 \, \forall k \quad t > 0.
\end{aligned} \tag{5.40}$$

5.5 Handling Multi-period Data

Most DEA models focus on the efficiency measure of DMUs observed in the same time period. In this section, consider the case of multi-period data. A first approach is to use the so-called inter-temporal approach and consider all the observations, whatever the time period, as belonging to the same PPS. In that case the efficiency of each DMU can be computed using a standard approach that benchmarks all the observations.

Another alternative is the so-called contemporaneous approach (Tulkens and Vanden Eeckaut 1995) in which it is assumed that for each time period $t = 1, 2, \ldots, L$ there exists a different PPS T^t which is inferred from the observations (X_j^t, Y_j^t) corresponding to that time period.

Another approach is the window analysis (WA) technique (Cooper et al. 2000), in which the analyst sets an arbitrary number of periods $\tau < L$, and solves a number of $w = L - \tau + 1$ different DEA models with $n \cdot \tau$ DMUs each. The first model considers a PPS inferred from the observations in periods $t = 1, 2, \ldots, \tau$ $\left(X_j^1, Y_j^1\right), \left(X_j^2, Y_j^2\right), \ldots, \left(X_j^\tau, Y_j^\tau\right)$, the second model considers a PPS inferred from the observations in periods $t = 2, 3, \ldots, \tau + 1$ $\left(X_j^2, Y_j^2\right), \left(X_j^3, Y_j^3\right), \ldots, \left(X_j^{\tau+1}, Y_j^{\tau+1}\right)$ and the last model considers a PPS inferred from the observations in periods $t = L - \tau + 1, L - \tau + 2, \ldots, L$ $\left(X_j^{L-\tau+1}, Y_j^{L-\tau+1}\right), \left(X_j^{L-\tau+2}, Y_j^{L-\tau+2}\right), \ldots, \left(X_j^L, Y_j^L\right)$. Once the corresponding DEA models are solved, WA provides $n \cdot \tau \cdot w$ measures of efficiency that can be displayed as a table. Table 5.2 shows an example for $n = 3$ DMUs (labeled A, B, and C), $L = 7$ time periods and a window size of $\tau = 3$ time periods. For example, ρ_{4B}^2 represents the efficiency computed by the second window model for the observation corresponding to DMU_B in period 4. Note that the first window model includes the three DMUs (labeled A, B, and C) in periods 1–3, the second model include the DMUs in periods 2–4, and so on. This analysis allows computing the average of the efficiencies obtained ($\bar{\rho}$) and the average of the variances ($\bar{\sigma}^2$) for each DMU.

A different analysis can be carried out with multi-period data and consist in measuring productivity change between periods and decomposing it into efficiency change and technical change components. This can be done using two alternative metrics: the Malmquist productivity index (Färe et al. 1994) and the Malmquist-Luenberger index (Chambers et al. 1996).

Under the assumption of CRS technology, the Malmquist input-based productivity index (MI) for DMU 0 is

$$MI_0(X^t, Y^t, X^{t+1}, Y^{t+1}) = \left(\frac{\theta_t^t}{\theta_{t+1}^t} \cdot \frac{\theta_t^{t+1}}{\theta_{t+1}^{t+1}}\right)^{\frac{1}{2}}. \tag{5.41}$$

where $\theta_t^t, \theta_{t+1}^t, \theta_t^{t+1}, \theta_{t+1}^{t+1}$ are the solutions of the following four DEA models:

Table 5.2 WA example

DMU	Efficiency scores							Mean	Variance
	1	2	3	4	5	6	7		
A	ρ_{1A}^1	ρ_{2A}^1	ρ_{3A}^1						
		ρ_{2A}^2	ρ_{3A}^2	ρ_{4A}^2				$\bar{\rho}_{\mathrm{A}}$	$\bar{\sigma}_{\mathrm{A}}^2$
			ρ_{3A}^3	ρ_{4A}^3	ρ_{5A}^3				
				ρ_{4A}^4	ρ_{5A}^4	ρ_{6A}^4			
					ρ_{5A}^5	ρ_{6A}^5	ρ_{7A}^5		
B	ρ_{1B}^1	ρ_{2B}^1	ρ_{3B}^1						
		ρ_{2B}^2	ρ_{3B}^2	ρ_{4B}^2				$\bar{\rho}_{\mathrm{B}}$	$\bar{\sigma}_{\mathrm{B}}^2$
			ρ_{3B}^3	ρ_{4B}^3	ρ_{5B}^3				
				ρ_{4B}^4	ρ_{5B}^4	ρ_{6B}^4			
					ρ_{5B}^5	ρ_{6B}^5	ρ_{7B}^5		
C	ρ_{1C}^1	ρ_{2C}^1	ρ_{3C}^1						
		ρ_{2C}^2	ρ_{3C}^2	ρ_{4C}^2				$\bar{\rho}_{\mathrm{C}}$	$\bar{\sigma}_{\mathrm{C}}^2$
			ρ_{3C}^3	ρ_{4C}^3	ρ_{5C}^3				
				ρ_{4C}^4	ρ_{5C}^4	ρ_{6C}^4			
					ρ_{5C}^5	ρ_{6C}^5	ρ_{7C}^5		

$$\begin{aligned}
&Min \quad \theta_t^t \\
&s.t. \\
&\sum_{j=1}^{n} \lambda_j X_j^t \leq \theta_t^t X_0^t \\
&\sum_{j=1}^{n} \lambda_j Y_j^t \geq Y_0^t \\
&\lambda_j \geq 0 \, \forall j \quad \theta_t^t \; free.
\end{aligned} \tag{5.42}$$

$$\begin{aligned}
&Min \quad \theta_{t+1}^t \\
&s.t. \\
&\sum_{j=1}^{n} \lambda_j X_j^t \leq \theta_{t+1}^t X_0^{t+1} \\
&\sum_{j=1}^{n} \lambda_j Y_j^t \geq Y_0^{t+1} \\
&\lambda_j \geq 0 \, \forall j \quad \theta_{t+1}^t \; free.
\end{aligned} \tag{5.43}$$

$$\begin{aligned}
&Min \quad \theta_t^{t+1} \\
&s.t. \\
&\sum_{j=1}^{n} \lambda_j X_j^{t+1} \leq \theta_t^{t+1} X_0^t \\
&\sum_{j=1}^{n} \lambda_j Y_j^{t+1} \geq Y_0^t \\
&\lambda_j \geq 0 \, \forall j \quad \theta_t^{t+1} \; free.
\end{aligned} \tag{5.44}$$

$$\begin{aligned}
&Min \quad \theta_{t+1}^{t+1} \\
&s.t. \\
&\sum_{j=1}^{n} \lambda_j X_j^{t+1} \leq \theta_{t+1}^{t+1} X_0^{t+1} \\
&\sum_{j=1}^{n} \lambda_j Y_j^{t+1} \geq Y_0^{t+1} \\
&\lambda_j \geq 0 \, \forall j \quad \theta_{t+1}^{t+1} \; free.
\end{aligned} \tag{5.45}$$

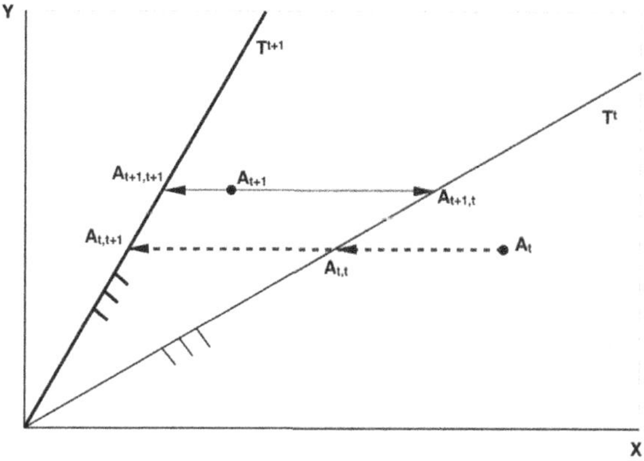

Fig. 5.8 Malmquist input-based productivity index

Thus, the notation θ_a^b (where $a, b \in \{t, t+1\}$) means computing the efficiency of DMU 0 in period a using the PPS corresponding to the observations in period b. The expression (5.41) can be decomposed into the following two terms:

$$MI_0(X^t, Y^t, X^{t+1}, Y^{t+1}) = \frac{\theta_t^t}{\theta_{t+1}^{t+1}} \cdot \left(\frac{\theta_{t+1}^{t+1}}{\theta_{t+1}^t} \cdot \frac{\theta_t^{t+1}}{\theta_t^t} \right)^{\frac{1}{2}} = ECI_0 \cdot TCI_0. \quad (5.46)$$

where ECI_0 captures the changes in efficiency of DMU 0 between periods t and $t + 1$, (efficiency improvement when $ECI_0 < 1$ and worsening if $ECI_0 > 1$); and TCI_0 the technical change measured by the shift in the frontier between t and $t + 1$ (technical progress if $TCI_0 < 1$ and regress if $TCI_0 > 1$).

Figure 5.8 shows a single-input–single-output CRS case in which the observed operation point of DMU A in periods t and t + 1 (A_t and A_{t+1}, respectively) are projected onto the efficient frontier of the technology of each period (T^t and T^{t+1}).

The projections obtained by models (5.42), (5.43), (5.44), and (5.45) are respectively $A_{t,t}$; $A_{t+1,t}$; $A_{t,t+1}$; and $A_{t+1,t+1}$, and the values of the objective functions can be used to obtain the MI_A substituting in formula (5.41).

The Malmquist output-based productivity index is analogous:

$$MO_0(X^t, Y^t, X^{t+1}, Y^{t+1}) = \left(\frac{\gamma_t^t}{\gamma_{t+1}^t} \cdot \frac{\gamma_t^{t+1}}{\gamma_{t+1}^{t+1}} \right)^{\frac{1}{2}}. \quad (5.47)$$

where, as before, γ_a^b means computing the (inverse) efficiency of DMU 0 in period a using the PPS corresponding to the observations in period b and are computed using the following DEA models:

$$\begin{aligned} &Max \quad \gamma_t^t \\ &s.t. \end{aligned}$$

$$
\begin{aligned}
&\sum_{j=1}^{n} \lambda_j X_j^t \le X_0^t \\
&\sum_{j=1}^{n} \lambda_j Y_j^t \ge \gamma_t^t Y_0^t \\
&\lambda_j \ge 0 \, \forall j \quad \gamma_t^t \ free.
\end{aligned} \tag{5.48}
$$

$$
\begin{aligned}
&Min \quad \gamma_{t+1}^t \\
&s.t. \\
&\sum_{j=1}^{n} \lambda_j X_j^t \le X_0^{t+1} \\
&\sum_{j=1}^{n} \lambda_j Y_j^t \ge \gamma_{t+1}^t Y_0^{t+1} \\
&\lambda_j \ge 0 \, \forall j \quad \gamma_{t+1}^t \ free.
\end{aligned} \tag{5.49}
$$

$$
\begin{aligned}
&Min \quad \gamma_t^{t+1} \\
&s.t. \\
&\sum_{j=1}^{n} \lambda_j X_j^{t+1} \le X_0^t \\
&\sum_{j=1}^{n} \lambda_j Y_j^{t+1} \ge \gamma_t^{t+1} Y_0^t \\
&\lambda_j \ge 0 \, \forall j \quad \gamma_t^{t+1} \ free.
\end{aligned} \tag{5.50}
$$

$$
\begin{aligned}
&Min \quad \gamma_{t+1}^{t+1} \\
&s.t. \\
&\sum_{j=1}^{n} \lambda_j X_j^{t+1} \le X_0^{t+1} \\
&\sum_{j=1}^{n} \lambda_j Y_j^{t+1} \ge \gamma_{t+1}^{t+1} Y_0^{t+1} \\
&\lambda_j \ge 0 \, \forall j \quad \gamma_{t+1}^{t+1} \ free.
\end{aligned} \tag{5.51}
$$

Finally, MO_0 can be decomposed into efficiency change and technical change components ECO_0 and TCO_0, respectively, as follows:

$$MO_0(X^t, Y^t, X^{t+1}, Y^{t+1}) = \frac{\gamma_t^t}{\gamma_{t+1}^{t+1}} \cdot \left(\frac{\gamma_{t+1}^{t+1}}{\gamma_{t+1}^t} \cdot \frac{\gamma_t^{t+1}}{\gamma_t^t} \right)^{\frac{1}{2}} = ECO_0 \cdot TCO_0. \quad (5.52)$$

The interpretation of ECO_0 and TCO_0 are analogous to those made in the input measure case, taken into account that in case of efficiency improvement and technical progress correspond to $ECO_0 > 1$ and $TCO_0 > 1$, with values lower than one implying efficiency worsening and technical regress.

The Malmquist-Luenberger index measures the change of productivity between two periods using DDF instead of a radial oriented model (as in the Malmquist productivity index). Similar to models (5.42–5.45), the following model allows computing the DDF of DMU 0 in period a using the PPS in period b (with $a, b \in \{t, t+1\}$):

$$\begin{aligned}
&D_b\left(X_0^a, Y_0^a, g^x, g^y\right) = Max \quad \beta \\
&s.t. \\
&\sum_{j=1}^{n} \lambda_j X_j^b = X_0^a - \beta g^x \\
&\sum_{j=1}^{n} \lambda_j Y_j^b = Y_0^a + \beta g^y \\
&\lambda_j \geq 0 \quad \forall j.
\end{aligned} \quad (5.53)$$

Chambers et al. (1996) defined the Malmquist-Luenberger Productivity Index as the arithmetic mean between the productivity change between t and $t + 1$ measured with respect to the PPS of periods t and $t + 1$:

$$ML_0\left(Z^t, Z^{t+1}\right) = \frac{1}{2} \cdot \left[D_{t+1}\left(Z^t, g\right) - D_{t+1}\left(Z^{t+1}, g\right) + D_t\left(Z^t, g\right) - D_t\left(Z^{t+1}, g\right)\right]. \quad (5.54)$$

where $Z^t = \left(X^t, Y^t\right)$, $Z^{t+1} = \left(X^{t+1}, Y^{t+1}\right)$, and $g = (g^x, g^y)$. Note that ML_0 takes a positive value in case the productivity of DMU 0 grows and a negative value when the productivity of DMU 0 declines.

Figure 5.9 shows the observed operation point of DMU A in periods t (A_t) and t + 1 (A_{t+1}), and their projections onto the efficient frontiers of each period using the directional vector g, represented by its components g^x and g^y. Note that $D_t(A_t, g)$; $D_{t+1}(A_t, g)$; $D_t(A_{t+1}, g)$; and $D_{t+1}(A_{t+1}, g)$ represent the optimal value of the objective function of model (5.53) when parameters a and b take each of the possible combinations of t and t + 1. It is worthy to mention that, in Fig. 5.9, $D_t(A_{t+1}, g) < 0$, as the projection of A_t onto the efficient frontier in t + 1 corresponds to increasing input X and reducing output Y. However, using (5.54), ML_A takes a positive value, since $D_{t+1}(A_t, g) > D_{t+1}(A_{t+1}, g)$ and $D_t(A_{t+1}, g) < 0$, as mentioned above.

Similar to the Malmquist index, ML_0 can be additively decomposed as follows:

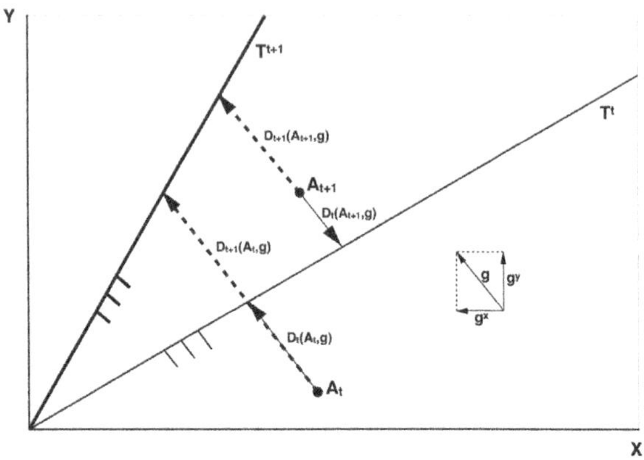

Fig. 5.9 Malmquist-Luenberger productivity index

$$ML_0(Z^t, Z^{t+1}) = \left[D_t(Z^t, g) - D_{t+1}(Z^{t+1}, g)\right] + \frac{1}{2} \cdot \left[D_{t+1}(Z^{t+1}, g) - D_t(Z^{t+1}, g) + D_{t+1}(Z^t, g) - D_t(Z^t, g)\right]. \tag{5.55}$$

The first term represents the efficiency change of DMU 0 between t and $t+1$ while the second term measures the technical change between those two time periods.

5.6 Handling Ratio Variables

There are many examples reported in the literature in which the data include ratio measures (growth rates, proportions, averages, …). When this happens, the variables can be divided into volume and ratio inputs and outputs: $(X, Y) = (X^V, X^R, Y^V, Y^R)$. The standard DEA models cannot be used directly to handle ratio variables, because ratio measures are inconsistent with the convexity assumption usually considered when inferring the PPS (Olesen and Petersen, 2006). Therefore, except for T^{FDH} (which does not assume convexity), DEA models need to be modified when ratio variables are considered.

For ratio variables, Emrouznejad and Amin (2009) proposed to treat the numerator and the denominator as volume input and volume output separately. Therefore, for input ratios, the numerator will be treated as an input, and the denominator as an output, and the opposite happens for output ratios. An important limitation of this method is the need for knowing the values of both the numerator and denominator to apply it. Unfortunately, in many cases, this information is unavailable.

Olesen et al. (2015) has proposed a new methodology that allows the inclusion of ratio measures as inputs and outputs defining the T^{R-VRS} and the T^{R-CRS} technologies. To do so, it is necessary to define upper bounds for the ratio inputs $X^R \leq \bar{X}^R$ and for the ratio outputs $Y^R \leq \bar{Y}^R$. The T^{R-VRS} technology satisfies axiom A1 (presented in

Sect. 5.2), and modifies the free disposability axiom (A2) and the convexity axiom (A3) into (A′2) and (A′3):

A′2. Free disposability of inputs and outputs:

$$(X, Y) \in T, \hat{X} \geq X, \hat{Y} \leq Y;^{R} \hat{X}^{R} \leq \bar{X}^{R}, \hat{Y}^{R} \leq \bar{Y}^{R} \Rightarrow \left(\hat{X}, \hat{Y}\right) \in T.$$

A′3. Selective convexity:

$$(X, Y), \left(\hat{X}, \hat{Y}\right) \in T, X^{R} = \hat{X}^{R} \Rightarrow \left(\lambda X + (1-\lambda)\hat{X}, \lambda Y + (1-\lambda)\hat{Y}\right) \in T, \forall\lambda \in [0, 1].$$

Hence, the Ratio VRS (R-VRS) technology corresponds to assuming A1 + A′2 + A′3

$$T^{R-VRS} = \left\{ \begin{array}{l} (X, Y) : \sum_{j=1}^{n} \lambda_j X_j^V \leq X^V, \sum_{j=1}^{n} \lambda_j Y_j^V \geq Y^V, \lambda_j \left(X_j^R - X^R\right) \leq 0 \; \forall j, \\ \lambda_j \left(Y_j^R - Y^R\right) \geq 0 \; \forall j, \sum_{j=1}^{n} \lambda_j = 1, \lambda_j \geq 0 \; \forall j \end{array} \right\}. \tag{5.56}$$

In Olesen et al. (2017) the authors formulated DEA models with T^{R-VRS} technology for some orientations. For instance, the radial output-orientated DEA model would be

$$\begin{aligned} & Max \quad \gamma \\ & \sum_{j=1}^{n} \lambda_j X_j^V \leq X_0^V \\ & \sum_{j=1}^{n} \lambda_j Y_j^V \geq \gamma Y_0^V \\ & \lambda_j \left(X_j^R - X_0^R\right) \leq \mathbf{0} \; \forall j \\ & \lambda_j \left(Y_j^R - \gamma Y_0^R\right) \geq \mathbf{0} \; \forall j \\ & \sum_{j=1}^{n} \lambda_j = 1 \\ & \lambda_j \geq 0 \forall j \quad \gamma \; free. \end{aligned} \tag{5.57}$$

This is a non-linear optimization model that can be transformed into a mixed-integer linear program (MILP), replacing the constraints $\lambda_j \left(Y_j^R - \gamma Y_0^R\right) \geq \mathbf{0}$ with

$$\lambda_j \leq \delta_j \quad \forall j$$

$$\gamma Y_0^R - Y_j^R \leq \mathbf{M}(1-\delta_j) \quad \forall j$$
$$\delta_j \in \{0,1\} \; \forall j. \tag{5.58}$$

where $\mathbf{M}$ is a vector of the same dimension as Y^R, with large positive components.

For the definition of T^{R-CRS}, it is necessary to introduce some additional concepts and modify the two axioms relating to the scalability assumptions (A4 and A5 in Sect. 5.2). From now on, we will consider the following types of ratio variables: $(X^F, X^P, X^D, X^U, Y^F, Y^P, Y^D)$, where:

(X^F, Y^F), fixed ratio measures: when the volume inputs and outputs are scaled by $\lambda \geq 0$ their values remain constant.
(X^P, Y^P), proportional ratio measures: when the volume inputs and outputs are scaled by $\lambda \geq 0$ their values change in the same proportion.
X^D, downward proportional ratio inputs: when the volume inputs and outputs are scaled by λ, X^D takes the following values:

$$X^D = \begin{cases} \lambda X_0^D + (1-\lambda)A & if\, 0 \leq \lambda < 1 \\ X_0^D & if\, \lambda \geq 1. \end{cases}$$

where $A \geq X_0^D$ is an assumed value.

Y^D, downward proportional ratio outputs: when the volume inputs and outputs are scaled by λ, Y^D takes the following values:

$$Y^D = \begin{cases} \lambda Y_0^D + (1-\lambda)B & if\, 0 \leq \lambda < 1 \\ Y_0^D & if\, \lambda \geq 1. \end{cases}$$

where $B \leq Y_0^D$ is an assumed value.

X^U, upward proportional ratio inputs: when the volume inputs and outputs are scaled by λ, X^U is constant for $0 \leq \lambda < 1$, and increases proportionally when $\lambda \geq 1$.

As in the R-VRS technology, upper bounds for every type of ratio measures are defined as $(\bar{X}^R, \bar{Y}^R) = (\bar{X}^F, \bar{X}^P, \bar{X}^D, \bar{X}^U, \bar{Y}^F, \bar{Y}^P, \bar{Y}^D)$.

Once the different types of ratio variables have been defined, the modified scalability axioms, labeled A′4 and A′5, can be stated as

A′4. Selective proportional contraction:

$$(X,Y) = (X^V, X^F, X^P, X^D, X^U, Y^V, Y^F, Y^P, Y^D) \in T$$
$$\Rightarrow \left(\alpha X^V, X^F, \alpha X^P, \alpha X^D + (1-\alpha)A, X^U, \alpha Y^V, Y^F, \alpha Y^P, \alpha Y^D + (1-\alpha)B\right) \in T,\; 0 \leq \alpha \leq 1.$$

A′5. Selective proportional expansion:

$$(X,Y) = (X^V, X^F, X^P, X^D, X^U, Y^V, Y^F, Y^P, Y^D) \in T$$
$$\Rightarrow \forall \beta \geq 1 : \beta X^U \leq \bar{X}^U, \beta X^P \leq \bar{X}^P$$

$$\left(\beta X^V, X^F, \beta X^P, \beta X^D, X^U, \beta Y^V, Y^F, min\left\{\beta Y^P, \bar{Y}^P\right\}, \beta Y^D\right) \in T.$$

The Ratio CRS (R-CRS) technology corresponds to assuming A1 + A′2 + A′3 + A′4 + A′5:

$$T^{R-CRS} = \left\{ \begin{array}{l} (X,Y) : \sum_{j=1}^{n} \lambda_j \alpha_j \beta_j X_j^V \leq X^V, \sum_{j=1}^{n} \lambda_j \alpha_j \beta_j Y_j^V \geq Y^V, \\ if \lambda_j > 0, \\ \left(X_j^F - X^F\right) \leq 0, \\ \left(Y_j^F - Y^F\right) \geq 0, \\ \alpha_j X_j^D + \left(1 - \alpha_j\right) A \leq X^D, \\ \alpha_j Y_j^D + \left(1 - \alpha_j\right) B \geq Y^D, \\ \beta_j X_j^U \leq X^U, \\ \alpha_j \beta_j X_j^P \leq X^P, \\ \alpha_j \beta_j Y_j^P \geq Y^P, \\ \sum_{j=1}^{n} \lambda_j = 1, \\ \alpha_j \leq 1, \beta_j \geq 1, \forall j \\ \lambda_j, \alpha_j \geq 0, \forall j \end{array} \right\}. \tag{5.59}$$

The DEA models that assume the R-CRS technology are generally non-linear but they can be written in an equivalent form with computational advantages by means of the following change of variables (see Olesen et al. 2015):

$$\begin{aligned} \mu_j &= \lambda_j \alpha_j \left(\beta_j - 1\right) \quad \forall j \\ v_j &= \lambda_j \left(1 - \alpha_j\right) \quad \forall j. \end{aligned} \tag{5.60}$$

5.7 Handling Undesirable Outputs

It happens often, especially in eco-efficiency and environmental efficiency assessment applications, that some of the outputs of the input–output transformation process are undesirable. The most typical example is pollution (e.g., SO_2, NO_x) generated by coal-fired power plants or greenhouse gases emissions produced by a country or economic sector. Other examples of undesirable outputs are noise and delays in the case of airports, accidents in the case of bus transport, or non-performing loans in the case of banks.

When undesirable outputs are considered then the input–output vector of each DMU j is denoted $\left(X_j, Y_j, Z_j\right)$, where $Z_j = \left(z_{1j}, \ldots, z_{qj}\right) \in \mathbb{R}_+^q$ represent the undesirable outputs. Undesirable outputs differ from desirable outputs in that the

DMU does not want to increase them but to reduce them, much like in the case of inputs. This similarity with inputs moved some researchers to model these variables as if they were inputs. However, in DEA inputs are generally freely disposable, something which does not occur in the case of undesirable outputs. The recommended way of modeling undesirable outputs is defining an appropriate so-called environmental technology that handles undesirable outputs specifically.

Thus, while inputs and outputs are assumed to be freely disposable (Axiom A2 in Sect. 5.2) for the undesirable outputs the following two assumptions can be made

A2′. Joint weak disposability of undesirable and desirable outputs:

$$(X, Y, Z) \in T \Rightarrow (X, \lambda Y, \lambda Z) \in T, \forall\lambda \in [0, 1].$$

A2″. Null-jointness of desirable and undesirable outputs:

$$(X, Y, 0) \in T \Rightarrow Y = \mathbf{0}_s.$$

The interpretation of axiom A2′ is that it is possible to uniformly reduce (i.e., scale down) the undesirable outputs provided that the desirable outputs are also simultaneously reduced. This is different from the free disposability of the desirable outputs, which can be reduced without any limitation. As regards axiom A2″ it simply means that the only way to produce no undesirable outputs is by producing zero desirable outputs.

Assuming A1 + A2 + A2′ + A2″ + A3 + A4, and applying the Minimum Extrapolation Principle, the following CRS environmental technology results (Färe et al. 1994)

$$T^{EnvCRS} = \left\{ (X, Y, Z) : \sum_{j=1}^{n} \lambda_j X_j \leq X, \sum_{j=1}^{n} \lambda_j Y_j \geq Y, \sum_{j=1}^{n} \lambda_j Z_j = Z, \lambda_j \geq 0 \forall j \right\}. \tag{5.61}$$

Once the PPS is defined it is easy to formulate DEA models with this DEA technology. Thus, for example, the corresponding DDF model for projecting DMU 0 using direction vector $g = (g^x, g^y, g^z)$ would be

$$\begin{aligned}
&Max \quad \beta \\
&s.t. \\
&\sum_{j=1}^{n} \lambda_j X_j \leq X_0 - \beta \cdot g^x \\
&\sum_{j=1}^{n} \lambda_j Y_j \geq Y_0 + \beta \cdot g^y
\end{aligned}$$

$$\sum_{j=1}^{n}\lambda_j Z_j = Z_0 - \beta \cdot g^z$$
$$\lambda_j \geq 0\ \forall j. \tag{5.62}$$

Dropping the scalability axiom A4 and applying the Minimum Extrapolation Principle assuming just A1 + A2 + A2′ + A2″ + A3 leads to the VRS environmental technology (Kuosmanen 2005)

$$T^{EnvVRS} = \left\{ \begin{array}{c} (X, Y, Z) : \sum_{j=1}^{n} \left(\lambda_j + \mu_j\right) \cdot X_j \leq X,\ \sum_{j=1}^{n} \lambda_j Y_j \geq Y,\ \sum_{j=1}^{n} \lambda_j Z_j = Z, \\ \lambda_j, \mu_j \geq 0\ \forall j,\ \sum_{j=1}^{n} \left(\lambda_j + \mu_j\right) = 1 \end{array} \right\}. \tag{5.63}$$

Formulating DEA models with this technology is done as before. Thus, for example, the corresponding DDF model for projecting DMU 0 using direction vector $g = (g^x, g^y, g^z)$ would be

$$\begin{aligned} &Max\ \ \beta \\ &s.t. \\ &\sum_{j=1}^{n} \left(\lambda_j + \mu_j\right) \cdot X_j \leq X_0 - \beta \cdot g^x \\ &\sum_{j=1}^{n} \lambda_j Y_j \geq Y_0 + \beta \cdot g^y \\ &\sum_{j=1}^{n} \lambda_j Z_j = Z_0 - \beta \cdot g^z \\ &\sum_{j=1}^{n} \left(\lambda_j + \mu_j\right) = 1 \\ &\lambda_j, \mu_j \geq 0\ \forall j. \end{aligned} \tag{5.64}$$

5.8 Handling Non-discretionary Variables

It is often the case that some of the input and output variables are not-discretionary (ND). For example, the population and GDP inputs considered when assessing the performance of nations at the Olympics are non-discretionary. The key feature of a non-discretionary input is that, unlike a discretionary one, the DMU cannot reduce it because it is impossible, impractical, or just unreasonable. Another example is the runway length, terminal area, boarding gates and similar inputs considered when assessing the efficiency of airports. It does not make sense for an airport to try to achieve efficiency by reducing those inputs. Similarly, a non-discretionary output cannot be increased by the DMU. Sometimes this corresponds to exogenous variables like, for example, the number of competitors in the vicinity of a supermarket or a retail store. That type of variables is considered a non-discretionary output beyond the control of management. Similarly, if the demand faced by a DMU is inelastic then the corresponding sales output can be considered non-discretionary.

Camanho et al. (2009) distinguish between ND factors that are internal to the production process but cannot be controlled by the DMU and ND factors that characterize the external conditions under which the DMU operates. A classical approach for continuous internal ND variables is the one proposed in Banker and Morey (1986). Thus, let I^{ND} and O^{ND} be the sets of non-discretionary inputs and outputs, respectively, and I^D and O^D, the corresponding discretionary variables. For the VRS case, instead of (5.6), the constraints corresponding to the ND variables would be

$$\hat{x}_i = x_{i0} \geq \sum_{j=1}^{n} \lambda_j x_{ij} \quad \forall i \in I^{ND} \quad \hat{y}_k = y_{k0} \leq \sum_{j=1}^{n} \lambda_j y_{kj} \quad \forall k \in O^{ND}. \tag{5.65}$$

These constraints imply that (1) for the non-discretionary inputs and outputs, the target value is equal to the observed value and should be bounded by a linear combination of the observations, and (2) the operating point resulting from the linear combination of the observations should dominate the DMU also in those dimensions that are non-discretionary and cannot be improved. The rationale is that the feasible operating point computed by the linear combination of the observations produces more of the discretionary outputs and consumes less of the discretionary inputs with a value of each non-discretionary input that is no larger than that of DMU 0 and a value of each non-discretionary output that is no lower than that of DMU 0.

In the CRS case, if the ND variables are scale dependent (i.e., volume measures) then for the ND variables (5.65) can be used. However, in case the ND variables cannot be scaled up or down then the corresponding constraints would be

$$\sum_{j=1}^{n} \lambda_j x_{ij} \leq \sum_{j=1}^{n} \lambda_j x_{i0} \quad \forall i \in I^{ND} \quad \sum_{j=1}^{n} \lambda_j y_{kj} \geq \sum_{j=1}^{n} \lambda_j y_{k0} \quad \forall k \in O^{ND}. \tag{5.66}$$

In all cases, for the non-discretionary variables there are neither slack variables, nor radial nor non-radial multiplicative factors. This is so because, as mentioned above, the DEA model will not try to reduce those inputs nor increase those outputs. Thus, for example, for the input-oriented radial model (CCR-I/BCC-I) the input reduction constraints would only affect the discretionary inputs

$$\hat{x}_i = \theta x_{i0} - s_i^- \quad \forall i \in I^D. \tag{5.67}$$

and the objective function of the corresponding slacks-maximizing phase II would only involve the slacks of the discretionary inputs and outputs

$$Max \sum_{i \in I^D} s_i^- + \sum_{k \in O^D} s_k^+. \tag{5.68}$$

For external ND variables there is the approach proposed in Ruggiero (1998), which in the case of a single non-discretionary variable z (whose value is supposed to be such that a larger value implies a more favorable environment), VRS, radial metric and an input orientation can be formulated as

$$\begin{aligned}
&Min \;\; \theta \\
&s.t. \\
&\sum_{j=1}^{n} \lambda_j x_{ij} \leq \theta x_{i0} \quad \forall i \\
&\sum_{j=1}^{n} \lambda_j y_{kj} \geq y_{k0} \quad \forall k \\
&\sum_{j=1}^{n} \lambda_j = 1 \\
&\lambda_j \geq 0 \, if \; z_j > z_0 \quad j = 1, 2, \ldots n \\
&\lambda_j \geq 0 \forall j \quad \theta \; free.
\end{aligned} \tag{5.69}$$

The idea is that the DMUs that operate in a more favorable environment than DMU 0 cannot be used to construct the benchmark. In other words, the exogenous variable induces a conditional PPS (and a conditional efficient frontier) that for each level of the exogenous variable excludes those observations with a larger value of the exogenous variable.

In the case of multiple exogenous variables, Ruggiero (1998) suggest a three-stage approach. In the first stage the technical efficiency of the different DMUs θ_j^{stageI} are computed ignoring the exogenous variables. In the second stage these efficiency scores are regressed on the q exogenous variables, i.e.,

$$\theta^{stageI} = \alpha + \beta_1 z_1 + \beta_2 z_2 + \ldots + \beta_q z_q + \varepsilon. \tag{5.70}$$

Finally, in stage III for each DMU j an aggregated exogenous variable is computed with the regression coefficients of (5.70), i.e.,

$$z_j = \beta_1 z_{1j} + \beta_2 z_{2j} + \ldots + \beta_q z_{qj}. \tag{5.71}$$

and model (5.69) is solved for each DMU 0.

5.9 Handling Integer Variables

It happens sometimes that one or more variables are restricted to take integer values, for example, because it is measured in units that cannot be fractioned. An example is the number of machines in a manufacturing cell or the number of Automated Guided Vehicles (AGVs) in Flexible Manufacturing Systems (FMS). Sometimes although a variable is an integer relaxing it and considering as continuous is acceptable. That happens, for example, to a variable such as number of employees. The variable can be modeled as continuous and the real-valued targets can be interpreted in Full Time Equivalent (FTE) terms. Another example is the number of passengers that use an airport. Since in that case, the figures are generally large, they can be measured in thousand persons, and thus the variable can have fractions and be treated as continuous. Alternatively, when the figures are large, treating the integer variable as continuous and rounding the target value up or down to the nearest integer does not introduce a significant error.

There are occasions, however, in which the integer character of a variable cannot be ignored and has to be taken explicitly into account in the model. This leads to integer DEA models (e.g., Lozano and Villa 2006; Kuosmanen and Kazemi Matin 2009). We will follow the same path as before, i.e., enunciate the axioms that fit this situation, use the Minimum Extrapolation Principle to derive the PPS and then we are ready to formulate the corresponding DEA models.

The axiomatic derivation of the integer DEA technologies for different returns to scale axioms is due to Kazemi Matin and Kuosmanen (2009). Let I^{ND} and O^{ND} be the sets of integer inputs and outputs, respectively. For those inputs, instead of the free disposability and convexity axioms (A2 and A3 in Sect. 5.2), these other two axioms can be used

B2. Natural disposability of integer inputs and outputs:

$$\begin{aligned}&(X, Y) \in T,\ \Delta X \geq 0_m,\ \Delta x_i \in \mathbb{Z}_+\ \forall i \in I^{\text{int}},\\&Y \geq \Delta Y \geq 0_s,\ \Delta y_k \in \mathbb{Z}_+\ \forall k \in O^{\text{int}} \Rightarrow (X + \Delta X, Y - \Delta Y) \in T.\end{aligned}$$

B3. Natural convexity of integer inputs and outputs:

$$\begin{aligned}&(X, Y), \left(\hat{X}, \hat{Y}\right) \in T, \lambda \in [0, 1],\\&\lambda x_i + (1-\lambda)\hat{x}_i \in \mathbb{Z}_+\ \forall i \in I^{\text{int}}\\&\lambda y_k + (1-\lambda)\hat{y}_k \in \mathbb{Z}_+\ \forall k \in O^{\text{int}} \Rightarrow \left(\lambda X + (1-\lambda)\hat{X}, \lambda Y + (1-\lambda)\hat{Y}\right) \in T.\end{aligned}$$

The interpretation of these two axioms is that free disposability and convexity apply but in a restricted way that is compatible with the integrality of the integer variables. Therefore, assuming A1 + B2 + B3, and applying the Minimum Extrapolation Principle, the following VRS integer DEA technology results

$$T^{IntVRS} = \left\{ \begin{array}{c} (X,Y): \sum_{j=1}^{n} \lambda_j X_j \leq X, \sum_{j=1}^{n} \lambda_j Y_j \geq Y, \sum_{j=1}^{n} \lambda_j = 1, \lambda_j \geq 0\, \forall j, \\ x_i \in \mathbb{Z}_+ \, \forall i \in I^{\text{int}}, y_k \in \mathbb{Z}_+ \, \forall k \in O^{\text{int}} \end{array} \right\}. \tag{5.72}$$

In order to derive the CRS integer DEA technology, the scalability axiom (A4 in Sect. 5.2) should be replaced by its natural analogue:

B4. Natural scalability of integer inputs and outputs:

$$(X,Y) \in T,\ \exists \lambda \geq 0 : \lambda x_i \in \mathbb{Z}_+ \forall i \in I^{\text{int}}, \lambda y_k \in \mathbb{Z}_+ \forall k \in O^{\text{int}} \Rightarrow (\lambda X, \lambda Y) \in T.$$

Assuming A1 + B2 + B3 + B4, and applying the Minimum Extrapolation Principle, the following CRS integer PPS results

$$T^{IntCRS} = \left\{ \begin{array}{c} (X,Y): \sum_{j=1}^{n} \lambda_j X_j \leq X, \sum_{j=1}^{n} \lambda_j Y_j \geq Y, \lambda_j \geq 0\, \forall j, \\ x_i \in \mathbb{Z}_+ \, \forall i \in I^{\text{int}}, y_k \in \mathbb{Z}_+ \, \forall k \in O^{\text{int}} \end{array} \right\}. \tag{5.73}$$

The derivation of the NIRS and NDRS integer DEA technologies involves splitting axiom B4 into

B5. Natural divisibility of integer inputs and outputs:

$$(X,Y) \in T,\ \exists \lambda \in [0,1] : \lambda x_i \in \mathbb{Z}_+ \forall i \in I^{\text{int}}, \lambda y_k \in \mathbb{Z}_+ \forall k \in O^{\text{int}} \Rightarrow (\lambda X, \lambda Y) \in T.$$

B6. Natural augmentability of integer inputs and outputs:

$$(X,Y) \in T,\ \exists \lambda \geq 1 : \lambda x_i \in \mathbb{Z}_+ \forall i \in I^{\text{int}}, \lambda y_k \in \mathbb{Z}_+ \forall k \in O^{\text{int}} \Rightarrow (\lambda X, \lambda Y) \in T.$$

These two axioms correspond to the standard downward and upward scalability (axioms A5 and A6 in Sect. 5.2) but in a restricted way that is compatible with the integrality of the integer variables. Therefore, assuming A1 + B2 + B3 + B5 leads to the corresponding NIRS integer PPS

$$T^{IntNIRS} = \left\{ \begin{array}{c} (X,Y): \sum_{j=1}^{n} \lambda_j X_j \leq X, \sum_{j=1}^{n} \lambda_j Y_j \geq Y, \sum_{j=1}^{n} \lambda_j \leq 1, \lambda_j \geq 0\, \forall j, \\ x_i \in \mathbb{Z}_+ \, \forall i \in I^{\text{int}}, y_k \in \mathbb{Z}_+ \, \forall k \in O^{\text{int}} \end{array} \right\}. \tag{5.74}$$

while assuming A1 + B2 + B3 + B6 leads to the NDRS integer PPS

$$T^{IntNDRS} = \left\{ \begin{array}{c} (X,Y): \sum_{j=1}^{n} \lambda_j X_j \leq X, \sum_{j=1}^{n} \lambda_j Y_j \geq Y, \sum_{j=1}^{n} \lambda_j \geq 1, \lambda_j \geq 0\, \forall j, \\ x_i \in \mathbb{Z}_+ \, \forall i \in I^{\text{int}}, y_k \in \mathbb{Z}_+ \, \forall k \in O^{\text{int}} \end{array} \right\}. \tag{5.75}$$

In any case, formulating DEA models with any of these integer DEA technologies can be done as always: looking for a feasible (i.e., belonging to the corresponding PPS) that maximizes the input and output improvements. This can be done using an oriented or non-oriented way and using a radial or a non-radial metric. Thus, for example, the radial input-oriented CRS DEA model would be

$$\begin{aligned}
&Min \quad \theta - \varepsilon \cdot \left(\sum_{i=1}^{n} s_i^- + \sum_{k=1}^{s} s_k^+ \right) \\
&s.t. \\
&\sum_{j=1}^{n} \lambda_j X_j \leq \hat{X} = \theta X_0 - S^- \\
&\sum_{j=1}^{n} \lambda_j Y_j \geq \hat{Y} = Y_0 + S^+ \\
&\hat{x}_i \in \mathbb{Z}_+ \ \forall i \in I^{\text{int}},\ \hat{y}_k \in \mathbb{Z}_+ \ \forall k \in O^{\text{int}} \\
&\lambda_j \geq 0 \ \forall j.
\end{aligned} \tag{5.76}$$

5.10 Other DEA Topics and DEA Limitations

This chapter has presented an introduction to DEA and although it has tried to give a broad overview of the methodology there are many DEA topics that have not been covered. Thus, all the DEA models presented in this chapter are of envelopment type and have corresponding dual multiplier formulations. Multiplier DEA models are widely used, e.g., cross-efficiency, common sets of weights, etc. There are also Network DEA models which do not model the input–output transformation as a single process but as multiple stages or, more generally, as a network of processes. Different centralized DEA approaches have also been proposed. Another interesting feature of DEA is its ability to handle uncertainty, e.g., interval or fuzzy data. And let us not forget the most important feature of all and it is the wide applicability of the methodology. DEA has been applied in practically all sectors: transportation, healthcare, energy, environment, telecommunications, education, banking, agriculture, sports, government, logistics, tourism, etc.

Our final words are for acknowledging that DEA has also limitations and drawbacks. Actually, it is a wonder that in spite of its limitations the methodology has proved itself so applicable and so useful. Thus, DEA is sensitive to outliers and noisy data, it is affected by missing data and it loses discriminant power when the number of DMUs is small relative to the number of variables. Another problem, not generally acknowledged, is the absence of an objective way of validating the selection of input and output variables. This is normally done ad hoc and subject to data availability. Another issue is the assumption, taken for granted, that the observed DMUs are homogeneous, something which may not happen always and which, in any case, we do not know very well how to check. A final limitation, this well-acknowledged, is that DEA can provide improvement targets and benchmarks but it does not provide any guide about what are exactly the best practices of those benchmarks that should be learnt by the DMU or the cost and difficulty of, and the time required for, importing those practices.

Acknowledgements This research was carried out with the financial support of the Spanish Ministry of Science and the European Regional Development Fund (ERDF), grant DPI2017-85343-P.

References

Aparicio, J., Ruiz, J. L., & Sirvent, I. (2007). Closest targets and minimum distance to the Pareto-efficient frontier in DEA. *Journal of Productivity Analysis, 28,* 209–218.

Aparicio, J. (2016). A survey on measuring efficiency through the determination of the least distance in data envelopment analysis. *Journal of Centrum Cathedra, 9,* 143–167.

Banker, R. D., Charnes, A., & Cooper, W. W. (1984). Some models for estimating technical and scale inefficiencies in data envelopment analysis. *Management Science, 30,* 1078–1092.

Banker, R. D., & Morey, R. (1986). Efficiency analysis for exogenously fixed inputs and outputs. *Operations Research, 34,* 513–521.

Camanho, A. S., Portela, M. C., & Vaz, C. B. (2009). Efficiency analysis accounting for internal and external non-discretionary factors. *Computers & Operations Research, 36,* 1591–1601.

Chambers, R. G., Chung, Y., & Färe, R. (1996). Benefit and distance functions. *Journal of Economic Theory, 70,* 407–419.

Charnes, A., Cooper, W. W., Golany, B., Seiford, L., & Stutz, J. (1985). Foundations of data envelopment analysis for Pareto-Koopmans efficient empirical production functions. *Journal of Econometrics, 30,* 91–107.

Charnes, A., Cooper, W. W., & Rhodes, E. (1978). Measuring the efficiency of decision making units. *European Journal of Operational Research, 2,* 429–444.

Cooper, W. W., Pastor, J. T., Borras, F., Aparicio, J., & Pastor, D. (2011). BAM: A bounded adjusted measure of efficiency for use with bounded additive models. *Journal of Productivity Analysis, 35,* 85–94.

Cooper, W. W., Seiford, L., & Tone, K. (2000). *Data envelopment analysis: A comprehensive text with models, application, references, and DEA-Solver software*. Norwell, Massachusetts: Kluwer Academic Publishers.

Cook, W. D., Tone, K., & Zhu, J. (2014). Data envelopment analysis: Prior to choosing a model. *Omega, 44,* 1–4.

Emrouznejad, A., & Amin, G. R. (2009). DEA models for ratio data: Convexity consideration. *Applied Mathematical Modelling, 33,* 486–498.

Färe, R., Grosskopf, S., & Lovell, C. A. K. (1985). The measurement of efficiency of production. Kluwer Nijhof Publishing.

Färe, R., Grosskopf, S., & Lovell, C. A. K. (1994). Production frontiers. Cambridge University Press.

Färe, R., Mardaritis, D., Rouse, P., & Roshdi, I. (2016). Estimating the hyperbolic distance function: A directional distance function approach. *European Journal of Operational Research, 254,* 312–319.

Kazemi Matin, R., & Kuosmanen, T. (2009). Theory of integer-valued data envelopment analysis under alternative returns to scale axioms. *Omega, 37,* 988–995.

Kuosmanen, T. (2005). Weak disposability in nonparametric production analysis with undesirable outputs. *American Journal of Agricultural Economics, 87,* 1077–1082.

Kuosmanen, T., & Kazemi Matin, R. (2009). Theory of integer-valued data envelopment analysis. *European Journal of Operational Research, 192,* 658–667.

Lozano, S., & Villa, G. (2004). Centralized resource allocation using data envelopment analysis. *Journal of Productivity Analysis, 22,* 143–161.

Lozano, S., & Villa, G. (2006). Data envelopment analysis of integer-valued inputs and outputs. *Computers & Operations Research, 33,* 3004–3014.

Lozano, S., & Villa, G. (2010). DEA-based pre-merger planning tool. *Journal of the Operational Research Society, 61,* 1485–1497.

Lozano, S. (2012). Information sharing in DEA: A cooperative game theory approach. *European Journal of Operational Research, 222,* 558–565.

Lozano, S. (2013a). DEA production games. *European Journal of Operational Research, 231,* 405–413.

Lozano, S. (2013b). Using DEA to find the best partner for a horizontal cooperation. *Computers & Industrial Engineering, 66,* 286–292.

Lozano, S. (2014). Company-wide production planning using a multiple technology DEA approach. *Journal of the Operational Research Society, 65,* 723–734.

Olesen, O. B., & Petersen, N. C. (2006). Controlling for socioeconomic characteristics in DEA. North American productivity workshop. New York: Stern School of Business, New York University.

Olesen, O. B., Petersen, N. C., & Podinovski, V. V. (2015). Effciency analysis with ratio measures. *European Journal of Operational Research, 245,* 446–462.

Olesen, O. B., Petersen, N. C., & Podinovski, V. V. (2017). Effciency measures and computational approaches for data envelopment analysis models with ratio inputs and outputs. *European Journal of Operational Research, 261,* 640–655.

Pastor, J. T. (1996). Translation invariance in data envelopment analysis: A generalization. *Annals of Operations Research, 66,* 93–102.

Pastor, J. T., Ruiz, J. L., & Sirvent, I. (1999). An enhanced DEA Russell graph efficiency measure. *European Journal of Operational Research, 115,* 596–607.

Ruggiero, J. (1998). Non-discretionary inputs in data envelopment analysis. *European Journal of Operational Research, 111,* 461–469.

Seiford, L. M., & Zhu, J. (2002). Modeling undesirable factors in efficiency evaluation. *European Journal of Operational Research, 142,* 16–20.

Sharp, J. A., Liu, W. B., & Meng, W. (2006). A modified slacks-based measure model for data envelopment analysis with natural negative outputs and inputs. *Journal of the Operational Research Society, 57*(1), 1–6.

Silva Portela, M. C. A., Thanassoulis, E., & Simpson, G. (2004). Negative data in DEA: A directional distance approach applied to bank branches. *Journal of the Operational Research Society, 55,* 1111–1121.

Tone, K. (2001). A slacks-based measure of efficiency in data envelopment analysis. *European Journal of Operational Research, 130,* 498–509.

Tone, K. (2002). A slacks-based measure of super-efficiency in data envelopment analysis. *European Journal of Operational Research, 143,* 32–41.

Tulkens, H., & Vanden Eeckaut, P. (1995). Non-parametric efficiency, progress and regress measures for panel data: Methodological aspects. *European Journal of Operational Research, 80,* 474–499.

Zofio, J. L., Pastor, J. T., & Aparicio, J. (2013). The directional profit efficiency measure: on why profit inefficiency is either technical or allocative. *Journal of Productive Analysis, 40,* 257–266.

Chapter 6
The Measurement of Firms' Efficiency Using Parametric Techniques

Luis Orea

Abstract In this chapter we summarize the main features of the standard econometric approach to measuring firms' inefficiency. We provide guidance on the options that are available using the Stochastic Frontier Analysis (SFA), the most popular parametric frontier technique. We start this chapter summarizing the main results of production theory using the concept of distance function. Next, we outline the most popular estimation methods: maximum likelihood, method-of-moments and distribution-free approaches. In the last section we discuss more advance topics and extend the previous models. For instance, we examine how to control for observed environmental variables or endogeneity issues. We also outline several empirical strategies to control for unobserved heterogeneity in panel data settings or using latent class and spatial stochastic frontier models. The last topics examined are dynamic efficiency measurement, production risk and uncertainty, and the decomposition of Malmquist productivity indices.

6.1 Introduction

The rapid development of information technology has boosted competitive pressure in many markets all over the world. In such an environment, firms and organizations must improve their productive and economic performance in order to survive. On the other hand, the availability of large size databases permits conducting comprehensive analysis of firms' performance. This chapter outlines some of the econometric methods that allow managers and researchers to identify gains in firms' efficiency and productivity that remain untapped. In this sense, this chapter combines two different but related literatures: (i) *Data Science*, which comprises a collection of scientific methods aiming to extract useful information from available data; and (ii) *Frontier Analysis*, which aims to measure firms' efficiency using real data, in contrast to engineering-based techniques based on simulated data.

L. Orea (✉)
Oviedo Efficiency Group, Departamento de Economía, Universidad de Oviedo, Oviedo, Spain
e-mail: lorea@uniovi.es

V. Charles et al. (eds.), *Data Science and Productivity Analytics*,
International Series in Operations Research & Management Science 290,
https://doi.org/10.1007/978-3-030-43384-0_6

In this chapter we summarize the main features of the standard econometric approach to measuring firms' inefficiency (and productivity). Given that the efficient production/cost of each firm is not directly observed, it must be inferred from real data using *frontier* models that involve the estimation of both the technological parameters and the parameters of firms' inefficiency. In this chapter we provide guidance on the options that are available in order to successfully undertake research in this field using the so-called Stochastic Frontier Analysis (SFA) models, the most popular parametric frontier technique.[1] For a comprehensive survey of this literature, we recommend the following references: Kumbhakar and Lovell (2000), Parmeter and Kumbhakar (2014), and Kumbhakar et al. (2015). Some of the following sections are inspired in two previous manuscripts written by the author of this chapter where we examine common issues in SFA and DEA approaches to measuring firms' production performance (see Orea and Zofío 2017, 2019).

We start this chapter summarizing in Sect. 6.2 the main results of production theory; particularly the possibility of characterizing the behaviour of the firm from the primal–technological perspective. As firms produce multiple outputs using multiple inputs, the primal representation of the technology relies on the concept of distance function, which is also interpreted as a measure of productive performance. We discuss in this section the choice of functional forms when representing firms' technology and examine the advantages and drawbacks of the so-called *flexible* functional forms.

Section 6.3 outlines the most popular estimation methods available to undertake SFA efficiency analyses. The most popular estimation method is *maximum likelihood*, where the parameters of the distance (production) function and the random term capturing firms' inefficiency are estimated in a single stage. A second method is the *method-of-moments* approach, where the distance function is first estimated using standard econometric techniques and distributional assumptions are only invoked in a second stage to estimate the parameter(s) describing the structure of the error components. Unlike the two abovementioned methods, firms' efficiency scores can also be computed without making specific distributional assumptions using the so-called *distribution-free approach.*

Section 6.4 discusses more advance topics and extends somehow the basic models introduced in the previous section. In particular, we examine here how to control for environmental or contextual variables that do not fall within managerial discretion. We also present a series of recent models that deal with addressing endogeneity issues or are able to control for *unobserved* differences in firms' technology or environmental conditions. This subsection summarize several frontier models that incorporate the dynamic nature of the decision-making process into efficiency analysis or are able to account for production risk, stochastic technologies and uncertainty. This subsection concludes with the popular (parametric) Malmquist productivity index

[1]The individual results from parametric and non-parametric methods will generally differ. However, the difference between the two methods is less pronounced nowadays than they used to be because both approaches now benefit from recent advances that address their shortcomings.

and its decomposition into several terms explaining productivity change (efficiency change, technical change…) based on estimated production and distance functions.

This chapter finishes with Sect. 6.5 that includes a set of concluding remarks.

6.2 Theoretical Background

The point of departure of any theoretical and empirical study of efficiency and productivity is whether it is merely concerned with technical performance from an engineering perspective or it has an economic dimension. The technical or engineering approach is the only available choice when prices are unavailable (for example, the public sector provision of some public goods and services), or when they simply do not exist (for example, undesirable by-products such as waste and pollution). In this case we presume that the objective of the firm is technological, based on quantities only, and technology must be inferred using a primal approach, such as production or distance functions. On the contrary, as would be the case of firms in an industry, if market prices for inputs and outputs are available, then we can extend our engineering analysis to the firm's market environment. In this case we presume that the objective of the firm is economic, and its analysis requires data on quantities and prices and dual representations of firms' technology such as cost and profit functions.

6.2.1 Primal Approach: Production and Distance Functions

6.2.1.1 Definitions and Properties

For the single output case, the technology can be represented by the production function defined as the maximum amount of output that can be obtained from any combination of inputs:

$$f(x) = max\{y: (x, y) \in T\} \tag{6.1}$$

where T is the technology set. In the multi-output case, a suitable representation of the technology is given by the distance function introduced by Shephard (1970). This representation can be made from alternative orientations including the following output and input-oriented distance functions:

$$D_O(x, y) = min\{\theta: (x, y/\theta) \in T\} \tag{6.2}$$

$$D_I(x, y) = max\{\lambda: (x/\lambda, y) \in T\} \tag{6.3}$$

If the technology satisfies the customary axioms, the *output* distance function, ODF, has the range $0 \leq D_O(x, y) \leq 1$. It is homogeneous of degree one in outputs, non-decreasing in outputs and non-increasing in inputs. Notice that the advantage of this interpretation is that it leaves room for technical inefficiency when $D_O(x, y) < 1$. In this case, value of the output distance function can be directly interpreted as a measure of firms' technical efficiency, that is $ET = D_O$. In contrast, the *input* distance function, IDF, has the range $D_I(x, y) \geq 1$. It is homogeneous of degree one in inputs, non-decreasing in inputs, and non-increasing in outputs. A firm is inefficient when $D_I(x, y) > 1$. Therefore, firms' technical efficiency can be measured as $ET = 1/D_I$.[2]

More recent and flexible characterizations are the additive *directional* distance functions that can be defined as:

$$\vec{D}(x, y, -g_x, g_y) = max\{\tau: (x - \tau g_x, y + \tau g_y) \in T\} \tag{6.4}$$

The directional distance function, DDF, measures the simultaneous maximum reduction in inputs and expansion in outputs given a pre-specified directional vector defined by $g = (g_x, g_y)$ and the actual technology. The properties of this function are presented in Chambers et al. (1996, 1998). Just mention here that this function nests Shephard's input and output distance functions depending on the specific values of the directional vector.

6.2.1.2 The Importance of Imposing Theoretical Properties

Notice that, at first sight, the distance functions in (6.2)–(6.4) dependent on the same vector of inputs and outputs. Thus, if we were able to estimate a function of inputs and outputs, say $D(x, y)$, how do we ensure that we have estimated our preferred choice, say, an output distance function, and not an input distance function? For identification purposes we need to take advantage of one of the properties of distance functions. In particular, the key property for identification is the homogeneity condition for the input and output distance functions and the translation property for the directional distance functions. The latter property is the additive analogy to the multiplicative homogeneity property of Shephard's distance functions. Identification works because each homogeneity condition involves different sets of variables.[3]

For instance, in an ***output distance function***, the linear homogeneity condition of the distance function implies that $D(x, \lambda y) = \lambda D(x, y)$. If we assume that

[2]I have often been asked whether it is possible to compute the elasticity of a specific input (output) variable using distance functions due to their radial definition involves the whole set of inputs (outputs). The answer to this question is: yes. In Appendix A we show how to compute relevant economic properties of the multi-input multi-output distance function such as specific input and output elasticities or marginal effects, and the scale elasticity regardless we use an input or output-oriented distance functions.

[3]Interestingly, although the underlying technology is the same, the coefficients of each distance function differ.

$\lambda = 1/y_M$, we get after taking logs that:

$$lnD = lnD(x, y/y_M) + lny_M \tag{6.5}$$

The term measuring firms' inefficiency (i.e. lnD) is not observed by the researcher and thus it cannot be used as a proper dependent variable. However, the linear homogeneity condition immediately "produces" an observed dependent variable for the above model if we rewrite (6.5) as:

$$-lny_M = lnD(x, y/y_M) + u \tag{6.6}$$

where $u = -lnD$, or

$$lny_M = -lnD(x, y/y_M) - u \tag{6.7}$$

Note that this ODF collapses to a standard production function if M = 1, and that we have reversed the signs of all the coefficients of $lnD(\cdot)$. Therefore, the estimated parameters can be interpreted as the coefficients of a (multi-output) production function. A similar expression can be obtained if we impose the linear homogeneity condition in inputs rather than in outputs, and an ***input distance function*** is estimated instead.[4]

The choice of orientation should be determined, at least partially, by the capability of firms to adjust their inputs and outputs in order to become fully efficient. However, Kumbhakar et al. (2007) show that, once the distance function is known, input (output)-oriented inefficiency scores can be obtained from output (input) distance functions. To see this clearly, assume that we want to estimate the output distance function (6.5) but using an input-oriented measure of firms' efficiency. The equation to be estimated can be written as:

$$0 = \ln D\left(xe^{-\eta}, y/y_M\right) + lny_M \tag{6.8}$$

where now η measures firms' efficiency in terms of input reductions, conditional on the observed output vector. The choice of orientation is a relevant issue because the efficiency and technological results will differ due to endogeneity issues, and for the "complexity" of the stochastic part of the SFA model. For instance, Kumbhakar and Tsionas (2006) show that the standard Maximum Likelihood (ML) method cannot be applied to estimate input-oriented production functions. This issue is examined later once several functional forms for the distance functions have been introduced.

[4]The linear homogeneity condition in inputs yields the following IDF:

$$lnx_J = f(x/x_J, y) + u$$

where now $u = lnD \geq 0$ measures firms' inefficiency in terms of inputs, and $f(x/x_J, y)$ is non-increasing in inputs, and non-decreasing in outputs. Therefore $f(x/x_J, y)$ can be interpreted as an input requirement function.

Regarding the ***directional distance function***, while its general specification is given in (6.4), quite often the directional vector is set to $(g_x, g_y) = (1, 1)$. In this case, this function can be written as:

$$\vec{D} = \vec{D}(x, y; -1, 1) = \vec{D}(x, y) \tag{6.9}$$

If the above directional distance function satisfies the *translation property* that says that if output is expanded by α and input is contracted by α, then the resulting value of the distance function is reduced by α:

$$\vec{D}(x - \alpha, y + \alpha,) = \vec{D}(y, x) - \alpha \tag{6.10}$$

Thus, replacing above $\vec{D}(x, y)$ with $\vec{D}(x - \alpha, y + \alpha) + \alpha$, we get:

$$-\alpha = \vec{D}(x - \alpha, y + \alpha,) - u \tag{6.11}$$

where now $u = \vec{D}$. We obtain variation on the left-hand side by choosing an α that is specific to each firm. For instance, $\alpha = y_M$.

6.2.1.3 Functional Forms

The initial and most commonly employed distance functions (or, equivalently, their corresponding production functions in the single output case), i.e. Cobb–Douglas (CD) or Constant Elasticity of Substitution (CES), as well as their associated dual functions, place significant restrictions on the technology (e.g. all elasticities are common to all firms or returns to scale do not vary with firms' size).[5] However, these functions are "well-behaved" in the sense that they are continuous and twice differentiable, and the Shephard and Hotelling's lemmas allow the recovery of the demand and supply equations. The so-called second order flexible functional forms (see Diewert 1971) emerged in the 70s permit a more general representation of the production technology. While the Quadratic, Leontief or Translog forms can be seen as second order Taylor-series mathematical expansions around different points with different transformations of the variables, other flexible forms based on Laurent and Fourier expansions (Thompson 1988) provide global rather than local approximations to the underlying technology; but they are much less popular since its econometric estimation and parameter interpretation are more demanding.

The fact that the number of parameters to be estimated increases exponentially with the number of variables included in the functional form, empirical research is de facto restricted to the quadratic approximation. If a large sample cannot be collected, degrees of freedom can be easily exhausted, and a general practice is to aggregate commodities and prices; but consistent aggregation is only possible under strong

[5]The limitations of the Cobb–Douglas functions when testing the neoclassical theory of the firm constituted the basis for newer, less restrictive functional forms (Zellner and Revankar 1969).

restrictions on the underlying technology—e.g. separability. The properties of flexible functional forms ultimately determine whether they are globally well-behaved in the presence of large data variability. For instance, the Quadratic and Translog specifications fail to satisfy the regularity conditions over the entire range of sample observations. However, how to test those global properties and impose regularity conditions globally remains unclear because imposing regularity conditions globally often comes at the cost of limiting the flexibility of the functional form. Given this trade-off, the common practice is to evaluate the estimated functions at the sample mean, rather than at each individual observation.[6]

Despite these caveats, flexible functional forms are useful and have become standard in empirical studies. To exemplify their capabilities when testing functional, we show two representative specifications. The first one makes use of the *Translog* formulation to specify the output distance function, and the second one corresponds to the *Quadratic* directional distance function.

As for the Translog output distance function with output-oriented inefficiency, the specification corresponds to:

$$\begin{aligned}\ln y_M = {} & \beta_0 + \sum\nolimits_{j=1}^{J} \beta_j \ln x_j + \sum\nolimits_{m=1}^{M-1} \beta_m \ln y_m^* + \frac{1}{2}\sum\nolimits_{j=1}^{J} \beta_{jj} \ln x_j^2 \\ & + \frac{1}{2}\sum\nolimits_{m=1}^{M-1} \beta_{mm} \ln y_m^{*2} \\ & + \sum\nolimits_{j=1}^{J} \sum\nolimits_{k\neq j}^{K} \beta_{jk} \ln x_j \ln x_k + \frac{1}{2}\sum\nolimits_{m=1}^{M-1} \sum\nolimits_{n\neq m}^{N} \beta_{mn} \ln y_m^* \ln y_n^* \\ & + \sum\nolimits_{m=1}^{M-1} \sum\nolimits_{j=1}^{J} \beta_{mj} \ln y_m^* \ln x_j - u \end{aligned} \tag{6.12}$$

where $\ln y_m^* = \ln y_m - \ln y_M$. Note that the output-oriented inefficiency term appears above as an additive term. Therefore, the above parameters can be easily estimated using the standard maximum likelihood (ML) techniques because the typical distributional assumptions for u provide a closed-form for the distribution of the error term. If instead we are willing to a Translog output distance function using an input-oriented measure of firms' efficiency, the model to be estimated is:

$$\begin{aligned}\ln y_M = {} & \beta_0 + \sum\nolimits_{j=1}^{J} \beta_j \ln\left(x_j e^{-\eta}\right) + \sum\nolimits_{m=1}^{M-1} \beta_m \ln y_m^* + \frac{1}{2}\sum\nolimits_{j=1}^{J} \beta_{jj} \ln\left(x_j e^{-\eta}\right)^2 \\ & + \frac{1}{2}\sum\nolimits_{m=1}^{M-1} \beta_{mm} \ln y_m^{*2} + \sum\nolimits_{j=1}^{J} \sum\nolimits_{k\neq j}^{K} \beta_{jk} \ln\left(x_j e^{-\eta}\right) \ln\left(x_k e^{-\eta}\right) \\ & + \frac{1}{2}\sum\nolimits_{m=1}^{M-1} \sum\nolimits_{n\neq m}^{N} \beta_{mn} \ln y_m^* \ln y_n^* + \sum\nolimits_{m=1}^{M-1} \sum\nolimits_{j=1}^{J} \beta_{mj} \ln y_m^* ln\left(x_j e^{-\eta}\right) \end{aligned} \tag{6.13}$$

[6]It should be pointed out, however, that it is possible to maintain *local* flexibility using Bayesian techniques. See Griffiths et al. (2000) and O'Donnell and Coelli (2005).

Assuming one input, the model can be written as:

$$\ln y_M = -lnD(x, y/y_M) - \left[\beta_j + \beta_{jj} + \sum_{m=1}^{M-1} \beta_{mj} \ln y_m^*\right]\eta + \beta_{jj}\eta^2 \quad (6.14)$$

The presence of the η^2 term makes the derivation of a closed likelihood function impossible, and this precludes using standard ML techniques. Similar comments can be made if we were to use a directional distance function. In all cases where we have intractable likelihood functions, they can be maximized by simulated maximum likelihood.[7] A final important remark regarding Eq. (6.14) is that the output orientation of the distance function does not force the researcher to use an input-oriented measure of firms' inefficiency. We first do it just for simplicity, and in doing so are likely to attenuate endogeneity problems as well.

As for the directional distance function, the reason why the quadratic formulation is the best choice is that the translation property can be easily imposed on this specification–just as the homogeneity properties corresponding to the radial input or output distance functions can be easily imposed on the Translog specification. Once the translation property is imposed using $\alpha = y_M$, the quadratic specification of (6.11) can be written as:

$$\begin{aligned} -y_M = \beta_0 + \sum_{j=1}^{J} \beta_j x_j^* + \frac{1}{2}\sum_{j=1}^{J}\sum_{k=1}^{K} \beta_{jk} x_j^* x_k^* + \sum_{m=1}^{M-1} \beta_m y_m^* \\ + \frac{1}{2}\sum_{m=1}^{M-1}\sum_{n=1}^{N} \beta_{mn} y_m^* y_n^* + \sum_{m=1}^{M-1}\sum_{j=1}^{J} \beta_{mj} y_m^* x_j^* - u \end{aligned} \quad (6.15)$$

where $x_j^* = x_j - y_M$ and $y_m^* = y_m + y_M$. It is worth mentioning that inefficiency is measured here in physical units. However, as the Quadratic specification is normally estimated once the variables are normalized with the sample means (see Färe et al. 2005; p. 480), the normalized variables are unit free, and thus in practice the estimated inefficiency scores can be interpreted as proportional changes in outputs and inputs, in the same fashion as in the standard radial distance functions.

6.2.2 *Dual Approach: Cost Functions*

We introduce here a cost minimization objective in order to discuss the duality framework allowing for an overall economic efficiency analysis. Based on the previous primal representations of the technology, and considering the vectors of input prices, w, the following cost function can be defined:

$$C(y, w) = \min_x \{wx: (x, y) \in T\} \quad (6.16)$$

[7]As shown by Parmeter and Kumbhakar (2014; p. 52) using a Translog cost function, if the production technology is homogeneous in outputs, the model can be estimated using simple ML techniques.

The cost function represents the minimum cost of producing a given amount of outputs, and assuming the necessary derivative properties–including continuity and differentiability, yields the input demand functions by applying Shephard's lemma.[8] If the technology satisfies the customary axioms, the cost function (6.16) is homogeneous of degree one in input prices, and non-decreasing in outputs and in input prices. Chambers et al. (1998) prove the duality between the input distance functions and its associated cost function. Unlike the distance function that only provides a measure of technical efficiency, the above definition leaves room for both technical and allocative inefficiency. However, Kumbhakar et al. (2015) point out that outputs and input prices are endogenous if firms are allocative inefficient because in this case the traditional u term depends on y and w.

Regarding the functional forms, the Translog cost function corresponds to:

$$\begin{aligned} ln\left(\frac{C}{w_J}\right) &= \beta_0 + \sum_{j=1}^{J-1} \beta_j \ln\left(\frac{w_j}{w_J}\right) + \sum_{m=1}^{M} \beta_m \ln y_m + \frac{1}{2}\sum_{j=1}^{J-1} \beta_{jj} \ln\left(\frac{w_j}{w_J}\right)^2 \\ &+ \frac{1}{2}\sum_{m=1}^{M} \beta_{mm} \ln y_m^2 \\ &+ \sum_{j=1}^{J-1}\sum_{k\neq j}^{K-1} \beta_{jk} \ln\left(\frac{w_j}{w_J}\right)\ln\left(\frac{w_k}{w_J}\right) + \sum_{m=1}^{M}\sum_{n\neq m}^{N} \beta_{mn} \ln y_m \ln y_n \\ &+ \sum_{m=1}^{M}\sum_{j=1}^{J-1} \beta_{mj} \ln y_m \ln\left(\frac{w_j}{w_J}\right) + u \end{aligned} \tag{6.17}$$

where u measures firms' technical and allocative inefficiency in terms of cost increases. Notice that we have already imposed linear homogeneity in input prices in the above cost function, and that the input-oriented inefficiency term appears above as an additive term. Again, this implies that the above parameters can be estimated by ML. Applying the Shephard's lemma in (6.17), we get the following cost share equations:

$$\begin{aligned} S_1 &= \beta_1 + \beta_{11} \ln\left(\tfrac{w_1}{w_J}\right) + \textstyle\sum_{k\neq 1}^{J-1} \beta_{1k} \ln\left(\tfrac{w_k}{w_J}\right) + \sum_{m=1}^{M} \beta_{m1} \ln y_m \\ &\vdots \\ S_{J-1} &= \beta_{J-1} + \beta_{J-1,J-1} \ln\left(\tfrac{w_{J-1}}{w_J}\right) + \textstyle\sum_{k\neq J-1}^{J-2} \beta_{J-1k} \ln\left(\tfrac{w_k}{w_J}\right) + \sum_{m=1}^{M} \beta_{mJ-1} \ln y_m \end{aligned} \tag{6.18}$$

[8]It is also possible to define shadow cost functions $C(y, w^s)$ constituting the dual representation of the technology for non-market-oriented (i.e. non-profit) organizations (e.g. public goods such as the provision of health and education services). In this case, for instance, the so-called shadow prices w^s rationalize the observed input quantity vector x as a cost-minimizing choice for the observed output vector y. If the minimum-cost condition is satisfied, the shadow price vector equals the market price vector. Rodríguez-Álvarez and Lovell (2004) show that these vectors may differ as a result of utility maximizing behavior on the part the bureaucrat, restricted by a budget constraint.

In principle, estimating the cost system (6.17)–(6.18) is more efficient from a statistical perspective because no additional parameters are added to the model.[9] However, Kumbhakar et al. (2015) clearly show that estimating a cost system using (6.17) and (6.18) is problematic, except with input-oriented technical inefficiency and zero allocative inefficiency. For this reason, they strongly prefer estimating *primal* system of equations consisting of a stochastic production (distance) function and a set of first-order conditions for cost minimization.

6.3 Estimation Methods

In this section we outline the most popular parametric frontier techniques aiming to measure both firms' inefficiency and technology. For notational ease, we develop this and next sections for cross-sectional data, except when it is compulsory to use a panel data framework. We also confine our discussion to the estimation of technical efficiency using output distance functions because they can be interpreted as a traditional but multi-output production function.[10] Thus, firm performance is evaluated by means of the following distance function:

$$lny_{Mi} = -\ln D\left(x_i, \frac{y_i}{y_{Mi}}, \beta\right) + v_i - u_i \tag{6.19}$$

where the subscript i stands for firm, β is now a vector of technological parameters, v_i is a two-sided noise term, and $u_i = -lnD_i \geq 0$ is a one-sided term capturing firms' inefficiency. In Eq. (6.19) we specify the distance function as being stochastic in order to capture random shocks that are not under the control of the firm. It can also be interpreted as a specification error term that appears when the researcher tries to model the firm's technology. Note also that this model can be immediately estimated econometrically once a particular functional form is chosen for $\ln D(x_i, y_i/y_{Mi}, \beta)$, and u_i is properly modelled.[11]

Note also that the composed error term $\varepsilon_i = v_i - u_i$ in (6.19) comprises two independent parts, a noise term and an inefficiency term. It is conventionally assumed that v_i follows a symmetric distribution since random shocks and specification errors might take both positive and negative values. In contrast, u_i is assumed to be non-negative (and asymmetrically) distributed because inefficient performance always produces a contraction in firms' output.

The estimation of the model involves both the parameters of the distance (production) function and the inefficiency. Both technological parameters of the distance

[9]This happens if we do not allow for non-zero mean values of the error terms traditionally added to each cost share equation in (6.18).

[10]Although most early SFA applications used production functions, the distance function became as popular as the production functions since Coelli and Perelman (1996), who helped practitioners to estimate and interpret properly the distance functions.

[11]The input distance functions as well as the directional distance function deserve similar comments.

function (β) and the structure of the two error components (i.e. the variance of v_i and u_i) are estimated simultaneously in a single stage using *maximum likelihood* (ML) once particular (perhaps strong) distributional assumptions on both random terms are made.

A second method that we can choose is the *method-of-moments* (MM) approach, where all technological parameters of the distance function are first estimated using standard econometric techniques (e.g. OLS, IV or GMM). This stage is independent of distributional assumptions in respect of either error component. Distributional assumptions are invoked in a second stage to obtain ML estimates of the variances of v_i and u_i. The most comprehensive SFA versions of the MM estimator are becoming increasingly popular among researchers because it allows for instance dealing with endogenous variables (see Guan et al. 2009), or distinguishing between transient and permanent efficiency (Filippini and Greene 2016).

The next step is to obtain the efficiency values for each firm. They are often estimated by decomposing the estimated residuals of the production function. Following Jondrow et al. (1982), both the mean and the mode of the conditional distribution of u_i given the composed error term ε_i can be used as a point estimate of u_i. Firms' efficiency scores can also be computed without making specific distributional assumptions on the error components using the so-called *distribution-free approach*. This approach includes the well-known COLS method for cross-sectional data, and the SS and CSS methods for panel data settings. As Kumbhakar et al. (2015; p. 49) remark, the drawback of this approach is that the statistical properties of the estimator of u_i may not be readily available.

6.3.1 ML Estimation

6.3.1.1 Single Equation Models

In order to estimate Eq. (6.19) using ML, we are forced to choose a distribution for v_i and u_i. The noise term is often assumed to be normally distributed with zero mean and constant standard deviation, i.e. $v_i \sim N(0, \sigma_v)$, with the following density function:

$$f(v_i) = \phi\left(\frac{v_i}{\sigma_v}\right) = \frac{1}{\sqrt{2\pi}\sigma_v}\exp\left(-\frac{v_i^2}{2\sigma_v^2}\right) \tag{6.20}$$

Note that $v_i = \varepsilon_i + u_i$, where $\varepsilon_i = lny_{Mi} - X'_{it}\beta$, and $X'_{it}\beta$ is the log of the frontier production (distance) function (e.g. Translog). So,

$$f(v_i) = f(\varepsilon_i + u_i) = \frac{1}{\sqrt{2\pi}\sigma_v}\exp\left(-\frac{(\varepsilon_i + u_i)^2}{2\sigma_v^2}\right) \tag{6.21}$$

The most popular distribution for inefficiency term is by far the half-normal (Aigner et al. 1977), which is the truncation (at zero) of a normally distributed random variable with zero mean and constant standard deviation, that is $u_i \sim N^+(0, \sigma_u)$.[12] Note that, for notational ease, we use σ_u to indicate hereafter the standard deviation of the pre-truncated normal distribution, and not the standard deviation of the post-truncated variable. If u_i follows a (homoscedastic) half-normal distribution, its density function can be written as:

$$f(u_i) = \frac{2}{\sigma_u}\phi\left(\frac{u_i}{\sigma_u}\right) = \frac{2}{\sqrt{2\pi}\cdot\sigma_u}\exp\left(-\frac{u_i^2}{2\sigma_u^2}\right) \tag{6.22}$$

Assuming that of v_i and u_i are distributed independently, the density function of the composed error term $\varepsilon_i = v_i - u_i$ can be written as:

$$f(\varepsilon_i) = \int_0^\infty f(\varepsilon_i + u_i)\cdot f(u_i)du_i \tag{6.23}$$

Given the assumed distributions, the above integration can be computed analytically. The density function of the composed error term of a normal-half-normal model is:

$$f(\varepsilon_i) = \frac{2}{\sigma}\phi\left(\frac{\varepsilon_i}{\sigma}\right)\Phi\left(-\frac{\varepsilon_i\lambda}{\sigma}\right) = \frac{2}{\sigma}\left[\frac{1}{\sqrt{2\pi}}\exp\left(-\frac{\varepsilon_i^2}{2\sigma^2}\right)\right]\Phi\left(-\frac{\varepsilon_i\lambda}{\sigma}\right) \tag{6.24}$$

where $\sigma = \sqrt{\sigma_u^2 + \sigma_v^2}$ and $\lambda = \sigma_u/\sigma_v$. Therefore, the log-likelihood function for the whole sample (assuming N observations) can be written as:

$$lnLF = \frac{N}{2}\ln(2/\pi) - Nln\sigma - \sum_{i=1}^N \frac{\varepsilon_i^2}{2\sigma^2} + \sum_{i=1}^N ln\Phi\left(-\frac{\varepsilon_i\lambda}{\sigma}\right) \tag{6.25}$$

The standard distributional assumptions for v_i and u_i provide a closed-form for the distribution of the composed error term, making the direct application of ML straightforward. The model is simply estimated by choosing the parameters that maximize the likelihood function (6.25). Newer models are appearing in the literature that do not yield tractable likelihood functions and must be estimated by simulated maximum likelihood. See Parmeter and Kumbhakar (2014, Sect. 7) for an excellent review of recent contributions dealing with this issue. To catch an idea about how this approach works, let us point out that the model can be estimated if we integrate out u_i from $f(\varepsilon_i + u_i)$ in (6.21):

$$f(\varepsilon_i) = \int_0^\infty \frac{1}{\sqrt{2\pi}\sigma_v}\exp\left(-\frac{(\varepsilon_i + u_i)^2}{2\sigma_v^2}\right)f(u_i)du_i \tag{6.26}$$

[12]Other distributions proposed in the literature are the exponential (Meeusen and van den Broeck 1977), the truncated normal (Stevenson 1980) and gamma (Greene 1990) distributions.

Notice that the integral can be viewed as an expectation, which we can evaluate through ***simulation*** as opposed to ***analytically***. Taking many draws, the above integral can be approximated as:

$$f(\varepsilon_i) \approx \frac{1}{R}\sum\nolimits_{r=1}^{R} \frac{1}{\sqrt{2\pi}\sigma_v}\exp\left(-\frac{(\varepsilon_i + \sigma_u|U_r|)^2}{2\sigma_v^2}\right) \tag{6.27}$$

The final task is obtaining the efficiency scores for each firm. As the procedure to get these scores is the same in ML and MM, it is explained later on in Sect. 6.3.2.

6.3.1.2 System Models

The previous discussion is concerned with the technical side of the firm. The allocation of inputs in a production model, or in our output distance function, is assumed to be either 100% efficient or they are assumed to be exogenously given. Recent developments in duality theory allow the decomposition of overall economic efficiency into technical and allocative terms in a consistent way. Parmeter and Kumbhakar (2014) prefer estimating primal system of equations consisting of a stochastic distance (production) function and a set of first-order conditions (FOC) for cost minimization. Following Kumbhakar et al. (2015, pp. 210–223), this system of equations can be written using a two-input Cobb–Douglas distance function as:

$$\begin{aligned} \ln y_{Mi} &= \alpha_0 + \alpha_1 \ln x_{1i} + \alpha_2 \ln x_{2i} + f\left(\frac{y_i}{y_{Mi}}, \beta\right) + v_i - u_i \\ \ln(\alpha_2/\alpha_1) &- \ln(w_{2i}/w_{1i}) - \ln x_{2i} + \ln x_{1i} = \xi_{2i} \end{aligned} \tag{6.28}$$

where $v_i \sim N(0, \sigma_v)$, $u_{it} \sim N^+(0, \sigma_u)$, and $\xi_i \sim N(\rho, \sigma_\xi)$.[13] The likelihood function of the whole system is:

$$LF_i = g(v_i - u_i) \cdot d(\xi_i) \cdot |J_i| \tag{6.29}$$

where $g(v_i - u_i)$ is the density function of a normal-half-normal random variable, $d(\xi_i)$ is the probability density function for ξ_i, and $|J_i|$ is the determinant of the Jacobian matrix:

$$|J_i| = \left|\frac{\partial(v_i - u_i, \xi_i)}{\partial(\ln x_{1i}, \ln x_{2i})}\right| \tag{6.30}$$

After estimating the parameters of the model by ML, firm-specific efficiency scores can be computed using the Jondrow et al. (1982) formula. Allocative inefficiency can be obtained from the residuals of the FOCs. If $\xi_i < 0$, input x_{2i} is ***overused*** relative to input x_{1i}, ***underused*** otherwise.

[13]If there are more than two inputs, $\xi_i = (\xi_{2i}, \ldots, \xi_{Ji})$ follows a multivariate normal distribution.

6.3.2 *MM Estimation*

The MM approach involves three stages. In the ***first stage***, we ignore the structure of the composed error term and estimate the frontier parameters using OLS if the explanatory variables are exogenous or GMM if they are endogenous. Taking expectations in (6.19), the model to be estimated in the first stage can be written as:

$$lny_{Mi} = E\left(lny_{Mi}|x_i, \frac{y_i}{y_{Mi}}; \beta \right) + \varepsilon_i = X'_{it}\beta + v_i - u_i \tag{6.31}$$

The endogeneity of some regressors will lead to OLS being biased and inconsistent. This source of inconsistency can be dealt with by using GMM. However, the parameter estimates can still be inconsistent if u_i is heteroskedastic itself. To achieve consistent estimates, it is critical to ensure that chosen instruments do not include determinants of u_i. Suppose that we can find a vector of instruments M_i that satisfy the following moment condition:

$$E[M_i \cdot \varepsilon_i] = E\left[M_i \cdot \left(lny_{Mi} - X'_{it}\beta\right)\right] = E[m_i(\beta)] = 0 \tag{6.32}$$

The efficient two-step GMM estimator is then the parameter vector that solves:

$$\hat{\beta} = \arg\min\left[\sum_{i=1}^{N} m_i(\beta)\right]' C^{-1}\left[\sum_{i=1}^{N} m_i(\beta)\right] \tag{6.33}$$

where C is an optimal weighting matrix obtained from a consistent preliminary GMM estimator. This optimal weighting matrix can take into account both heteroskedasticity and autocorrelation of the error term.[14]

In the *second* stage of the estimation procedure, distributional assumptions are invoked to obtain consistent estimates of the parameter(s) describing the standard deviations of v_i and u_i, conditional on the first-stage estimated parameters. Given that we are going to assume a particular distribution for the inefficiency term, both variances can be estimated using ML. The ML estimators are obtained by maximizing the likelihood function associated to the error term $\hat{\varepsilon}_i = lny_{Mi} - X'_{it}\beta$ that can be obtained from an estimate of the first-stage production Eq. (6.31). However, it should be pointed out that $\hat{\varepsilon}_i$ is a biased estimate of ε_i because $E(u_i) > 0$. We have two options to control for this bias. First, we can estimate the following (unrestricted) ML model:

$$\hat{\varepsilon}_i = \gamma_0 + v_i - u_i \tag{6.34}$$

[14] If we allow v_i or u_i be heteroscedastic, an efficient GMM estimator is needed.

where $\hat{\gamma}_0$ is an estimate of $E(u_i)$. If we assume that u_i follows a half-normal distribution, its mean value is equal to $\sqrt{2/\pi}\sigma_u$. Thus, the second option is estimating the above ML model with the following restriction $\gamma_0 = \sqrt{2/\pi}\sigma_u$.[15]

In the third stage we obtain the efficiency scores for each firm. From previous stages we have estimates of $\varepsilon_i = v_i - u_i$, which obviously contain information on u_i. The problem is to extract the information that ε_i contains on u_i. Jondrow et al. (1982) propose using the *conditional* distribution of the asymmetric random term u_i given the composed error term ε_i. The best predictor of u_i is the conditional expectation $E(u_i|\varepsilon_i)$ (see Kumbhakar and Lovell 2000). Given our distributional assumptions, $E(u_i|\varepsilon_i)$ can be written as follows:

$$\hat{u}_i = E(u_i|\varepsilon_i) = \mu_* + \frac{\sigma_* \phi\left(\frac{\mu_{*i}}{\sigma_*}\right)}{\Phi\left(\frac{\mu_{*i}}{\sigma_*}\right)} = E\left(u_i^*|\varepsilon_i\right) \tag{6.35}$$

where $\mu_* = -\varepsilon_i\sigma_u^2\left(\sigma_v^2 + \sigma_u^2\right)^{-1}$ and $\sigma_*^2 = \sigma_v^2\sigma_u^2\left(\sigma_v^2 + \sigma_u^2\right)^{-1}$. One might be tempted to validate the chosen specification of the inefficiency term by simply comparing the observed distribution of $\hat{u}_i$ to the assumed distribution for u_i. Wang and Schmidt (2009) show that this is not a good idea. To carry out this test we should compare the distribution of $\hat{u}_i$ and $E(u_i|\varepsilon_i)$. In this sense, they propose non-parametric Chi-square and Kolmogorov–Smirnov type statistics to perform this test properly. These authors also point out that, although $\hat{u}_i$ is the minimum mean squared error estimate of u_i, and it is unbiased in the unconditional sense $E\left(\hat{u}_i - u_i\right) = 0$, it is a shrinkage of u_i towards its mean. An implication of shrinkage is that on average we will overestimate u_i when it is small and underestimate u_i when it is large. This result, however, simply reflects the familiar principle that an optimal (conditional expectation) forecast is less variable than the term being forecasted.

Two comments are in order to conclude this section. First, it should be pointed out that using OLS or GMM does not let researchers dispense with distributional assumptions altogether because we still need the distributional assumptions to calculate the JLMS-type efficiency estimates based on Jondrow et al. (1982) formula. Second, using a Hausman test of the difference between the ML and first-stage OLS equation to test distributional assumptions might not be a good idea. In principle the ML estimator should be more efficient because it uses the distributional information, and the first-stage OLS estimator is likely to be consistent regardless of whether or not the inefficiency term follows a particular (homoscedastic) distribution. However, as Orea and Álvarez (2019) point out, it is not clear whether the ML estimator is still consistent if the assumed distribution of the inefficiency term is not correct. In the case that both estimators are consistent, we can use a Hausman test, but it will not necessarily show power if the ML is consistent too.

---

[15]The second-stage model can be also estimated by MM that relies on the second and third moments of the error term $\hat{\varepsilon}_i$ in Eq. (6.31).

6.3.3 *Distribution-Free Approaches*

Firms' efficiency scores can also be computed without making specific distributional assumptions on the error components using the so-called *distribution-free approach*. In the following paragraphs, we present three methods that do not make distributional assumptions on either v_i or u_i.

6.3.3.1 COLS Method

The Corrected Ordinary Least Squares (COLS) method was proposed by Winsten (1957) and can be used with cross-sectional or panel data sets. The estimation proceeds in two stages. In the first stage, we estimate the frontier parameters of (6.31) using OLS if the explanatory variables are exogenous, or GMM if they are endogenous. At this stage, we obtain the zero-mean first-stage residuals as $\hat{\varepsilon}_i = lny_{Mi} - X'_{it}\hat{\beta}$. The value of $\hat{\varepsilon}_i$ can be greater, equal to, or less than zero. At the second stage, the estimated function is shifted upward to the extent that the function after the adjustment bounds all observations below. Once the residuals are adjusted upward, the frontier model becomes:

$$lny_{Mi} = \max(\hat{\varepsilon}_i) + X'_{it}\hat{\beta} - \hat{u}_i \tag{6.36}$$

and the inefficiency term is computed as:

$$\hat{u}_i = \max_i(\hat{\varepsilon}_i) - \hat{\varepsilon}_i \geq 0 \tag{6.37}$$

Notice that frontier model in (6.36) is deterministic in nature because any deviation from the frontier is now interpreted as inefficiency. This limitation can be addressed if panel data is available and we use the following SS and CSS methods.

6.3.3.2 SS Method

A fixed-effect estimator can be used to estimate the frontier model if panel data is available. In this case, it is possible to compute firm-specific efficiency scores without making specific distributional assumptions on the error components and using a stochastic or non-deterministic frontier framework.

Schmidt and Sickles (1984) assumed a production (distance) model with firm-specific intercepts that can be written as:

$$lny_{Mit} = \beta_0 + X'_{it}\beta + v_{it} - u_i = \alpha_i + X'_{it}\beta + v_{it} \tag{6.38}$$

where $\alpha_i = \beta_0 - u_i$ are firm-specific intercepts that are to be estimated along with the parameter vector β, and $X'_{it}\beta$ is the log of the frontier production (distance) function. Schmidt and Sickles (1984) showed that we can apply standard FE panel data estimation methods to estimate the firm-specific effects. Once $\hat{\alpha}_i$ are available, the following transformation is used to get time-invariant inefficiency scores for each firm:

$$\hat{u}_i = \max_i(\hat{\alpha}_i) - \hat{\alpha}_i > 0 \tag{6.39}$$

6.3.3.3 CSS Method

To make the inefficiency term time-varying, Cornwell et al. (1990) suggest replacing α_i by $\alpha_{it} = \alpha_{0i} + \alpha_{1i}t + \alpha_{2i}t^2$. The model can be estimated using OLS if a set of firms' dummies and their interaction with t and t^2 are added to the model:

$$lny_{Mit} = \sum_{i=1}^{N}(\alpha_{0i} + \alpha_{1i}t + \alpha_{2i}t^2)D_i + X'_{it}\beta + v_{it} \tag{6.40}$$

Finally, $\hat{u}_{it}$ is obtained by:

$$\hat{u}_{it} = \hat{\alpha}_t - \hat{\alpha}_{it} = \max_i(\hat{\alpha}_{it}) - \hat{\alpha}_{it} \geq 0 \tag{6.41}$$

Notice that we can rewrite (6.40) using (6.41) as $lny_{Mit} = \hat{\alpha}_t + X'_{it}\beta + v_{it} + v_{it} - \hat{u}_{it}$. As $\hat{\alpha}_t$ changes over time, the CSS model allows implicitly for technical change, and the rate of technical change can be computed as $TC = \hat{\alpha}_t - \hat{\alpha}_{t-1}$.

6.4 More (Advanced) Topics and Extensions

6.4.1 Observed Environmental Conditions

The concern about the inclusion of environmental variables (also called *contextual* or z-variables) has generated the development of several models either using parametric, nonparametric or semi-parametric techniques. Here we only mention the methods most frequently applied that include z-variables as determinants of firms' inefficiency.

The first methodological choice is whether we should incorporate the z-variables as either frontier determinants, determinants of firms' inefficiency or as determinants of both the frontier and the inefficiency term. As Orea and Zofío (2019) point out, the key question that should be responded in order to include the z-variables as frontier determinants is whether a fully efficient firm will need to use more inputs to provide the same services or produce the same output level if an increase in a contextual variable represents a deterioration in the environment where it operates.

In general, we should include as frontier drivers those variables that are fundamental to production. If they in addition make it more difficult or easier to manage the firm, they should be also treated as determinants of firms' inefficiency.

Early papers proceeded in two steps. In the first step, a homoscedastic SFA model is estimated ignoring the z-variables. In the second step, the first-step efficiency scores are regressed against the z-variables. Wang and Schmidt (2002) and other authors showed that this two-step procedure might be seriously biased. The solution to this bias is a one-step procedure based on a heteroscedastic SFA model. Once heteroscedastic SFA models are to be estimated, a second methodological choice appears: how to do it. Summaries of this literature can be found in Kumbhakar and Lovell (2000) and Parmeter and Kumbhakar (2014). The available options can be discussed using a general specification of the inefficiency term:

$$u_i \sim N^+(\mu(z_i), \sigma_u(z_i)) \tag{6.42}$$

where both the pre-truncation mean and standard deviation of the distribution might depend on the z-variables. According to this model, Álvarez et al. (2006) divide most heteroscedastic SFA models into three groups. In the *mean-oriented* models, it is assumed that the variance of the pre-truncated normal variable is homoscedastic and, thus, the contextual variables are introduced here through the pre-truncated mean. Following Battese and Coelli (1995), this specification can be written as:

$$u_i \sim N^+\left(\theta_0 + z_i'\theta, e^{\gamma_0}\right) \tag{6.43}$$

In contrast, in the *variance-oriented* models, it is assumed that the mean of the pre-truncated normal variable is homoscedastic and, hence, the environmental variables are treated as determinants of the variance of the pre-truncated normal variable. Following Caudill et al. (1995), this specification can be illustrated as:

$$u_i \sim N^+\left(0, e^{\gamma_0 + z_i'\gamma}\right) \tag{6.44}$$

In more *general* models, the contextual variables are introduced through both the mean and variance of the pre-truncated normal distributed random variable. Álvarez et al. (2006) and Lai and Huang (2010) proposed, respectively, exponential and lineal specifications for this model:

$$u_i \sim N^+\left(e^{\theta_0 + z_i'\theta}, e^{\gamma_0 + z_i'\gamma}\right) \tag{6.45}$$

$$u_i \sim N^+\left(\theta_0 + z_i'\theta, e^{\gamma_0 + z_i'\gamma}\right) \tag{6.46}$$

Some of the above models satisfy the so-called *scaling property* in the sense that the inefficiency term can be written as a deterministic (scaling) function of a set of efficiency covariates (h_i) times a one-sided random variable (u_i^*) that does not depend on any efficiency determinant. That is:

$$u_i = h(z_i'\gamma) \cdot u_i^* = h_i u_i^* \tag{6.47}$$

where e.g. u_i^* might follow a truncated normal or a half-normal distribution.[16] For instance, the variance-oriented model in (6.44) has the scaling property due it can be written as:

$$u_i = e^{z_i'\gamma} \cdot u_i^* \tag{6.48}$$

where $h_i = e^{z_i'\gamma}$ and $u_i^* \sim N^+(0, e^{\gamma_0})$. As Parmeter and Kumbhakar (2014) point out, the ability to reflect the scaling property requires that both the mean and the variance of the truncated normal are parameterized identically and with the same parameters in each parameterization. In this sense, the general model introduced by Álvarez et al. (2006) also has the scaling property if we impose in (6.45) that $\theta = \gamma$. In this case, $h_i = e^{z_i'\gamma}$ and $u_i^* \sim N^+(e^{\theta_0}, e^{\gamma_0})$.

As noted by Simar et al. (1994), Wang and Schmidt (2002) and Álvarez et al. (2006), the most fundamental benefit of the scaling property from a statistical point of view is that the stochastic frontier and the deterministic component of inefficiency can be recovered without requiring a specific distributional assumption on u_i^*.[17] Indeed, if we take into account our specification of firms' inefficiency in (6.48) and define $\mu^* = E(u_i^*) \geq 0$, then taking expectations in (6.31) yields:

$$lny_{Mi} = X_i'\beta - h(z_i'\gamma) \cdot \mu^* + \varepsilon_i^* \tag{6.49}$$

where again and $X_i'\beta$ is the log of the frontier production (distance) function, and

$$\varepsilon_i^* = v_i - h(z_i'\gamma)[u_i^* - \mu^*] \tag{6.50}$$

The parameters in (6.50) can be estimated using nonlinear least squares as[18]:

$$\left(\hat{\beta}, \hat{\gamma}, \hat{\mu}^*\right) = \arg min \frac{1}{N} \sum_{i=1}^{N} [lny_{Mi} - X_i'\beta + h(z_i'\gamma)\mu^*]^2 \tag{6.51}$$

Given that ε_i^* is heteroscedastic, robust standard errors should be constructed to ensure valid inferences. The presence of μ^* in (6.51) implies that one cannot include

16 As pointed out Álvarez et al. (2006), u_i^* can be viewed as a measure of basic inefficiency which captures things like the managers' natural skills, which we view as random. How well these natural skills are exploited to manage the firm efficiently depends on other variables z_i, which might include the manager's education or experience, or measures of the environment in which the firm operates.

17 Orea and Zofío (2017) point out that the scaling property has been also useful to remove individual fixed effects and still get a closed-form for the likelihood function (Wang and Ho 2010), to address endogeneity issues (Griffiths and Hajargasht 2016) or to relax the zero-rebound effect assumption in traditional demand frontier models (Orea et al. 2015).

18 To impose $\mu^* \geq 0$ in practice, we could replace μ^* in (6.49) with e^{θ_μ}.

a constant in h_i as this leads to identification issues (see Parmeter and Kumbhakar 2014, p. 88). In a second stage, distributional assumptions are invoked to estimate σ_v and σ_u. Notice that, if we assume that $u_i^* \sim N^+(0, \sigma_u)$, we have already got an estimate of σ_u using the first-stage estimate of μ^* as follows: $\hat{\sigma}_u = \hat{\mu}^* \sqrt{\pi/2}$. Thus, only σ_v should be estimated in the second stage of the procedure. In a third stage, we can obtain the estimates of efficiency for each firm using the conditional expectation $E\left(u_i|\varepsilon_i^*\right)$.

All heteroscedastic frontier models above can be used to examine exogenous (marginal) effects on firm's expected inefficiency. These effects can be easily computed if the inefficiency term has the scaling property. For instance, assume u_i follows the heteroscedastic half-normal distribution in (6.44). In this case, the conditional expectation $E(u_i|z_i)$ is equal to $h_i \cdot E\left(u_i^*\right) = e^{z_i'\gamma} \cdot \left[\sqrt{2/\pi} e^{\gamma_0}\right]$. Thus, the marginal effect of z_i on $E(u_i|z_i)$ is:

$$\frac{\partial E(u_i|z_i)}{\partial z_i} = \gamma \cdot e^{z_i'\gamma}\left[\sqrt{2/\pi} e^{\gamma_0}\right] \tag{6.52}$$

In order to get non-monotonic effects, we could include quadratic terms, or estimate more general models with both heteroscedastic mean and variance. However, in the latter case, Wang (2002) shows that the marginal effects are complex functions of both γ and θ parameters.

6.4.2 *Endogeneity Issues*

Endogeneity problems can arise if both inputs and/or outputs levels and inputs and/or outputs ratios are decided by the firms. Dealing with the endogeneity issue is relatively more complicated in a SFA framework than in standard regression models due to the special nature of the error term. Several authors have recently proposed alternative empirical strategies to account for endogenous regressors in SFA settings. In the next paragraphs we closely follow Orea and Zofío (2017, 2019) to outline the main features of these methods, trying to identify their relative advantages and disadvantages.

Let us first assume that we are interested in estimating the following production model with endogenous regressors:

$$lny_{Mi} = X_i'\beta + v_i - u_i \tag{6.53}$$

$$X_i = z_i'\delta + \eta_i \tag{6.54}$$

where X_i is a vector of endogenous production drivers, and z_i is a vector of exogenous or instrumental variables. Equation in (6.54) can be viewed as a reduced-form expression that links the endogenous variables with the set of instruments. The endogeneity problem arises if η_i is correlated with either v_i or u_i. In order to estimate consistently the frontier model (6.53), Guan et al. (2009) propose a two-step MM estimation strategy. In the first step, they suggest estimating the frontier parameters using a GMM estimator as long as valid instruments are found. In the second step, σ_v and σ_u are estimated using ML, conditional on the first-stage estimated parameters.

Instead of introducing instruments for these endogenous variables in an ad hoc fashion (e.g. temporal lags of inputs and outputs), Kumbhakar et al. (2013) and Malikov et al. (2015) bring additional equations for the endogenous variables from the first-order conditions of profitability (cost) maximization (minimization). They advocate using a system approach for two reasons. First, estimates of allocative inefficiencies can be obtained from the residuals of the first-order conditions. Second, since the first-order conditions contain the same technology parameters, their estimates are likely to be more precise (efficient). However, estimation of such a system requires availability of input and output prices.

Other authors address the endogeneity problem using the joint distribution of the stochastic frontier and the associated reduced form equations in (6.54). For instance, Karakaplan and Kutlu (2013) assume that the error terms in (6.53) and (6.54) satisfy the following[19]:

$$\begin{pmatrix} \Omega_\eta^{-1/2}\eta_i \\ v_i \end{pmatrix} \sim N\left(\begin{bmatrix} 0 \\ 0 \end{bmatrix}, \begin{bmatrix} I_p & \rho\sigma_v \\ \rho'\sigma_v & \sigma_v^2 \end{bmatrix}\right) \tag{6.55}$$

where Ω_η is the variance-covariance matrix of η_i and ρ is a correlation vector between v_i and η_i. Based on (6.55), the equations in (6.53) and (6.54) can be written as:

$$lny_{Mi} = X_i'\beta + \tau\left(X_i - z_i'\delta\right) + \omega_i - u_i \tag{6.56}$$

where $\omega_i = \left(1 - \rho'\rho\right)v_i$ and $\tau = \sigma_v\rho'\Omega_\eta^{-1/2}$, which can be viewed as a correction term for bias. Conditional on X_i and z_i, the distribution of the composed error term in (6.56) is exactly the same as their traditional counterparts from the stochastic frontier literature. They then show that the joint log-likelihood function of lny_{Mi} and X_i is given by:

$$lnLF = lnLF_{y|X} + lnLF_X \tag{6.57}$$

The first part of the log-likelihood function, $lnLF_{y|X}$, is almost the same as that of a traditional stochastic frontier model where the residual is adjusted by the $\tau\left(X_i - z_i'\delta\right)$ factor. The second part, $lnLF_X$, is just the likelihood function of the reduced form equations in (6.54), that is the likelihood function of a multivariate normal variable. The likelihood function (6.57) can be maximized by brute force in a single step, one can use a two-step maximum likelihood estimation method if computational difficulties appear. In the first stage, $lnLF_X$ is maximized with respect to Ω_η and δ.

[19] In his model, the distribution of u_i is not allowed to have efficiency determinants.

In the second stage, the rest of the parameters are estimated by maximizing $lnLF_{y|X}$ taking the estimates of Ω_η and δ as given.[20]

The abovementioned ML model does not address the potential correlation with the inefficiency term. Amsler et al. (2016) is the first paper to allow endogeneity of the inputs with respect to statistical noise and inefficiency separately using a copula specification of the joint distribution of the three random variables. One obvious difficulty with this approach is the need to specify a copula. Another difficulty of this approach is that it may be computationally challenging. Tran and Tsionas (2015) also use a Gaussian copula function to directly model the dependency of the endogenous regressors and the composed error without using instrumental variables. Consistent estimates can be obtained by maximizing the likelihood function in a two-step procedure. The first step requires, however, using numerical integration as in Amsler et al. (2016).[21]

6.4.3 Unobserved Heterogeneity

Many industries worldwide are incentive regulated. The aim is to provide firms with incentives to improve their efficiency and to ensure that consumers benefit from the gains. As regulators reward or penalize firms in line with their respective (in)efficiency levels, the reliability of these scores is crucial for the fairness and effectiveness of the regulatory framework. Obtaining reliable measures of firms' inefficiency requires controlling for the different environmental conditions under which each firm operates. This is particularly important in the case of benchmarking of electricity, gas and water networks where the results of efficiency analysis have important financial implications for the firms. However, there are many characteristics (e.g. geography, climate or network characteristics) that affect firms' production (costs) but which are *unobserved* or *omitted* variables.

Several statistical methods have been developed in the SFA literature to address this issue. A simple or naïve strategy is the sample separation approach. Estimation of the technology is carried out in two stages. First, the sample observations are classified into several groups. In the second stage, separate analyses are carried out for each class, conditional on the first-stage (maybe ad hoc) sample separation. More sophisticated and popular approaches to deal with omitted variables use panel data, random coefficients, latent class models or spatial econometrics.

---

[20] However, the standard errors from this two-stage method are inconsistent because the estimates are conditional on the estimated error terms from the first stage. Kutlu (2010) suggests using a bootstrapping procedure in order to get the correct standard errors.

[21] In the abovementioned papers, there were no environmental variables determining firms' inefficiency. Amsler et al. (2017) provides a systematic treatment of endogeneity in stochastic frontier models and allows environmental variables to be endogenous because they are correlated with either the statistical noise or the basic inefficiency term or both.

6.4.3.1 Panel Data Models

For instance, the True Fixed/Random Effects models introduced by Greene (2005) capture the unobserved heterogeneity through a set of firm-specific intercepts α_i:

$$lny_{Mit} = \alpha_i + X'_{it}\beta + v_{it} - u_{it} \tag{6.58}$$

If we treat α_i as fixed parameters which are not part of inefficiency, then the above model becomes the "True Fixed Effects" (TFE) panel stochastic frontier model. The model is labelled as "True Random Effects" model when α_i is treated as a time-invariant random variable. Estimation of the model in (6.58) is not easy. When the number of firms is too large, the model encounters the incidental parameter problem. This problem appears when the number of parameters to be estimated increases with the number of cross-sectional observations in the data. In this situation, consistency of the parameter estimates is not guaranteed even if $N \to \infty$.

Wang and Ho (2010) solve the problem in Greene (2005) using temporal transformations of (6.58). In order to remove time-invariant firm-specific effects, they carried out first-differences and within transformations of the model. Their within transformation of the inefficiency term (see Eq. 6.24) is reproduced below with our notation:

$$u_{it}^{w} = u_{it} - \frac{1}{T}\sum\nolimits_{t=1}^{T} u_{it} = u_{it} - u_{i.} \tag{6.59}$$

As u_{it}^{w} is the difference of "two" one-sided error terms, the distribution of u_{it}^{w} is not known if u_{it} is independently distributed over time. To get a closed-form for the likelihood function, they assumed that the inefficiency term u_{it} possesses the scaling property so that it can be multiplicatively decomposed into two components as follows:

$$u_{it} = h(z_{it}, \delta) \cdot u_i^* = h_{it} \cdot u_i^* \tag{6.60}$$

where $h_{it} \geq 0$ is a function of firm exogenous variables, and $u_i^* \geq 0$ is a firm-specific and *time-invariant* inefficiency term which captures aspects such as the manager's natural skills which are viewed as random. This implies that the within-transformed inefficiency term in (6.59) can be rewritten as:

$$u_{it}^{w} = \left(h_{it} - \frac{1}{T}\sum\nolimits_{t=1}^{T} h_{it}\right) \cdot u_i^* \tag{6.61}$$

Note that the distribution of u_i^* is not affected by the within-transformation. This key aspect of their model enabled them to get a tractable likelihood function for their transformed model.

Both TFE and WH models capture the unobserved heterogeneity through a set of firm-specific intercepts α_i. They are not linked to firms' inefficiency. However, if we use the SS method, the adjusted individual effects are used to provide a measure of persistent (time-invariant) inefficiency. In order to separate persistent inefficiency from both time-invariant unobserved heterogeneity and transient (time-varying) inefficiency, Filippini and Greene (2016) propose estimating a model with four error terms:

$$lny_{Mit} = \beta_0 + X'_{it}\beta + \alpha_i + v_{it} - (u_{\text{i}} + \tau_{it}) \tag{6.62}$$

where α_i captures time-invariant unobserved heterogeneity, v_{it} is the traditional noise term that follows a normal distribution, u_{i} is one-sided error term capturing persistent inefficiency, and τ_{it} is one-sided error term capturing transient inefficiency. Estimation of the model in (6.62) can be undertaken in a single stage ML method based on the distributional assumptions on the four error terms (Colombi et al. 2014).[22] Kumbhakar et al. (2015) consider a simpler multi-stage procedure if the model is rewritten as:

$$lny_{Mit} = \beta_0^* + X'_{it}\beta + \alpha_i^* + \omega_{it} \tag{6.63}$$

where $\beta_0^* = \beta_0 - E(u_{\text{i}}) - E(\tau_{it}), \alpha_i^* = \alpha_i - u_{\text{i}} + E(u_{\text{i}})$ and $\omega_{it} = v_{it} - \tau_{it} + E(\tau_{it})$. This model can be easily estimated in three stages. In the first stage, we estimate (6.63) using a FE or RE estimator and get the first-stage fixed effects ($\hat{\alpha}_i^*$) and residuals ($\hat{\omega}_{it}$). In a second stage, we estimate a standard SFA model regressing $\hat{\omega}_{it}$ on an intercept, which can be interpreted as an estimate of $E(\tau_{it})$. Using the Jondrow et al. (1982) formula, we decompose $\left(\hat{\omega}_{it} - \hat{E}(\tau_{it})\right)$ into $\hat{v}_{it}$ and $\hat{\tau}_{it}$. In the third stage, we estimate a SF model regressing $\hat{\alpha}_i^*$ on an intercept, which can be interpreted as an estimate of $E(u_{\text{i}})$. Using again Jondrow et al. (1982), we next decompose $\left(\alpha_i^* - \hat{E}(u_{\text{i}})\right)$ into $\hat{\alpha}_i$ and $\hat{u}_i$.

To conclude this subsection, it is worth mentioning that the above panel data models only use the temporal (i.e. within) variation contained in the data to estimate the coefficients of the main production drivers. This is quite problematic in many applications because many important determinants of firm costs (production) are persistent or slow changing variables (such as the energy delivered or number of customers in electricity distribution).

6.4.3.2 Latent Class Models

Possible differences among firms associated with their use of different technologies are also often addressed using latent class models. The latent class stochastic frontier (hereafter LCSF) models combine the stochastic frontier approach with a latent

[22]Filippini and Greene (2016) proposed a simulation-based procedure to circumvents many of the challenges that appears when estimating the model by brute force maximization.

class structure (see Orea and Kumbhakar 2004; Greene 2005, for some applications). A conventional LCSF model assumes there is a finite number of technologies (classes) underlying the data and allocates probabilistically each firm in the sample to a particular technology.[23]

Let us first assume that there are J different technologies, and that each firm belongs to one, and only one, of these technologies. Conditional on technology $j(=1, \ldots, J)$, the general specification of the LCSF model can be written as follows:

$$lny_{Mi} = X_i'\beta_j + v_{i|j} + u_{i|j} \tag{6.64}$$

where $v_{i|j} \sim N(0, \sigma_{vj})$ is a noise term that follows a normal distribution, and $u_{i|j} \sim N^+(0, \sigma_{uj})$ is a one-sided error term capturing firms' inefficiency.[24] Given that the researcher lacks knowledge as to whether a particular firm belongs to class j or another, the class-membership probability should be estimated simultaneously alongside other parameters of the model. Following Greene (2005), the class probabilities are parameterized as a multinomial logit function:

$$\Pi_{ij}(\gamma_j) = \frac{exp(q_i'\gamma_j)}{1 + exp(q_i'\gamma_j)}, \quad j = 1, \ldots, J-1 \tag{6.65}$$

where q_i is a vector of firm-specific variables. The last probability is obtained residually taking into account that the sum of all probabilities should be equal to one. The *unconditional* likelihood for firm i is obtained as the weighted sum of their technology-specific likelihood functions, where the weights are probabilities of technology-class membership, Π_{ij}. That is:

$$LF_i(\theta) = \sum_{j=1}^{J} LF_{i|j}(\beta_j, \sigma_{vj}, \sigma_{uj})\Pi_{ij}(\gamma_j) \tag{6.66}$$

where θ encompasses all parameters. The overall likelihood function can be written as:

$$ln\, LF(\theta) = \sum_{i=1}^{N} ln\, LF_i(\theta) = \sum_{i=1}^{N} ln\left\{\sum_{j=1}^{J} LF_{i|j}(\beta_j, \sigma_{vj}, \sigma_{uj})\Pi_{ij}(\gamma_j)\right\} \tag{6.67}$$

[23]The LCSF model is similar to the stochastic frontier model with random coefficients introduced by Tsionas (2002), in the sense that a latent class model can be viewed as a discrete approximation to a (continuous) random coefficient model (see Greene 2005, p. 287).

[24]Orea and Jamasb (2017) assumed the existence of two behavioral classes: fully efficiency and inefficient. While in the "inefficient" class it is assumed that $u_{i|j}$ follows a half-normal distribution, the "fully efficient" class is defined by imposing that the variance of the pre-truncated normal distribution is zero, i.e. $\sigma_{uj}^2 = 0$.

Maximizing the above maximum likelihood function gives asymptotically efficient estimates of all parameters. The estimated parameters can then be used to compute (unconditional) posterior class membership probabilities for each technology. The posterior probabilities can be used to allocate each firm to a technology-class with highest probability.

6.4.3.3 Spatial Frontier Models

A common feature of the above approaches is that they ignore the spatial structure of the data. Orea et al. (2018) advocate using a different empirical strategy to account for the unobserved differences in environmental conditions based on firms' geographic location. Indeed, as many unobservable variables are likely to be spatially correlated (such as weather and geographic conditions, population structure, electricity demand patterns, input prices, etc.), an alternative empirical strategy emerges. Their spatial-based approach can be used in panel data settings. Indeed, as they utilize different (spatial vs. temporal) dimensions of our data, they can be viewed as *complementary* approaches to deal with unobserved variables.

Orea et al. (2018) proposed a frontier model with cross-sectional correlation in the noise term, which can be written assuming a single input as:

$$lny_{Mi} = X_i'\beta + v_i - u_i = X_i'\beta + (z_i + \omega_i) - u_i \tag{6.68}$$

$$z_i = \lambda W_i z \tag{6.69}$$

where z_i represent *unobserved* environmental variables that are spatially correlated, and ω_i is the traditional non-spatially correlated noise term, z is a vector of $Nx1$ unobserved environmental variables, W_i is a known $1xN$ spatial weight vector with elements that are equal to unity if firms i and j operate in neighbouring areas (markets), and equal to zero otherwise. The term λ is a coefficient that measures the degree of spatial correlation between the unobserved environmental variables. Hence the spatial effects estimated in this model lack an economic interpretation as they are completely "spurious". Equation (6.68) can be alternatively rewritten as follows:

$$lny_{Mi} = X_i'\beta + \lambda W_i lny_M + W_i X'(-\lambda\beta) + \tilde{\omega}_i - \tilde{u}_i \tag{6.70}$$

where $\tilde{\omega}_i = \omega_i - \lambda W_i \omega$, $\tilde{u}_i = u_i - \lambda W_i u$, lny_M is a vector of $Nx1$ production levels, X is a vector of $Nx1$ explanatory variables, u is $Nx1$ vectors of the firms' inefficiency terms, and ω is again $Nx1$ vectors of the firms' non-spatially correlated noise terms.

Equation (6.70) resembles a conventional spatial econometric model. However, only one additional coefficient is estimated here, i.e. the coefficient of the spatially lagged dependent variable. Unlike other similar spatial econometric models, this coefficient should not be interpreted as the effect of neighbours' production on the production of a particular firm. Rather, λ is measuring the spatial correlation between the unobserved or omitted variables in our sample. On the other hand, it is worth mentioning that (6.70) is similar to the Durbin Stochastic Frontier (SDF) model introduced recently by Glass et al. (2016) in which they propose estimating the following model:

$$lny_{Mi} = X_i'\beta + \lambda W_i lny_M + W_i X'\theta + v_i - u_i \tag{6.71}$$

Notice that θ in the above SDF model is not restricted to be equal to $-\lambda\beta$. In this sense, our spatial model in (6.70) is nested in the SDF model. However, no spatially correlated omitted (random) variables are explicitly modelled in the SDF model. While the spatial spillovers in Glass et al. (2016) have an economic or causal interpretation, the spatial spillovers in (6.70) are simply associated with the omitted variables.

Orea et al. (2018) discuss how to estimate (6.70) taking into account that this model includes two spatially correlated error terms. They propose a two-step procedure. In the first step, Eq. (6.70) is estimated ignoring the spatial and frontier structure of the composed error term. The degree of spatial correlation of omitted variables (i.e. parameter λ) and other coefficients of the frontier model are estimated using GMM because the spatially lagged dependent variable is endogenous. The estimated λ parameter is then used to get a predicted value for z_i. In the second step, they estimate (6.68) once the original omitted variable z_i is replaced with its predicted counterpart.

Orea and Álvarez (2019) develop a cross-sectional (spatial) frontier model that explicitly allows for cross-sectional (spatial) correlation in both noise and inefficiency terms. Their model can be written as:

$$lny_{Mi} = X_i'\beta + \tilde{v}_i(\rho) + \tilde{u}_i(\tau) \tag{6.72}$$

where both error terms are cross-sectionally correlated using spatial moving average (SMA) or spatial autoregressive (SAR) spatial stochastic processes. The coefficients ρ and τ measure the degrees of cross-sectional (spatial) correlation between firms' noise and inefficiency terms, respectively. In a SMA specification of the model, the noise and inefficiency terms are defined as $\tilde{v}_i = v_i + \rho W_i v$, and $\tilde{u}_i = u_i + \tau W_i u$. A SAR specification for the two error terms can be expressed as: $\tilde{v}_i = v_i + \rho W_i \tilde{v}$, and $\tilde{u}_i = u_i + \tau W_i \tilde{u}$.

Note that (6.72) has the structure of a traditional SFA model as it includes a noise term ($\tilde{v}_i$) and an inefficiency term ($\tilde{u}_i$). However, the above model cannot be estimated using full maximum likelihood because the distribution of $\tilde{u}_i$ is *generally* not known *if* we assume that u_i is independently distributed across firms (see, for instance, Wang 2003). To address this issue, Areal et al. (2012), Tsionas and Michaelides (2016) and Schmidt et al. (2009) proposed several computational algorithms based on Gibbs sampling or simulated ML. In contrast, Orea and Álvarez (2019) assumed that the basic inefficiency term u_i possesses the scaling property, but they replace Wang and Ho (2010)'s firm-specific term u_i^* with an industry-specific term u^*:

$$u_i = h(z_i, \delta) \cdot u^* = h_i \cdot u^* \tag{6.73}$$

where $h_i \geq 0$ is again function of firm exogenous variables, and $u^* \geq 0$ is an industry-specific inefficiency term. For simplicity, Orea and Álvarez (2019) assume that $u^* \sim N^+(0, \sigma_u)$.[25] The above specification of u_i implies that the SMA-transformed inefficiency term can be written as:

$$\tilde{u}_i = u_i + \tau W_i u = \left(h_i + \tau \frac{1}{n_i} \sum\nolimits_{j \in A_i} h_j\right) u^* = \tilde{h}_i \cdot u^* \tag{6.74}$$

or, in simpler notation:

$$\tilde{u} = (I_N + \tau W)u = M_\tau u = M_\tau h u^* = \tilde{h} \cdot u^* \tag{6.75}$$

where $h = (h_1, \ldots, h_N)$, and $\tilde{h} = \left(\tilde{h}_1, \ldots, \tilde{h}_N\right)$ are N × 1 vectors of *idiosyncratic* and *generalized* scaling functions, respectively. If the inefficiency term instead follows a SAR process, we just need to replace $M_\tau = I_N + \tau W$ with $M_\tau = (I_N - \tau W)^{-1}$. Regardless of whether SMA or SAR processes are assumed, the half-normal distribution of u^* is not affected by the cross-sectional transformation. This is the crucial aspect of the model that enables Orea and Álvarez (2019) to get a tractable likelihood function that can be maximized using standard software. In this sense, the proposed model can be viewed as a new application of the scaling property in SFA analyses. Moreover, some portions of the model can also be estimated using non-linear least squares (NLLS).

6.4.4 Dynamic Efficiency

Two different approaches have been used in the literature to incorporate the dynamic nature of the decision-making process into efficiency analyses: reduced-form models and structural models.[26]

6.4.4.1 Reduced-Form Models

The *reduced-form models* do not define explicitly a mathematical representation of dynamic behaviour of the firm but recognize a persistence effect of firms' inefficiency over time and specify its evolution as an autoregressive process. For instance, Tsionas (2006) departs from a typical stochastic production frontier of the following form:

[25] As the random inefficiency component in (6.73) does not vary across firms, consistency of σ_u can be obtained if we use a panel data set and $T \rightarrow \infty$.

[26] For a more comprehensive review of this literature see Emvalomatis (2009).

$$lny_{Mit} = X'_{it}\beta + v_{it} + lnET_{it} \tag{6.76}$$

where $ET_{it} = e^{-u_{it}} \leq 1$ is the usual technical efficiency of firm i in period t. To avoid the complications inherent in the specification of autoregressive processes on non-negative variables, Tsionas (2006) converts the technical efficiency term into an autoregressive form using $s_{it} = \ln(-lnET_{it})$ instead of directly $lnET_{it}$[27]:

$$s_{it} = z'_{it}\delta + \rho s_{it-1} + \xi_{it} \tag{6.77}$$

The distinguishing feature of (6.59) is that past values of efficiency determine the value of ET_{it}. Estimating the above dynamic stochastic frontier model is far from simple. While Tsionas (2006) estimate the model using Bayesian techniques, Emvalomatis et al. (2011) use Kalman filtering techniques and proceed to estimation by maximum likelihood.

6.4.4.2 Structural Models

The *structural models* that make explicit assumptions regarding the objective of the firm. For instance, the objective of the firm is often assumed to be the maximization of the following intertemporal problem (see Emvalomatis 2009; p. 30)[28]:

$$\begin{aligned} J = \max_{I,x} E^t\left\{\textstyle\int_0^\infty e^{-\rho t}\pi(y,x,k,I)dt\right\} \\ s.t. \qquad \dot{K} = I - \delta K \\ \vec{D} = \vec{D}(y,K,x,I,-g_x,g_I) \\ given\ K \end{aligned} \tag{6.78}$$

In this formulation, the objective of the firm is to maximize the expected discounted flow of an instantaneous profit function over time. The choice variables are the levels of variable inputs (x) to be employed and the level of investment in quasi-fixed inputs (I). While the first restriction describes the evolution of capital through time, the second restriction is a *dynamic* representation of technology in terms of a directional distance function. Given the level of quasi-fixed inputs, this function describes the vectors of outputs that can be produced from a given vector of variable inputs and gross investment. In the presence of adjustment costs in quasi-fixed inputs, the above dynamic specification indicate that static measures do not correctly reflect inefficiency.

Serra et al. (2011) and Tovar and Wall (2014) used the adjustment cost framework of Silva and Lansink (2013), but instead of DEA they carried out a parametric

[27] Alternatively, Emvalomatis et al. (2011) define $s_{it} = \ln(ET_{it}/(1 - ET_{it}))$ as the latent-state variable. In this specification, ρ measures the percentage change in the efficiency to inefficiency ratio that is carried from one period to the next.

[28] This subsection heavily relies on Emvalomatis' (2009) thesis.

estimation generalizing the static input-oriented directional distance function introduced by Färe et al. (2005). They use a quadratic functional form for the directional distance function because it is easy to impose the above translation property. Setting $(g_x, g_I) = (1, 1)$, their dynamic directional distance function can be written as:

$$\vec{D}(y_{it}, K_{it}, x_{it} - \alpha_{it}, I_{it} + \alpha_{it}) = \vec{D}(y_{it}, K_{it}, x_{it}, I_{it}) - \alpha_{it} \tag{6.79}$$

This simply states that if investment is expanded by α_{it} and input contracted $-\alpha_{it}$, the value of the distance function will be reduced by α_{it}. The above papers set $\alpha_{it} = I_{it}$. Stochastic estimation is accomplished by maximum likelihood procedures in Tovar and Wall (2014). Estimating the above directional distance function only provides estimates of technical inefficiency. To get cost efficiency scores in a dynamic framework, Serra et al. (2011) and Tovar and Wall (2014) propose estimating the following (quadratic) cost frontier model:

$$C_{it} = rW(y_{it}, K_{it}, w_{it}) - W_K(\cdot)\dot{K}_{it} + v_{it} + u_{it} \tag{6.80}$$

where C_{it} is observed cost (normalized by a variable input price), $W(y_{it}, K_{it}, w_{it})$ is optimum cost, $W_K(\cdot)$ is its derivative with respect to the capital stock; v_{it} is white noise and u_{it} is a one-sided term measuring firms' cost inefficiency. The dynamic directional distance function (6.79) allows estimating technical inefficiency of both variable and quasi-fixed inputs. The parametric dynamic cost model (6.80) allows estimating the dynamic cost inefficiency defined as the difference between the observed shadow cost of input use and the minimum shadow cost. Finally, an allocative inefficiency score can be obtained as the difference between dynamic cost inefficiency and dynamic technical inefficiency.

6.4.5 Production Risk

Most of the literature measuring firms' production performance lacks an explicit recognition that production takes place under conditions of uncertainty. Although SFA models are stochastic, their stochastic elements arise primarily from econometric concerns (measurement error, missing variables) and not as an endogenous response to the stochastic environment in which firms operate. Ignoring uncertainty in efficiency and productivity analyses may have remarkable welfare and policy implications, which serve to jeopardize our interpretation of the efficiency measures and also bias our representation of the stochastic technology. This may be a serious issue in many applications, such as agriculture, fishing or banking where production uncertainty is relatively high.

Early papers (e.g. Just and Pope 1978) used a simple production function with heteroskedastic error terms to represent production risk. Kumbhakar (2002), among others, extended this framework and proposed estimating the following single output SFA model:

$$y_i = f(x_i, \beta) + g(x_i, \lambda)\{v_i - u_i\} \tag{6.81}$$

where $g(x_i, \lambda)$ is the output risk function. If the variance of the composed random term is normalized to 1, the variance of output is therefore $g(x_i, \lambda)$. In this framework, an input is risk-increasing (reducing) (neutral) according to $\partial g(x_i, \lambda)/\partial x_i >$ $(<)(=)0$. Kumbhakar assumed later on that producers maximize the expected utility of anticipated profits. Assuming a single input, the first-order condition of the above problem can be expressed as:

$$\frac{\partial f(x_i, \beta)}{\partial x_i} = w - \theta(\cdot)\frac{\partial g(x_i, \beta)}{\partial x_i} \tag{6.82}$$

where w is the input price relative to the output price, and $\theta(\cdot)$ is a risk preference function that measures firms' risk aversion.[29] Orea and Wall (2012) used the above framework to show that productivity and welfare changes do not necessarily follow the same path under conditions of uncertainty. Although risk aversion coefficients can be estimated from this equation (or a system of equations in the case of more inputs), it is difficult to derive an algebraic form of the risk preference function that keeps the model simple for estimation purposes.

A common feature of the previous model is that it is developed using standard stochastic frontier models that are too simple to account properly for the stochastic elements of the producer decision environment. In this sense, O'Donnell et al. (2010) show that the application of standard methods of efficiency analysis to data arising from production under uncertainty may give rise to spurious findings of efficiency differences between firms. Chambers and Quiggin (2000) proposed to model uncertainty in terms of a state-contingent technology to deal with this issue. Empirical application of the state-contingent approach has proved difficult because most of the data needed to estimate these models are lost in unrealized states of nature (i.e. outputs are typically observed only under one of the many possible states of nature). O'Donnell and Griffiths (2006) show how to estimate state-contingent models using a latent class model approach if the technology is "output-cubical" in the terminology of Chambers and Quiggin (2000).[30] In this case, the production technology can be described by the set of state-contingent production functions:

$$lny_i = \alpha_s + f(x_i, \beta) + v_{is} - u_{is} \tag{6.83}$$

where α_s is a state-varying intercept that allows expected log-output to vary across the states of nature. The standard deviation of v_{is} is assumed state-dependent. Technically

[29]The coefficient of risk aversion in this equation can be viewed as a measure of overall risk preferences regarding both noise and inefficiency terms.

[30]Chavas (2008) proposes a method that allows the researcher to test whether or not the state-contingent technology is "output-cubical" if the states are independently distributed across observations. The main limitation of this method is that it focuses exclusively on the observed outputs. As such, the approach neglects the potential outputs that could have been obtained had nature selected different states.

inefficiency will also be expected to differ across states. The above model can be viewed as a conventional SFA model with state-specific parameters. As the state of nature that has produced each observation is not observed, O'Donnell and Griffiths (2006) nest the above model into a latent class model (LCM) structure, where both state-specific production functions and the probabilities for the realization of each state are estimated simultaneously by ML techniques.[31]

6.4.6 Total Factor Productivity Decomposition

An estimated distance function can constitute the building block for the measurement of productivity change and its decomposition into its basic sources. First, let us add a t superscript to all variables of the output distance function (6.19) and a time trend to capture technological changes over time. Taking into account that $u_{it} = -lnD_{it}$, the distance function in period t can be rewritten as:

$$lnD_{it} = lny_{Mit} + \ln D\left(x_{it}, \frac{y_{it}}{y_{Mit}}, t, \beta\right) + v_{it} = \ln D(x_{it}, y_{it}, t, \beta) + v_{it} \quad (6.84)$$

If we take first differences, we get:

$$\Delta lnD_{it} = \Delta \ln D(x_{it}, y_{it}, t, \beta) + \Delta v_{it} \quad (6.85)$$

As the average change in the noise term tend to vanish over time, we hereafter ignore Δv_{it} for notational ease. We next assume that the above distance function has a Translog form. Since the Translog distance function is ***quadratic*** in logs, the change in the value of the distance function can be decomposed as:

$$\begin{aligned}\Delta \ln D(x_{it}, y_{it}, t, \beta) &= \frac{1}{2}\sum\nolimits_{m=1}^{M}(\varepsilon_{mi}(t) + \varepsilon_{mi}(t-1))\Delta lny_{mit} \\ &+ \frac{1}{2}\sum\nolimits_{j=1}^{J}\big(\varepsilon_{ji}(t) + \varepsilon_{ji}(t-1)\big)\Delta lnx_{jit} + \frac{1}{2}(\varepsilon_t(t) + \varepsilon_t(t-1))\end{aligned} \quad (6.86)$$

[31]Note that the elasticity of expected output with respect to the input in (6.64) is state-invariant. This property may be implausible in some production contexts (e.g. irrigation in rainy and dry seasons). If we allow the slope coefficients in (6.64) to vary across states of nature, an identification (or labelling) problem arises. If there are only two different states of nature, which class should be labelled as a "bad" or "good" state? To solve the identification problem, O'Donnell and Griffiths (2006) suggest scaling the inputs so that $x_i = 0$ at the sample mean. The state with the lowest (highest) α_s will be labelled as "bad" ("good") state. In this case, however, this labelling is local in the sense that it is only valid for the "representative" firm. O'Donnell and Griffith (2006) rely on Bayesian estimation to address the identification problem and impose the labelling restriction globally.

where $D(t)$ is short for $D(x_{it}, y_{it}, t, \beta)$, $\varepsilon_{mi}(t) = \frac{\partial lnD(t)}{\partial lny_{mi}}$ is the elasticity of the distance function with respect to y_{mi}, $\varepsilon_{ji}(t) = \frac{\partial lnD(t)}{\partial lnx_{ji}}$ is the elasticity of the distance function with respect to x_{ji}, and $\varepsilon_t(t) = \frac{\partial lnD(t)}{\partial t}$ is the rate of technical change evaluated at period t. In order to measure total factor productivity changes, Orea (2002) proposed the following Generalized Malmquist Productivity Index:

$$lnG_{t,t-1} = \frac{1}{2}\sum\nolimits_{m=1}^{M}\left(\frac{\varepsilon_{mi}(t)}{\sum_{m=1}^{M}\varepsilon_{mi}(t)} + \frac{\varepsilon_{mi}(t-1)}{\sum_{m=1}^{M}\varepsilon_{mi}(t-1)}\right)\cdot \Delta lny_{mit}$$
$$-\frac{1}{2}\sum\nolimits_{j=1}^{J}\left(\frac{\varepsilon_{ji}(t)}{\sum_{j=1}^{J}\varepsilon_{ji}(t)} + \frac{\varepsilon_{ji}(t-1)}{\sum_{j=1}^{J}\varepsilon_{ji}(t-1)}\right)\cdot \Delta lnx_{jit} \quad (6.87)$$

Notice that in (6.87) we have not imposed the linear homogeneity in outputs of the distance function $\sum_{m=1}^{M}\varepsilon_{mi}(t) = 1$ in order to show that it can also be used with input distance functions. Inserting (6.67) into (6.66), Orea (2002) obtained the following parametric decomposition of the Malmquist productivity index (6.66)[32]:

$$lnG_{t,t-1} = \Delta lnD_{it} - \frac{1}{2}(\varepsilon_t(t) + \varepsilon_t(t-1))$$
$$+\frac{1}{2}\sum\nolimits_{j=1}^{J}\left(\frac{\varepsilon_{ji}(t)}{\sum_{j=1}^{J}\varepsilon_{ji}(t)}EE(t) + \frac{\varepsilon_{ji}(t-1)}{\sum_{j=1}^{J}\varepsilon_{ji}(t-1)}EE(t-1)\right)\cdot \Delta lnx_{jit} \quad (6.88)$$

where $EE(t) = -\sum_{j=1}^{J}\varepsilon_{ji}(t) - 1$ is a measure of firms' economies of scale. Equation (6.88) provides a meaningful decomposition of a total factor productivity indicator into changes in technical efficiency (TE), technical change (TC) and a scale effect (SE) that vanishes under the assumption of constant returns to scale or constant input quantities.

It should be pointed out that the above decomposition does not individualize any output or input mix effect. However, an input mix effect can be easily obtained if we measure the scale effect with respect to the *average* input change, instead of the change of each input. In this case, the scale effect in (6.88) can be in turn decomposed in a *pure* scale effect and a term measuring relative changes in the input mix:

$$SE = \left\{\frac{1}{2}\sum\nolimits_{j=1}^{J}\left(\frac{\varepsilon_{ji}(t)}{\sum_{j=1}^{J}\varepsilon_{ji}(t)}EE(t) + \frac{\varepsilon_{ji}(t-1)}{\sum_{j=1}^{J}\varepsilon_{ji}(t-1)}EE(t-1)\right)\right\}\cdot \Delta\ln\bar{x}_{it}$$
$$+\frac{1}{2}\sum\nolimits_{j=1}^{J}\left(\frac{\varepsilon_{ji}(t)}{\sum_{j=1}^{J}\varepsilon_{ji}(t)}EE(t) + \frac{\varepsilon_{ji}(t-1)}{\sum_{j=1}^{J}\varepsilon_{ji}(t-1)}EE(t-1)\right)\cdot \Delta\ln\tilde{x}_{jit} \quad (6.89)$$

[32] A similar decomposition can be obtained from a parametric directional distance function using a Luenberger productivity index (see Färe et al. 2008; p. 593).

where $\ln \bar{x}_{it} = \frac{1}{J}\sum_{j=1}^{J} lnx_{jit}$ and $\ln \tilde{x}_{jit} = lnx_{jit} - \ln \bar{x}_{it}$. A similar output mix effect can be obtained if we decompose the output growth in Eq. (6.87) taking into account the *average* change in outputs.

6.5 Concluding Remarks

This chapter serves as guide to efficiency evaluation from an econometric perspective. The analytical framework relies on the most general parametric models and up to date representations of the production technology through Translog and Quadratic distance functions. We conclude this chapter emphasizing the importance of choosing a suitable analytical framework that is in accordance with the industry characteristics and the restrictions faced by the firm, most particularly the relative discretion that managers have over output production and input usage. This sets the stage for the economic objective of the firm that often is assumed to maximize profits (profitability) or minimize cost. The next question that scholars face is the choice of methods that are available to study variability in firm performance.

The dispersion of results obtained with the methods surveyed in this chapter is a general matter of concern that has been addressed by several authors, who employing the same datasets resort to compare the similarity of the distributions of the efficiency scores (see, e.g. Cummins and Zi 1998; Bauer et al. 1998). Ultimately, what matters is the ability to provide reliable results on individual performance, not only for the managers of the firms operating within an industry, but also for stakeholders and government agencies involved in regulation, competition and general policy analysis. In general, the higher the consistency of efficiency results in terms of rankings, temporal stability, etc., the more confidence regulators and competition authorities will have on the conclusions derived from them, and the intended effect of their policy decisions.

We thus conclude emphasizing the relevance of the methods surveyed in this chapter in unveiling the economic performance of firm in terms of technical (and allocative) efficiencies. Many challenges are still ahead, but cross fertilization of ideas with other research fields will result in a better understanding of the ultimate causes and consequences of inefficient economic performance.

Appendix A

Assume that we have estimated the following multi-input multi-output distance function:

$$\ln D = \ln D\left(x, y, \hat{\beta}\right) \tag{A.1}$$

In order to examine relevant features of firms' technology we should first notice that they must be computed once we assume that the observation belongs to the frontier, i.e. that D = 1. Next, we must differentiate the distance function taking into account that $dD = 0$ as we are moving over the frontier. After some simple manipulations we get:

$$0 = \frac{dD}{D} = \sum_{j=1}^{J} \frac{\partial D}{\partial x_j} \cdot \frac{x_j}{D} \cdot \frac{dx_j}{x_j} + \sum_{m=1}^{M} \frac{\partial D}{\partial y_m} \cdot \frac{y_m}{D} \cdot \frac{dy_m}{y_m} \tag{A.2}$$

or

$$0 = \sum_{j=1}^{J} \varepsilon_{Dj} \cdot dlnx_j + \sum_{m=1}^{M} \varepsilon_{Dm} \cdot dlny_m \tag{A.3}$$

where $\varepsilon_{Dj} = \partial lnD/\partial lnx_j$ and $\varepsilon_{Dm} = \partial lnD/\partial lny_m$. The elasticity of output m with respect to input j can be computed once we assume above that $dlnx_k = 0 \forall k \neq j$ and $dlny_n = 0 \forall n \neq m$, that is:

$$0 = \varepsilon_{Dj} dlnx_j + \varepsilon_{Dm} dlny_m \tag{A.4}$$

This yields the following expression for this specific elasticity:

$$\varepsilon_{mj} = \frac{dlny_m}{dlnx_j} = -\frac{\varepsilon_{Dj}}{\varepsilon_{Dm}} \tag{A.5}$$

Notice that the above elasticity can be computed from both input, output and directional distance function. A return to scale measure (RTS) can be obtained using a similar fashion. In this case, we are interested in the proportional change in outputs caused by a proportional change in all inputs. This implies that $lnx_j = dlnx \forall j = 1, \ldots, J$, and $dlny_m = dlny \forall m = 1, \ldots, M$.

$$0 = \sum_{j=1}^{J} \varepsilon_{Dj} dlnx + \sum_{m=1}^{M} \varepsilon_{Dm} dlny \tag{A.6}$$

This yields the following expression for the RTS measure:

$$RTS = \frac{dlny}{dlnx} = -\frac{\sum_{j=1}^{J} \varepsilon_{Dj}}{\sum_{m=1}^{M} \varepsilon_{Dm}} \tag{A.7}$$

Again, the above scale elasticity can be computed from both input, output and directional distance function. However, it can be simplified if we take into account that they satisfy the corresponding homogeneity or transition properties. For instance, if an input distance function has been estimated, $\sum_{j=1}^{J} \varepsilon_{Dj} = 1$, and hence the RTS in (A.7) is equal to:

$$RTS = -\left(\sum\nolimits_{m=1}^{M} \varepsilon_{Dm}\right)^{-1} \tag{A.8}$$

If instead an output distance function has been estimated, $\sum\nolimits_{m=1}^{M} \varepsilon_{Dm} = 1$, and hence the RTS in (A.7) collapses to:

$$RTS = -\sum\nolimits_{j=1}^{J} \varepsilon_{Dj} \tag{A.9}$$

References

Aigner, D. J., Lovell, C. A., & Schmidt, P. (1977). Formulation and estimation of stochastic frontier production functions. *Journal of Econometrics, 6*(1), 21–37.

Álvarez, A., Amsler, C., Orea, L., & Schmidt, P. (2006). Interpreting and testing the scaling property in models where inefficiency depends on firm characteristics. *Journal of Productivity Analysis, 25,* 201–212.

Amsler, C., Prokhorov, A., & Schmidt, P. (2016). Endogeneity in stochastic frontier models. *Journal of Econometrics, 190*(2), 280–288.

Amsler, C., Prokhorov, A., & Schmidt, P. (2017). Endogenous environmental variables in stochastic frontier models. *Journal of Econometrics, 199,* 131–140.

Areal, F. J., Balcombe, K., & Tiffin, R. (2012). Integrating spatial dependence into stochastic frontier analysis. *Australian Journal of Agricultural and Resource Economics, 56*(4), 521–541.

Battese, G. E., & Coelli, T. J. (1995). A model for technical inefficiency effects in a stochastic frontier production function for panel data. *Empirical Economics, 20*(1), 325–332.

Bauer, P. W., Berger, A. N., Ferrier, G. D., & Humphrey, D. B. (1998). Consistency conditions for regulatory analysis of financial institutions: A comparison of frontier efficiency methods. *Journal of Economics and Business, 50*(2), 85–114.

Caudill, S. B., Ford, J. M., & Gropper, D. M. (1995). Frontier estimation and firm-specific inefficiency measures in the presence of heteroscedasticity. *Journal of Business, 13,* 105–111.

Chambers, R., & Quiggin, J. (2000). *Uncertainty, production, choice and agency: The state-contingent approach.* New York: Cambridge University Press.

Chambers, R. G., Chung, Y., & Färe, R. (1996). Benefit and distance functions. *Journal of Economic Theory, 70,* 407–419.

Chambers, R. G., Chung, Y., & Färe, R. (1998). Profit, directional distance functions and Nerlovian efficiency. *Journal of Optimization Theory and Applications, 95*(2), 351–364.

Chavas, J. P. (2008). A cost approach to economic analysis under state-contingent production uncertainty. *American Journal of Agricultural Economics, 90*(2), 435–446.

Coelli, T., & Perelman, S. (1996). *Efficiency measurement, multiple-output technologies and distance functions: With application to European railways.* No. DP 1996/05. CREPP.

Colombi, R., Kumbhakar, S., Martini, G., & Vittadini, G. (2014). Closed-skew normality in stochastic frontiers with individual effects and long/short-run efficiency. *Journal of Productivity Analysis, 42*(2), 123–136.

Cornwell, C., Schmidt, P., & Sickles, R. C. (1990). Production frontiers with cross-sectional and time-series variation in efficiency levels. *Journal of Econometrics, 46*(1–2), 185–200.

Cummins, J. D., & Zi, H. (1998). Comparison of frontier efficiency methods: An application to the U.S. life insurance industry. *Journal of Productivity Analysis, 10,* 131–152.

Diewert, W. E. (1971). An application of the Shephard duality theorem: A generalized Leontief production function. *Journal of Political Economy, 79,* 461–507.

Emvalomatis, G. (2009). *Parametric models for dynamic efficiency measurement*. Unpublished thesis.
Emvalomatis, G., Stefanou, S. E., & Lansink, A. O. (2011). A reduced-form model for dynamic efficiency measurement: Application to dairy farms in Germany and the Netherlands. *American Journal of Agricultural Economics, 93*(1), 161–174.
Färe, R., Grosskopf, S., & Margaritis, D. (2008). Efficiency and productivity: Malmquist and more. In H. Fried, C. A. Lovell, & S. S. Schmidt (Eds.), *The measurement of productive efficiency and productivity growth*. New York: Oxford University Press.
Färe, R., Grosskopf, S., Noh, D. W., & Weber, W. (2005). Characteristics of a polluting technology: Theory and practice. *Journal of Econometrics, 126,* 469–492.
Filippini, M., & Greene, W. (2016). Persistent and transient productive inefficiency: A maximum simulated likelihood approach. *Journal of Productivity Analysis, 45*(2), 187–196.
Glass, A. J., Kenjegalieva, K., & Sickles, R. C. (2016). A spatial autoregressive stochastic frontier model for panel data with asymmetric efficiency spillovers. *Journal of Econometrics, 190*(2), 289–300.
Greene, W. (1990). A Gamma-distributed stochastic frontier model. *Journal of Econometrics, 46*(1–2), 141–164.
Greene, W. (2005). Reconsidering heterogeneity in panel data estimators of the stochastic frontier model. *Journal of Econometrics, 126,* 269–303.
Griffiths, W. E., & Hajargasht, G. (2016). Some models for stochastic frontiers with endogeneity. *Journal of Econometrics, 190*(2), 341–348.
Griffiths, W. E., O'Donnell, C. J., & Tan- Cruz, A. (2000). Imposing regularity conditions on a system of cost and factor share equations. *Australian Journal of Agricultural and Resource Economics, 44*(1), 107–127.
Guan, Z., Kumbhakar, S. C., Myers, R. J., & Lansink, A. O. (2009). Measuring excess capital capacity in agricultural production. *American Journal of Agricultural Economics, 91,* 765–776.
Jondrow, J., Lovell, C. A., Materov, S., & Schmidt, P. (1982). On the estimation of technical efficiency in the stochastic frontier production function model. *Journal of Econometrics, 19*(2–3), 233–238.
Just, R. E., & Pope, R. D. (1978). Stochastic specification of production functions and economic implications. *Journal of Econometrics, 7,* 67–86.
Karakaplan, M. U., & Kutlu, L. (2013). *Handling endogeneity in stochastic frontier analysis: A solution to endogenous education cost frontier models*. Working paper, Department of Economics.
Kumbhakar, S. C. (2002). Specification and estimation of production risk, risk preferences and technical efficiency. *American Journal of Agricultural Economics, 84,* 8–22.
Kumbhakar, S. C., & Lovell, C. A. (2000). *Stochastic frontier analysis*. Cambridge: Cambridge University Press.
Kumbhakar, S. C., & Tsionas, E. G. (2006). Estimation of stochastic frontier production functions with input-oriented technical efficiency. *Journal of Econometrics, 133*(1), 71–96.
Kumbhakar, S. C., Asche, F., & Tveteras, R. (2013). Estimation and decomposition of inefficiency when producers maximize return to the outlay: An application to Norwegian fishing trawlers. *Journal of Productivity Analysis, 40,* 307–321.
Kumbhakar, S. C., Hung-Jen, W., & Horncastle, A. P. (2015). *A practitioner's guide to stochastic frontier analysis using stata*. Cambridge: Cambridge University Press.
Kumbhakar, S. C., Orea, L., Rodríguez-Álvarez, A., & Tsionas, E. G. (2007). Do we have to estimate an input or an output distance function? An application of the mixture approach to European railways. *Journal of Productivity Analysis, 27*(2), 87–100.
Kutlu, L. (2010). Battese-Coelli estimator with endogenous regressors. *Economic Letters, 109,* 79–81.
Lai, H. P., & Huang, C. J. (2010). Likelihood ratio tests for model selection of stochastic frontier models. *Journal of Productivity Analysis, 34,* 3–13.

Malikov, E., Kumbhakar, S. C., & Tsionas, M. G. (2015). A cost system approach to the stochastic directional technology distance function with undesirable outputs: The case of US banks in 2001–2010. *Journal of Applied Econometrics, 31*(7), 1407–1429.
Meeusen, W., & Van Den Broeck, J. (1977). Efficiency estimation from Cobb-Douglas production functions with composed error. *International Economic Review, 18*(2), 435–444.
O'Donnell, C. J., & Coelli, T. J. (2005). A Bayesian approach to imposing curvature on distance functions. *Journal of Econometrics, 126*(2), 493–523.
O'Donnell, C. J., & Griffiths, W. E. (2006). Estimating state-contingent production frontiers. *American Journal of Agricultural Economics, 88*(1), 249–266.
O'Donnell, C. J., Chambers, R. G., & Quiggin, J. (2010). Efficiency analysis in the presence of uncertainty. *Journal of Productivity Analysis, 33*(1), 1–17.
Orea, L. (2002). A parametric decomposition of a generalized Malmquist productivity index. *Journal of Productivity Analysis, 18,* 5–22.
Orea, L., & Álvarez, I. (2019). A new stochastic frontier model with cross-sectional effects in both noise and inefficiency terms. *Journal of Econometrics, 213*(2), 556–577.
Orea, L., Álvarez, I., & Jamasb, T. (2018). A spatial stochastic frontier model with omitted variables: Electricity distribution in Norway. *Energy Journal, 39*(3), 93–116.
Orea, L., & Kumbhakar, S. (2004). Efficiency measurement using stochastic frontier latent class model. *Empirical Economics, 29*(1), 169–183.
Orea, L., & Jamasb, T. (2017). Regulating heterogeneous utilities: A new latent class approach with application to the Norwegian electricity distribution networks. *Energy Journal, 38*(4), 101–128.
Orea, L., Llorca, M., & Filippini, M. (2015). A new approach to measuring the rebound effect associated to energy efficiency improvements: An application to the US residential energy demand. *Energy Economics, 49,* 599–609.
Orea, L., & Wall, A. (2012). Productivity and producer welfare in the presence of production risk. *Journal of Agricultural Economics, 63*(1), 102–118.
Orea, L., & Zofío, J. L. (2017). *A primer on the theory and practice of efficiency and productivity analysis*. Efficiency Series Paper 5/2017, Oviedo Efficiency Group, Department of Economics, University of Oviedo.
Orea, L., & Zofío, J. L. (2019). Common methodological choices in nonparametric and parametric analyses of firms' performance. In T. ten Raa & W. Greene (Eds.), *The Palgrave handbook of economic performance analysis*. Cham: Palgrave Macmillan.
Parmeter, C. F., & Kumbhakar, S. C. (2014). Efficiency analysis: A primer on recent advances. *Foundations and Trends in Econometrics, 7*(3–4), 191–385.
Rodríguez-Álvarez, A., & Lovell, C. A. (2004). Excess capacity and expense preference behavior in national health systems: An application to the Spanish public hospitals. *Health Economics, 13*(2), 157–169.
Schmidt, A. M., Moreira, A. R. B., Helfand, S. M., & Fonseca, T. C. O. (2009). Spatial stochastic frontier models: Accounting for unobserved local determinants of inefficiency. *Journal of Productivity Analysis, 31,* 101–112.
Schmidt, P., & Sickles, R. C. (1984). Production frontiers and panel data. *Journal of Business and Economic Statistics, 2*(4), 367–374.
Serra, T., Oude Lansink, A., & Stefanou, S. E. (2011). Measurement of dynamic efficiency: A directional distance function parametric approach. *American Journal of Agricultural Economics, 93*(3), 756–767.
Shephard, R. W. (1970). *Theory of cost and production functions*. Princeton, New Jersey: Princeton University Press.
Silva, E., & Oude Lansink, A. (2013). *Dynamic efficiency measurement: A directional distance function approach*. Centro de Economia e Finanças da UPorto, cef. upworking paper 2013-07.
Simar, L., Lovell, C. A., & van den Eeckaut, P. (1994). *Stochastic frontiers incorporating exogenous influences on efficiency*. Discussion paper no. 9403, Institut de Statistique, Université Catholique de Louvain.

Stevenson, R. E. (1980). Likelihood functions for generalized stochastic frontier estimation. *Journal of Econometrics, 13*(1), 57–66.

Thompson, G. D. (1988). Choice of flexible functional forms: Review and appraisal. *Western Journal of Agricultural Economics, 13*(2), 169–183.

Tovar, B., & Wall, A. (2014). The impact of demand uncertainty on port infrastructure costs: Useful information for regulators? *Transport Policy, 33,* 176–183.

Tran, K. C., & Tsionas, E. G. (2015). Endogeneity in stochastic frontier models: Copula approach without external instruments. *Economics Letters, 133,* 85–88.

Tsionas, E. G. (2002). Stochastic frontier models with random coefficients. *Journal of Applied Econometrics, 17*(2), 127–147.

Tsionas, E. G. (2006). Inference in dynamic stochastic frontier models. *Journal of Applied Econometrics, 21*(5), 669–676.

Tsionas, E. G., & Michaelides, P. G. (2016). A spatial stochastic frontier model with spillovers: Evidence for Italian regions. *Scottish Journal of Political Economy, 63*(3), 243–257.

Wang, H. J. (2002). Heteroscedasticity and non-monotonic efficiency effects of a stochastic frontier model. *Journal of Productivity Analysis, 18*(3), 241–253.

Wang, H. J. (2003). A stochastic frontier analysis of financing constraints on investment: The case of financial liberalization in Taiwan. *Journal of Business and Economic Statistics, 21,* 406–419.

Wang, H. J., & Ho, C. W. (2010). Estimating fixed-effect panel stochastic frontier models by model transformation. *Journal of Econometrics, 157*(2), 286–296.

Wang, H. J., & Schmidt, P. (2002). One-step and two-step estimation of the effects of exogenous variables on technical efficiency levels. *Journal of Productivity Analysis, 18,* 129–144.

Wang, W. S., & Schmidt, P. P. (2009). On the distribution of estimated technical efficiency in stochastic frontier models. *Journal of Econometrics, 148,* 36–45.

Winsten, C. B. (1957). Discussion on Mr. Farrell's paper. *Journal of the Royal Statistical Society, 120*(3), 282–284.

Zellner, A., & Revankar, N. S. (1969). Generalized production functions. *Review of Economics and Statistics, 36*(2), 241–250.

Chapter 7
Fair Target Setting for Intermediate Products in Two-Stage Systems with Data Envelopment Analysis

Qingxian An, Haoxun Chen, Beibei Xiong, Jie Wu, and Liang Liang

Abstract In a two-stage system with two divisions connected in series, fairly setting the target outputs for the first stage or equivalently the target inputs for the second stage is critical, in order to ensure that the two stages have incentives to collaborate with each other to achieve the best performance of the whole system. Data envelopment analysis (DEA) as a non-parametric approach for efficiency evaluation of multi-input, multi-output systems has drawn a lot of attention. Recently, many two-stage DEA models were developed for studying the internal structures of two-stage systems. However, there was no work studying the fair setting of the target intermediate products (or intermediate measures) although unreasonable setting will result in unfairness to the two stages because setting higher (fewer) intermediate measures means that the first (second) stage must make more efforts to achieve the overall production plan. In this chapter, a new DEA model taking account of fairness in the setting of the intermediate products is proposed, where the fairness is interpreted based on the Nash bargaining game model, in which the two stages negotiate their target efficiencies in the two-stage system based on their individual efficiencies. This approach is illustrated by an empirical application to insurance companies.

Keywords Data envelopment analysis · Efficiency analysis · Intermediate products · Fairness concern · Nash bargaining game

---

Q. An
School of Business, Central South University, Changsha 410083, China

H. Chen
Industrial Systems Optimization Laboratory, University of Technology of Troyes, 10004 Troyes, France

B. Xiong (✉)
School of Business Administration, Hunan University, Changsha 410082, Hunan, China
e-mail: xiongbeibei20@163.com

J. Wu · L. Liang
School of Management, University of Science and Technology of China, Hefei 230026, China

V. Charles et al. (eds.), *Data Science and Productivity Analytics*,
International Series in Operations Research & Management Science 290,
https://doi.org/10.1007/978-3-030-43384-0_7

7.1 Introduction

In a two-stage system such as an enterprise with a production division and a distribution division, the collaboration and coordination between the two divisions is critical for the enterprise to achieve its highest efficiency. Normally, they are managed as two separate business units or two profit centers, whose profits or losses are calculated separately, and their performances are evaluated individually. In such a system, since the interaction between the two divisions (stages) is realized through intermediate products, i.e., the outputs for the first stage or equivalently the inputs for the second stage, the enterprise may set a target (goal) for the intermediate products. For a system to achieve target outputs with the available resources, setting higher (fewer) intermediate products means that the first (second) stage must make more efforts to achieve the overall production plan. Then, the first (second) stage may think the intermediate products set is unfair to it, this will affect its morale in cooperation with the other stage to achieve the target outputs. Since the setting of target intermediate products has an important management implication as it provides a direction (benchmark) for the two stages to achieve, thus, fairly setting intermediate products is an important issue for the two-stage system.

It is clear that the setting of the intermediate products will directly affect the target efficiency of each stage (division) in the system. When a two-stage system uses the available inputs to produce the target outputs, higher intermediate products set for the first stage means higher target efficiency for this stage to achieve whereas higher intermediate products set for the second stage means lower target efficiency for this stage to achieve. That is, setting higher target efficiency for the first stage means that this stage should produce more intermediate products with the available inputs and setting higher target efficiency for the second stage means that this stage should consume few intermediate products to produce the target outputs. Thus, the fair intermediate products can be determined by setting appropriate target efficiencies for two stages. When does a stage feel unfair? Taking a non-life insurance company as an example which has a weak premium acquisition ability in the first stage but a strong profit generation ability in the second stage, setting higher intermediate products, direct written premiums and reinsurance premiums, is unfair to the first stage, and is also unfavorable for the company to realize its target outputs. Therefore, the setting of an appropriate target for the intermediate products of the system should consider the production abilities' difference of its two divisions. In other words, the setting of the expected (target) efficiency for each stage should consider its individual ability, i.e., the performance of the stage is compared with its homogenous divisions (stages) in other similar two-stage systems.

Data Envelopment Analysis (DEA) is a non-parametric mathematical programming approach for evaluating the relative efficiency of a set of homogeneous decision-making units (DMUs) with multiple inputs and multiple outputs (Charnes et al. 1978; Khezrimotlagh et al. 2019; Chen et al. 2020). It has been extensively developed and applied in the performance evaluation of multi-input multi-output complex systems

(Cook and Seiford 2009). The conventional DEA model can well deal with single-stage systems, but it cannot be used to evaluate more complex systems, such as two-stage systems, because the traditional DEA model considers the internal structure of a system as a black-box. Recently, two-stage DEA approaches were used to evaluate two-stage network structures in which the outputs of the first stage become the whole or part of the inputs of the second stage (Liang et al. 2008; Tone and Tsutsui 2014; Halkos et al. 2014; An et al. 2016). Following the review made by Cook et al. (2010), we classify the previous studies on two-stage DEA approaches into four categories: standard DEA approaches, efficiency decomposition approaches, game-theoretic approaches, and network DEA approaches. In the standard approaches, the conventional DEA methodology is applied separately to the first and second stage without considering possible conflicts between the two stages (Seiford and Zhu 1999; Sexton and Lewis 2003). The first efficiency decomposition approach considers the multiplicative or additive relationship between two stages by assuming that the weights of the output products of the first stage are identical to the weights of the input products of the second stage, such as Chen et al. (2009a, b) with additive efficiency decomposition and Kao and Hwang (2008) with multiplicative efficiency decomposition. For more recent works see Kao and Hwang (2011), Wang et al. (2014), and Despotis et al. (2016). Game-theoretic approaches model the performance evaluation of a two-stage system as a non-cooperative or cooperative game. Among them, Liang et al. (2006) proposed a cooperative game DEA model to calculate the efficiency of a two-stage system, where the two stages have the same bargaining power and cooperate with each other to jointly maximize their total efficiency. Other works in this category include Liang et al. (2008), Du et al. (2011), and Li et al. (2012). Network DEA approaches are related to the network DEA concept narrowly defined in Cook et al. (2010) for two-stage systems. One important work in this category is that of Färe and Grosskopf (1996) who investigated DMUs with two-stage structure as a network DEA. A number of studies have been reported following this work, such as Lewis and Sexton (2004), Tone and Tsutsui (2014), Liu et al. (2013), and Lim and Zhu (2019). These studies considered the two stages of a system, respectively, when building its two-stage model, and established the relationships between the two stages through intermediate products.

In general, all these previous works on two-stage DEA models focus on the overall efficiency of a two-stage system or the efficiency of each stage in the system. Few works studied the frontier projection of a two-stage system except Chen et al. (2010), Chen et al. (2013), and Lim and Zhu (2016). Chen et al. (2010) are the first ones to use an envelopment DEA model to produce the frontier projection through determining the optimal values for the intermediate measures. Later, Chen et al. (2014) stated that multiple and envelopment DEA models are dual models under the standard DEA, but there is a pitfall that these two types of models should be used, respectively, in deriving information for divisional efficiency and frontier projections (i.e., projected points on the production frontier). Lim and Zhu (2016) used a linear program to calculate the overall and divisional efficiencies and frontier projections simultaneously. However, they pointed out "*possible multiple optimal solutions exist. Therefore, the frontier projections and divisional efficiency scores are not necessarily unique. In fact, we show that a range of projections for the intermediate measures can be obtained for the frontier projections*". That is, there are usually a large range of frontier projections for the intermediate measures that the DMU has to choose as its production targets. As we have explained above, setting higher (fewer) intermediate measures as the projection means that the first (second) stage must make more efforts than the second (first) stage to achieve the overall production plan. It is very necessary to consider the fairness between the two stages in the setting of target intermediate products. In this chapter, by constructing a Nash bargaining game, we build a new DEA model with fairness concern to address this issue. The two stages are considered as two players in our model who bargain (negotiate) their target efficiencies in the two-stage system. Based on this model and its Nash bargaining solution, we not only fairly set the target efficiencies of the two stages, but also obtain fair target intermediate products of the two-stage system. Moreover, the production frontier point of each stage in the system can be obtained. With the fair setting of target intermediate products, the corresponding frontier projections are more easily accepted by the two stages because they are treated equally in the system.

The rest of this chapter is organized as follows. Section 7.2 briefly reviews some DEA models for two-stage systems. In Sect. 7.3, we present our approach for fairly setting the target intermediate products and determining frontier points through a Nash bargaining model. An application of the approach to 24 Taiwanese non-life insurance companies is given in Sect. 7.4. Some remarks for future research are given in the conclusion section.

7.2 Two-Stage DEA Models for Two-Stage Systems

Assume that there are n DMUs to be evaluated, where each DMU as shown in Fig. 7.1 contains s different outputs, t intermediate products, and m different inputs. Denote the ith input, dth intermediate product, and rth output for DMUj $(j = 1, 2, \ldots, n)$ as $x_{ij}(i = 1, 2, \ldots, n)$, $z_{dj}(d = 1, \ldots, t)$, and $y_{rj}(r = 1, 2, \ldots, m)$, respectively. $x_{ij} \geq 0$, $y_{rj} \geq 0$, $z_{dj} \geq 0$, and each DMU must have at least one positive input, one positive intermediate product, and one positive output value. In the first stage of the system, $X_j(x_{1j}, \ldots, x_{mj})$ are used as "inputs" to produce the intermediate products (outputs for the first stage) $Z_j(z_{1j}, \ldots, z_{tj})$. In the second stage, intermediate products $Z_j(z_{1j}, \ldots, z_{tj})$ are used as "inputs" to produce the outputs $Y_j(y_{1j}, \ldots, y_{sj})$. We denote the DMU being evaluated as DMU0 hereafter.

Färe and Grosskopf (1996) proposed an equivalent network DEA model for measuring the efficiency of the same system.

Fig. 7.1 Two-stage system

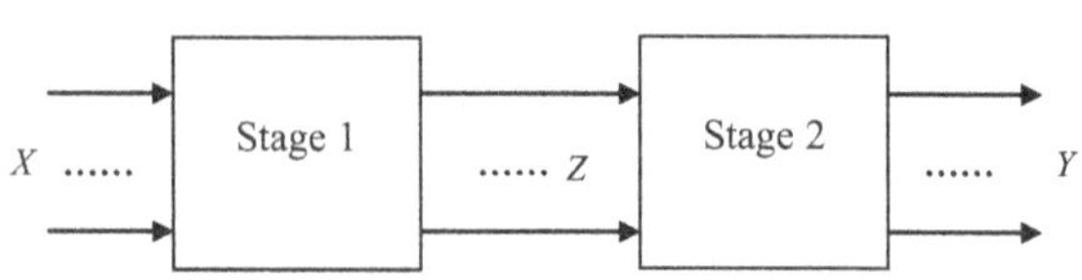

$$\min\ \phi$$
$$\textit{subject to}$$
$$\sum_{j=1}^{n} \gamma_j x_{ij} \le \phi x_{i0}, i = 1, \ldots, m, \tag{7.1a}$$

$$\sum_{j=1}^{n} \gamma_j z_{dj} \ge \tilde{z}_{d0}, d = 1, \ldots, t, \tag{7.1b}$$

$$\gamma_j \ge 0, j = 1, \ldots, n, \tag{7.1c}$$

$$\sum_{j=1}^{n} \pi_j z_{dj} \le \tilde{z}_{d0}, d = 1, \ldots, t, \tag{7.1d}$$

$$\sum_{j=1}^{n} \pi_j y_{nj} \ge y_{r0}, r = 1, \ldots, s, \tag{7.1e}$$

$$\pi_j \ge 0, j = 1, \ldots, n. \tag{7.1f}$$

where $\gamma(\gamma_1, \ldots, \gamma_n)$, $\pi(\pi_1, \ldots, \pi_n)$ are the intensity vectors corresponding to the first stage and the second stage, respectively. Denote the optimal value of model (7.1a, 7.1b, 7.1c, 7.1d, 7.1e, 7.1f) by ϕ^*, which represents the efficiency of the system. ϕx_{i0}, $i = 1, \ldots, m$ indicates how many inputs DMU0 should take to produce its given outputs. The constraints of (7.1a)–(7.1c) are imposed on the first stage, whereas the constraints of (7.1d)–(7.1f) are imposed on the second stage. $\tilde{z}_{d0}$, $d = 1, \ldots, t$ are the variables representing the target intermediate products that DMU0 should have in order to achieve the overall efficiency of the whole system. If $\phi^* = 1$, the two-stage system, DMU0, is overall efficient; otherwise, it is not overall efficient (Färe and Grosskopf 1996). It should be noted that the overall efficient here is actually weakly overall efficient. In this chapter, we follow this handling that overall efficient all refers to weakly overall efficient without special illustration.

Chen and Zhu (2004) firstly proposed a DEA model that can both evaluate the efficiency and obtain the production frontier points of a two-stage system. The model under the assumption of constant returns to scale is presented as follows:

$$\min\ w_1\alpha - w_2\beta$$
$$\textit{subject to}$$
$$\sum_{j=1}^{n} \gamma_j x_{ij} \le \alpha x_{i0}, i = 1, \ldots, m, \tag{7.2a}$$

$$\sum_{j=1}^{n} \gamma_j z_{dj} \ge \tilde{z}_{d0}, d = 1, \ldots, t, \tag{7.2b}$$

$$\alpha \le 1, \tag{7.2c}$$

$$\gamma_j \geq 0, j = 1, \ldots, n, \tag{7.2d}$$

$$\sum_{j=1}^{n} \pi_j z_{dj} \leq \tilde{z}_{d0}, d = 1, \ldots, t, \tag{7.2e}$$

$$\sum_{j=1}^{n} \pi_j y_{nj} \geq \beta y_{r0}, r = 1, \ldots, s, \tag{7.2f}$$

$$\beta \geq 1, \tag{7.2g}$$

$$\pi_j \geq 0, j = 1, \ldots, n. \tag{7.2h}$$

where γ_j, π_j, $\tilde{z}_{d0}$, α, and β are variables. Among the constraints, the constraints (7.2a)–(7.2d) are imposed on the first stage, whereas the constraints (7.2e)–(7.2h) are imposed on the second stage. The two stages are linked by target intermediate products $\tilde{z}$. Chen and Zhu (2004) indicated that w_1, w_2 are user-specified weights which reflect the preference over the two stages' performances. In Sect. 7.4, we will approach this model which often results in extreme optimal values of α, β no matter what the weights w_1, w_2 are chosen. Denote by (γ_j^*, π_j^*, $\tilde{z}_{d0}^*$, α^*, β^*) the optimal solution of model (7.2a, 7.2b, 7.2c, 7.2d, 7.2e, 7.2f, 7.2g, 7.2h). Then, $\alpha^* x$ indicates the (minimum) inputs that are required to produce the intermediate products $\tilde{z}_{d0}^*$, thus α^* can be considered as the target efficiency of stage 1. Analogously, $\beta^* y$ indicates the (maximum) outputs that can be produced by stage 2 using the intermediate products $\tilde{z}_{d0}^*$, and this implies that stage 2 can use the intermediate products $(1/\beta^*)\tilde{z}_{d0}^*$ to produce the outputs y under the assumption of constant returns to scale. Thus $1/\beta^*$ can be considered as the target efficiency of stage 2.

Later, Chen et al. (2009a, b) considered a special case of their model by setting $w_1 = 1$, $w_2 = 1$, i.e., by assuming that the user (decision maker) has the same preference over the performances of the two stages. They proved that the efficiency α^* of the first stage is always equal to 1. As α^* is always equal to 1, it can be replaced by 1 in the model. This model can thus be equivalently transformed into the output-oriented DEA model of Chen et al. (2009a, b) for determining the production frontier of a two-stage system.

Both Chen et al. (2009a, b) and Cook et al. (2010) pointed out that the overall efficiency of a two-stage system obtained by using the approach of Färe and Grosskopf (1996) is equal to that obtained by the model of Kao and Hwang (2008), the centralized model of Liang et al. (2008), and the model of Chen et al. (2010) under constant returns to scale, thus all these models are equivalent in the evaluation of the overall efficiency of a two-stage system. Furthermore, according to the study of Chen et al. (2009a, b), the frontier projections for intermediate products can be directly obtained by the envelopment DEA model. In this chapter, we use the envelopment DEA model to determine the target intermediate products of the system and obtain its frontier projections while measuring the overall efficiency of a two-stage system.

7.3 Methodology

In this section, we will propose a new non-oriented network DEA model with consideration of fairness between two stages for setting target intermediate products and determining frontier projections for a two-stage system. Firstly, a multi-objective linear programming model is built for analyzing the properties of two stages' target efficiencies of the system. Based on the properties, a Nash bargaining game model is established for setting the fair intermediate products of the system.

7.3.1 Multi-objective Linear Programming Model for Analyzing Target Efficiencies

As we know, the setting of target intermediate products directly affects the target efficiency of each stage; inversely, the setting of the target efficiencies of the two stages can influence the target intermediate products, which can be found in the analysis of Chen and Zhu (2004). However, the DEA model proposed by Chen and Zhu (2004) usually generates the extreme values of the target efficiencies. Thus, instead of considering a single objective function obtained by a linear combination of the two target efficiencies, we first provide the following multi-objective linear programming model to formulate the efficiencies of the two stages in a two-stage system.

$$
\begin{aligned}
& min\, \alpha \\
& min\, 1/\beta \\
& subject\, to\, (stage\, 1) \\
& \qquad \textstyle\sum_{j=1}^{n} \gamma_j x_{ij} \le \alpha x_{i0},\, i = 1, \ldots, m, \\
& \qquad \textstyle\sum_{j=1}^{n} \gamma_j z_{dj} \ge \tilde{z}_{d0},\, d = 1, \ldots, t, \\
& \qquad \alpha \le 1. \\
& \qquad \gamma_j \ge 0,\, j = 1, \ldots, n, \\
& \qquad (stage\, 2) \\
& \qquad \textstyle\sum_{j=1}^{n} \pi_j z_{dj} \le \tilde{z}_{d0},\, d = 1, \ldots, t, \\
& \qquad \textstyle\sum_{j=1}^{n} \pi_j y_{rj} \ge \beta y_{r0},\, r = 1, \ldots, s, \\
& \qquad \beta \ge 1. \\
& \qquad \pi_j \ge 0,\, j = 1, \ldots, n.
\end{aligned}
\tag{7.3}
$$

In model (7.3), each stage wants to optimize (minimize) its target efficiency so that it can make less effort to achieve its production task assigned by the underlying two-stage system. For the multi-objective optimization model, its Pareto-optimal solutions which are generally not unique are defined as follows.

Definition 7.1 For a feasible solution $p^* = (\alpha^*, \beta^*, \gamma^*, \tilde{z}_0^*, \pi^*)$ of model (7.3), if there is no other feasible solution $p' = (\alpha', \beta', \gamma', \tilde{z}_0', \pi')$ such that $\alpha' \le \alpha^*$, $1/\beta' \le$

$1/\beta^*$ and at least one of the two inequalities must be strict, then p^* is a Pareto-optimal solution of the model.

Let us denote the set of all Pareto-optimal solutions of model (7.3) by P. For any $p^* = (\alpha^*, \beta^*, \gamma^*, \tilde{z}_0^*, \pi^*) \in P$, α^* is the target efficiency of the first stage and $1/\beta^*$ is the target efficiency of the second stage. As soon as p^* is chosen from set P, the target intermediate products, i.e., the value $\tilde{z}_0^*$ of variables $\tilde{z}_0$ in model (7.3), can be determined, which can also be called projections of intermediate products. For simplicity, some components of a solution may be omitted when it is cited hereafter.

Theorem 7.1 *Model* (7.3) *has at least one Pareto-optimal solution.*

See Appendix 7.1 for the proof.

The linear combination of inputs, intermediate products and outputs whose weights are given by any Pareto-optimal solution of model (7.3) forms a projected point for the evaluated DMU, $\left(\sum_{j=1}^{n} \gamma_j^* x_{ij}, \tilde{z}_{d0}^*, \sum_{j=1}^{n} \pi_j^* y_{rj}\right)$. Before further analysis, we first give two definitions as follows.

Definition 7.2 A DMU (x_{i0}, z_{d0}, y_{r0}) is called overall efficient if and if only there is no other real or virtual DMU, DMU0' $\left(x'_{i0}, z'_{d0}, y'_{r0}\right)$ that satisfies $x'_{i0} \leq x_{i0}, i = 1, \ldots, m$ and $y'_{r0} \geq y_{r0}, r = 1, \ldots, s$, where $x'_{i0} = \sum_{j=1}^{n} \gamma_j x_{ij}, y'_{r0} = \sum_{j=1}^{n} \pi_j z_{dj}$ $\gamma_j \geq 0, \pi_j \geq 0$.

Definition 7.3 The projected point of a decision-making unit is called a projected DMU.

The Pareto-optimal solutions of model (7.3) have the following property:

Theorem 7.2 *Consider the projected point of DMU0,* $\left(\sum_{j=1}^{n} \gamma_j^* x_{ij}, \tilde{z}_{d0}^*, \sum_{j=1}^{n} \pi_j^* y_{rj}\right)$, *determined by a Pareto-optimal solution* $(\alpha^*, \beta^*, \gamma^*, \tilde{z}_0^*, \pi^*)$ *of model* (7.3). *The projected DMU* $\left(\sum_{j=1}^{n} \gamma_j^* x_{ij}, \tilde{z}_{d0}^*, \sum_{j=1}^{n} \pi_j^* y_{rj}\right)$, *created by replacing the DMU under evaluation with its projection while keeping the data for other DMUs unchanged, is overall efficient.*

See Appendix 7.2 for the proof.

Based on the above theorem, we know that the projected DMU, i.e., $\left(\sum_{j=1}^{n} \gamma_j^* x_{ij}, \tilde{z}_{d0}^*, \sum_{j=1}^{n} \pi_j^* y_{rj}\right)$, is overall efficient. So, we consider it as the benchmark of the evaluated DMU. As we know, the target intermediate products for DMU0 are determined by the two stages' target efficiencies, α^* and $1/\beta^*$, which reflect how many inputs can be reduced when keeping outputs unchanged for two stages. However, fairness is not ignored in this model. Moreover, this model cannot be solved directly and may have multiple Pareto-optimal solutions because it has two objectives. By introducing two new variables $\vartheta = 1/\beta$ and $\tau_j = \pi_j \vartheta$, we can transform model (7.3) into a new non-linear multi-objective programming problem.

$$
\begin{aligned}
& min\,\alpha \\
& min\,\vartheta \\
& subject\ to\,(stage\ 1) \\
& \qquad \textstyle\sum_{j=1}^{n} \gamma_j x_{ij} \le \alpha x_{i0}, i = 1, \ldots, m, \\
& \qquad \textstyle\sum_{j=1}^{n} \gamma_j z_{dj} \ge \tilde{z}_{d0}, d = 1, \ldots, t, \\
& \qquad \alpha \le 1. \\
& \qquad \gamma_j \ge 0, j = 1, \ldots, n, \\
& \qquad (stage\ 2) \\
& \qquad \textstyle\sum_{j=1}^{n} \tau_j z_{dj} \le \vartheta \tilde{z}_{d0}, d = 1, \ldots, t, \\
& \qquad \textstyle\sum_{j=1}^{n} \tau_j y_{rj} \ge y_{r0}, r = 1, \ldots, s, \\
& \qquad \vartheta \le 1, \\
& \qquad \tau_j \ge 0, j = 1, \ldots, n.
\end{aligned}
\tag{7.4}
$$

The solutions of model (7.4) have the following property:

Theorem 7.3 *The product of any Pareto-optimal objective values α^* and ϑ^* of model* (7.4) *is equal to the system's overall efficiency ϕ^* obtained by model* (7.1a, 7.1b, 7.1c, 7.1d, 7.1e, 7.1f).

See Appendix 7.3 for the proof.

The relationship between two values α^* and ϑ^* of a Pareto-optimal solution is illustrated in Fig. 7.2.

The X-axis represents the value of the target efficiency of the first stage α^* while the Y-axis represents the value of the target efficiency of the second stage ϑ^*. $\underline{\alpha}$ represents the minimum value of the target efficiency of the first stage, which can be obtained from model (7.4) by setting the value of ϑ to 1. Similarly, $\underline{\vartheta}$ represents the minimum value of the target efficiency of the second stage. As $(\underline{\alpha}, 1)$, $(1, \underline{\vartheta})$ are two Pareto-optimal solution of model (7.4), thus $\underline{\alpha} * 1 = \phi^*$, $1 * \underline{\vartheta} = \phi^*$, so $\underline{\alpha} = \phi^*$ and $\underline{\vartheta} = \phi^*$. Because $(\phi^*,\ 1)$ is a Pareto-optimal solution of model (7.4), any combination $(\phi^*/b,\ \ b * 1)$ with variable b satisfying $\underline{\alpha} \le \phi^*/b \le 1$; $\underline{\vartheta} \le$

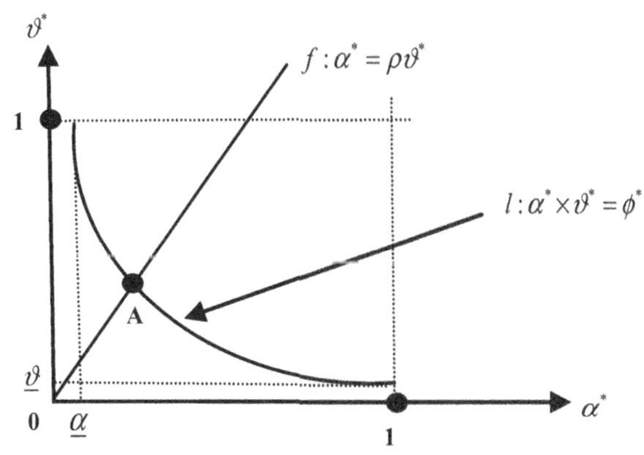

Fig. 7.2 Relationship between two divisional target efficiencies

$b \leq 1$ is also a Pareto-optimal solution of model (7.4). Thus, any point on the curve $l : \alpha^* \times \vartheta^* = \phi^*$, with $\underline{\alpha} \leq \alpha^* \leq 1$; $\underline{\vartheta} \leq \vartheta^* \leq 1$, is a Pareto-optimal solution of model (7.4). The two values α^*, ϑ^* have competition because if one becomes larger, the other will become smaller. Also, we know that α^*, ϑ^*, and ϕ^* are within (0, 1], which is consistent with the concept of efficiency.

7.3.2 Nash Bargaining Game for Fairly Setting the Target Intermediate Products

It is clear that a different combination of α^* and ϑ^* in model (7.4) may lead to different target intermediate products. If either α^* or ϑ^* is set larger, the corresponding stage will be asked to have a higher efficiency, that is, this stage must make more effort to realize its target efficiency. Different from those of Chen and Zhu (2004) and Chen et al. (2009a, b), we set the intermediate products of a two-stage system based on its Nash bargaining game model. In the model, the two stages of the system are considered two players who bargain (negotiate) their target efficiencies. Then, the Nash bargaining solution of the model can be used to fairly set the intermediate products. The Nash bargaining game is a simple two-player game in which two players bargain (negotiate) with each other for the division of an item (e.g., an apple pie, a payoff) between them (Nash 1950, 1953). This solution, called the Nash bargaining solution of the game, satisfies four axioms (invariant to affine transformations, Pareto optimality, independence of irrelevant alternatives, and symmetry). It is thus considered a fair division of the item between the two players.

Let d_1 and d_2 represent the disagreement points of player 1 and player 2, respectively. The disagreement point of a player is the value (the portion) it can expect to receive if the negotiation with the other player breaks down. A pair of payoffs (u_1^*, u_2^*) is the Nash bargaining solution of the game if it solves the following optimization problem:

$$\begin{aligned} &\max_{t_1, t_2} \quad (u_1 - d_1)(u_2 - d_2) \\ &subject \ (u_1, u_2) \in U \end{aligned} \tag{7.5}$$

where u_1 and u_2 represent the payoff of player 1 and player 2, respectively, U denotes the set of feasible payoffs (possible bargaining agreements) of the two players.

Assume that the overall efficiency of the two-stage system is ϕ^*, which is obtained by model (7.1a, 7.1b, 7.1c, 7.1d, 7.1e, 7.1f). Denote the target efficiency of the first stage and the second stage by α and ϑ, respectively. Then, we have $\alpha * \vartheta = \phi^*$. By fairly setting the target efficiency α for stage 1 and the target efficiency ϑ for stage 2, the intermediate products of the system can then be fairly determined. Higher target efficiency of one stage implies that it must make more efforts to use fewer inputs to produce more outputs. Each stage wants its target efficiency to be set as low as possible.

Denote by α' and ϑ' are the disagreement efficiencies of stage 1 and stage 2, respectively, i.e., the efficiency of each stage without cooperation of the other stage, which corresponds to the individual efficiency of the stage when it is regarded as a DMU. Such efficiency can be obtained by considering the stage as a DMU and comparing it with homogenous stages (DMUs) in similar two-stage systems according to their past production by applying the conventional CCR model. In the Nash bargaining game, two players demand a portion of an item (good or amount of money). That is, $u = u_1 + u_2$, where u is the total amount of good or money, u_i is the portion obtained by player i, $i = 1, 2$, $\underline{u} \geq 0$, $u_1 \geq 0$, $u_2 \geq 0$. In order to apply the Nash bargaining game model to fairly set the target efficiencies of the two stages for a two-stage system, we have to transform the equation $\alpha * \vartheta = \phi^*$ into an equivalent one of type $u = u_1 + u_2$ by changing variables. This can be achieved by first taking the logarithmic operation and then the negation operation on both sides of the first equation, leading to $(-\log(\alpha)) + (-\log(\vartheta)) = -\log(\phi^*)$. Since $-\log(\phi^*) \geq 0$, $-\log(\alpha) \geq 0$, and $-\log(\vartheta) \geq 0$, we can regard the two stages as two players, $-\log(\phi^*)$ as the total amount of an item to be divided between the two players, $-\log(\alpha)$ and $-\log(\vartheta)$ as the portions of the item obtained by player 1 and player 2, respectively. With such analogy, we can change variables α and ϑ to two new variables u_1 and u_2 by defining $u_1 = -\log(\alpha)$, $u_2 = -\log(\vartheta)$, $d_1 = -\log(\alpha')$, and $d_2 = -\log(\vartheta')$. This leads to $u = u_1 + u_2$ as required by the Nash bargaining game. Accordingly, $d_1 = -\log(\alpha')$ and $d_2 = -\log(\vartheta')$ can be regarded as the disagreement points (portions) of the two stages (players), respectively, in the game. A fair allocation of the overall efficiency of the two-stage system between its two stages can thus be determined by model (7.5) that defines the Nash bargaining solution. This model maximizes $(u_1 - d_1)(u_2 - d_2)$, i.e., maximizes $(-\log(\alpha) - (-\log(\alpha'))) * (-\log(\vartheta) - (-\log(\vartheta')))$ or equivalently maximizes $(\log(\alpha) - \log(\alpha')) * (\log(\vartheta) - \log(\vartheta'))$, subject to the constraint $(u_1, u_2) \in U$, i.e.,$(-\log(\alpha), -\log(\vartheta)) \in U$. The remaining thing is to determine the set U for the two-stage system. Obviously, any feasible payoffs of the two stages must satisfy the constraint $\alpha * \vartheta = \phi^*$ or equivalently the constraint $\log(\alpha) + \log(\vartheta) = \log(\phi^*)$ and all constraints in model (7.4) according to Theorem 7.3, but additional constraints are required for defining U.

For the two-stage system, two cases may happen. The first case is that its overall efficiency is set no higher than the product of the individual efficiencies of its two stages. This case occurs when a two-stage system sets its efficiency target within the two stages' capabilities measured by their individual efficiencies. In this case, no stage wants to have its efficiency set higher than its individual efficiency in the two-stage system. Thus, we can view the individual efficiency of each stage as the maximum acceptable efficiency that the stage is willing to achieve without changing its current internal structure (i.e., using its currently available resources to produce its outputs) in the two-stage system. In this case, the individual efficiency of each stage can be regarded as its disagreement efficiency, and the additional constraints for defining U are $\alpha \leq \alpha'$ and $\vartheta \leq \vartheta'$, or equivalently $\log(\alpha) \leq \log(\alpha')$ and $\log(\vartheta) \leq \log(\vartheta')$. Note that in this case $\phi^* \leq \alpha' \times \vartheta'$, i.e., $\log(\phi^*) \leq \log(\alpha') + \log(\vartheta')$. The second case is that the overall efficiency of the two-stage system is set higher than the product of the individual efficiencies of its two stages. This case occurs when the two-stage system sets its target efficiency exceeding the two stages' current capabilities measured by their individual efficiencies. Such case is possible if the two stages make more efforts to produce more outputs by using the current inputs or to produce the current outputs by consuming fewer inputs, or to produce more outputs by consuming fewer inputs. In this case, to be fair, each stage must make more efforts than its individual efficiency in order to achieve the overall efficiency. Thus, we can view the individual efficiency of each stage as the "minimum acceptable efficiency" that the stage is expected (by the other stage) to achieve in the two-stage system. In this case, the individual efficiency of each stage can still be regarded as its disagreement efficiency, but the additional constraints for defining U become $\alpha \geq \alpha'$ and $\vartheta \geq \vartheta''$, or equivalently $\log(\alpha) \geq \log(\alpha')$ and $\log(\vartheta) \geq \log(\vartheta')$. Note that in this case $\phi^* \geq \alpha' \times \vartheta'$, i.e., $\log(\phi^*) \geq \log(\alpha') + \log(\vartheta')$.

According to the above discussion, the problem of fairly setting the target intermediate products in the two-stage system can be formulated as the following optimization model:

$$\begin{aligned} max \quad & \left(\log(\alpha) - \log(\alpha')\right) * \left(\log(\vartheta) - \log(\vartheta')\right) \\ subject\ to \quad & (stage\ 1) \\ & \sum_{j=1}^{n} \gamma_j x_{ij} \leq \alpha x_{i0}, i = 1, \ldots, m, \end{aligned} \tag{7.6.1}$$

$$\sum_{j=1}^{n} \gamma_j z_{dj} \geq \tilde{z}_{d0}, d = 1, \ldots, t, \tag{7.6.2}$$

$$\alpha \leq 1, \tag{7.6.3}$$

$$\gamma_j \geq 0, j = 1, \ldots, n, \tag{7.6.4}$$

$$(stage\ 2)$$

$$\sum_{j=1}^{n} \tau_j z_{dj} \leq \vartheta \tilde{z}_{d0}, d = 1, \ldots, t, \tag{7.6.5}$$

$$\sum_{j=1}^{n} \tau_j y_{rj} \geq y_{r0}, r = 1, \ldots, s, \tag{7.6.6}$$

$$\vartheta \leq 1, \tag{7.6.7}$$

$$\tau_j \geq 0, j = 1, \ldots, n. \tag{7.6.8}$$

$$\log(\alpha) + \log(\vartheta) = \log(\phi^*) \tag{7.6.9}$$

$$\log(\alpha) \le \log(\alpha'), \log(\vartheta) \le \log(\vartheta'), \text{ if } \log(\phi^*) \le \log(\alpha') + \log(\vartheta') \quad (7.6.10a)$$

$$or\ \log(\alpha) \ge \log(\alpha'), \log(\vartheta) \ge \log(\vartheta'), \text{ if } \log(\phi^*) \ge \log(\alpha') + \log(\vartheta') \quad (7.6.10b)$$

Note that in the above model constraints (7.6.10a) and (7.6.10b) are exclusive, i.e., either (7.6.10a) or (7.6.10b) is satisfied. From constraints (7.6.10a) and (7.6.10b), it is easy to prove that the optimal objective value of the model is nonnegative. As a result, constraints (7.6.10a) and (7.6.10b) are redundant. Because if $\log(\phi^*) \le \log(\alpha') + \log(\vartheta')$, the optimal solution of the model must satisfy (7.6.10a), otherwise either $\log(\alpha) > \log(\alpha')$ or $\log(\vartheta) > \log(\vartheta')$. According to Theorem 7.3, $\log(\alpha) + \log(\vartheta) = \log(\phi^*)$ always holds in this model. If $\log(\alpha) > \log(\alpha')$, then $\log(\vartheta) = \log(\phi^*) - \log(\alpha) \le \log(\alpha') + \log(\vartheta') - \log(\alpha) < \log(\vartheta')$. This leads to that the objective value of the model is negative, so (α, ϑ) is not an optimal solution of the model. Similarly, if $\log(\phi^*) \ge \log(\alpha') + \log(\vartheta')$, the optimal solution must also satisfy (7.6.10b). Therefore, model (7.6.1, 7.6.2, 7.6.3, 7.6.4, 7.6.5, 7.6.6, 7.6.7, 7.6.8, 7.6.9, 7.6.10a, 7.6.10b) is equivalently converted into the following model.

$$\begin{aligned}
\max \quad & (\log(\alpha) - \log(\alpha')) * (\log(\vartheta) - \log(\vartheta')) \\
& subject\ to\ (stage\ 1) \\
& \quad \textstyle\sum_{j=1}^{n} \gamma_j x_{ij} \le \alpha x_{i0}, i = 1, \ldots, m, \\
& \quad \textstyle\sum_{j=1}^{n} \gamma_j z_{dj} \ge \tilde{z}_{d0}, d = 1, \ldots, t, \\
& \quad \alpha \le 1, \\
& \quad \gamma_j \ge 0, j = 1, \ldots, n, \\
& \quad (stage\ 2) \\
& \quad \textstyle\sum_{j=1}^{n} \tau_j z_{dj} \le \vartheta \tilde{z}_{d0}, d = 1, \ldots, t, \\
& \quad \textstyle\sum_{j=1}^{n} \tau_j y_{rj} \ge y_{r0}, r = 1, \ldots, s, \\
& \quad \vartheta \le 1, \\
& \quad \tau_j \ge 0, j = 1, \ldots, n.
\end{aligned} \quad (7.7)$$

Model (7.7) is a non-linear programming problem. However, since $\log(\alpha) + \log(\vartheta) = \log(\phi^*)\ (\alpha \times \vartheta = \phi^*)$ always holds in this model, it can be solved by the following line search algorithm to obtain a solution very close to the optimal solution.

Algorithm for solving model (7.7):

Step 1. Replace $\log(\vartheta)$ with $\log(\vartheta) = \log(\phi^*) - \log(\alpha)$, ϑ with ϕ^*/α, and transform model (7.7) into a non-linear programming model with a single variable α and the objective function $(\log(\alpha) - \log(\alpha')) * (\log(\phi^*/\alpha) - \log(\vartheta')) = (\log(\alpha) - \log(\alpha'))* (\log(\phi^*) - \log(\vartheta') - \log(\alpha))$.

Step 2. For the single variable non-linear programming model, apply the line search of α from $\underline{\alpha}$ to 1. Since the interval contains an infinite number of real values between $\underline{\vartheta}$ and 1, we cannot consider all the values. Instead, we discretize

the interval by only considering a finite number of values $\vartheta_k = \underline{\vartheta} + k \times \Delta, k = 0, 1, \ldots, K$, where K is the largest integer equal to $(1 - \underline{\vartheta})/\Delta$, Δ is a step size.
Step 3. Compare the objective values of model (7.7) for each ϑ_k value to obtain the maximum objective value and its corresponding ϑ_k^*. ϑ_k^* is very close to the optimal solution of the model if we take Δ sufficiently small ($\Delta = 0.00001$ for example).

Model (7.7) has the following properties:

Theorem 7.4 *For any optimal solution* $(\gamma_j^*, \tilde{z}_{d0}^*, \tau_j^*, \alpha^*, \vartheta^*)$ *of model* (7.7), *the projected DMU* $(\sum_{j=1}^n \gamma_j^* x_{ij}, \tilde{z}_{d0}^*, 1/\vartheta * \sum_{j=1}^n \pi_j^* y_{rj})$, *created by replacing a DMU under evaluation with its projection while keeping the data for the other DMUs unchanged, is overall efficient.*

Proof Firstly, we know that any optimal solution of model (7.7) $(\gamma_j^*, \tilde{z}_{d0}^*, \tau_j^*, \alpha^*, \vartheta^*)$ is a Pareto-optimal solution of model (7.4). Because Model (7.4) is multiple-objective programming model which has the same constraints as model (7.7) and pursues the minimization of both α and ϑ, the projected DMU $(\sum_{j=1}^n \gamma_j^* x_{ij}, \tilde{z}_{d0}^*, 1/\vartheta * \sum_{j=1}^n \pi_j^* y_{rj})$ obtained by the optimal solution created from model (7.7) is obviously a projected DMU in model (7.4). According to Theorem 7.2, the projected DMU from model (7.7) must be overall efficient. □

Theorem 7.5 *Model* (7.7) *provides unique divisional efficiencies.*

Proof For any optimal solution α, ϑ of model (7.7), we have $\alpha * \vartheta = \phi^*$, where ϕ^* is constant. Then, $\log(\alpha) + \log(\vartheta) = \log(\phi^*)$. Substituting $\log(\vartheta) = \log(\phi^*) - \log(\alpha)$ into the objective function of model (7.7), the function becomes $(\log(\alpha) - \log(\alpha')) * (\log(\phi^*) - \log(\alpha) - \log(\vartheta'))$. Denote $\log(\alpha)$ by x. The objective function can be rewritten as $f(x) = (x - \log(\alpha')) * (\log(\phi^*) - x - \log(\vartheta'))$. Since the second derivative $f''(x) = -2 < 0$, $f(x)$ is a quadratic concave function with respect to x. Moreover, the constraints of model (7.7) become the constraints of model (7.3) after substituting τ_j by $\vartheta\pi_j$ and ϑ by $1/\beta$ while keeping variable α unchanged. As the feasible region of α defined by the constraints of model (7.7) is the same as that defined by the constraints of model (7.3), which are linear, we analyze the feasible region of α based on the constraints of model (7.3). α_1 and α_2 denote the two feasible values of α which satisfy these constraints. Then there are two feasible solutions $(\gamma_{1j}, \pi_{1j}, \beta_1)$ and $(\gamma_{2j}, \pi_{2j}, \beta_2)$ such that

$$
\begin{array}{lcl}
\sum_{j=1}^n \gamma_{1j} x_{ij} \le \alpha_1 x_{i0}, i = 1, \ldots, m, & & \sum_{j=1}^n \gamma_{2j} x_{ij} \le \alpha_2 x_{i0}, i = 1, \ldots, m, \\
\sum_{j=1}^n \gamma_{1j} z_{dj} \ge \tilde{z}_{1d0}, d = 1, \ldots, t, & & \sum_{j=1}^n \gamma_{2j} z_{dj} \ge \tilde{z}_{2d0}, d = 1, \ldots, t, \\
\alpha_1 \le 1, & & \alpha_2 \le 1, \\
\gamma_{1j} \ge 0, j = 1, \ldots, n, & \text{and} & \gamma_{2j} \ge 0, j = 1, \ldots, n, \\
\sum_{j=1}^n \pi_{1j} z_{dj} \le \tilde{z}_{1d0}, d = 1, \ldots, t, & & \sum_{j=1}^n \pi_{2j} z_{dj} \le \tilde{z}_{2d0}, d = 1, \ldots, t, \\
\sum_{j=1}^n \pi_{1j} y_{rj} \ge \beta_1 y_{r0}, r = 1, \ldots, s, & & \sum_{j=1}^n \pi_{2j} y_{rj} \ge \beta_2 y_{r0}, r = 1, \ldots, s, \\
\beta_1 \ge 1, & & \beta_2 \ge 1, \\
\pi_{1j} \ge 0, j = 1, \ldots, n. & & \pi_{2j} \ge 0, j = 1, \ldots, n.
\end{array}
$$

Then any convex combination of α_1 and α_2, i.e., $\alpha_3 = w\alpha_1 + (1-w)\alpha_2, 0 \le w \le 1$, satisfies these constraints too. Because by defining $\gamma_{3j} = w\gamma_{1j} + (1-w)\gamma_{2j}$, $\pi_{3j} = w\pi_{1j} + (1-w)\pi_{2j}$, $\beta_3 = w\beta_1 + (1-w)\beta_2$, and $\tilde{z}_{3d0} = w\tilde{z}_{1d0} + (1-w)\tilde{z}_{2d0}$, we can easily obtain

$$\begin{aligned}
&\sum_{j=1}^{n} \gamma_{3j} x_{ij} \le \alpha_3 x_{i0}, i = 1, \ldots, m, \\
&\sum_{j=1}^{n} \gamma_{3j} z_{dj} \ge \tilde{z}_{3d0}, d = 1, \ldots, t, \\
&\alpha_3 \le 1, \\
&\gamma_{3j} \ge 0,\ j = 1, \ldots, n, \\
&\sum_{j=1}^{n} \pi_{3j} z_{dj} \le \tilde{z}_{3d0}, d = 1, \ldots, t, \\
&\sum_{j=1}^{n} \pi_{3j} y_{rj} \ge \beta_3 y_{r0}, r = 1, \ldots, s, \\
&\beta_3 \ge 1, \\
&\pi_{3j} \ge 0,\ j = 1, \ldots, n.
\end{aligned}$$

This implies that α_3 is a feasible solution of these constraints too. Therefore, the feasible region of α under the constraints of model (7.4) is convex. Since the variable α is one dimensional, so its feasible region can be written as $[\underline{\alpha}, 1]$, so the feasible region of x is $[\log(\underline{\alpha}), 0]$ since $\log(\alpha)$ is a continuous increasing function of α. So the feasible region of $x (x = \log(\alpha))$ under these constraints is also convex. This implies model (7.7) is a concave programming with respect to the single variable x within $[\log(\underline{\alpha}), 0]$, so its optimal solution, denoted by x^*, is unique. As $x* = \log(\vartheta^*)$, ϑ^* is unique as well. Moreover, because $\alpha^* * \vartheta^* = \phi^*$, α^* is also unique. □

7.4 An Empirical Application to Insurance Industry

In the last two decades, financial liberalization has resulted in substantial changes in the insurance industry all over the world (Huang and Eling 2013). As insurance markets play a significant role in national economics, the study of insurance systems has drawn a lot of attention in the academic community (Biener and Eling 2012; Kao and Wang 2014; Ilyas and Rajasekaran 2019). In this section, 24 Taiwanese non-life insurance companies studied in Kao and Hwang (2008) are used to illustrate our approach. Following Kao and Hwang, the financial process of the non-life insurance industry is divided into two stages: premium acquisition stage and profit generation stage. This process can be seen as a financial supply chain. "Operation expenses" (OE) and "insurance expenses" (IE) are used as the two inputs in the first stage. "Direct written premiums" (DWP) and "reinsurance premiums" (RP) are the intermediate products which are taken as both the outputs of the first stage and the inputs of the second stage. "Underwriting profits" (UP) and "investment profits" (IP) are taken as the outputs of the second stage. See Kao and Hwang (2008) for a detailed discussion on these measures. The detailed data can be obtained from Table 7.2 of Kao and Hwang (2008).

Moreover, we use $\tilde{z}_1$ and $\tilde{z}_2$ to denote target intermediate products. Firstly, we obtain the individual efficiency of each stage through input-oriented CCR model by its historical data. Notably, we should use the data of two stages when they worked independently. However, the accurate information on these data is unavailable, thus we take the above historical data as the independent scenario data, which doesn't affect us to illustrate our approach. These efficiencies are given in Table 7.1 by its second and third column, respectively. Assume that two stages work together and the overall manager sets the target of using initial inputs X to produce the final output Y. By substituting the individual efficiencies of two stages obtained from Table 7.1 into model (7.7), we can obtain the overall efficiency and divisional target efficiencies of DMUs in the last three columns as shown in Table 7.1.

Table 7.1 Individual efficiencies, overall and divisional efficiencies of DMUs

DMU	Individual		Overall and divisional efficiency		
	Stage 1	Stage 2	Overall efficiency	Stage 1	Stage 2
1	0.993	0.713	0.699	0.986	0.709
2	0.998	0.627	0.626	0.998	0.627
3	0.69	1	0.69	0.69	1
4	0.724	0.432	0.304	0.714	0.426
5	0.838	1	0.767	0.802	0.957
6	0.964	0.406	0.39	0.962	0.405
7	0.752	0.538	0.277	0.622	0.445
8	0.726	0.511	0.275	0.625	0.44
9	1	0.292	0.223	0.876	0.255
10	0.862	0.674	0.467	0.773	0.604
11	0.741	0.327	0.164	0.61	0.269
12	1	0.76	0.76	1	0.76
13	0.811	0.543	0.209	0.558	0.374
14	0.725	0.518	0.289	0.636	0.454
15	1	0.705	0.614	0.934	0.658
16	0.907	0.385	0.32	0.87	0.368
17	0.723	1	0.36	0.511	0.705
18	0.794	0.374	0.259	0.742	0.349
19	1	0.416	0.411	0.996	0.413
20	0.933	0.901	0.547	0.753	0.727
21	0.751	0.28	0.201	0.735	0.273
22	0.59	1	0.59	0.59	1
23	0.85	0.56	0.42	0.799	0.526
24	1	0.335	0.135	0.636	0.213

From Table 7.1, we can see that no non-life insurance company is overall efficient. For the first stage, only DMU 12 is efficient, whereas for the second stage, only DMU 3 and 22 are efficient. Although the output-oriented model and input-oriented model of Chen et al. (2010) can be used for obtaining the target frontier projections of intermediate products, they are not applicable for the estimation of the divisional efficiency scores (Lim and Zhu 2016). In Table 7.2, the target efficiencies obtained by our model are compared with those obtained by Lim and Zhu (2016).

In Table 7.2, α represents the target efficiency of the first stage, and $1/\beta$ represents the target efficiency of the second stage. Because $\vartheta = 1/\beta$ in model (7.7), we can obtain the value of $1/\beta$ by ϑ. From Table 7.2, we can find the difference between the stage efficiencies of our model and those of Lim and Zhu (2016) is significant. That's

Table 7.2 Comparison of stage efficiencies by our approach and by Lim and Zhu (2016)

DMU	Our model		Lim and Zhu (2016)	
	α	$1/\beta$	α	$1/\beta$
1	0.986	0.709	0.993	0.705
2	0.998	0.627	0.998	0.627
3	0.69	1	0.690	1
4	0.714	0.426	0.724	0.42
5	0.802	0.957	0.831	0.923
6	0.962	0.405	0.961	0.406
7	0.622	0.445	0.671	0.412
8	0.625	0.44	0.663	0.415
9	0.876	0.255	1	0.223
10	0.773	0.604	0.862	0.541
11	0.61	0.269	0.647	0.253
12	1	0.760	1	0.760
13	0.558	0.374	0.672	0.310
14	0.636	0.454	0.670	0.431
15	0.934	0.658	1	0.614
16	0.87	0.368	0.886	0.361
17	0.511	0.705	0.628	0.574
18	0.742	0.349	0.794	0.326
19	0.996	0.413	1	0.411
20	0.753	0.727	0.933	0.586
21	0.735	0.273	0.732	0.274
22	0.59	1	0.59	1
23	0.799	0.526	0.843	0.499
24	0.636	0.213	0.429	0.315

because our approach has considered the fairness issue between the two stages based on their individual efficiencies when setting the target intermediate products.

The results of the target intermediate products (i.e., projections of intermediate products) obtained by our approach and the two models of Chen et al. (2010) are given in Table 7.3. As the projection for intermediate measure obtained by Lim and Zhu (2016) can be any choice in an interval, we omit the comparison between Lim and Zhu (2016) with our model.

In Table 7.3, $\tilde{z}_1$ and $\tilde{z}_2$ represent the target "underwriting profits" and "investment profits" that non-life insurance companies should have in order to achieve the overall efficiency of the whole system. The first two columns show the intermediate products obtained by our approach, the middle two columns show the results obtained by

Table 7.3 The intermediate products of the non-life insurance companies

DMU	Our model		Output-oriented		Input-oriented	
	$\tilde{z}_1$	$\tilde{z}_2$	$\tilde{z}_1$	$\tilde{z}_2$	$\tilde{z}_1$	$\tilde{z}_2$
1	7,235,646.9	949,892.9	7,335,749	963,015.6	5,129,409	673,373.7
2	10,032,690.9	1,325,308.8	10,063,742	1,324,944	6,287,502	827,782.2
3	4,776,548.0	560,244	6,922,331	811,924.1	4,776,548	560,244
4	3,127,616.7	408,841.3	4,379,619	572,502.8	1,332,365	174,166.4
5	31,482,466.3	4,364,920.1	39,280,374	5,446,233	30,127,364	4,177,166
6	9,400,413.5	1,075,046.3	9,769,924	1,117,304	3,807,167	435,393.8
7	8,406,613.9	1,470,434.1	13,516,290	2,364,791	3,738,287	654,045.2
8	12,620,488.3	2,293,197.7	20,180,210	3,666,832	5,553,015	1,009,007
9	8,496,376.3	1,379,582.3	9,703,191	1,575,536	2,166,576	351,793.5
10	7,318,522.6	1,115,896	9,480,468	1,440,327	4,417,507	671,133.1
11	3,503,854.2	980,342.7	5,745,796	1,608,421	941,872.50	263,658.4
12	9,447,561.1	1,112,683.1	9,434,406	1,118,489	7,166,191	849,582.4
13	7,097,107.9	1,726,606.7	12,748,722.	3,100,926	2,649,297	644,399.8
14	6,056,718.7	1,001,514.8	9,526,595	1,575,280	2,749,750	454,687.7
15	8,613,660.8	964,960.5	9,226,915	1,033,948	5,663,750	634,667.6
16	5,161,403.2	511,148.2	5,932,758	587,537.6	1,899,396	188,102.5
17	4,971,489.5	1,028,755.1	9,735,516	2,014,580	3,504,900	725,272.3
18	3,888,100.9	643,310.1	5,242,330	867,375.7	1,356,947	224,515.2
19	1,149,636.7	262,087.2	1,154,679	263,236.6	474,800	108,242
20	417,314.5	88,002.7	554,326.4	117,026.5	302,964.9	63,960.38
21	250,168.5	35,204.9	340,151.7	47,867.8	68,296	9610.947
22	52,063	14,574	88,313.7	24,721.7	52,063	14,574
23	261,767.9	30,702.9	327,564.3	38,420.2	137,689.90	16,149.72
24	750,944.6	158,607.8	1,184,206	244,366.5	159,644.20	32,943.34

output-oriented model, and the last two columns show the results obtained by input-oriented model. From Table 7.3, we can observe that the intermediate products of all DMUs on the production frontier obtained by model (7.7) are smaller than that obtained by output-oriented model but larger than that obtained by input-oriented model. The target intermediate products set by the output-oriented model is unfair to the first stage because it should make more efforts to realize the targets than the second stage, while the target intermediate products set by the input-oriented model is unfair to the second stage because it should make more efforts than the first stage. Compared to these two models, the target intermediate products set according to our model are fairer to the two stages because they should make the same efforts. Thus, the two stages would be more willing to accept the target intermediate products by our approach.

As we know, the intermediate products are the outputs of the first stage whose inputs are given. According to the value of target intermediate products in Table 7.3, the efficiency of the first stage of each DMU obtained by model (7.7) is smaller than that obtained by input-oriented model but larger than that obtained by output-oriented model. Similarly, the target efficiency of the second stage of each DMU obtained by model (7.7) is larger than that obtained by input-oriented model but smaller than that obtained by output-oriented model. Table 7.2 clearly shows this phenomenon about the target efficiencies of two stages in three models.

In order to evaluate the size of change for original intermediate products to achieve the target, a new relative measure of "the mean of absolute percentage change" I defined as $I = |(\tilde{Z} - Z)/Z|$ is introduced, where $\tilde{Z}$ is the value of one target intermediate product obtained by model (7.7), output-oriented model and input-oriented model, and Z is the original value of the corresponding intermediate product. A larger value of I means that a larger change should be made in the corresponding intermediate product to achieve its target for the evaluated DMU. The values of I for three models are given in Table 7.4.

In Table 7.4, I_1 and I_2 represent the mean of absolute percentage changes of "underwriting profits" and "investment profits". I in bold indicates that the corresponding intermediate product should be increased by a certain proportion to achieve its target while other I indicates that the corresponding intermediate product should be decreased by a certain proportion. Taking DMU 6 in our model, for example, we should decrease its direct written premiums by 3.56 % and increase its reinsurance premiums by 12.89 % in order to achieve its targets. As the mean of value of I of all DMUs can reflect the effort of DMUs for improving intermediate products in order to be efficient under an approach, we define it as a new index, MI. The standard deviation (SD) of value I is used to quantify the amount of variation or dispersion of values I. The MI of direct written premiums and reinsurance premiums are, respectively, 16.78 % and 62.17 % for model (7.7), 28.68 % and 113.15 % for output oriented model, and 49.74 % and 47.27 % for input-oriented model. We can see that the mean of the values of I_1 for model (7.7) is the smallest among all three models, and the mean of the values of I_2 for model (7.7) is much smaller than that for output-oriented model and a little bit larger than that for input-oriented model. This means that the average change of intermediate products for a company to be efficient by our model

Table 7.4 The percentage decrease of original intermediate products

DMU	Our model		Output-oriented		Input-oriented	
	I_1 (%)	I_2 (%)	I_1 (%)	I_2 (%)	I_1 (%)	I_2 (%)
1	2.90	**10.87**	1.56	**12.41**	31.17	21.40
2	**0.12**	26.90	**0.43**	26.92	37.25	54.34
3	0.00	0.00	**44.92**	**44.92**	0.00	0.00
4	1.49	**9.94**	**37.95**	**53.96**	58.03	53.16
5	15.81	**148.88**	**5.05**	**210.54**	19.43	**138.18**
6	3.56	**12.89**	**0.23**	**17.32**	60.94	54.28
7	21.33	**128.54**	**26.49**	**267.54**	65.02	**1.65**
8	26.91	**102.12**	**16.87**	**223.18**	67.84	11.07
9	25.95	**152.51**	15.43	**188.38**	81.12	35.61
10	10.86	**121.18**	**15.47**	**185.48**	46.20	**33.02**
11	51.49	**52.42**	20.44	**150.07**	86.96	59.01
12	**0.14**	0.52	0.00	0.00	24.04	24.04
13	49.02	**112.81**	8.42	**282.20**	80.97	20.58
14	18.11	**115.14**	**28.80**	**238.40**	62.82	2.32
15	17.35	**28.68**	11.47	**37.88**	45.66	15.37
16	7.93	**26.87**	**5.83**	**45.83**	66.12	53.31
17	35.40	**200.38**	**26.51**	**488.22**	54.45	**111.77**
18	**7.07**	35.39	**44.36**	12.88	62.63	77.45
19	**0.67**	45.77	**1.11**	45.53	58.42	77.60
20	**31.72**	33.29	**74.96**	11.29	4.38	51.52
21	**10.75**	13.16	**50.58**	**18.07**	69.77	76.29
22	0.00	0.00	**69.63**	**69.63**	0.00	0.00
23	**6.45**	38.43	**33.20**	22.95	44.01	67.61
24	**57.62**	75.40	**148.56**	62.10	66.49	94.89
MI	16.78	62.17	28.68	113.15	49.74	47.27
SD	17.47	58.09	33.34	122.87	25.22	36.87

MI represents the mean of values I in each column. SD is the standard deviation of values I

is smaller than that of the other models. The standard deviation of values I_1 in direct written premiums by model (7.7) is the smallest among all three models. The SD of I_2 for model (7.7) is much smaller than that for output-oriented model and a little bit larger than that for input-oriented model. This means that the variation of the change of intermediate products for the two stages of all companies by our model is relatively small. From these observations, we can say the target intermediate products set by our approach are relatively fair because the targets of intermediate products can be more easily accepted by the managers of the two stages (divisions) in these insurance companies. Besides, we can also obtain the benchmark (frontier projection) of each company through our model which is omitted here for saving the space.

7.5 Conclusions

Systems with a two-stage structure are very common in production or service organizations. Although studies on two-stage systems are numerous, they are mainly focused on the efficiency evaluation of the systems. Only a few studies have considered how to determine a target (projection) for the intermediate products of a two-stage system. But these works ignore the fairness issue between two stages when setting the target intermediate products. In this study, we address the fair setting of the target intermediate products for such a system. An approach is proposed based on a new DEA model and a Nash bargaining game to determine the target intermediate products and further the production frontier projections of all DMUs. Our approach was illustrated and validated by an empirical example of insurance companies. The numerical results show that the target intermediate products set by using our approach are more reasonable than those set by using two existing approaches and more easily accepted by divisional managers of these companies. Because of the fair setting of the target intermediate products by our approach, the two stages would to cooperate with each other to realize the targets of the whole system than the earlier approaches.

In this chapter, the fairness in the setting of the intermediate products is interpreted based on the Nash bargaining game model in which the individual efficiencies of two stages in a two-stage system are taken as their disagreement efficiencies. It should be noted that other disagreement efficiencies may also be considered in the model, since the choice of the disagreement efficiencies may be not unique. For example, the full efficiency, i.e., the efficiency of 1, may also be taken as the disagreement efficiency of each stage, because this corresponds to the worst target efficiency of each stage in the fair setting of the intermediate products in a two-stage system. Obviously, this disagreement efficiency for each stage leads to the target efficiency of each stage equal to the square root of the efficiency of the whole system. However, no matter what kind of disagreement efficiency is chosen, the approach for fairly setting the intermediate products proposed in this chapter is always valid.

Although this chapter only considers DMUs with constant returns to scale (CRS), our proposed approach may be extended to DMUs with variable returns to scale (VRS). Under the CRS assumption, we can obtain an important property that the product of any Pareto-optimal objective values of model (7.4) is equal to the system's overall efficiency obtained by model (7.1a, 7.1b, 7.1c, 7.1d, 7.1e, 7.1f). Under the VRS assumption, this property may not be valid. However, the idea of setting the target intermediate products for a two-stage system based on the individual efficiencies of its two stages is also suitable for a system under the VRS assumption. That is, we should keep the ratio of the target efficiencies of the two stages in the system as close to the ratio of their individual efficiencies as possible. However, more work is required to make such extension, because under the VRS assumption, the overall efficiency of a two-stage system may be affected by its intermediate products since the input-oriented efficiency of the system doesn't have the reciprocal relationship with its output-oriented efficiency. For this reason, the study of fairly setting intermediate products under VRS is our future work. In future, we will investigate the influences of various technological assumptions on setting the target intermediate products.

This study also opens two research directions. Firstly, the models presented in this chapter may be extended to systems with more than two stages or more complex networks. Secondly, in this chapter, all inputs, intermediate products and outputs are considered discretionary and desirable, so an extension may also be made to consider both discretionary and non-discretionary inputs, and also consider both desirable and undesirable outputs. If some variables are non-discretionary, we may need to fix them in the model. If some intermediate products are undesirable, these intermediate products may be required as lower as possible by both the first stage and the second stage. How to deal with fairness in this complex scenario is also a future research topic.

Appendices

Appendix 7.1. Proof of Theorem 1

Proof It can be easily proved because we can find that $(\alpha', \beta', \gamma', \tilde{z}'_0, \pi')$ where $\alpha' = \phi^*$, (ϕ^* is the optimal value of model (7.1a, 7.1b, 7.1c, 7.1d, 7.1e, 7.1f), $\beta' = 1$, $\gamma'_0 = 1$, $\gamma'_j = 1$, $j \neq j_0$, $\tilde{z}'_0 = z_0$, $\pi'_0 = 1$, $\pi'_j = 1$, $j \neq j_0$ must be a Pareto solution of model (7.3).

As ϕ^* is the optimal value of model (7.1a, 7.1b, 7.1c, 7.1d, 7.1e, 7.1f), thus ϕ^* must be the minimum value of α can be obtained from model (7.3) when setting $\beta = 1$. This also proves that $(\alpha', \beta', \gamma', \tilde{z}'_0, \pi')$ is a feasible solution of model (7.3). Thus, the optimal value of β must not be smaller than 1.

Then, we prove 1 is the maximum value of β when setting $\alpha = \phi^*$ in model (7.3). If it is not equal to 1, then the optimal value of β model (7.3) must be larger than 1 when setting $\alpha = \phi^*$. By the variable substitution of $\bar{\gamma}_j = \gamma_j/\beta$, $\bar{z}_{d0} = \tilde{z}_{d0}/\beta$, $\bar{\pi}_j = \pi_j/\beta$, $\beta = \phi^*$, model (7.3) can be converted into

$$
\begin{aligned}
& min\, \alpha = \phi^* \\
& min\, 1/\beta \\
& subject\, to\, (stage\, 1) \\
& \qquad \sum_{j=1}^{n} \bar{\gamma}_j x_{ij} \le (\phi^*/\beta) x_{i0}, i = 1, \ldots, m, \\
& \qquad \sum_{j=1}^{n} \bar{\gamma}_j z_{dj} \ge \bar{z}_{d0}, d = 1, \ldots, t, \\
& \qquad \alpha \le 1. \\
& \qquad \bar{\gamma}_j \ge 0, j = 1, \ldots, n, \\
& \qquad (stage\, 2) \\
& \qquad \sum_{j=1}^{n} \bar{\pi}_j z_{dj} \le \bar{z}_{d0}, d = 1, \ldots, t, \\
& \qquad \sum_{j=1}^{n} \bar{\pi}_j y_{rj} \ge y_{r0}, r = 1, \ldots, s, \\
& \qquad \pi_j \ge 0, j = 1, \ldots, n.
\end{aligned}
\tag{A.1}
$$

Then, as $\beta > 1$, then $\phi^*/\beta < \phi^*$. This is contradicted with $\phi^* and$ is the optimal value of model (7.1a, 7.1b, 7.1c, 7.1d, 7.1e, 7.1f). So $(\alpha', \beta', \gamma', \tilde{z}_0', \pi')$ is a Pareto-optimal solution of model (7.3). Thus, model (7.3) has at least one Pareto-optimal solution. □

Appendix 7.2. Proof of Theorem 7.2

Proof Let us evaluate the efficiency of the projected DMU $\left(\sum_{j=1}^{n} \gamma_j^* x_{ij}, \tilde{z}_{d0}^*, \sum_{j=1}^{n} \pi_j^* y_{rj}\right)$. If this DMU is not overall efficient, there must be a feasible solution $p' = (\alpha', \beta', \gamma', \tilde{z}_0', \pi')$ which makes $(\sum_{j=1}^{n} \gamma_j' x_{ij}, -\sum_{j=1}^{n} \pi_j' y_{rj}) < (\sum_{j=1}^{n} \gamma_j^* x_{ij}, -\sum_{j=1}^{n} \pi_j^* y_{rj})$. That is, $\sum_{j=1}^{n} \gamma_j' x_{ij} < \sum_{j=1}^{n} \gamma_j^* x_{ij}$, $\sum_{j=1}^{n} \pi_j' y_{rj} > \sum_{j=1}^{n} \pi_j^* y_{rj}$. Because model (7.3) has the constraints of $\sum_{j=1}^{n} \gamma_j x_{ij} \le \alpha x_{i0}$, $\sum_{j=1}^{n} \pi_j y_{rj} \ge \beta y_{r0}$ and its two objectives are minimizing α, $1/\beta$, then for the solution p', it must have $\sum_{j=1}^{n} \gamma_j' x_{ij} \le \alpha' x_{i0}$, $\sum_{j=1}^{n} \pi_j' y_{rj} \ge \beta' y_{r0}$, and there are at least one input i' and one output r' such that $\sum_{j=1}^{n} \gamma_j' x_{i'j} = \alpha' x_{i'0}$ and $\sum_{j=1}^{n} \pi_j' y_{r'j} = \beta' y_{r'0}$. And for the solution p^*, it must have $\sum_{j=1}^{n} \gamma_j^* x_{ij} \le \alpha^* x_{i0}$, $\sum_{j=1}^{n} \pi_j^* y_{rj} \ge \beta^* y_{r0}$ for all inputs and outputs. As $\sum_{j=1}^{n} \gamma_j' x_{ij}$ is strictly smaller than $\sum_{j=1}^{n} \gamma_j^* x_{ij}$ for any input i and $\sum_{j=1}^{n} \pi_j' y_{rj}$ is strictly larger than $\sum_{j=1}^{n} \pi_j^* y_{rj}$ for any output r. So for input i' and output r', $\alpha' x_{i'0} = \sum_{j=1}^{n} \gamma_j' x_{i'j} < \sum_{j=1}^{n} \gamma_j^* x_{i'j} \le \alpha^* x_{i'0}$ and $\beta' y_{r'0} = \sum_{j=1}^{n} \pi_j' y_{r'j} > \sum_{j=1}^{n} \pi_j^* y_{r'j} \ge \beta^* y_{r'0}$. Therefore, the optimal objective values will satisfy $(\alpha', 1/\beta') < (\alpha^*, 1/\beta^*)$. This leads to a contradiction with the assumption that p^* is a Pareto-optimal solution of model (7.3). Hence, the projected DMU $\left(\sum_{j=1}^{n} \gamma_j^* x_{ij}, \tilde{z}_{d0}^*, \sum_{j=1}^{n} \pi_j^* y_{rj}\right)$ is overall efficient. □

Appendix 7.3. Proof of Theorem 7.3

Proof Let $\bar{z}_{d0} = \vartheta \tilde{z}_{d0}$, $\bar{\gamma}_j = \vartheta \gamma_j$, then programming (7.4) can be converted into the following programming:

$$
\begin{aligned}
& min\, \alpha \\
& min\, \vartheta \\
& subject\, to\, (stage\, 1) \\
& \qquad \textstyle\sum_{j=1}^{n} \bar{\gamma}_j x_{ij} \le \alpha\vartheta x_{i0}, i = 1, \ldots, m, \\
& \qquad \textstyle\sum_{j=1}^{n} \bar{\gamma}_j z_{dj} \ge \tilde{z}_{d0}, d = 1, \ldots, t, \\
& \qquad \alpha \le 1. \\
& \qquad \gamma_j \ge 0, j = 1, \ldots, n, \\
& \qquad (stage\, 2) \\
& \qquad \textstyle\sum_{j=1}^{n} \tau_j z_{dj} \le \tilde{z}_{d0}, d = 1, \ldots, t, \\
& \qquad \textstyle\sum_{j=1}^{n} \tau_j y_{rj} \ge y_{r0}, r = 1, \ldots, s, \\
& \qquad \vartheta \le 1. \\
& \qquad \tau_j \ge 0, j = 1, \ldots, n.
\end{aligned} \tag{A.2}
$$

For any pair of Pareto-optimal objective values (α^*, ϑ^*) of model (7.4), it is also a pair of Pareto-optimal objective values of model (A.2). Denote by $\underline{\phi}$ the minimum value of $\alpha\vartheta$ that can be obtained among all pairs of Pareto-optimal objective values (α, ϑ) of the model, we must have $\alpha^*\vartheta^* = \underline{\phi}$. Because if $\alpha^*\vartheta^* > \underline{\phi}$, there is α' such that $\alpha^*\vartheta^* > \alpha'\vartheta^* = \underline{\phi}$, where $0 \le \alpha' < \alpha^* \le 1$. Since $\underline{\phi}$ the minimum value, there is a Pareto-optimal solution $\left(\underline{\alpha}, \underline{\vartheta}, \underline{\gamma}, \underline{\tilde{z}}_0, \underline{\tau}\right)$ of model (A.2) such that $\sum_{j=1}^{n} \underline{\gamma}_j x_{ij} \le \underline{\phi} x_{i0}, i = 1, \ldots, m$, where $\underline{\phi} = \underline{\alpha\vartheta}$. From $\alpha'\vartheta^* = \underline{\phi}$, we can construct a new feasible solution $(\alpha', \vartheta^*, \underline{\gamma}, \underline{\tilde{z}}_0, \underline{\tau})$ of model (A.2), this is contradictory with the assumption that (α^*, ϑ^*) is a pair of Pareto-optimal objective values of the model, so $\alpha^*\vartheta^* = \underline{\phi}$ is proved. Moreover, if we take $\alpha = 1$ in model (A.2), it is reduced to model (7.1a, 7.1b, 7.1c, 7.1d, 7.1e, 7.1f) by replacing variable ϑ with ϕ. This implies that $(1, \phi^*)$ is a feasible pair of objective values of model (A.2). This pair is Pareto-optimal, because otherwise there is a pair of Pareto-optimal objective values (α^*, ϑ^*) such that $\alpha^* < 1$ and $\vartheta^* \le \phi^*$, or $\alpha^* \le 1$ and $\vartheta^* < \phi^*$. For both case, $\alpha^*\vartheta^* < \phi^*$ and we have $\sum_{j=1}^{n} \underline{\gamma}_j x_{ij} \le \alpha^*\vartheta^* x_{i0}, i = 1, \ldots, m$, this implies that we can find a feasible solution of model (7.1a, 7.1b, 7.1c, 7.1d, 7.1e, 7.1f) with objective value $\alpha^*\vartheta^*$ by letting $\phi = \alpha^*\vartheta^*$ and the values of other variables the same as those of the corresponding variables in model (A.2). This is contradictory with the assumption that ϕ^* is the minimum objective value of model (7.1a, 7.1b, 7.1c, 7.1d, 7.1e, 7.1f), so $(1, \phi^*)$ is Pareto-optimal. This implies $\underline{\phi} = \phi^*$, i.e., the product of any Pareto-optimal values α^* and ϑ^* of model (7.4) is equal to the overall efficiency obtained by model (7.1a, 7.1b, 7.1c, 7.1d, 7.1e, 7.1f). □

References

An, Q., Yan, H., Wu, J., & Liang, L. (2016). Internal resource waste and centralization degree in two-stage systems: An efficiency analysis. *Omega, 61,* 89–99.

Biener, C., & Eling, M. (2012). Organization and efficiency in the international insurance industry: A cross-frontier analysis. *European Journal of Operational Research, 221*(2), 454–468.

Charnes, A., Cooper, W. W., & Rhodes, E. (1978). Measuring the efficiency of decision making units. *European Journal of Operational Research, 2*(6), 429–444.

Chen, L., Huang, Y., Li, M. J., & Wang, Y. M. (2020). Meta-frontier analysis using cross-efficiency method for performance evaluation. *European Journal of Operational Research, 280*(1), 219–229.

Chen, Y., Cook, W. D., Li, N., & Zhu, J. (2009a). Additive efficiency decomposition in two-stage DEA. *European Journal of Operational Research, 196*(3), 1170–1176.

Chen, Y., Liang, L., & Zhu, J. (2009b). Equivalence in two-stage DEA approaches. *European Journal of Operational Research, 193*(2), 600–604.

Chen, Y., Cook, W. D., & Zhu, J. (2010). Deriving the DEA frontier for two-stage processes. *European Journal of Operational Research, 202*(1), 138–142.

Chen, Y., Cook, W. D., Kao, C., & Zhu, J. (2013). Network DEA pitfalls: divisional efficiency and frontier projection under general network structures. *European Journal of Operational Research, 226*, 507–515.

Chen, Y., Cook, W. D., Kao, C., & Zhu, J. (2014). Network DEA pitfalls: Divisional efficiency and frontier projection. In *Data envelopment analysis* (pp. 31–54). Springer, Boston, MA.

Chen, Y., & Zhu, J. (2004). Measuring information technology's indirect impact on firm performance. *Information Technology and Management, 5*(1–2), 9–22.

Cook, W. D., & Seiford, L. M. (2009). Data envelopment analysis (DEA)—Thirty years on. *European Journal of Operational Research, 192*(1), 1–17.

Cook, W. D., Liang, L., & Zhu, J. (2010). Measuring performance of two-stage network structures by DEA: A review and future perspective. *Omega, 38*(6), 423–430.

Despotis, D. K., Koronakos, G., & Sotiros, D. (2016). Composition versus decomposition in two-stage network DEA: A reverse approach. *Journal of Productivity Analysis, 45*(1), 71–87.

Du, J., Liang, L., Chen, Y., Cook, W. D., & Zhu, J. (2011). A bargaining game model for measuring performance of two-stage network structures. *European Journal of Operational Research, 210*(2), 390–397.

Färe, R., & Grosskopf, S. (1996). Productivity and intermediate products: A frontier approach. *Economics Letters, 50*(1), 65–70.

Halkos, G. E., Tzeremes, N. G., & Kourtzidis, S. A. (2014). A unified classification of two-stage DEA models. *Surveys in Operations Research and Management Science, 19*(1), 1–16.

Huang, W., & Eling, M. (2013). An efficiency comparison of the non-life insurance industry in the BRIC countries. *European Journal of Operational Research, 226*(3), 577–591.

Ilyas, A. M., & Rajasekaran, S. (2019). An empirical investigation of efficiency and productivity in the Indian non-life insurance market. *Benchmarking—An International Journal, 26*(7), 2343–2371.

Kao, C., & Hwang, S. N. (2008). Efficiency decomposition in two-stage data envelopment analysis: An application to non-life insurance companies in Taiwan. *European Journal of Operational Research, 185*(1), 418–429.

Kao, C., & Hwang, S. N. (2011). Decomposition of technical and scale efficiencies in two-stage production systems. *European Journal of Operational Research, 211*(3), 515–519.

Kao, C., & Hwang, S. N. (2014). Multi-period efficiency and Malmquist productivity index in two-stage production systems. *European Journal of Operational Research, 232*(3), 512–521.

Khezrimotlagh, D., Zhu, J., Cook, W. D., & Toloo, M. (2019). Data envelopment analysis and big data. *European Journal of Operational Research, 274*(3), 1047–1054.

Lewis, H. F., & Sexton, T. R. (2004). Network DEA: efficiency analysis of organizations with complex internal structure. *Computers & Operations Research, 31*(9), 1365–1410.

Liang, L., Cook, W. D., & Zhu, J. (2008). DEA models for two-stage processes: Game approach and efficiency decomposition. *Naval Research Logistics (NRL), 55*(7), 643–653.

Liang, L., Yang, F., Cook, W. D., & Zhu, J. (2006). DEA models for supply chain efficiency evaluation. *Annals of Operations Research, 145*(1), 35–49.

Li, Y., Chen, Y., Liang, L., & Xie, J. (2012). DEA models for extended two-stage network structures. *Omega, 40*(5), 611–618.

Lim, S., & Zhu, J. (2016). A note on two-stage network DEA model: Frontier projection and duality. *European Journal of Operational Research, 248*(1), 342–346.

Lim, S., & Zhu, J. (2019). Primal-dual correspondence and frontier projections in two-stage network DEA models. *Omega, 83,* 236–248.

Liu, J. S., Lu, L. Y., Lu, W. M., & Lin, B. J. (2013). Data envelopment analysis 1978–2010: A citation-based literature survey. *Omega, 41*(1), 3–15.

Nash, J. (1950). The bargaining problem. *Econometrica: Journal of the Econometric Society*, 155–162.

Nash, J. (1953). Two-person cooperative games. *Econometrica: Journal of the Econometric Society*, 128–140.

Seiford, L. M., & Zhu, J. (1999). Profitability and marketability of the top 55 US commercial banks. *Management Science, 45*(9), 1270–1288.

Sexton, T. R., & Lewis, H. F. (2003). Two-stage DEA: An application to major league baseball. *Journal of Productivity Analysis, 19*(2–3), 227–249.

Tone, K., & Tsutsui, M. (2014). Dynamic DEA with network structure: A slacks-based measure approach. *Omega, 42*(1), 124–131.

Wang, K., Huang, W., Wu, J., & Liu, Y. N. (2014). Efficiency measures of the Chinese commercial banking system using an additive two-stage DEA. *Omega, 44,* 5–20.

Chapter 8
Fixed Cost and Resource Allocation Considering Technology Heterogeneity in Two-Stage Network Production Systems

Tao Ding, Feng Li, and Liang Liang

Abstract Many studies have concentrated on fixed cost allocation and resource allocation issues by using data envelopment analysis (DEA). Existing approaches allocate fixed cost and resource primary based on the efficiency maximization principle. However, due to the existing of technology heterogeneity among DMUs, it is impractical for all the DMUs to achieve a common technology level, especially when some DMUs are far from the efficient frontier. In this chapter, under the centralized decision environment, we present a new approach to deal with fixed cost and resource allocation issues for a two-stage production system by considering the factor of technology heterogeneity. Specifically, technology difference is analyzed in the performance evaluation framework firstly. Then, by taking the technology heterogeneity into account, the two-stage DEA-based fixed cost allocation and resource allocation models are proposed. In addition, two illustrated examples are calculated to show the feasibility of the two proposed models. Finally, this chapter is concluded.

Keywords Data envelopment analysis · Fixed cost allocation · Resource allocation · Two-stage · Technology heterogeneity

This chapter is an extended work based on Ding, T., Chen, Y., Wu, H., & Wei, Y. (2018). Centralized fixed cost and resource allocation considering technology heterogeneity: A DEA approach. Annals of Operations Research, 268(1–2), 497–511.

T. Ding (✉)
School of Economics, Hefei University of Technology, 485 Danxia Road, Hefei 230601, China
e-mail: hfutdingtao@126.com

F. Li
School of Business Administration, Southwestern University of Finance and Economics, Chengdu 611130, Sichuan Province, China
e-mail: lifeng1990@swufe.edu.cn

L. Liang
School of Management, Hefei University of Technology, 193 Tunxi Road, Hefei 230009, China
e-mail: lliang@hfut.edu.cn

V. Charles et al. (eds.), *Data Science and Productivity Analytics*,
International Series in Operations Research & Management Science 290,
https://doi.org/10.1007/978-3-030-43384-0_8

8.1 Introduction

Data envelopment analysis (DEA) has become an effective tool for performance evaluation and benchmarking since it is first introduced by Charnes et al. (1978). Following the first CCR model, various kinds of DEA models have been proposed (Banker et al. 1984; Cook and Seiford 2009) and widely applied to different fields (Cooper et al. 2007, 2011). Recently, two of the most popular DEA applications are fixed cost allocation and resource allocation (Cooper et al. 2011).

Many DEA-based approaches have been proposed to deal with the fixed cost allocation and resource allocation problems (Cook and Kress 1999; Cook and Zhu 2005). The former problem expects the cost allocated to each DMU to be as less as possible Si et al. (2013). Conversely, the latter expects the resource allocated to each DMU to be as more as possible (Lozano et al. 2009; Bi et al. 2011; Fang 2013, 2015). All the approaches mentioned above are based on two principles. One is efficiency invariance principle which means all DMUs keep their efficiencies without any change after allocation. For example, Cook and Kress (1999) first introduce the DEA technique to obtain a cost allocation plan which is based on two principles: efficiency invariance and Pareto-minimality. Following them, Cook and Zhu (2005) further extend their approach under both input and output orientations, and provide a linear programming for calculating the cost allocation. In consideration of the non-uniqueness problem in Cook and Kress (1999), Lotfi et al. (2012) propose a fixed allocation mechanism that is based on a common dual weights approach. Lin (2011) proposes an approach for allocating fixed cost in such a way that the relative efficiency of every DMU remains unchanged. Furthermore, Lin et al. (2016) propose a new proportional sharing model to determine a unique fixed cost allocation under two assumptions: efficiency invariance and zero slack. The other principle is efficiency maximized principle that means all DMUs maximized their efficiencies to be efficient after allocation. For example, Beasley (2003) develops a non-linear resource allocation model by maximizing the average efficiency of the DMUs in an organization. Korhonen and Syrjänen (2004) develop an interactive formal approach based on DEA and multiple-objective linear programming (MOLP) to obtain the most optimal allocation plan. Lozano and Villa (2004) suggested allocating resource by a two-phase method, minimizing the total input consumption under final output constraints. After these operations, all the DMUs were projected onto a region of the efficient frontier. Different from Lozano et al. (2004), Asmild et al. (2009) develop centralized BCC models to allocate resources, in which only adjustments of previously inefficient units are considered. Li et al. (2009) consider a situation that the fixed cost is a complement of other inputs rather than an extra independent input. And they investigate the relationship between the allocated cost and the efficiency score. Li et al. (2013) further propose a maximization model and a corresponding algorithm to generate a unique fixed cost allocation based on satisfaction degree. Du et al. (2014) propose a DEA cross-efficiency to distribute fixed cost, and ensure that all DMUs become efficient after the fixed cost is allocated as an additional input measure. Li et al. (2018) proposed a cooperative game cross-efficiency approach that maximized the cross-efficiency and determined the allocation plan using the Shapley value. Recently, Li et al. (2019b) suggested a non-egoistic principle which states that each DMU should propose its allocation proposal in such a way that the maximal cost would be allocated to itself. Chu and Jiang (2019) suggested generating the allocation scheme by maximizing the minimum utility across all DMUs based on a common set of weights, which also implicitly maximized the efficiency scores. Amirteimoori and Tabar (2010) present an approach to fixed resource allocation and target setting such that each DMU has an efficiency score one.

To summarize, existing approaches to deal with fixed cost and resource allocation problems primary based on two efficiency principles. In reality, however, both two principles are difficult to realize because of the limitations of technology progress. To be specific, for efficiency invariance principle, it is not rational for all DMUs without any improvement in efficiencies after increasing resources or decreasing costs. For efficiency maximized principle, it may not be suitable in many cases because of the neglect of possible technology heterogeneity among the DMUs. Conventional efficiency maximization allocation plans automatically assume that all DMUs are equipped with the same level of production technology. However, the assessed DMUs usually have different production technologies due to differences in geographical location, national policy, and socioeconomic conditions (Chen and Song 2006). For example, the industrial technology level in western provinces of China is far below that in eastern provinces. It is impractical and irrational to make an allocation plan such that every province can achieve the common frontier (Wang et al. 2013). Moreover, the DMU with a very low efficiency level is unable to be efficient just after an allocation. Therefore, it could lead to an unfair allocation result ignoring the technology heterogeneity.

In order to take the technology heterogeneity into account in an analysis framework, Battese and Rao (2002) propose the meta-frontier production function framework, which utilizes stochastic frontier approach (SFA) to investigate the environmental efficiency of firms in groups that have different technologies. In another work, O'Donnell et al. (2008) construct the meta-frontier based on DEA instead of SFA. However, these studies still evaluate the DMUs on the basis of technology homogeneity. As far as we know, few literature has considered the technology heterogeneity factor in DEA-based fixed cost and resource allocation problems.

Recently, Ding et al. (2018) developed a novel new approach to deal with fixed cost and resource allocation problems by considering the factor of technology heterogeneity under the centralized decision environment. Both the concepts of meta-efficiency and group efficiency as well as meta-technology ratio were introduced to reflect the technology level of the DMUs, and then two centralized DEA models considering technology heterogeneity were proposed to allocate fixed cost and resources, respectively. However, the aforementioned studies as well as Ding et al. (2018) deal with the one-stage DEA structure, in which the internal structures of DMUs are ignored and the units are treated as 'black boxes' (Lewis and Sexton 2004). However, in many cases, the production process usually has intermediate measures, and more specifically there exists a two-stage network process in which outputs generated from the first stage will be input to the second stage. Such two-stage network structures are common in many real applications and have been extensively studied in the DEA literature. Recently, Yu et al. (2016) extended the game cross-efficiency approach of Du et al. (2014) to two-stage network structures, where the intermediate outputs produced from the first stage were only inputs to the second stage. Afterwards, Ding et al. (2019) and Zhu et al. (2019) extended the satisfaction degree method of Li et al. (2013), and some variants of the two-stage structures and satisfaction degree concepts were also studied in the two-stage fixed cost allocation problem. Recently, Li et al. (2019b) studied the same problem, of which the most significant feature of the two-stage fixed cost allocation approaches was that it can obtain a unique allocation scheme through an iterative procedure. In addition, Li et al. (2019b) took the operation sizes of DMUs into account, and the final allocation scheme was supposed to reflect the current input usage and output production from a size perspective. Chu et al. (2019) studied the same problem using a satisfaction degree bargaining game approach, and their approach was able to theoretically guarantee the uniqueness of allocation plans. An et al. (2019) also studied the two-stage FCA problem and the efficiency invariance principle was considered for both the whole system and the individual stage.

In this chapter, we aim to present more fair approaches to the fixed cost and resource allocation problems for reducing the negative influence of the inconsideration of technology gaps among the DMUs with two-stage network production systems. Under a centralized decision-making environment, this chapter first introduces the concept of meta-frontier and group frontier to analyze the two-stage technology heterogeneity. With the assumption that DMUs in the same group have similar technology level, both subjective and objective grouping methods are presented. And then the meta-technology ratio is calculated to reflect the group technology level. Afterwards, considering the technology heterogeneity factor, we develop centralized DEA-based approaches to allocate fixed cost and resources across DMUs with two-stage production systems, respectively. Specifically, from a conventional point of view, the allocated cost is still treated as a new input in our model. Take the technology heterogeneity factor into consideration, the efficiency of each DMU should be non-decreasing and should not exceed the meta-technology ratio. Our new approach has two main principles. First, the efficiencies of all DMUs in the next period should be greater than or equal to their relative efficiencies in the current period. Second, the efficiencies of all DMUs in the next period should not exceed their technology level. A distinct feature of the current work is that it focuses not only on the efficiency principle but also on practical feasibility.

The structure of this chapter is organized as follows: Sect. 8.2 introduces the meta-frontier and group-frontier concepts to analyze a DMU's two-stage technology heterogeneity. By incorporating the meta-technology ratio into analysis, Sect. 8.3 proposes the two-stage DEA fixed cost allocation and resource allocation approaches, respectively. Two numerical examples are given to illustrate the feasibility of the proposed methods. Conclusions are provided in the last section.

8.2 Technology Difference Analysis

8.2.1 The Meta-frontier and Group Frontier

When using the traditional DEA models to evaluate the efficiency of each DMU, it is usually assumed that all DMUs possess the same level of production technology. However, in many situations, the assessed DMUs have different production technologies due to differences in geographical location, national policy, production scale, and other factors. Herein, O'Donnell et al. (2008) apply the meta-frontier concept to estimating DEA efficiency. Particularly, they gauge a meta-frontier through the use of overall samples, divide the DMUs and estimate the group frontiers of group samples. Here in this chapter, we limit our focus on the most typical two-stage production system, in which the intermediate outputs generated in the first stage will be unique inputs that will be used to produce final outputs in the second stage (Kao and Hwang 2008; Chen et al. 2010).

Suppose that there are n DMUs under evaluation, with each having a two-stage network production system as depicted in Fig. 8.1. More specifically, $DMU_j(j = 1, \ldots, n)$ will consume m inputs $x_{ij}(i = 1, \ldots, m)$ to generate q intermediate outputs $z_{pj}(p = 1, \ldots, q)$, which are further used to generate s outputs $y_{rj}(r = 1, \ldots, s)$. According to the differences in production technologies, all DMUs can be divided into $J(J > 1)$ groups. The number of DMUs in the kth group is N_k such that $\sum_{k=1}^{J} N_k = n$.

The two-stage production technology of the kth group frontier is defined by

$$T^k = \left\{ \left(x_i^k, z_p^k, y_r^k\right) \middle| \begin{array}{l} \sum_{j=1}^{N_k} \lambda_j^k x_{ij}^k \le x_i^k, i = 1, \ldots, m \\ \sum_{j=1}^{N_k} \lambda_j^k z_{pj}^k \ge z_p^k, p = 1, \ldots, q \\ \sum_{j=1}^{N_k} \eta_j^k z_{pj}^k \le z_p^k, p = 1, \ldots, q \\ \sum_{j=1}^{N_k} \eta_j^k y_{rj}^k \ge y_r^k, r = 1, \ldots, s \\ \lambda_j^k, \eta_j^k \ge 0, j = 1, \ldots, N_k \end{array} \right. \quad (8.1)$$

The production technology of the meta-frontier (T^{meta}) is defined by

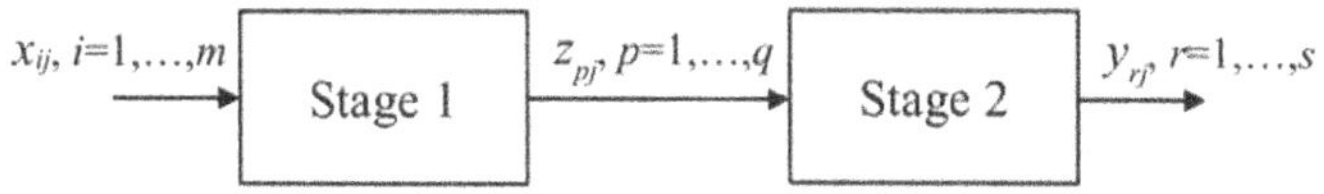

Fig. 8.1 Two-stage network structure

$$T^{meta} = \left\{ (x_i, z_p, y_r) \middle| \begin{array}{l} \sum_{k=1}^{J} \sum_{j=1}^{N_k} \lambda_j^k x_{ij}^k \le x_i, i = 1, \ldots, m \\ \sum_{k=1}^{J} \sum_{j=1}^{N_k} \lambda_j^k z_{pj}^k \ge z_p, p = 1, \ldots, q \\ \sum_{k=1}^{J} \sum_{j=1}^{N_k} \eta_j^k z_{pj}^k \le z_p, p = 1, \ldots, q \\ \sum_{k=1}^{J} \sum_{j=1}^{N_k} \eta_j^k y_{rj}^k \ge y_r, r = 1, \ldots, s \\ \lambda_j^k, \eta_j^k \ge 0, k = 1, \ldots, J; j = 1, \ldots, N_k \end{array} \right. \tag{8.2}$$

In Eqs. (8.1) and (8.2), T^k and T^{meta} represent the specific technologies of the meta-frontier and group frontier, respectively. And they hold on the following properties:

(1) If $(x_i, z_p, y_r) \in T^k$, then $(x_i, z_p, y_r) \in T^{meta}$ for any k;
(2) If $(x_i, z_p, y_r) \in T^{meta}$, then $(x_i, z_p, y_r) \in T^k$ for some k.

In addition, we make the following assumption:

Assumption *For two given DMU_f and DMU_g, it is needed that:*

(1) *If DMU_f and DMU_g belong to the same group, then they have the same or similar production technology;*
(2) *If DMU_f and DMU_g belong to different groups, then they have different production technologies.*

8.2.2 Calculation of Meta-group Efficiency

Considering the two-stage series structure in Fig. 8.1, the relative efficiency of the whole system can be computed as the weighted sum of two substage efficiency (Chen et al. 2010). Under the meta-frontier, for a given DMU_d *and* DMU_g its two-stage relative meta-efficiency (ME) can be computed by the following standard CCR model:

$$\begin{aligned} \theta_d^{ME} &= Max(w_1\theta_{d1} + w_2\theta_{d2}) \\ s.t.\theta_{j1} &= \frac{\sum_{p=1}^{q} \varphi_p z_{pj}}{\sum_{i=1}^{m} v_i x_{ij}} \le 1, j = 1, \ldots, n \\ \theta_{j2} &= \frac{\sum_{r=1}^{s} u_r y_{rj}}{\sum_{p=1}^{q} \varphi_p z_{pj}} \le 1, j = 1, \ldots, n \\ u_r, \varphi_p, v_i &\ge 0, r = 1, \ldots, s; p = 1, \ldots, q; i = 1, \ldots, m. \end{aligned} \tag{8.3}$$

Model (8.3) encapsulates the idea that the considered DMU can select a set of relative weights and a feasible allocation plan to maximize its possible relative efficiency. In addition, w_1 and w_2 are important indexes of two substages such that $w_1 + w_2 = 1$. According to Chen et al. (2010), a rational method to determine the value of w_1 and w_2 is the proportion of resources consumed by each substage. Therefore, we have

$$w_1 = \frac{\sum_{i=1}^{m} v_i x_{id}}{\sum_{i=1}^{m} v_i x_{id} + \sum_{p=1}^{q} \varphi_p z_{pd}}, \; w_2 = \frac{\sum_{p=1}^{q} \varphi_p z_{pd}}{\sum_{i=1}^{m} v_i x_{id} + \sum_{p=1}^{q} \varphi_p z_{pd}}. \tag{8.4}$$

Thus, the overall efficiency of the whole two-stage system for DMU_d is calculated by solving the following fractional model under CRS assumption:

$$\begin{aligned} \theta_d^{ME} &= Max \frac{\sum_{p=1}^{q}\varphi_p z_{pd} + \sum_{r=1}^{s} u_r y_{rd}}{\sum_{i=1}^{m} v_i x_{id} + \sum_{p=1}^{q}\varphi_p z_{pd}} \\ s.t.\, \theta_{j1} &= \frac{\sum_{p=1}^{q}\varphi_p z_{pj}}{\sum_{i=1}^{m} v_i x_{ij}} \le 1, j = 1, \ldots, n \\ \theta_{j2} &= \frac{\sum_{r=1}^{s} u_r y_{rj}}{\sum_{p=1}^{q}\varphi_p z_{pj}} \le 1, j = 1, \ldots, n \\ u_r, \varphi_p, v_i &\ge 0, r = 1, \ldots, s; p = 1, \ldots, q; i = 1, \ldots, m. \end{aligned} \tag{8.5}$$

The fractional program (8.5) can be equivalently converted into a linear model (8.6) by using the Charnes–Cooper transformation (Charnes and Cooper 1962).

$$\begin{aligned} \theta_d^{ME} &= Max\left(\sum_{p=1}^{q}\varphi_p z_{pd} + \sum_{r=1}^{s} u_r y_{rd}\right) \\ s.t.\, &\sum_{i=1}^{m} v_i x_{id} + \sum_{p=1}^{q}\varphi_p z_{pd} = 1 \\ &\sum_{p=1}^{q}\varphi_p z_{pj} - \sum_{i=1}^{m} v_i x_{ij} \le 0, j = 1, \ldots, n \\ &\sum_{r=1}^{s} u_r y_{rj} - \sum_{p=1}^{q}\varphi_p z_{pj} \le 0, j = 1, \ldots, n \\ &u_r, \varphi_p, v_i \ge 0, r = 1, \ldots, s; p = 1, \ldots, q; i = 1, \ldots, m. \end{aligned} \tag{8.6}$$

Solving the above model (8.6) can determine an optimal meta-efficiency θ_d^{ME}. Similarly, under the group frontier, for a given DMU_d belong to group *k*, its two-stage relative group efficiency (GE) can be computed by the following model (8.7):

$$\begin{aligned} \theta_d^{GE} &= Max\left(\sum_{p=1}^{q}\varphi_p z_{pd} + \sum_{r=1}^{s} u_r y_{rd}\right) \\ s.t.\, &\sum_{i=1}^{m} v_i x_{id} + \sum_{p=1}^{q}\varphi_p z_{pd} = 1 \\ &\sum_{p=1}^{q}\varphi_p z_{pj} - \sum_{i=1}^{m} v_i x_{ij} \le 0, j = 1, \ldots, N_k \\ &\sum_{r=1}^{s} u_r y_{rj} - \sum_{p=1}^{q}\varphi_p z_{pj} \le 0, j = 1, \ldots, N_k \\ &u_r, \varphi_p, v_i \ge 0, r = 1, \ldots, s; p = 1, \ldots, q; i = 1, \ldots, m. \end{aligned} \tag{8.7}$$

Because the meta-frontier envelops the *k* group frontier, the efficiency of a given DMU_j measured on the basic of the meta-frontier is less than that of the group frontiers, which can be described by $\theta_j^{ME} \le \theta_j^{GE}$. O'Donnell et al. (2008) state that technical efficiency under meta-frontier can be decomposed into the group efficiency (GE) as well as the technology gap ratio (TGR) denoted by $\beta_j = \frac{\theta_j^{ME}}{\theta_j^{GE}}$. GE indicates the relative efficiency of a given DMU under specific group-frontier technologies, whereas MTR indicates how close a group-frontier technology is to a meta-frontier technology. Since $\theta_j^{ME} \le \theta_j^{GE}$ always holds, the MTR is not greater than unity. The higher the MTR, the closer the group-frontier technology is to the meta-frontier technology. Apparently, when β_j is closer to 1, DMU_j has the smaller gap between the two technologies. Conversely, when β_j is closer to 0, DMU_j has the bigger gap between the two technologies.

8.2.3 Methods for Group Formulation

In this subsection, we present two ways for group formulation. One is subjective, i.e., the external characteristic approach. The other is objective, i.e., the extended context-dependent approach.

External characteristic approach
As mentioned above, the sources of technology heterogeneity are differences in geographical location, production scale, socioeconomic conditions, and some other inherent attributes. In many cases, the external characteristic differences among the DMUs are easy to be identified. It is practical to divide the DMUs into several groups according to the external characteristic. For example, considering the level of industrial development of China, each province can be regarded as a DMU. It is clear that the production technology in eastern provinces is higher than that in western provinces. According to the geographical region, all the provinces in China can be divided into three groups: the east region, the central region, and the west region.

However, in some conditions, it is not easy to find the external characteristic differences among the DMUs. A subjective division is no longer applied. Then, an objective method has to be proposed to group the DMUs.

Extended context-dependent approach
Seiford and Zhu (2003) present a context-dependent DEA approach to divide a set of DMUs into different levels of efficient frontiers. Specifically, the first-level efficient frontier is obtained by the CCR model such as model (8.6). Then removing the efficient DMUs, the remaining inefficient DMUs will shape a second-level efficient frontier, and so on. In this way, all the DMUs are grouped into several efficient frontiers. By using this method, every DMU is group efficient, namely $\theta_j^{GE} = 1$. And the meta-technology ratio comes to be $\beta_j = \theta_j^{ME}$.

However, this will also lead to some drawbacks to apply the context-dependent grouping approach. For example, too many groups are produced because DMUs are impossible to have strictly the same technologies in actual. Even a group has only one DMU in some situations. It may lead to a narrower applicability of the context-dependent approach. Moreover, the managers are unwilling to see a huge gap between group-frontier technology and meta-frontier technology. In these regards, we modify the context-dependent grouping method such that it is more closely to practical applications.

For a given set of DMUs denoted by $K^h = \{DMU_j, j = 1, \ldots, n\}$, K^{h+1} is defined as $K^h - G^h$ where $|G^h = \{DMU_k \in K^h | \psi(h, k) \geq \gamma\}$. Specially, γ is a threshold efficiency value given by the decision-maker with a range from [0, 1]. And $\psi(h, k)$ is the optimal value solved from the following model:

$$
\begin{gathered}
\psi(h,k) = Min\,\theta \\
s.t.\ \sum_{j\epsilon\Phi(K^h)} \lambda_j x_{ij} \le \theta x_{ik}, i = 1,\ldots,m \\
\sum_{j\epsilon\Phi(K^h)} \lambda_j z_{pj} - \sum_{j\epsilon\Phi(K^h)} \eta_j z_{pj} + \theta z_{pk} \ge z_{pk}, p = 1,\ldots,q \\
\sum_{j\epsilon\Phi(K^h)} \eta_j y_{rj} \ge y_{rk}, r = 1,\ldots,s \\
\lambda_j, \eta_j \ge 0, j\epsilon\Phi\left(K^h\right),
\end{gathered}
\tag{8.8}
$$

where $\Phi(\Delta)$ represents the subscript index set with respect to a DMU set. Model (8.8) is a dual formulation of model (8.6).

Similar to Seiford and Zhu's (2003) approach, by the calculation of model (8.8) and the definition of G^h, the DMUs are divided into several groups clearly. It should be point out that there is a negative correlation between the threshold value γ and number of groups. Specifically, the number of groups reduce to one when $\gamma = 0$. And the grouping result is the same to Seiford and Zhu's (2003) when $\gamma = 1$.

8.3 Proposed Fixed Cost and Resource Allocation Models

8.3.1 *The Fixed Cost Allocation Approach*

Using the methods in Sect. 8.2 as a backup, we propose a centralized DEA approach to allocate the fixed cost in this section. Suppose that a fixed cost R is to be assigned among n DMUs, and each DMU_j obtains a fixed cost R_j, i.e., $\sum_{j=1}^{n} R_j = R$. Because the whole process of DMU_j can freely allocate the fixed cost R_j between the first and second stages, the fixed cost can be viewed as an additional shared input (Yu et al. 2016). Without loss of generality, DMU_j allocates R_{1j} to the first stage and R_{2j} to the second stage. Taking the allocate cost into consideration, we can develop the following model (8.9) to evaluate the relative overall efficiency for DMU_d.

$$
\begin{gathered}
E_d = Max\left(w_1' \frac{\sum_{p=1}^{q} \varphi_p z_{pd}}{\sum_{i=1}^{m} v_i x_{id} + R_{1d}} + w_2' \frac{\sum_{r=1}^{s} u_r y_{rd}}{\sum_{p=1}^{q} \varphi_p z_{pd} + R_{2d}}\right) \\
s.t.\ \frac{\sum_{p=1}^{q} \varphi_p z_{pj}}{\sum_{i=1}^{m} v_i x_{ij} + R_{1j}} \le 1, j = 1,\ldots,n \\
\frac{\sum_{r=1}^{s} u_r y_{rj}}{\sum_{p=1}^{q} \varphi_p z_{pj} + R2_j} \le 1, j = 1,\ldots,n \\
\sum_{j=1}^{n} \left(R_{1j} + R_{2j}\right) = R \\
u_r, \varphi_p, v_i, R_{1j}, R_{2j} \ge 0, r = 1,\ldots,s;\ p = 1,\ldots,q;\ i = 1,\ldots,m;\ j = 1,\ldots,n.
\end{gathered}
\tag{8.9}
$$

Model (8.9) immediately takes the value of one for the relative weight attached to the allocated cost as it will not affect and change the results (Beasley 2003). Further, similar to the previous section, a reasonable weight choice of each stage is the proportion of total inputs devoted to each stage. Here, we have

$$w_1' = \frac{\sum_{i=1}^{m} v_i x_{id} + R_{1d}}{\sum_{i=1}^{m} v_i x_{id} + R_{1d} + \sum_{p=1}^{q} \varphi_p z_{pd} + R_{2d}},\ w_2' = \frac{\sum_{p=1}^{q} \varphi_p z_{pd} + R_{2d}}{\sum_{i=1}^{m} v_i x_{id} + R_{1d} + \sum_{p=1}^{q} \varphi_p z_{pd} + R_{2d}} \quad (8.10)$$

Thus, the overall efficiency of the whole two-stage network process with the allocated cost for a given DMU_d is assessed by solving the following model (8.11) and the relative efficiency of DMU_d can be denoted by $\frac{\sum_{p=1}^{q} \varphi_p z_{pd} + \sum_{r=1}^{s} u_r y_{rd}}{\sum_{i=1}^{m} v_i x_{id} + R_{1d} + \sum_{p=1}^{q} \varphi_p z_{pd} + R_{2d}}$.

$$\begin{aligned} E_d = Max & \frac{\sum_{p=1}^{q} \varphi_p z_{pd} + \sum_{r=1}^{s} u_r y_{rd}}{\sum_{i=1}^{m} v_i x_{id} + R_{1d} + \sum_{p=1}^{q} \varphi_p z_{pd} + R_{2d}} \\ s.t. & \frac{\sum_{p=1}^{q} \varphi_p z_{pj}}{\sum_{i=1}^{m} v_i x_{ij} + R_{1j}} \le 1,\ j = 1, \ldots, n \\ & \frac{\sum_{r=1} u_r y_{rj}}{\sum_{p=1}^{q} \varphi_p z_{pj} + R_{2j}} \le 1,\ j = 1, \ldots, n \\ & \sum_{j=1}^{n} (R_{1j} + R_{2j}) = R \\ u_r, \varphi_p, v_i, R_{1j}, R_{2j} \ge 0, & r = 1, \ldots, s;\ p = 1, \ldots, q;\ i = 1, \ldots, m;\ j = 1, \ldots, n. \end{aligned} \quad (8.11)$$

By applying the efficiency maximal principle, the fixed cost should be allocated such that each DMU_j has an efficient score 1. However, it is unfair and irrational in some conditions due to the ignorance of technological heterogeneity among the n DMUs. In other words, a low technology DMU is unlikely to achieve the efficient frontier through an allocation. Conversely, the efficiency invariance principle is also impractical because all DMUs are not allowed to have any improvement in efficiencies after increasing or decreasing fixed costs.

Following Wang et al. (2013), the inefficiency of the meat-frontier (MTI) can be divided into two parts: one is technology gap inefficiency (TGI) and the other is group-frontier managerial inefficiency (GMI), which are expressed in Eqs. (8.12) and (8.13).

$$TGI_j^k = GE_j^k (1 - TGR_j^k) = \theta_j^{GE} \left(1 - \frac{\theta_j^{ME}}{\theta_j^{GE}} \right) = \theta_j^{GE} - \theta_j^{ME} \quad (8.12)$$

$$GMI_j^k = 1 - GE_j^k = 1 - \theta_j^{GE} \quad (8.13)$$

The TGI_j^k represents the inefficiency of DMU_j belong to group k originating from the technical gap between the meta-frontier and the group-specific frontiers. The cause of inefficiency is attributed to technological differences, which cannot be eliminated in a short time. And the GMI_j^k represents the inefficiency of the DMU_j belong to group k originating from input excess and output shortfall. The reason for the inefficiency is attributed to managerial failure of DMU_j, which is able to reduce or even eliminate by managerial improvement in a short time.

Based on the above analysis, taking the technology gap between group-frontier technology and meta-frontier technology into consideration, we set up the following inequalities:

$$
\begin{cases}
\theta_j^{ME} \le \frac{\sum_{p=1}^{q}\varphi_p z_{pj}+\sum_{r=1}^{s}u_r y_{rj}}{\sum_{i=1}^{m}v_i x_{ij}+R_{1j}+\sum_{p=1}^{q}\varphi_p z_{pj}+R_{2j}} \le \theta_j^{GE}, j=1,\ldots,n \\
\frac{\sum_{p=1}^{q}\varphi_p z_{pj}}{\sum_{i=1}^{m}v_i x_{ij}+R_{1j}} \le 1, j=1,\ldots,n \\
\frac{\sum_{r=1}^{s}u_r y_{rj}}{\sum_{p=1}^{q}\varphi_p z_{pj}+R2_j} \le 1, j=1,\ldots,n \\
\sum_{j=1}^{n}\left(R_{1j}+R_{2j}\right)=R \\
u_r, \varphi_p, v_i, R_{1j}, R_{2j} \ge 0, r=1,\ldots,s; p-1,\ldots,q; i=1,\ldots,m; j=1,\ldots,n
\end{cases}
\tag{8.14}
$$

where $0 \le \theta_j^{ME} \le 1$ and $0 \le \theta_j^{GE} \le 1$ are two constant parameters taking the values given by the conventional two-stage CCR efficiency computed by model (8.6) and (8.7) in Sect. 8.2, respectively. The constraints indicate that each DMU's efficiency can be increased by reducing the group-frontier managerial inefficiency (GMI) without changing its inherent technology. It is more practical and reasonable because the managerial improvement can be achieved but the technology improvement cannot be achieved in a short time.

Property 1 *System of inequalities* (8.14) *is feasible.*

Proof Consider $u_s = \frac{R}{\sum_{j=1}^{n}\left(y_{sj}/\theta_j^{GE}\right)}$, $v_i = \varphi_p = u_r = 0(\forall i, p, r \ne s)$ and $R_{1j} = R_{2j} = \frac{R y_{sj}/\theta_j^{GE}}{2\sum_{j=1}^{n}\left(y_{sj}/\theta_j^{GE}\right)}(\forall j)$. Clearly, $\left(v_i, \varphi_p, u_r, R_{1j}, R_{2j}, \forall i, p, r, j\right)$ is a nonnegative solution. Furthermore, $\left(v_i, \varphi_p, u_r, R_{1j}, R_{2j}, \forall i, p, r, j\right)$ satisfies all constraints of system (8.14), as for any $DMU_j(j = 1,\ldots,n)$ it holds

$$
\frac{\sum_{p=1}^{q}\varphi_p z_{pj}+\sum_{r=1}^{s}u_r y_{rj}}{\sum_{i=1}^{m}v_i x_{ij}+R_{1j}+\sum_{p=1}^{q}\varphi_p z_{pj}+R_{2j}} = \frac{u_s y_{sj}}{R_{1j}+R_{2j}} = \frac{\frac{R y_{sj}}{\sum_{j=1}^{n}\left(y_{sj}/\theta_j^{GE}\right)}}{\left[\frac{R y_{sj}/\theta_j^{GE}}{2\sum_{j=1}^{n}\left(y_{sj}/\theta_j^{GE}\right)}+\frac{R y_{sj}/\theta_j^{GE}}{2\sum_{j=1}^{n}\left(y_{sj}/\theta_j^{GE}\right)}\right]} = \theta_j^{GE},
$$

$$
\frac{\sum_{p=1}^{q}\varphi_p z_{pj}}{\sum_{i=1}^{m}v_i x_{ij}+R_{1j}} = \frac{0}{R_{1j}} \le 1,
$$

$$
\frac{\sum_{r=1}^{s}u_r y_{rj}}{\sum_{p=1}^{q}\varphi_p z_{pj}+R2_j} = \frac{u_s y_{sj}}{R_{2j}} = \frac{\frac{R y_{sj}}{\sum_{j=1}^{n}\left(y_{sj}/\theta_j^{GE}\right)}}{\frac{R y_{sj}/\theta_j^{GE}}{2\sum_{j=1}^{n}\left(y_{sj}/\theta_j^{GE}\right)}} = \theta_j^{GE}/2 \le 1.
$$

Therefore, $\left(v_i, \varphi_p, u_r, R_{1j}, R_{2j}, \forall i, p, r, j\right)$ is a feasible solution to system (8.14) and this completes the proof.

Assuming that all DMUs are under the control of the centralized decision-maker (DM), the centralized DM aims to maximize the system efficiency of the organization. Many DEA applications such as bank branches, schools, hospitals, and so on fall into this situation where the organization controls the resources. Lozano and Villa (2004) In such a decision environment, the DM tends to pursue improvement in system efficiency rather than consider the optimal efficiency of an individual unit alone. In this subsection, the allocation plan lays more stress on practical feasibility. In order to calculate the post-allocation efficiency of the system, we propose the following mathematical model:

$$
\begin{aligned}
& Max \frac{\sum_{p=1}^{q}\varphi_p\sum_{j=1}^{n}z_{pj}+\sum_{r=1}^{s}u_r\sum_{j=1}^{n}y_{rj}}{\sum_{i=1}^{m}v_i\sum_{j=1}^{n}x_{ij}+\sum_{p=1}^{q}\varphi_p\sum_{j=1}^{n}z_{pj}+R} \\
s.t.\ \theta_j^{ME} \le & \frac{\sum_{p=1}^{q}\varphi_p z_{pj}+\sum_{r=1}^{s}u_r y_{rj}}{\sum_{i=1}^{m}v_i x_{ij}+R_{1j}+\sum_{p=1}^{q}\varphi_p z_{pj}+R_{2j}} \le \theta_j^{GE},\ j=1,\ldots,n \\
& \frac{\sum_{p=1}^{q}\varphi_p z_{pj}}{\sum_{i=1}^{m}v_i x_{ij}+R_{1j}} \le 1,\ j=1,\ldots,n \\
& \frac{\sum_{r=1}^{s}u_r y_{rj}}{\sum_{p=1}^{q}\varphi_p z_{pj}+R2_j} \le 1,\ j=1,\ldots,n \\
& R_j^- \le R_{2j}+R_{2j} \le R_j^+,\ j=1,\ldots,n \\
& \sum_{j=1}^{n}\left(R_{1j}+R_{2j}\right)=R \\
u_r, \varphi_p, v_i, R_{1j}, R_{2j} \ge 0,\ & r=1,\ldots,s;\ p=1,\ldots,q;\ i=1,\ldots,m;\ j=1,\ldots,n
\end{aligned}
\tag{8.15}
$$

in which R_j^- and R_j^+ denote the lower and upper bound of the fixed cost allocated to DMU_j given by the decision-maker, respectively. The constraint is added due to the consideration of practical feasibility. For example, the transferred range should not be too exorbitant in the short run so as to keep steady in fixed cost allocation applications.

Note that model (8.15) is a non-linear program because of its fractional form, and it can be equivalently converted into a linear program through the Charnes and Cooper (1962) transformation.

$$
\begin{aligned}
& Max \sum_{p=1}^{q}\varphi_p\sum_{j=1}^{n}z_{pj}+\sum_{r=1}^{s}u_r\sum_{j=1}^{n}y_{rj} \\
& s.t.\ \sum_{i=1}^{m}v_i\sum_{j=1}^{n}x_{ij}+\sum_{p=1}^{q}\varphi_p\sum_{j=1}^{n}z_{pj}+tR=1 \\
& \theta_j^{ME}\left(\sum_{i=1}^{m}v_i x_{ij}+R'_{1j}+\sum_{p=1}^{q}\varphi_p z_{pj}+R'_{2j}\right)-\left(\sum_{p=1}^{q}\varphi_p z_{pj}+\sum_{r=1}^{s}u_r y_{rj}\right)\le 0,\ j=1,\ldots,n \\
& \sum_{p=1}^{q}\varphi_p z_{pj}+\sum_{r=1}^{s}u_r y_{rj}-\theta_j^{GE}\left(\sum_{i=1}^{m}v_i x_{ij}+R'_{1j}+\sum_{p=1}^{q}\varphi_p z_{pj}+R'_{2j}\right)\le 0,\ j=1,\ldots,n \\
& \sum_{p=1}^{q}\varphi_p z_{pj}-\left(\sum_{i=1}^{m}v_i x_{ij}+R'_{1j}\right)\le 0,\ j=1,\ldots,n \\
& \sum_{r=1}^{s}u_r y_{rj}-\left(\sum_{p=1}^{q}\varphi_p z_{pj}+R'_{2j}\right)\le 0,\ j=1,\ldots,n \\
& tR_j^- \le R'_{1j}+R'_{2j}\le tR_j^+,\ j=1,\ldots,n \\
& \sum_{j=1}^{n}\left(R'_{1j}+R'_{2j}\right)=tR \\
& u_r, \varphi_p, v_i, R'_{1sj}, R'_{2j}\ge 0,\ r=1,\ldots,s;\ p=1,\ldots,q;\ i=1,\ldots,m;\ j=1,\ldots,n
\end{aligned}
\tag{8.16}
$$

Based on model (8.16), the optimal fixed cost for each substage as well as the whole DMU can be obtained.

8.3.2 The Resource Allocation Approach

Assuming that all the DMUs are under the control of the centralized decision-maker, the centralized DM aims to improve the system efficiency by reallocating inputs across all DMUs. The centralized resource allocation model developed by Mar-Molinero et al. (2014) can be written as follows:

$$
\begin{aligned}
& Max \frac{\sum_{r=1}^{s} u_r \sum_{j=1}^{n} y_{rj}}{\sum_{i=1}^{m} v_i \sum_{j=1}^{n} x_{ij}} \\
& s.t. \frac{\sum_{r=1}^{s} u_r y_{rj}}{\sum_{i=1}^{m} v_i x_{ij}} \le 1,\ j = 1,,,.n \\
& u_r, v_i \ge 0,\ r = 1, \ldots, s;\ i = 1, \ldots, m.
\end{aligned}
\tag{8.17}
$$

In this model, the whole organization is being treated as a macro unit that uses all of the inputs available in all of the units to generate all of the outputs that the system generates irrespective of the unit in which they are produced.

However, like many existing methods, the technology gap factor is not considered in this resource allocation model. We should point out that it is not practical in some conditions. Due to the existing of technology gap in different DMUs, some inefficient DMUs cannot be efficient by simple resource allocation. When using traditional allocation models, the heterogeneities of production technology among all DMUs may produce an unbiased resource allocation plan.

In this subsection, we suggest a centralized allocation plan considering the technology gap for two-stage production systems. The central authority still pursues improvement in system efficiency. Similar to model (8.17), we don't add the constraints of lower bound efficiency of each DMU for the pursuit of system efficiency. The centralized DEA model considering technology gaps is formulated as follows:

$$
\begin{aligned}
& Max \frac{\sum_{p=1}^{q} \varphi_p \sum_{j=1}^{n} j + \sum_{r=1}^{s} u_r \sum_{j=1}^{n} y_{rj}}{\sum_{i=1}^{m} v_i \sum_{j=1}^{n} x_{ij} + \sum_{p=1}^{q} \varphi_p \sum_{j=1}^{n} z_{pj}} \\
& s.t. \frac{\sum_{p=1}^{q} \varphi_p z_{pj} + \sum_{r=1}^{s} u_r y_{rj}}{\sum_{i=1}^{m} v_i x_{ij} + \sum_{p=1}^{q} \varphi_p z_{pj}} \le \theta_j^{GE},\ j = 1, \ldots, n \\
& u_r, \varphi_p, v_i \ge 0,\ r = 1, \ldots, s;\ p = 1, \ldots, q;\ i = 1, \ldots, m.
\end{aligned}
\tag{8.18}
$$

In practice, the resources can be reallocated in order to achieve a better system efficiency compared with the previous period. Obviously, each input for a given DMU can be added, reduced or maintained after the reallocation. We develop the following model (8.19) to maximize the system performance by reallocating input resources while considering managerial feasibility.

$$
\begin{aligned}
& Max \frac{\sum_{p=1}^{q} \varphi_p \sum_{j=1}^{n} z_{pj} + \sum_{r=1}^{s} u_r \sum_{j=1}^{n} y_{rj}}{\sum_{i=1}^{m} v_i \sum_{j=1}^{n} (x_{ij} + \Delta x_{ij}) + \sum_{p=1}^{q} \varphi_p \sum_{j=1}^{n} z_{pj}} \\
& s.t. \frac{\sum_{p=1}^{q} \varphi_p z_{pj} + \sum_{r=1}^{s} u_r y_{rj}}{\sum_{i=1}^{m} v_i (x_{ij} + \Delta x_{ij}) + \sum_{p=1}^{q} \varphi_p z_{pj}} \le \theta_j^{GE},\ j = 1, \ldots, n \\
& -\alpha x_{ij} \le \Delta x_{ij} \le \alpha x_{ij},\ i = 1, \ldots, m;\ j = 1, \ldots, n \\
& \sum_{j=1}^{n} \Delta x_{ij} = 0,\ i = 1, \ldots, m \\
& u_r, \varphi_p, v_i \ge 0,\ r = 1, \ldots, s;\ p = 1, \ldots, q;\ i = 1, \ldots, m.
\end{aligned}
\tag{8.19}
$$

In the above model, θ_j^{GE} is the group efficiency defined in Sect. 8.2. Δx_{ij} represents the amount of adjustment for DMU_j in input i. Since each input for DMU_j can be either increased or decreased, Δx_{ij} is free in sign. The constraint

$\sum_{j=1}^{n} \Delta x_{ij} = 0$ suggests that the total amount of each input across all DMUs remains unchanged. In general, a similar model can be developed if in the next time period total available input resources are changed. The constraint $\left|\Delta x_{ij}\right| \leq \alpha x_{ij}$ is added to guarantee practical feasibility, where α is an acceptable percentage of changes in resources according to specific applications. For example, we may add constraint $-0.1x_{ij} \leq \Delta x_{ij} \leq 0.1x_{ij}, i = 1, \ldots, m; j = 1, \ldots, n$ in model (8.19) to subject that input i for DMU_j could vary by at most 10% from its previous observed values.

Letting $\Delta' x_{ij} = v_i \Delta x_{ij}$, then the fractional program (8.19) can be converted into a linear model (8.20) by using the Charnes and Cooper (1962) transformation:

$$\begin{gathered} Max \sum_{p=1}^{q} \varphi_p \sum_{j=1}^{n} z_{pj} + \sum_{r=1}^{s} u_r \sum_{j=1}^{n} y_{rj} \\ s.t. \sum_{i=1}^{m} v_i \sum_{j=1}^{n} x_{ij} + \sum_{p=1}^{q} \varphi_p \sum_{j=1}^{n} z_{pj} = 1 \\ \sum_{p=1}^{q} \varphi_p z_{pj} + \sum_{r=1}^{s} u_r y_{rj} - \theta_j^{GE} \left(\sum_{i=1}^{m} v_i x_{ij} + \sum_{i=1}^{m} \Delta x_{ij} + \sum_{p=1}^{q} \varphi_p z_{pj} \right) \leq 0, j = 1, \ldots, n \\ -\alpha v_i x_{ij} \leq \Delta' x_{ij} \leq \alpha v_i x_{ij}, i = 1, \ldots, m; j = 1, \ldots, n \\ \sum_{j=1}^{n} \Delta' x_{ij} = 0, i = 1, \ldots, m \\ u_r, \varphi_p, v_i \geq 0, r = 1, \ldots, s; p = 1, \ldots, q; i = 1, \ldots, m. \end{gathered} \tag{8.20}$$

Based on model (8.20), we can obtain the optimal resources allocated for all DMUs.

8.4 Illustrations

In this section, two numerical examples from the previous study are presented to illustrate the feasibility and applicability of the proposed approaches in Sect. 8.3.

8.4.1 Illustration on Fixed Cost Allocation

We first apply our centralized DEA-based approach to fixed cost allocation using the numerical example proposed by Li et al. (2019a), which is a virtual dataset derived from Cook and Kress (1999) by randomly generating two intermediate outputs. As a result, this example includes 12 DMUs with three inputs (x_1, x_2, and x_3), two intermediate outputs (z_1 and z_2), and two final outputs (y_1 and y_2) as presented in Table 8.1. Further, it is assumed that a total shared cost $R = 100$ is to be covered by the set of all DMUs and accordingly their two substages.

Taking the technology heterogeneity into consideration, the 12 DMUs are divided into two groups A (DMUs 4, 5, 8, 9, 10, and 12) and B (DMUs 1, 2, 3, 6, 7, and 11). Reasonably, we assume that (1) the technology level of group A is higher than that of group B and (2) DMUs' technology level is similar in the same group. The original two-stage CCR efficiency and group efficiency for each DMU calculated by

Table 8.1 Numerical example

DMU	x_1	x_2	x_3	z_1	z_2	y_1	y_2
1	350	39	9	35	63	67	751
2	298	26	8	25	75	73	611
3	422	31	7	34	85	75	584
4	281	16	9	33	95	70	665
5	301	16	6	23	98	75	445
6	360	29	17	22	77	83	1070
7	540	18	10	30	57	72	457
8	276	33	5	40	57	78	590
9	323	25	5	27	63	75	1074
10	444	64	6	32	92	74	1072
11	323	25	5	24	62	25	350
12	444	64	6	35	91	104	1199

model (8.6) and model (8.7) are listed in the second and third column of Table 8.2, respectively. It can be seen that none of 12 DMUs is overall efficient about meta-efficiency, while two DMUs (7 and 12) are group efficient. Consequently, the meta-technology ratio of each DMU is listed in the last column of Table 8.2.

The last three columns of Table 8.3 present an allocation result determined by model (8.16). This numerical example is a special case of the proposed method because the bounded constraint of allocated cost is not needed. For comparison, the corresponding results from Cook and Kress (1999), Beasley (2003), Cook and Zhu

Table 8.2 Calculation results

DMU	Meta-efficiency	Group efficiency	Meta-technology ratio	Allocation		
				Stage 1	Stage 2	Whole
1	0.7253	0.8342	0.8695	1.4075	7.0812	8.4887
2	0.8174	0.9366	0.8727	0.5220	7.7153	8.2373
3	0.8011	0.8784	0.9120	0.0000	9.0240	9.0240
4	0.8270	0.8512	0.9716	1.2934	7.3982	8.6916
5	0.9072	0.9794	0.9262	0.1665	7.9267	8.0932
6	0.7542	0.9561	0.7888	0.0000	9.1752	9.1752
7	0.7496	1.0000	0.7496	0.0000	7.6096	7.6096
8	0.9588	0.9701	0.9883	0.0000	8.4978	8.4978
9	0.9513	0.9513	1.0000	0.0000	8.3323	8.3323
10	0.8594	0.8594	1.0000	1.2797	7.8210	9.1008
11	0.6239	0.7031	0.8874	0.0000	3.7579	3.7579
12	0.9944	1.0000	0.9944	0.0000	10.9917	10.9917

Table 8.3 Fixed cost allocation results

DMU	Proposed method	Cook and Kress (2009)	Beasley (2003)	Cook and Zhu (2005)	Du et al. (2014)	Ding et al. (2018)	Li et al. (2019b)
1	8.4887	14.52	6.78	11.22	5.79	4.69	6.5133
2	8.2373	6.74	7.21	0	7.95	10.28	8.9348
3	9.0240	9.32	6.83	16.95	6.54	8.93	8.6160
4	8.6916	5.6	8.47	0	11.1	6.39	9.6739
5	8.0932	5.79	7.48	0	8.69	11.62	10.6490
6	9.1752	8.15	10.06	15.43	13.49	0	9.2238
7	7.6096	8.86	5.09	0	7.1	2.75	8.9836
8	8.4978	6.26	7.74	0	6.83	14.13	9.0062
9	8.3323	7.31	15.11	17.62	16.68	12.68	10.9617
10	9.1008	10.08	10.08	21.15	5.42	9.69	5.4679
11	3.7579	7.31	1.58	17.62	0	0	2.0324
12	10.9917	10.08	13.97	0	10.41	18.85	9.9375

(2005), Du et al. (2014), Ding et al. (2018) and Li et al. (2019b) are listed in columns 3, 4, 5, 6, 7, and 8, respectively.

Cook and Kress (1999) allocate the fixed cost based on the principle of invariance and Pareto-minimality. As listed in column 3, DMUs 9 and 11, as well as 10 and 12 obtain the same allocated cost with identical inputs and different outputs. As stated by Beasley (2003), this is because the allocated cost depends completely on inputs values rather than outputs. In addition, due to the existing of multiple efficient DMUs, the allocation results are non-uniqueness.

To cope with the non-uniqueness problem in (Cook and Kress 1999; Beasley 2003) develops a multi-step approach to obtain a unique fixed cost allocation plan. Cook and Zhu (2005) extend the method of Cook and Kress (1999) by dividing the DMUs into two groups: efficient ones and inefficient ones. However, according to column 5, DMU 4, DMU 5, DMU 8, and DMU 12 are efficient but obtain no resource allocation at all. We should point out, Cook and Zhu's (2005) cost allocation only provides one feasible solution but not the optimal solution. To deal with the non-uniqueness problem in (Cook and Kress 1999; Lotfi et al. 2012) propose a common weights approach to allocate fixed cost. As listed in column 6, the DMUs with better performance in the production system tend to achieve much higher allocated fixed cost. Du et al. (2014) develop a DEA cross-efficiency-based approach to allocate fixed cost. According to their approach, all 12 DMUs become efficient when the allocated cost is considered as a new input. However, there exists some DMUs with poor efficiency such as DMU 3 and DMU 11. They achieve an efficient frontier together merely by a cost allocation. So, a rational question naturally arises. Ding et al. (2018) took technology heterogeneity into account, which is a basis of this paper, but Ding et al. (2018) remitted the responsibility of the total fixed costs (i.e., the allocated cost is zero) for some DMU 7 and DMU 11. Li et al. (2019b) suggested a non-egoistic principle to limit the feasible allocation space and further generate final allocation results by maximizing efficiency for all DMUs. The non-egoistic efficiency considered in Li et al. (2019b) might not be feasible considering different production technology.

Comparably, the gap of the values of allocated costs across the DMUs is 7.2338 (7.2338 = 10.9917 − 3.7579) for our proposed approach, which is the smallest gap among other approaches in Table 8.2, which are 8.92, 13.53, 21.15, 16.68, 18.85, and 8.9293, respectively. As a small gap between the largest value and the smallest value of the allocated fixed costs would cause less organizational resistance to and difficulty in implementing the allocation (Li et al. 2009), the allocation plan generated from our proposed approach will be more reasonable and acceptable for all DMUs involved. More importantly, under the control of the central authority, model (8.16) only needs to be solved one time to obtain the allocation results for the pursuit of system efficiency. By the way, the resulting weights are common weights, which are used to calculate the system efficiency as well as each DMU's efficiency. Moreover, DMUs' technology heterogeneity is considered in our model. It is easy to verify that in our approach, the two-stage CCR efficiency of each DMU after a cost allocation remains non-decreasing. Practically, the efficiency of inefficiency DMUs increases in a reasonable way rather than achieving the efficiency frontier together.

8.4.2 *Illustration on Resource Reallocation*

We now demonstrate our method to the resource reallocation problem using a data set of 27 bank branches from Li et al. (2019a). It was assumed that each bank branch has a two-stage structure, in which three original inputs (labors, fixed assets, Operation costs other than labor costs) are used to generate deposits and other raised funds as intermediate products in the first stage, and both deposits and other raised funds are further consumed to produce interest income, non-interest income as well as a jointly produced undesirable output, bad debts, as final outputs. Here in this chapter, we consider the dataset as a centralized two-stage resource reallocation problem, and the data is given in Table 8.4. In addition, the data transformation method used by Li et al. (2019a) was also considered in this chapter.

Under the control of central authority, input resources are allowed to reallocate to the 27 branches for the pursuit of improvement in system efficiency. Table 8.5 shows the computed results by our resource allocation method. Taking the technology heterogeneity factor into consideration, the 27 DMUs are divided into two groups by model (8.8), where the threshold efficiency value is specified as 0.80. The grouping results are listed in column 2. By solving model (8.6) and model (8.7), the original efficiency and group efficiency of each DMU are listed in column 3 and column 4. Accordingly, column 5 gives the meat-technology ratio for each DMU. Column 6 to column 8 reports the allocation results of the proposed model (8.20). The results are derived from an assumption that the parameter α is 20%, that is, through a resource

Table 8.4 Input and output data for 27 bank branches

DMU	x_1	x_2	x_3	z_1	z_2	y_1	y_2	y_3
1	25	619	538	2947	913	224	77237	34224
2	27	419	489	3138	478	516	88031	56559
3	40	1670	1459	5494	1242	877	164053	62776
4	42	2931	1497	3144	870	1138	145369	65226
5	52	2587	797	6705	854	618	166424	85886
6	45	2181	697	8487	1023	2096	215695	30179
7	33	989	1217	4996	767	713	114043	43447
8	107	6277	2189	21265	6282	6287	727699	294126
9	88	3197	949	8574	1537	1739	186642	53223
10	146	6222	1824	21937	5008	3261	614241	121784
11	57	1532	2248	8351	1530	2011	241794	83634
12	42	1194	1604	5594	858	1203	150707	57875
13	132	5608	1731	15271	4442	2743	416754	168798
14	77	2136	906	10070	2445	1487	276379	38763
15	43	1534	438	4842	1172	1355	133359	48239
16	43	1711	1069	6505	1469	1217	157275	27004
17	59	3686	820	6552	1209	1082	150827	60244
18	33	1479	2347	8624	894	2228	215012	78253
19	38	1822	1577	9422	967	1367	192746	76284
20	162	5922	2330	18700	4249	6545	533273	163816
21	60	2158	1153	10573	1611	2210	252568	77887
22	56	2666	2683	10678	1589	1834	269402	158835
23	71	2969	1521	8563	905	1316	197684	100321
24	117	5527	2369	15545	2359	2717	406475	106073
25	78	3219	2738	14681	3477	3134	371847	125323
26	51	2431	741	7964	1318	1158	190055	142422
27	48	2924	1561	11756	2779	1398	332641	94933

reallocation, the inputs for any DMU shouldn't change more than 10 percentage compared with the current period. We believe this is a reasonable assumption in the short run and under the spirit of slight changes. After the reallocation of input resources, the relative efficiency of each DMU is presented in the last column.

Comparably, the efficiency scores of some DMUs (1, 4, 5, 10, 11, 19, 21, and 27) will have a decreased efficiency score through our resource reallocation plan, while the efficiency scores of the remaining larger proportion of all DMUs are non-decreased. Furthermore, the average efficiency across all DMUs increased from 0.8321 to 0.8581. Naturally, we can deduce that the system efficiency has been improved. What's more, not all DMUs can achieve the efficient frontier, which is

Table 8.5 Resource allocation results

DMU	Group	Original efficiency	Group efficiency	Meta-technology ratio	Changes			New efficiency
					x_1	x_2	x_3	
1	*A*	0.8634	0.8634	1.0000	5.0000	123.8000	107.6000	0.8267
2	*A*	0.8798	0.8957	0.9822	5.4000	83.8000	97.8000	0.8798
3	*B*	0.7470	0.8936	0.8359	0.0000	−334.0000	−198.2861	0.8081
4	*B*	0.5726	0.7488	0.7647	8.4000	586.2000	299.4000	0.5286
5	*B*	0.7102	0.8640	0.8220	10.4000	517.4000	159.4000	0.6738
6	*A*	0.9304	0.9491	0.9803	3.8308	−436.2000	139.4000	0.9867
7	*B*	0.8152	0.9534	0.8551	−6.6000	−197.8000	−243.4000	0.8973
8	*A*	0.9618	0.9697	0.9919	21.4000	1255.4000	437.8000	0.9775
9	*B*	0.7399	0.8680	0.8525	−14.5059	−639.4000	189.8000	0.7661
10	*A*	0.9225	0.9411	0.9803	−2.0152	1244.4000	364.8000	0.8614
11	*A*	0.8770	0.9062	0.9678	−11.4000	231.4677	−167.9529	0.8336
12	*B*	0.7975	0.9144	0.8721	−8.4000	−238.8000	−320.8000	0.8712
13	*B*	0.8402	0.9714	0.8650	−26.4000	−1121.6000	−346.2000	0.9760
14	*A*	0.9499	0.9670	0.9823	−12.8719	427.2000	−181.2000	0.9777
15	*A*	0.9513	0.9548	0.9963	−8.0940	306.8000	−87.6000	0.9970
16	*B*	0.8090	0.9846	0.8216	−8.6000	−342.2000	−213.8000	0.9436
17	*B*	0.6934	0.8173	0.8485	11.8000	737.2000	−132.9141	0.7144
18	*A*	0.9876	1.0000	0.9876	−6.6000	−295.8000	−469.4000	0.9876
19	*A*	0.8937	0.8937	1.0000	7.6000	364.4000	315.4000	0.7979
20	*B*	0.8458	1.0000	0.8458	−32.4000	−1184.4000	−466.0000	1.0000

(continued)

Table 8.5 (continued)

DMU	Group	Original efficiency	Group efficiency	Meta-technology ratio	Changes			New efficiency
					x_1	x_2	x_3	
21	*A*	0.8850	0.9150	0.9672	12.0000	431.6000	230.6000	0.8530
22	*B*	0.7948	0.9314	0.8533	−11.2000	−533.2000	−139.9467	0.8274
23	*B*	0.6407	0.8311	0.7708	14.2000	−308.0677	304.2000	0.6864
24	*B*	0.6914	0.8638	0.8004	23.4000	−1105.4000	−108.4567	0.7751
25	*A*	0.9154	0.9169	0.9984	15.6000	−643.8000	265.7564	0.9342
26	*B*	0.8276	0.9561	0.8657	0.4563	486.2000	−148.2000	0.9007
27	*A*	0.9241	0.9393	0.9838	9.6000	584.8000	312.2000	0.8863

different from Du et al.'s (2014) results. We should point out that it is a reasonable result because we consider the factor of technology heterogeneity. In other words, those DMUs with low technology are impractical to be efficient only through resource allocation.

8.5 Conclusion

In order to solve the problem of fixed cost or resource allocation, existing DEA-based methods primary concentrate on the pursuit of efficiency. They usually assume that all the DMUs can achieve a common efficient frontier. However, taking the technology heterogeneity factor into consideration, it is unlikely to let those DMUs with low efficiency become efficient.

In this chapter, we address the fixed cost and resource allocation problem in a environment involved with two-stage series production systems. In contrast to existing studies, in this chapter, we present a new fixed cost and resource allocation approach under the environment of central authority. Two grouping methods are introduced to classify the DMUs. It is reasonable to allow the DMUs to achieve the group frontier rather than the common frontier together. In a centralized decision-making environment, our DEA-based allocation models improve the system efficiency for all DMUs rather than one specific DMU. Based on this principle, two centralized DEA models are proposed. In order to demonstrate the feasibility and superiority of our approaches, two numerical examples are presented.

In this chapter, we propose our allocation models under the assumption of constant returns to scale (CRS). Similarly, the proposed models can be extended to the variable returns to scale (VRS) situation. Moreover, the allocated fixed cost and resources in this chapter may not be unique even though we propose our DEA models to successfully allocate fixed cost and resources by considering technology heterogeneity. Hence, how to obtain a unique allocation can be further examined for future research.

References

Amirteimoori, A., & Tabar, M. M. (2010). Resource allocation and target setting in data envelopment analysis. *Expert Systems with Applications, 37*(4), 3036–3039.

An, Q., Wang, P., Emrouznejad., A., & Hu, J. (2019). Fixed cost allocation based on the principle of efficiency invariance in two-stage systems. *European Journal of Operational Research, 283*, 662–675.

Asmild, M., Paradi, J. C., & Pastor, J. T. (2009). Centralized resource allocation BCC models. *Omega, 37*(1), 40–49.

Battese, G. E., & Rao, D. S. P. (2002). Technology gap, efficiency, and a stochastic metafrontier function. *International Journal of Business and Economics, 1*(2), 87–93.

Banker, R. D., Charnes, A., & Cooper, W. W. (1984). Some models for estimating technical and scale inefficiencies in data envelopment analysis. *Management Science, 30*(9), 1078–1092.

Beasley, J. E. (2003). Allocating fixed costs and resources via data envelopment analysis. *European Journal of Operational Research, 147*(1), 198–216.
Bi, G. B., Ding, J. J., Luo, Y., & Liang, L. (2011). Resource allocation and target setting for parallel production system based on DEA. *Applied Mathematical Modelling, 35*(9), 4270–4280.
Charnes, A., & Cooper, W. W. (1962). Programming with linear fractional functionals. *Naval Research Logistics, 15,* 333–334.
Charnes, A., Cooper, W. W., & Rhodes, E. (1978). Measuring the efficiency of decision making units. *European Journal of Operational Research, 2*(78), 429–444.
Chen, Y., Du, J., David, H., & Zhu, J. (2010). DEA model with shared resources and efficiency decomposition. *European Journal of Operational Research, 207*(1), 339–349.
Chen, Z., & Song, S. (2006). Efficiency and technology gap in China's agriculture: A regional meta-frontier analysis. *China Economic Review, 19*(2), 287–296.
Chu, J., Wu, J., Chu, C., & Zhang, T. (2019). DEA-based fixed cost allocation in two-stage systems: leader-follower and satisfaction degree bargaining game approaches. *Omega*. https://doi.org/10.1016/j.omega.2019.03.012.
Cook, W. D., & Kress, M. (1999). Characterizing an equitable allocation of shared costs: A DEA approach. *European Journal of Operational Research, 119*(3), 652–661.
Cook, W. D., & Zhu, J. (2005). Allocation of shared costs among decision making units: A DEA approach. *Computers & Operations Research, 32*(8), 2171–2178.
Cook, W. D., & Seiford, L. M. (2009). Data envelopment analysis (DEA)—Thirty years on. *European Journal Operational Research, 192*(1), 1–17.
Cooper, W. W., Seiford, L. M., & Tone, K. (2007). *Data envelopment analysis: A comprehensive text with models, applications, references and DEA-solver software*. New York: Springer.
Cooper, W. W., Seiford, L. M., & Zhu, J. (2011). *Handbook on data envelopment analysis*. New York: Springer.
Ding, T., Chen, Y., Wu, H., & Wei, Y. (2018). Centralized fixed cost and resource allocation considering technology heterogeneity: A DEA approach. *Annals of Operations Research, 268*(1–2), 497–511.
Ding, T., Zhu, Q., Zhang, B., & Liang, L. (2019). Centralized fixed cost allocation for generalized two-stage network DEA. *INFOR: Information Systems and Operational Research, 57*(2), 123–140.
Du, J., Cook, W. D., Liang, L., & Zhu, J. (2014). Fixed cost and resource allocation based on DEA cross-efficiency. *European Journal of Operational Research, 235*(1), 206–214.
Fang, L. (2013). A generalized DEA model for centralized resource allocation. *European Journal of Operational Research, 228*(2), 405–412.
Fang, L. (2015). Centralized resource allocation based on efficiency analysis for step-by-step improvement paths. *Omega, 51,* 24–28.
Kao, C., & Hwang, S. N. (2008). Efficiency decomposition in two-stage data envelopment analysis: An application to non-life insurance companies in Taiwan. *European Journal of Operational Research, 185*(1), 418–429.
Korhonen, P., & Syrjanen, M. (2004). Resource allocation based on efficiency analysis. *Management Science, 50*(8), 1134–1144.
Lewis, H. F., & Sexton, T. R. (2004). Network DEA: Efficiency analysis of organizations with complex internal structure. *Computers & Operations Research, 31*(9), 1365–1410.
Li, F., Zhu, Q., & Liang, L. (2018). Allocating a fixed cost based on a DEA-game cross efficiency approach. *Expert Systems with Applications, 96,* 196–207.
Li, F., Zhu, Q., & Chen, Z. (2019a). Allocating a fixed cost across the decision making units with two-stage network structures. *Omega, 83,* 139–154.
Li, F., Zhu, Q., & Liang, L. (2019b). A new data envelopment analysis based approach for fixed cost allocation. *Annals of Operations Research, 274*(1–2), 247–272.
Li, Y. J., Yang, F., Liang, L., & Hua, Z. S. (2009). Allocating the fixed cost as a complement of other cost inputs: A DEA approach. *European Journal of Operational Research, 197*(1), 389–401.

Li, Y. J., Yang, M., Chen, Y., Dai, Q. Z., & Liang, L. (2013). Allocating a fixed cost based on data envelopment analysis and satisfaction degree. *Omega, 41*(1), 55–60.

Lin, R. Y. (2011). Allocating fixed costs and common revenue via data envelopment analysis. *Applied Mathematics and Computation, 218*(7), 3680–3688.

Lin, R., Chen, Z., & Li, Z. (2016). A new approach for allocating fixed costs among decision making units. *Journal of Industrial and Management Optimization, 12*(1), 211–228.

Lotfi, F. H., Hatami-Marbini, A., Agrell, P. J., Aghayi, N., & Gholami, K. (2012). Allocating fixed resources and setting targets using a common-weights DEA approach. *Computers & Industrial Engineering, 64*(2), 631–640.

Lozano, S., & Villa, G. (2004). Centralized resource allocation using data envelopment analysis. *Journal of Productivity Analysis, 22,* 143–161.

Lozano, S., Villa, G., & Brännlund, R. (2009). Centralized reallocation of emission permits using DEA. *European Journal of Operational Research, 193*(3), 752–760.

Mar-Molinero, C., Prior, D., Segovia, M. M., & Portillo, F. (2014). On centralized resource utilization and its reallocation by using DEA. *Annals of Operations Research, 221*(1), 273–283.

O'Donnell, C. J., Rao, D. S. P., & Battese, G. E. (2008). Metafrontier frameworks for the study of firm-level efficiencies and technology ratios. *Empirical Economics, 34*(2), 231–255.

Seiford, L. M., & Zhu, J. (2003). Context-dependent data envelopment analysis—Measuring attractiveness and progress. *Omega, 31*(5), 397–408.

Si, X. L., Liang, L., Jia, G. Z., Yang, L., Wu, H. Q., & Li, Y. J. (2013). Proportional sharing and DEA in allocating the fixed cost. *Applied Mathematics and Computation, 219*(12), 6580–6590.

Wang, Q. W., Zhao, Z. Y., Zhou, P., & Zhou, D. Q. (2013). Energy efficiency and production technology heterogeneity in China: A meta-frontier DEA approach. *Economic Modelling, 35*(5), 283–289.

Yu, M. M., Chen, L. H., & Hsiao, B. (2016). A fixed cost allocation based on the two-stage network data envelopment approach. *Journal of Business Research, 69*(5), 1817–1822.

Zhu, W., Zhang, Q., & Wang, H. (2019). Fixed costs and shared resources allocation in two-stage network DEA. *Annals of Operations Research, 278*(1–2), 177–194.

Chapter 9
Efficiency Assessment of Schools Operating in Heterogeneous Contexts: A Robust Nonparametric Analysis Using PISA 2015

Jose Manuel Cordero, Cristina Polo, and Rosa Simancas

Abstract The present study proposes an international comparison of education production efficiency using cross-country data on secondary schools from different countries participating in PISA 2015. Given that the context in which schools are operating might be heterogeneous, we need to account for those divergences in the environmental conditions when estimating the efficiency measures of school performance. In this way, each school can be benchmarked with units with similar characteristics regardless of the country they belong to. For this purpose, we use a robust nonparametric approach that allows us to clean the effect of contextual factors previously to the estimation of efficiency measures. Since this approach needs smoothing in the conditional variables in the middle of the sample and not at the frontier (where the number of units is smaller), it seems to be a better option than other nonparametric alternatives previously developed in the literature to deal with the effect of external factors. Likewise, by using this novel approach, we will also be able to explore how those contextual factors might affect both the attainable production set and the distribution of the efficiencies.

Keywords Education · Pure efficiency · Nonparametric · Cross-country comparison

9.1 Introduction

The participation of the majority of nations in international large-scale comparative studies in education like PISA (Programme for International Student Assessment), TIMSS (Trends in International Mathematics and Science Study) or PIRLS (Progress in International Reading Literacy Study) has provided researchers with rich and extensive cross-national databases that can be used to assess the performance and effectiveness of educational systems. As a result, comparative education studies have become increasingly popular in education sciences today (Gustafsson 2008), since

J. M. Cordero (✉) · C. Polo · R. Simancas
University of Extremadura, Badajoz, Spain
e-mail: jmcordero@unex.es

V. Charles et al. (eds.), *Data Science and Productivity Analytics*,
International Series in Operations Research & Management Science 290,
https://doi.org/10.1007/978-3-030-43384-0_9

researchers can look at the entire world as a natural laboratory to view the multiple ways in which societal factors, educational policies, and practices may vary across countries (Bray and Thomas 1995).

Most studies adopting a cross-country perspective are situated within the field of educational effectiveness research, i.e., they estimate an educational production function by means of an equation linking resource inputs with educational outcomes after controlling for various contextual characteristics to investigate the main factors influencing educational attainment (see Hanushek 1979; Todd and Wolpin 2003; Creemers and Kyriakides 2008). However, resource utilization is also a key matter of concern in science and technology management (David et al. 2000). In this sense, the existing constraints of resources faced by most countries and the great amount of national income devoted to educational costs, policy makers, and researchers have become increasingly concerned with developing guidelines for educational institutions to use resources more efficiently. As a result, the literature on school performance assessment is growing notably in recent years,[1] although most empirical studies have been restricted to schools operating in the same country or region.

This study proposes an international comparison of education production efficiency using cross-country data on secondary schools from different countries participating in PISA 2015. In this sense, it is worth mentioning that international comparisons are extremely challenging since schools are operating in very different contexts. Unless we consider the existing heterogeneity among schools, we would be implicitly assuming that all the schools are operating within the most favorable environment, which is unlikely in most cases. Some studies have addressed this problem by limiting the comparison group to similar countries (e.g., Bogetoft et al. 2015; Cordero et al. 2017; Dufrechou 2016). In this paper, however, our dataset includes a large sample of schools belonging to 67 countries with heterogeneous conditions, thus we need to take into account data about the diverse educational environments in which they are operating when estimating the efficiency measures of school performance. In this way, each unit can be benchmarked with other units from different countries provided that their operational environment is similar.

For this purpose, we rely on the most recent developments of the nonparametric conditional frontier literature (Daraio and Simar 2005, 2007a, b). These authors extend the probabilistic formulation of the production process proposed by Cazals et al. (2002) to account for the potential influence of heterogeneous contextual factors. Their approach is based on the estimation of conditional distribution functions (or conditional survival functions), where the conditioning is on the environmental factors Z. This procedure requires smoothing techniques for the environmental variables including a selection of optimal smoothing parameters (bandwidths) as those proposed by Badin et al. (2010) or, more recently, by Badin et al. (2019). Then, by exploiting the relationship between the conditional and unconditional measures, we can investigate the direction of their effect (favorable or unfavorable) on the production process. Furthermore, we can obtain cleaned efficiency scores by applying the

[1] See Worthington (2001) for an early review of this literature and Johnes (2015) or De Witte and López-Torres (2017) for updated revisions.

two-stage flexible location-scale regression model suggested by Badin et al. (2012) to eliminate the effect of contextual conditions and produce rankings of units assuming that they are operating in the same environmental conditions.

The main problem of this approach is that the estimated efficiency scores might be affected by the so-called "curse of dimensionality", especially when estimating the location-scale estimators if the number of environmental factors considered is relatively high. In order to avoid this potential problem, in this study, we apply the flexible location-scale model suggested by Florens et al. (2014), which can be interpreted as the aforementioned two-stage method, but the other way around. Thus, first, we eliminate the dependence of production inputs/outputs on external factors (Z) by means of nonparametric regression model and then, in a second step, we estimate the frontier and the efficiencies of the units using "pure" inputs and outputs, i.e., whitened from the influence of Z. We also estimate both the full frontier and their more robust versions using order-m partial frontiers. Finally, we also estimate the full and the order-m conditional frontier using the original input–output values in order to explore the influence of Z variables on both the attainable set and the distribution of efficiencies.

The main advantage of the approach used here is that it needs smoothing in the conditional variables in the center of the data cloud and not at the boundary in the space of the whitened inputs and outputs, where there are fewer observations. Therefore, is more robust than the two-stage approach proposed by Badin et al. (2012). We then obtain measures of pure efficiency which are more reliable to produce rankings or benchmarks of countries or educational systems around the world, according to the performance of schools participating in PISA in the hypothetical event that they all operate in a similar environment.

The remainder of the chapter is structured as follows. Section 9.2 reviews previous literature on cross-country studies using data from international large-scale assessments focusing on producing efficiency estimates of performance. Section 9.3 describes the methodology applied in our empirical analysis. Section 9.4 explains the main characteristics of the data and the variables selected for the empirical analysis. Section 9.5 discusses the main results compared with previous literature. Finally, Sect. 9.6 outlines some concluding remarks.

9.2 Literature Review

Since the publication of the pioneering work by Woessmann (2003) combining international student- and school-level microdata with several country-level indicators, multiple studies have adopted a cross-country approach to explore the main determinants of educational achievement from different perspectives (Ammermüller et al. 2005, Hanushek and Woessmann 2014; Le Donné 2014). These studies mainly use econometric techniques to identify relationships between student background, school-related variables, and educational outcomes (typically represented by test scores).

More recently, some of them have started to apply more sophisticated methods in order to identify causal relationships in the international data on educational achievement (see Cordero et al. 2018a for a review).

However, those studies do not consider the potential existence of an unexpected level of inefficiency in student, school or country performance (Levin 1974). Nevertheless, over the last few years, the number of studies devoted to the assessment of education systems using a cross-country approach has grown significantly. Among those works, the most common ones are those using cross-sectional data aggregated at a country level (Clements 2002; Afonso and St Aubyn 2006; Verhoeven et al. 2007; Giménez et al. 2007, 2017; Giambona et al. 2011; Thieme et al. 2012; Agasisti 2014; Coco and Lagravinese 2014; Aristovnik and Obadić 2014; Bogetoft et al. 2015). Nevertheless, we can also find studies comparing the performance of education systems in different countries using school-level data. For instance, Sutherland et al. (2009) study the performance of schools from 30 OECD countries participating in PISA 2003; Agasisti and Zoido (2018) derive efficiency measures for around 8,600 schools operating in 30 developed countries participating in PISA 2012; Aparicio et al. (2018) assess schools operating in the 34 OECD countries participating in PISA 2012 and identify different levels of inefficiency for reading and mathematics; Agasisti and Zoido (2019) assessed the performance of 6,800 schools from 28 developing countries using also PISA 2012 data. Finally, De Jorge and Santín (2010) and Deutsch et al. (2013) use student-level PISA data to estimate the efficiency of European Union and Latin American countries, respectively.

Most of the above studies use nonparametric techniques like DEA or FDH to estimate performance efficiency measures since they are flexible enough to adapt to the characteristics of public services provision,[2] especially to their multi-input multi-output nature. Moreover, in many cases, a two-stage procedure is also applied to examine the potential influence of contextual variables on efficiency estimates (e.g., Afonso and St Aubyn 2006; Verhoeven et al. 2007; De Jorge and Santín 2010; Agasisti 2014; Aparicio et al. 2018; Agasisti and Zoido 2018, 2019). The main problem with this procedure is that it assumes that environmental factors affect the shape of the distribution of inefficiencies (i.e., mean, variance, etc.) but not the attainable set or the estimated frontier (see Simar and Wilson 2007, 2011 for details). This is often unrealistic since contextual factors can be expected to be influencing both educational outcomes and the resources employed. Although the separability between the input–output space and the space of external variables can be tested in advance using the statistical tools developed by Daraio et al. (2015, 2018), none of the above empirical studies examined whether this assumption holds before applying this method.

Whenever the two-stage procedure is found to be inappropriate, the alternative option is to use the conditional nonparametric approach developed by Daraio and Simar (2005, 2007a, b), which allows for incorporating the effect of contextual factors into the estimation of efficiency scores without assuming the restrictive separability

[2]There are some exceptions using parametric methods (e.g., Deutsch et al. 2013; Sutherland et al. 2009).

condition. To the best of our knowledge, in previous literature, we can only find two empirical studies applying this method. Cordero et al. (2017) analyzed the effect of several contextual factors at school- and country-level on the performance of primary schools from 16 European countries participating in PIRLS 2011. In addition, they decomposed the estimated inefficiency levels between two different sources (school and country) using the metafrontier framework (O'Donnell et al. 2008). Subsequently, Cordero et al. (2018b) assess the performance of a set of more than 12,000 secondary schools from 36 countries participating in PISA 2012 exploring the influence of a wide range of contextual factors, including also variables at both school- and country-level. Moreover, they also obtain "whitened" efficiency scores by applying the second-stage approach suggested by Badin et al. (2012) to eliminate the effects of contextual conditions. However, in this study cleaning is conducted as the final step, while in the present work we estimate the frontier and the efficiencies of the units using "pre-whitened" inputs after cleaning the influence of environmental variables.

9.3 Methodology

Starting with the usual production process in which a set of inputs $X \in \mathbb{R}_+^p$ produces a set of outputs $Y \in \mathbb{R}_+^q$, the attainable set of feasible combinations of inputs and outputs can be defined by:

$$\Psi = \left\{(x, y) \in \mathbb{R}_+^{p+q} \mid x \text{ can produce } y\right\} \tag{9.1}$$

Following Cazals et al. (2002), this production function can be defined using a probabilistic formulation as:

$$H_{X,Y}(x, y) = \text{Prob}(X \leq x, Y \geq y) \tag{9.2}$$

In our context, we should note that school performance might be affected by the presence of contextual or exogenous factors that are beyond their control. Thus, we need to account for that environment in the efficiency analysis in order to make a fairer comparison among units. These factors can be included as a set of contextual variables that can have an impact on both, the input–output space (i.e., the frontier) and the distribution of the efficiencies. Following Daraio and Simar (2005, 2007b), those variables can be incorporated in the joint distribution of (X, Y) conditional on $Z = z$ as:

$$H_{X,Y|Z}(x, y \mid z) = \text{Prob}(X \leq x, Y \geq y|Z = z) \tag{9.3}$$

For an output conditional measure of efficiency, this function can be decomposed in two terms: the survival conditional function of outputs and the conditional distribution function of inputs, as follows:

$$H_{X,Y|Z}(x, y \mid z) = S_{Y \mid X,Z}(y \mid x, z) F_{X \mid Z}(x \mid z) \tag{9.4}$$

Thus, the conditional output-oriented efficiency measure can be defined as the proportionate increase in outputs required for the evaluated unit to have a zero probability of being dominated at the given input level and by other units facing the same environmental conditions z:

$$\lambda(x, y|z) = \sup\{\lambda > 0 \big| H_{X,Y|Z}(x, \lambda y|z)\big)0\} = \sup\{\lambda > 0 | S_{Y|X,Z}(\lambda y|x, z) > 0\} \tag{9.5}$$

In terms of the Farrell-Debreu efficiency scores, a value equal to one indicates that the unit is fully efficient and in the case of values higher than one, units are considered as inefficient (higher values imply higher distance to the frontier representing more inefficiency). By a plug-in rule, we can define the different conditional estimators of either the full frontier, such as the Free Disposal Hull (FDH) and Data Envelopment Analysis (DEA),[3] or partial frontiers as order-m and order-α.[4]

This approach is well-established in the literature, but presents some limitations that might be detrimental for schools' ranking purposes, which is one of the main objectives of this study. Particularly, the estimators defined in that way need smoothing in the conditional variables at the frontier where there are fewer units than inside of the data cloud. This fact makes the measures more sensitive to outliers and extreme data. On the other hand, the derived efficiency measures could suffer the problem of the endogeneity bias caused by reverse causality between external factors and input/output indicators. Moreover, this conditional model might be affected by the well-known curse of dimensionality problem due to the dimension of Z.[5]

In order to address these limitations, we propose to follow the methodology developed by Florens et al. (2014). These authors assume flexible nonparametric location-scale models that link the input–output space to the contextual variables and, subsequently, remove the dependence on Z in the original inputs and outputs, thus generating two new sets of "whitened" variables. Specifically, we suppose the following nonparametric location-scale regression model:

---

[3]While the conditional efficiency estimator of the FDH frontier was developed in Daraio and Simar (2005), the corresponding convex technology, i.e., the DEA estimator was established in Daraio and Simar (2007b).

[4]Robust order-m frontiers were presented by Cazals et al. (2002) and Daraio and Simar (2005). Daouia and Simar (2007a, b) developed the alpha-quantile conditional efficiency estimators.

[5]Florens et al. (2014) indicate that the rates of convergence are deteriorated by the smoothing in Z to get the different nonparametric estimators in the sense that n is to be replaced by $n \prod_{j=1}^{d_z} h_j$ (being h_j the corresponding bandwidth for each unit) when product kernels are used for smoothing the d_z components of Z (see Jeong et al. 2010 for details).

$$\begin{cases} X_i = \mu_x(Z_i) + \sigma_x(Z_i)\varepsilon_{x,i} \\ Y_i = \mu_y(Z_i) + \sigma_y(Z_i)\varepsilon_{y,i} \end{cases} \tag{9.6}$$

The first equation in (9.6) gathers p relations, one for each component of X, while the second one integrates q components as the dimension of Y. In the same way, μ_x, σ_x, and $\varepsilon_{x,i}$ have each p relations and μ_y, σ_y, and $\varepsilon_{y,i}$ have each q relations. All products of vectors are understood as componentwise. The terms associated with the residuals in this model (i.e., ε_x, ε_y) are considered as the "cleaned" inputs and outputs and they have mean zero and standard deviation equal to 1. The vectors μ_x and μ_y grasp the locations, as well as σ_x and σ_y capture the scale effects, all of them conditioned to Z s.

As pointed out in Florens et al. (2014), that mathematical specification of the methodology has some very relevant features. First, the model is built keeping its nonparametric character, since any hypothesis is not assumed for the distributions of ε_x and ε_y. Furthermore, the procedure removes any dependence among the exogenous variables and these vectors, so they can be accepted as the "whitened" version of X and Y.[6] Additionally, this well-defined model allows us to derive a "pure" efficiency measure that seems to be more robust than previous conditional nonparametric techniques which account for the impact of external factors. This fact is due to the smoothing is done in the center of the sample where there are more observations than those that form the boundary (see Badin et al. 2012 for details).

From model (9.6), we can estimate the location and scale vectors by running a double nonparametric regression. In the first of them, the location functions ($\mu_x(Z_i)$, $\mu_y(Z_i)$) are estimated based on a local linear model. The square residuals resulting from this first regression are used to estimate the scale functions ($\hat{\sigma}_x^2(Z_i)$, $\hat{\sigma}_x^2(Z_i)$) from a subsequent local constant model to assure positive values of the variances. Therefore, we can derive the residuals:

$$\hat{\varepsilon}_{x,i} = \frac{X_i - \hat{\mu}_x(Z_i)}{\hat{\sigma}_x(Z_i)} \tag{9.7}$$

$$\hat{\varepsilon}_{y,i} = \frac{Y_i - \hat{\mu}_y(Z_i)}{\hat{\sigma}_y(Z_i)} \tag{9.8}$$

These expressions represent the inputs and outputs for which the influence of the exogenous variables has been eliminated.[7] According to Mastromarco and Simar (2018), the assumption of independence among ε_x, ε_y, and Z s is asymptotically verified in this model since $\mathrm{Cov}(X_i, \hat{\varepsilon}_{x,i}) \to 0$ and $\mathrm{Cov}(Y_i, \hat{\varepsilon}_{y,i}) \to 0$ as $N \to \infty$.

By cleaning the inputs and outputs from the effect of the Z s, two of the main aforementioned limitations of previous conditional estimators are solved, i.e., the

[6]In the case of X and Y were independent of the exogenous variables, the vectors ε_x and ε_y would directly be the standardized inputs and outputs (Florens et al. 2014).

[7]Florens et al. (2014) propose a bootstrap based procedure to test the independence between the whitened inputs and outputs and the Zs. In this paper, we present the evidence of that independence as in Mastromarco and Simar (2017).

endogeneity bias caused by reverse causality between production process and the environment as well as the curse of dimensionality problem due to the size of Z.

The following step would be to estimate a pure measure of efficiency by using the whitened estimations ε_x and ε_y. To do this, we must transform Eq. (9.1) in such a way that the attainable set of ε_x and ε_y is now defined:

$$\Psi_\varepsilon = \left\{ \left(e_x, e_y\right) \in \mathbb{R}^{p+q} \middle| H_{\varepsilon_x,\varepsilon_y}\left(e_x, e_y\right) = \text{Prob}\left(\varepsilon_x \le e_x, \varepsilon_y \ge e_y\right) \rangle 0 \right\} \tag{9.9}$$

The different efficiency estimators as DEA or FDH can be obtained by replacing the empirical counterparts of $\hat{H}_{\varepsilon_x,\varepsilon_y}\left(e_x, e_y\right)$. Nonetheless, since pure inputs and outputs have mean zero, the measures based on radial distances from each observation to the efficient frontier are inappropriate. In that scenario, the directional distance functions seem to be more suitable,[8] therefore we apply:

$$\delta\left(e_x, e_y; d_x, d_y\right) = \sup\left\{ \gamma \middle| H_{\varepsilon_x,\varepsilon_y}\left(e_x - \gamma d_x, e_y + \gamma d_y\right) \rangle 0 \right\} \tag{9.10}$$

The terms d_x, d_y in (9.10) represent the desired direction of the projection over the efficient frontier. In this case, we select the output direction ($d_x = 0$ and $d_y = 1$) since schools are supposed to be interested in improving their performance by achieving better student results and not by reducing their use of resources. Then, the nonparametric pure efficiency estimator in the output direction can be obtained as:

$$\hat{\delta}\left(\hat{\varepsilon}_{x,i}, \hat{\varepsilon}_{y,i}; 0, 1\right) = \hat{\varphi}\left(\hat{\varepsilon}_{x,i}, \hat{\varepsilon}_{y,i}\right) - \hat{\varepsilon}_{y,i} \tag{9.11}$$

where $\hat{\varphi}\left(e_x, e_y\right) = \max\limits_{\left\{i \mid \hat{\varepsilon}_{x,i} \le e_x\right\}} \left\{ \min\limits_{j=1,\ldots,q} \left(\frac{\hat{\varepsilon}^j_{y,i}}{e^j_y} \right) \right\}$ is clearly identified as the output-oriented FDH estimator of the pure efficient frontier.

The efficiency derived from this estimator could be affected by outliers or extreme data presented in the sample. To avoid this problem, robust versions of the proposed estimator have been developed in the literature. One of the most used approach is the order-m estimator, which defines a partial frontier by using only a set of m observations randomly drawn from the population of units using less inputs than x. Following Florens et al. (2014), the pure version of that estimator is the expected value of the maximum of the outputs of the set with length m such that $\varepsilon_{x,i} \le e_x$:

$$\hat{\varphi}_m\left(e_x, e_y\right) = \hat{E}\left[\max_{i=1,\ldots,m} \left\{ \min_{j=1,\ldots,q} \left(\frac{\hat{\varepsilon}^j_{y,i}}{e^j_y} \right) \right\} \right] \tag{9.12}$$

We rely on these measures of pure efficiency described from the perspective of both full and partial frontier estimators, since they allow us to mitigate potential problems related to endogeneity bias as well as the dimensionality problem that

[8] For a more detailed explanation of directional distance functions (DDF), see Färe and Grosskopf (2000), Simar and Vanhems (2012), Daraio and Simar (2014, 2016) or Daraio et al. (2019).

might arise in traditional conditional models. This is possible because the influence of contextual variables has been removed from the original variables and hence from the estimations. Nonetheless, if we are interested on assessing the impact of the external variables Z on the shape of the frontier and also on the distribution of the inefficiencies, we need to resort to the conditional efficiencies estimated using the original values of the inputs and the outputs, i.e., X and Y. Specifically, we can explore this influence following the procedure proposed in Badin et al. (2012),[9] which allows us to distinguish both effects by calculating the ratio between the conditional efficiency estimates and the unconditional efficiency measures estimated considering only information about original inputs and outputs (X, Y):

$$\hat{R}(x, y|z) = \frac{\hat{\lambda}(x, y|z)}{\hat{\lambda}(x, y)} \quad (9.13)$$

$$\hat{R}_m(x, y|z) = \frac{\hat{\lambda}_m(x, y|z)}{\hat{\lambda}_m(x, y)} \quad (9.14)$$

The ratios in Eq. (9.13) correspond to the full frontiers estimators, whereas the ratios in (9.14) are obtained from the partial frontier estimators. Since the selected orientation of the approach is to maximize the outputs, the ratios from the full frontier estimators are below 1, representing exactly this value the coincidence of the marginal and the conditional frontiers. Nonetheless, by construction of the partial frontier estimators, the corresponding ratios are not bounded by 1, being possible to found values above 1.

We can explore the impact of Z s on the shape of the frontier by using the full ratios. These could be estimated in two ways, either with FDH scores or with the extreme order-m measures, i.e., when $m \rightarrow \infty$,[10] which can be understood as a robustness check of the full frontier estimation. Additionally, by estimating the ratios from a partial frontier that is built in the middle of the data cloud, we can assess the impact of the exogenous variables on the shift inside the attainable sets. In this particular case, the value $m = 1$ returns an average production function (see Badin et al. 2012 for details).

The interpretation of the direction of the impact produced by Zs in an output-oriented model is as follows. An upward trend of the ratio when the conditioning variables increase would indicate a favorable effect (the conditional frontier approximates the marginal frontier, so the variables act as freely available inputs), whereas a downward trend denotes an unfavorable impact (the conditional frontier moves away from the unconditional frontier when the variables increase, so the Zs act as undesirable outputs).

[9]Mastromarco and Simar (2017, 2018) also apply this procedure after using the pre-whitening approach suggested by Florens et al. (2014). These works extend the method to a dynamic framework in which time (the database is of type data panel) plays a role as an additional exogenous variable.

[10]In practice, this is equivalent to $m = \text{n}$ (total number of units).

9.4 Data and Variables

The data used in this study come from the Programme for International Student Assessment (from now on, PISA) designed by the OECD in the late 1990s as a comparative, international, and continuous study of certain characteristics and skills of students aged between 15 and 16 years. PISA assesses three core competencies: mathematics, reading, and science.[11] PISA also gathers information about students' backgrounds and school environments through different questionnaires completed by students, parents, teachers, and school principals. The first PISA survey took place in the year 2000 and since then, it has been repeated every 3 years. Although the three core competencies mentioned above are always assessed, each wave focuses on one domain. The latest wave available is PISA 2015, with a total of 72 participating countries (35 OECD members and 31 partners), being science the main domain, as well as in 2006 (OECD 2017).

This survey uses a two-stage stratified design sampling (Willms and Smith 2005). In the first stage of sampling, schools having age-eligible students are sampled systematically with probabilities proportional to the school size. A minimum of 150 schools is selected in each country. Subsequently, 42 15-year-old students are randomly selected from each school to participate in the survey.[12] Such two-stage sampling is convenient from a logistic point of view, but the accuracy can be deteriorated due to intraclass correlation (students within schools are more similar than students across schools).[13]

One of the main advantages of using PISA data is that this study does not evaluate cognitive abilities or skills through using one single score but each student receives five different scores (plausible values) that represent the range of abilities that a student might reasonably have (see OECD 2016 for details).[14] Specifically, the dataset provides measures on students' performance based upon pupils' responses to different test booklets, each of which includes only a limited number of test questions. Thus, it is difficult to make claims about individual performance with great accuracy. Using a complex process based on item response theory model (Rasch 1960/1980), the survey organizers produce test scores for participants taking into account the difficulty of each test question.[15] Plausible values can, therefore, be defined as random values drawn from this distribution of proficiency estimates.

In addition, the survey collects a great volume of data about other factors potentially related to those results, such as variables representing the student's background, school environment, or educational provision. This information comes from the

[11] In the most recent waves of this survey, PISA also evaluates other innovative skills such as collaborative problem-solving or financial literacy.

[12] Only 35 students for those countries where PISA assessment was administered in paper-based mode.

[13] If 42 students within a school are selected, they do not provide as much "information" as 42 students randomly selected from all schools (Wu 2010).

[14] For more detailed information about plausible values see Mislevy et al. (1992) or Wu (2005).

[15] See Von Davier and Sinharay (2013) for further details.

responses given to different questionnaires completed by students and school principals. From these data, it is possible to collect a great amount of information referred to the main determining factors of educational performance.

Our dataset comprises a total number of 16,643 schools distributed across 67 countries as reported in Table 9.1. However, those countries represent more than 85% of the world economy (see Fig. 9.1). As explained above, the minimum number of participating schools in each country must be 150, although in our sample we have some exceptional cases with a lower number of observations because the number of schools in the country is limited (e.g., Luxembourg, Malta or Montenegro). Likewise, in several countries, the sample is very large due to the existence of representative samples for different regions within the country (e.g., Australia, Brazil, Canada, Italy or the United Kingdom).

Although it is difficult to empirically quantify the education received by students, there is a broad consensus in the literature about considering the results from standardized tests as educational outputs in empirical studies (e.g., Cherchye et al.

Table 9.1 Dataset composition: number of schools per country

Country	n	Country	n	Country	n	Country	n
Algeria	161	France	252	Lithuania	311	Slovak Republic	290
Australia	758	Georgia	262	Luxembourg	44	Vietnam	188
Austria	269	Germany	256	Macao	45	Slovenia	333
Belgium	288	Greece	211	Malta	59	Spain	201
Brazil	841	Hong Kong	138	Mexico	275	Sweden	202
Bulgaria	180	Hungary	245	Moldova	229	Switzerland	227
Canada	759	Iceland	124	Montenegro	64	Thailand	273
Chile	227	Indonesia	236	Netherlands	186	Trinidad and Tobago	149
Chinese Taipei	214	Ireland	167	New Zealand	183	United Arab Emirates	473
Colombia	372	Israel	173	Norway	229	Tunisia	165
Costa Rica	205	Italy	474	Peru	281	Turkey	187
Croatia	160	Japan	198	Poland	169	Macedonia	106
Czech Republic	344	Jordan	250	Portugal	246	United Kingdom	550
Denmark	333	Korea	168	Qatar	167	United States	177
Dominican Republic	194	Kosovo	224	Romania	182	Uruguay	220
Estonia	206	Lebanon	270	Russian Federation	210	B-S-J-G (China)	268
Finland	168	Latvia	250	Singapore	177	**Total**	**16,643**

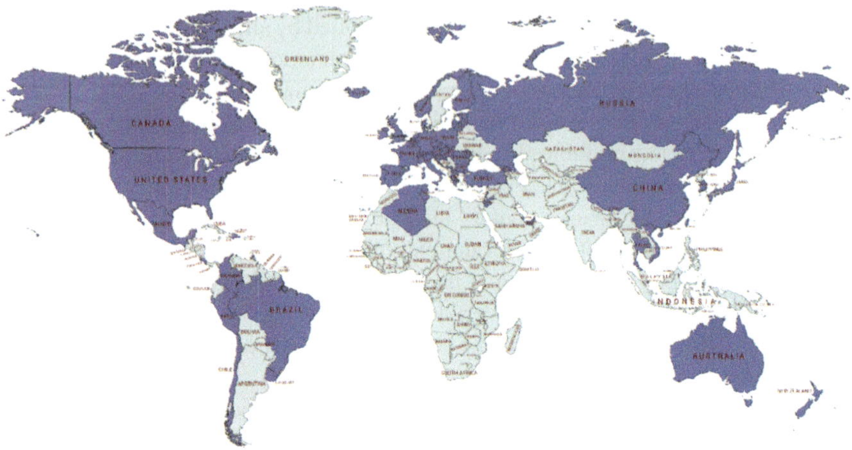

Fig. 9.1 Countries included in the sample

2010),[16] thus we have selected the average plausible values of students belonging to the same school in the three evaluated competences. For the sake of simplicity, in our analysis, we only consider a single plausible value (the first one) for each subject (PVMATH, PVREAD, and PVSCIE) since using one plausible value or more (there are ten available in the dataset) does not really make a substantial difference on large samples (see OECD 2009, p. 44 for details).

The selection of inputs is a tough decision since in the PISA database there is an extensive list of potential indicators that can be considered. Given that the literature does not provide an explicit rule to discriminate between them, we have based our decision on the following criteria. On one hand, input variables must fulfill the requirement of monotonicity (i.e., ceteris paribus, more input implies an equal or higher level of output). Thus, inputs should present a significant positive correlation with outputs. On the other, input variables should be objective measures of educational resources involved in the learning process. In our case, we have selected three variables that meet these requirements and are in line with previous works attempting to measure the efficiency of schools (e.g., Worthington 2001; De Witte and Lopez-Torres (2017). First, we select one indicator representing the economic, social, and cultural status of students in the school (ESCS), since students are the "raw material" to be transformed through the learning process.[17] This is an index created by PISA analysts that includes the highest educational level of any of the student's parents, the highest labor occupation of any of the student's parents and

[16]However, some authors have highlighted that when test scores are used as proxies of educational outcomes, other dimensions of learning such as social skills, attitudes, personal maturity, or moral values are ignored, even though they are crucial for individual development (Levin 2012).

[17]This is a common practice in several recent papers attempting to measure the efficiency of schools (e.g., Thieme et al. 2013; Agasisti 2014; Crespo-Cebada et al. 2014; Aparicio et al. 2017, 2018; Agasisti and Zoido 2018).

an index of educational possessions related to the household economy. Second, as a proxy for human resources, we included the number of teachers per hundred of students (TSRATIO), i.e., the inverse of the student–teacher ratio provided by PISA. The third input is an index representing the quality of school educational resources (SCHRES) computed as the inverse of EDUSHORT (original PISA index which measures the principals' perceptions of potential factors hindering the provision of instruction at school).[18]

Finally, we have also selected some continuous and dummy variables representing the educational environment in which they are operating as well as factors related to the type of school management. Specifically, we included the proportion of fully certified teachers with respect to the total number of teachers (PROPCERT), the total number of students enrolled per school (SCHSIZE), the level of responsibility that school staff have in allocating resources (RESPRES), the percentage of immigrants students (%IMMIG), the percentage of students who have repeated at least one grade (%REPEAT), and the percentage of students who had skipped a whole school day (%TRUANCY). As dummy variables, we considered whether the school is placed in a rural area (city with less than 15,000 inhabitants) (RURAL), the type of school ownership (PRIVATE), if the school has less than 20 students per class (SMCLASS), the school accountability policies, i.e., whether the school makes publicly available its students average achievement (ACCOUNT) and, finally, if the school distributes their students across classes based on their ability (ABGROUP). The main descriptive statistics for all these variables are summarized in Table 9.2.

9.5 Results

This section shows the results obtained for the dual efficiency analysis applied to the database described in the previous section. First of all, we eliminate the dependence of production inputs/outputs on external factors (Z) by means of the nonparametric location-scale model. By using the model described in (9.6) and the Eqs. (9.7) and (9.8), we derive the cleaned values for these new "pure" inputs and outputs. Table 9.3 shows the main descriptive statistics for those residuals. Here we can notice that, although there are some divergences in the dispersions of the new variables, all of them contain both positive and negative values, confirming the predicted results for model (9.6).

Before estimating the "pure" efficiencies from the new cleaned variables, we must check if whitening has been correctly conducted. Following Mastromarco and Simar (2017), we calculate the Pearson and Spearman rank correlations coefficients between the vectors $\hat{\varepsilon}_x$, $\hat{\varepsilon}_y$ and the Zs, which we report in Table 9.4. As expected,

[18]The original values of EDUSHORT and ESCS were rescaled to show positive values by adding up the minimum value to all the original values of the variables. This transformation does not alter the efficient frontier (or empirical production function), and hence the associated DEA model is translation invariant.

Table 9.2 Descriptive statistics of variables included in the analysis

Variables		Mean	SD	Min	1Q	Median	3Q	Max
Outputs	PVMATH	455.31	78.01	163.89	398.33	461.90	510.55	717.38
	PVREAD	456.51	79.16	136.74	401.66	464.93	514.76	690.08
	PVSCIE	461.50	75.40	166.68	405.38	465.33	515.97	723.98
Inputs	ESCS	4.56	0.80	0.01	4.07	4.66	5.15	6.44
	TSRATIO	9.46	9.43	1.00	6.04	7.74	10.00	100.00
	SCHRES	3.59	1.07	0.02	2.97	3.78	4.39	4.97
Continuous	PROPCERT	0.80	0.30	0.00	0.78	0.95	1.00	1.00
	SCHSIZE	752.43	541.18	10.00	341.00	651.00	985.00	2,503
	RESPRES	0.80	1.04	0.01	0.19	0.46	0.81	3.62
	%IMMIG	0.05	0.12	0.00	0.00	0.00	0.05	1.00
	%REPEAT	0.41	0.42	0.00	0.00	0.24	0.96	1.00
	%TRUANCY	0.39	0.32	0.00	0.14	0.31	0.57	1.00

		% Level 0	% Level 1
Dummies	RURAL	67.35	32.65
	PRIVATE	83.08	16.92
	SMCLASS	79.71	20.29
	ACCOUNT	63.86	36.14
	ABGROUP	61.86	38.14

Table 9.3 Descriptive statistics of whitened inputs and outputs: $\hat{\varepsilon}_x, \hat{\varepsilon}_y$

	Mean	SD	Min	1Q	Median	3Q	Max
Pure PVMATH	0.0080	0.3438	−1.0317	−0.1750	0.0036	0.1809	1.0906
Pure PVREAD	0.0139	0.4212	−1.3506	−0.1874	0.0123	0.2281	1.2252
Pure PVSCIE	0.0131	0.4522	−1.3721	−0.2057	−0.0011	0.2329	1.3882
Pure ESCS	2.2871	34.8066	−116.6218	−13.3132	2.0489	18.0496	117.9084
Pure TSRATIO	0.0215	5.0703	−17.1800	−0.9428	−0.1270	0.7235	22.3843
Pure SCHRES	2.6134	26.1132	−74.6512	−11.0938	2.6810	15.3160	155.7599

we find both positive and negative values, but the most relevant fact is that all of them are very low, thus we can conclude that the effect of the contextual variables has been totally removed from the original data. Therefore, we are assuming that all schools operate in the same environmental conditions regardless of the educational system they belong. Consequently, the possible existence of endogeneity bias and

Table 9.4 Pearson and Spearman correlations between $\hat{\varepsilon}_x$, $\hat{\varepsilon}_y$ and Z s

	PROPCERT	SCHSIZE	RESPRES	RURAL	PRIVATE	SMCLASS	%IMMIG	**%REPEAT**	**%TRUANCY**	**ACCOUNT**	**ABGROUP**
Pearson correlation											
Pure PVMATH	0.0174	0.0140	−0.0060	0.0198	0.0007	0.0053	0.0081	0.0000	−0.0259	0.0060	0.0042
Pure PVREAD	0.0144	0.0057	−0.0019	0.0153	0.0010	0.0077	0.0090	−0.0018	−0.0158	0.0136	0.0086
Pure PVSCIE	0.0173	0.0179	0.0010	0.0213	0.0025	0.0070	0.0024	0.0004	−0.0113	0.0147	0.0150
Pure ESCS	0.0159	0.0069	−0.0056	−0.0101	−0.0080	−0.0112	0.0188	−0.0154	−0.0307	0.0064	0.0012
Pure TSRATIO	−0.0103	0.1317	0.0125	−0.0112	0.0205	−0.0059	0.0185	−0.0003	−0.0111	0.0122	0.0004
Pure SCHRES	0.0028	0.0074	−0.0287	−0.0378	−0.0320	−0.0354	−0.0080	−0.0495	−0.0222	−0.0393	−0.0322
Spearman correlation											
Pure PVMATH	0.0234	0.0529	0.0937	0.0255	0.0105	0.0118	0.0520	0.0182	−0.0370	0.0054	0.0093
Pure PVREAD	0.0376	0.0559	0.0787	0.0173	0.0032	0.0062	0.0523	0.0044	−0.0354	0.0136	0.0122
Pure PVSCIE	0.0258	0.0655	0.0910	0.0247	0.0109	0.0052	0.0397	0.0096	−0.0233	0.0158	0.0179
Pure ESCS	0.0011	0.0422	0.1042	−0.0016	−0.0092	−0.0138	0.0976	−0.0158	−0.0308	0.0076	0.0091
Pure TSRATIO	−0.0246	0.0318	0.0453	−0.0209	0.0299	−0.0199	0.0545	−0.0001	−0.0151	0.0043	−0.0176
Pure SCHRES	−0.0061	0.0497	0.0563	−0.0233	−0.0216	−0.0286	0.0240	−0.0182	−0.0168	−0.0280	−0.0186

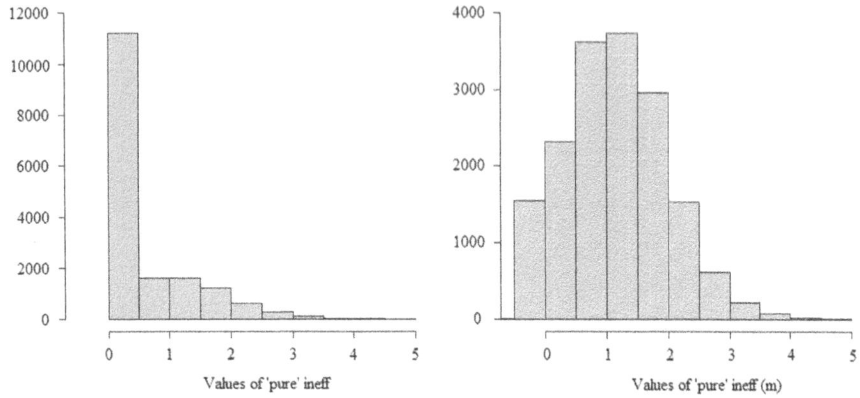

Fig. 9.2 Histograms of estimated "pure" inefficiencies relative to the full frontier $\hat{\varphi}$ (left panel) and the order-m frontier $\hat{\varphi}_m$ (right panel)

dimensionality problems due to the influence of different school environments should not affect the estimation of the efficiency scores and, subsequently, the construction of a ranking of countries according to the performance of their schools.

As we detailed in Sect. 9.3, once we have "whitened" input and output values, we estimate both FDH and order-m "pure" efficiencies. Following Daraio and Simar (2005), we determine the size of the partial frontier as the value of m for which the decrease in the number of super-efficient observations stabilizes. In our application, we have selected a value of $m = \sqrt[3]{N^2}$, as suggested by Tauchmann (2012), which corresponds to approximately 15% of the total number of schools. This implies that each school is compared to approximately 2,500 schools randomly drawn from observations in the whole dataset that consume at most the same amount of inputs. For statistical inference, we use 200 bootstrap replications. Figure 9.2 collects the two histograms of the distributions of the "pure" inefficiencies corresponding to both frontiers. For the full frontier, we can observe a potential problem of dimensionality, since the proportion of efficient schools is very high (10,224 out of 16,643). In contrast, the distribution of the inefficiencies in the case of the order-m is much more dispersed. The greater discrimination power of the latter approach allows us to better identify the best and worst performers among schools under the assumption of operating in the same environment. Specifically, we find 642 schools (around 4%) that are fully efficient and other 919 schools (around 5.5%) that are super-efficient (below 0), i.e., they perform better than the 15% schools against which they are benchmarked.

Table 9.5 reports the ranking of countries according to the average estimated "pure" efficiency order-m scores of their schools and the corresponding standard deviation.[19] Additionally, we also show the classification of countries according to

[19]We only present the classification by using the order-m estimations because they are more robust and present a higher level of discrimination power. The original values have been transformed into

Table 9.5 "Pure" efficiencies and results for science. Average scores and standard deviations by country

RK	Country	PureEff	SD	Country	PVSCIE	SD
1	Singapore	0.7370	0.3034	Singapore	544.35	64.79
2	Japan	0.6674	0.3003	Japan	537.07	64.34
3	Tunisia	0.6346	0.2942	Estonia	531.95	44.71
4	Slovenia	0.6083	0.2949	Finland	529.50	43.15
5	Korea	0.5875	0.3241	Chinese Taipei	529.29	60.32
6	Norway	0.5743	0.3418	Hong Kong	525.39	46.58
7	Greece	0.5578	0.2916	B-S-J-G (China)	523.55	71.71
8	New Zealand	0.5545	0.3152	Vietnam	517.37	51.09
9	Canada	0.5516	0.3328	Korea	513.35	49.68
10	Finland	0.5303	0.2167	Canada	513.24	44.25
11	Estonia	0.5207	0.2501	Macao	511.83	45.43
12	United Kingdom	0.5194	0.3083	Netherlands	507.46	78.43
13	Iceland	0.5155	0.3074	New Zealand	507.15	51.69
14	Austria	0.5129	0.3124	Poland	507.07	41.82
15	Turkey	0.5099	0.3095	Germany	503.59	72.08
16	Hong Kong	0.5084	0.2779	United Kingdom	502.80	50.15
17	United States	0.4980	0.3161	Australia	500.39	55.66
18	Israel	0.4842	0.3398	Norway	499.54	36.17
19	Italy	0.4634	0.2901	Ireland	499.25	38.36
20	Croatia	0.4629	0.2667	Switzerland	497.44	63.23
21	B-S-J-G (China)	0.4619	0.2974	Sweden	497.15	51.90
22	Brazil	0.4511	0.2948	Spain	495.15	35.79
23	France	0.4504	0.2950	Belgium	492.67	74.38
24	Switzerland	0.4491	0.3282	United States	492.53	49.94
25	Australia	0.4486	0.2993	Czech Republic	489.87	69.82
26	Russian Federation	0.4451	0.2771	Luxembourg	486.97	61.16
27	Poland	0.4427	0.2460	France	486.84	75.57
28	Czech Republic	0.4256	0.2725	Denmark	485.86	47.51
29	Vietnam	0.4016	0.2817	Latvia	483.67	41.10
30	Sweden	0.4010	0.2650	Austria	483.51	69.82
31	Romania	0.3996	0.2477	Portugal	482.33	54.23
32	Thailand	0.3958	0.2746	Russian Federation	482.01	43.90
33	Chinese Taipei	0.3953	0.2971	Italy	480.34	64.27
34	Moldova	0.3854	0.2389	Iceland	474.20	37.30
35	Jordan	0.3734	0.2729	Slovenia	473.38	72.82

(continued)

Table 9.5 (continued)

RK	Country	PureEff	SD	Country	PVSCIE	SD
36	Germany	0.3727	0.2486	Croatia	472.39	57.03
37	Algeria	0.3689	0.2681	Israel	463.57	69.49
38	Slovak Republic	0.3689	0.2497	Malta	461.20	79.34
39	Spain	0.3684	0.2615	Hungary	460.89	77.37
40	Denmark	0.3606	0.2335	Lithuania	456.09	55.18
41	Peru	0.3588	0.2506	Chile	451.96	68.30
42	Portugal	0.3455	0.2755	Slovak Republic	450.05	66.39
43	Indonesia	0.3423	0.2906	Greece	447.24	64.34
44	Costa Rica	0.3415	0.2728	Bulgaria	432.90	78.05
45	Belgium	0.3399	0.2647	Uruguay	430.95	56.09
46	Dominican Republic	0.3337	0.1600	United Arab Emirates	430.11	67.79
47	Hungary	0.3286	0.2038	Romania	429.81	51.93
48	Lithuania	0.3269	0.2210	Thailand	424.82	61.91
49	Bulgaria	0.3216	0.2246	Moldova	424.06	41.92
50	Qatar	0.3187	0.2264	Colombia	419.80	46.98
51	Montenegro	0.3172	0.2370	Costa Rica	417.33	39.05
52	Malta	0.3138	0.1803	Trinidad and Tobago	415.75	70.78
53	Colombia	0.3047	0.2867	Mexico	412.32	43.63
54	Georgia	0.3015	0.2494	Turkey	410.51	58.61
55	Netherlands	0.3015	0.2076	Georgia	407.56	47.62
56	Kosovo	0.3001	0.1965	Jordan	407.51	49.77
57	Macao	0.2983	0.2498	Montenegro	407.25	49.86
58	Ireland	0.2920	0.2033	Indonesia	405.40	44.37
59	Mexico	0.2859	0.2515	Qatar	402.22	66.36
60	Latvia	0.2783	0.2081	Brazil	388.94	56.33
61	Lebanon	0.2772	0.2208	Peru	388.33	51.41
62	United Arab Emirates	0.2624	0.2316	Tunisia	382.36	40.66
63	Chile	0.2426	0.2073	Macedonia	382.10	46.62
64	Macedonia	0.2423	0.2226	Lebanon	378.28	63.73
65	Trinidad and Tobago	0.2422	0.1742	Algeria	374.28	40.53
66	Uruguay	0.2385	0.2226	Kosovo	363.55	40.18
67	Luxembourg	0.1677	0.1026	Dominican Republic	324.23	47.65
	Total	**0.4058**	**0.2610**	**Total**	**461.37**	**55.56**

mean values of their results in science so that we can explore whether this classification changes or not with the consideration of inputs (and external factors).[20] At the top positions of both rankings, we can find several Asian countries (e.g., Singapore, Japan or Korea), some European countries such as Norway, Finland or Estonia, and other countries like Canada or New Zealand. However, we also notice some surprising results such as the high rank of several countries in the classification according to efficiency levels despite their results in science are below the average (e.g., Tunisia, Slovenia or Greece). This brings to light that the context in which these countries operate is highly affecting the performance of their schools, thus when we eliminate the dependence of production inputs on external factors (Z) they experience a remarkable improvement in their performance.

At the opposite end of the ranking, we identify several countries from Eastern Europe and Middle East (Georgia, Kosovo, Lebanon, and United Arab Emirates), and Latin American countries (Colombia, Chile, Trinidad and Tobago, and Uruguay) as the worst performers in terms of efficiency. Nevertheless, we can also find some unexpected cases, such as the Netherlands, Ireland, and Latvia, which are placed very below in the ranking in spite of having results in science above the average. Our interpretation of this result is that they are harmed by the consideration of inputs involved in the education process like the average socioeconomic status of students attending their schools and their favorable environment. Something similar, but in the opposite direction, might explain that some countries located at the bottom of the ranking of science results are placed in intermediate positions in the ranking based on efficiencies (e.g., Algeria and the Dominican Republic), or even clearly above the average like Brazil or Turkey.

In order to examine the influence of the above external factors on efficiency estimates, we have also estimated the unconditional and conditional efficiency scores with the original levels of inputs and outputs. The distributions for these measures corresponding to the full frontier (FDH) and the partial order-m frontier are represented in Fig. 9.3. Here, we can notice again a high concentration around the value 1 when the Z s are included in the estimation of FDH measures (green curves in the Figure). The curves corresponding to order-m efficiency scores present a higher distribution since they have a higher discrimination power.

Subsequently, we computed the full FDH ratios (9.13) and the partial order-m ratios (9.13) for $m = 16{,}643$ (robust version of the full ratios) to investigate the impact of Zs on the shape of the frontier and also for $m = 1$ to investigate the impact of Zs on the average of the inefficient distribution. Figure 9.4 displays the described ratios for each continuous exogenous variable. In order to facilitate the visual interpretation of these effects, we added a nonparametric regression line of

values between 0 and 1 in order to facilitate their interpretation (higher values indicate higher levels of efficiency).

[20]For comparative reasons, only average results in science are used, as this is the main competence assessed in PISA 2015 (OECD 2017). In any case, the three competencies assessed (science, mathematics, and reading) are highly correlated with each other.

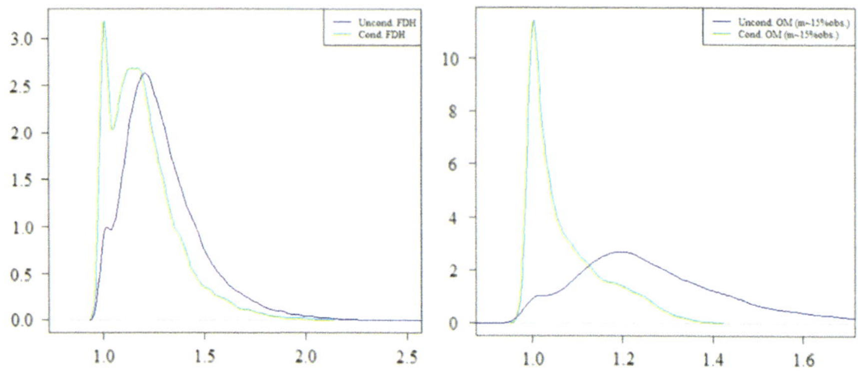

Fig. 9.3 Distributions of FDH scores (left panel) and the order-m scores (right panel)

the different ratios on Zs. For the dummy variables, we have summarized the global direction of the effect in Table 9.6.[21]

In Fig. 9.4, we observe different effects according to the plotted marginal views. The proportion of fully certified teachers seems to have a favorable effect on the shape of the frontier, although it can only be noticed for higher values (Fig. 9.4a). Nonetheless, the effect of this variable on the distribution of efficiencies seems to be slightly negative. The size of the school, in contrast, does not seem to have a clear impact on the frontier, but it presents a favorable effect on the center of the efficiencies, especially for lower values of the variable (Fig. 9.4b). A similar effect is noticed for the level of responsibility of school staff in the allocation of resources, although in this case, the effect is mainly on the shifts of the frontier instead of the efficiencies (Fig. 9.4c). In the case of the percentage of immigrants, we are not able to identify any clear effect in any graph (Fig. 9.4d), but in the last two scatterplots (Fig. 9.4e and f) we can visualize the negative impact of both the percentage of repeaters and the proportion of students who frequently skip classes on both the middle of the efficiency distribution and the frontier (only for lower values in the case of repeaters). These results are in line with previous evidence in empirical studies exploring the influence of these factors on efficiency measures of school performance (e.g., Agasisti and Zoido 2018; Aparicio et al. 2018); and also in the literature about the determinants of achievement (Jimerson 2001; Henry 2007).

With regard to the global impact of the dummy external variables on efficiency levels, the values reported in Table 9.6 indicate that the effect of being located in a rural area is unfavorable as well as in some previous studies based on PISA data (Aparicio et al. 2018). For the remaining four variables (being a private school, having

[21] The interpretation of the trend of the ratios for dummy variables could be confusing. Nonetheless, these graphs are available upon request.

less than 20 students per class, having systems of accountability, and using ability grouping between classes), the influence is positive, as we could expect according to previous evidence found in recent literature (e.g., Agasisti and Zoido 2018).

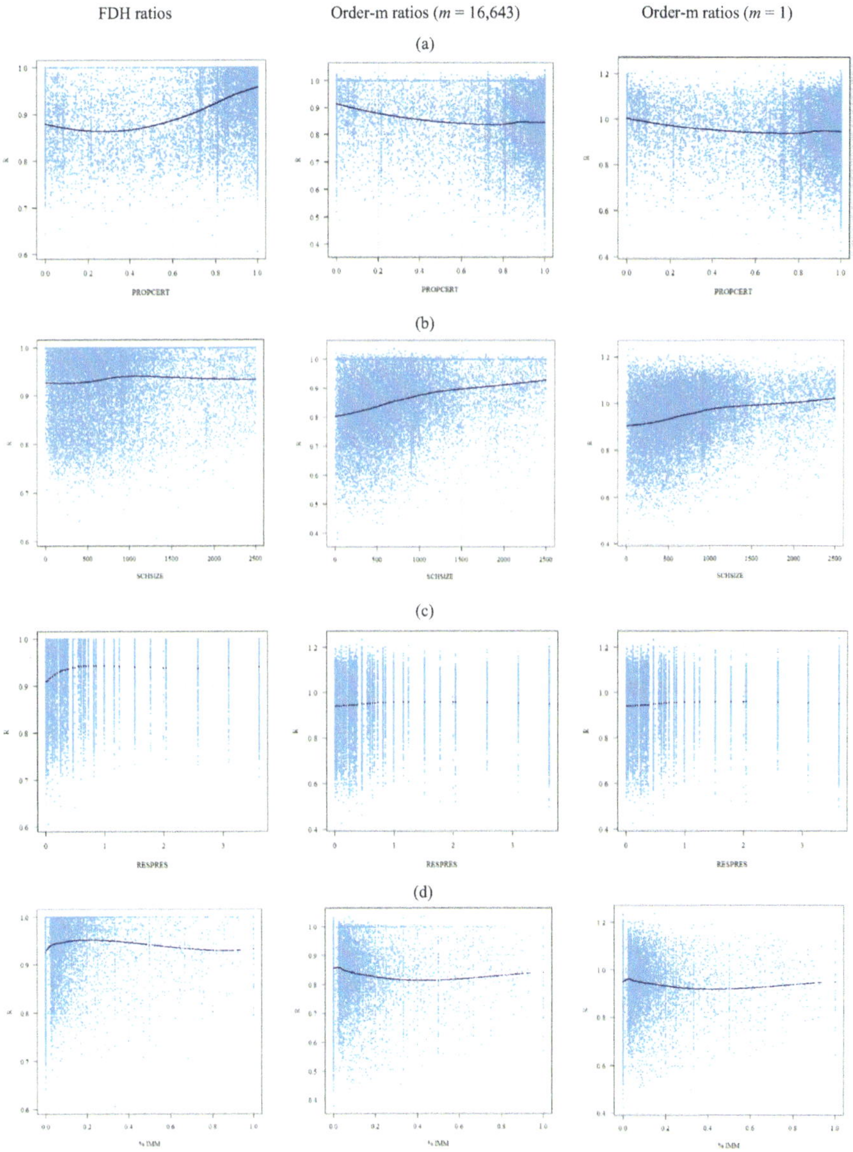

Fig. 9.4 Effect of continuous exogenous variables on the efficiency

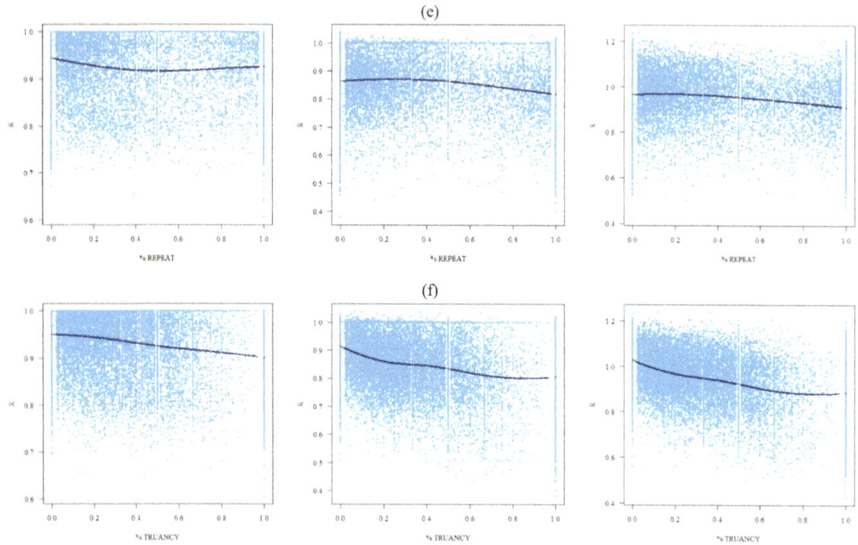

Fig. 9.4 (continued)

Table 9.6 Effect of dummy exogenous variables on efficiency estimates

Variable	RURAL	PRIVATE	SMCLASS	ACCOUNT	ABGROUP
Effect	Unfavorable	Favorable	Favorable	Favorable	Favorable

9.6 Conclusions

The analysis of the efficiency of educational systems and their schools is one of the hottest topics in the field of economics of education for two main reasons. On the one hand, the academic results obtained by students attending secondary schools is widely accepted as a measure of the quality of educational systems, which has a strong and stable association with greater economic growth rates (Hanushek and Kimko 2000). On the other hand, most countries have made a huge financial effort in providing resources for education in the last decades, although there is no straight positive correlation between higher per capita public expenditures on education and higher academic outcomes (Hanushek 2003). For these reasons benchmarking schools and analyzing their efficiency worldwide is one of the most promising tools to learn from best managerial practices. Moreover, this analysis may help policy makers to discard educational policies that do not work and reallocating this public expenditure in more promising alternatives.

In this paper, we applied some of the most recent nonparametric methods to assess the performance of a sample of secondary schools operating in 67 different countries using data from PISA 2015. Specifically, we apply the flexible location-scale model suggested by Florens et al. (2014), which allows us to eliminate the dependence of production inputs/outputs on external factors (Z) by means of nonparametric regression model, thus we can estimate pure or managerial efficiencies whitened from the influence of Z. The main advantage of this approach is that the estimated measures of performance are more reliable to produce rankings or benchmarks of countries or educational systems in the hypothetical case that schools were all operating in the same environment.

Our results reveal several interesting issues. First, we found that although there are some similarities in the classification of countries according to their school performance in terms of efficiency and the average results obtained in PISA, the consideration of inputs involved in the educational process together with the corrections made to equalize the conditions under which they are operating produces some substantial changes in the ranking of countries. Thus, we can find both nations with good results in PISA that are placed below the average in terms of efficiency (e.g., the Netherlands or Ireland) as well as countries with bad results in PISA that obtain average efficiency levels above the average (e.g., Tunisia, Brazil or Turkey). Second, we notice that the influence of the contextual factors included in our model seems to be in line with previous evidence in the literature of the determinants of efficiency and academic results. In particular, we observe a favorable effect for private ownership, small classes, the size of the school, using ability grouping, having accountability systems and having higher proportions of certified teachers and a negative impact of being located in a rural area, and having a high proportion of students who have repeated or who frequently skip classes.

Although these findings provide some interesting insights for the analysis of school efficiency, more research is still needed to further explore the results discussed here. For instance, we should also explore the potential influence of cross-country heterogeneity by incorporating some additional contextual factors at the country level, since some previous studies have suggested that those variables have a more relevant impact on efficiency measures than school environmental factors (Cordero et al. 2017, 2018b). Likewise, we should note that the results of the approach used in this paper cannot be interpreted causally since it would entail neglecting the potential presence of unobserved heterogeneity in school performance. For instance, we totally ignored the potential accumulative impact of inputs, since our results are based on cross-sectional data. In this sense, it is worth mentioning that some authors have developed methods to address the issue of unobserved heterogeneity and endogeneity in frontier estimations (Cazals et al. 2016; Simar et al. 2016), thus a potential area for further research could be applying these methods to derive more accurate estimates of performance that take those latent factors into account.

References

Afonso, A., & St Aubyn, M. (2006). Cross-country efficiency of secondary education provision: A semi-parametric analysis with non-discretionary inputs. *Economic Modelling, 23*(3), 476–491.

Agasisti, T. (2014). The efficiency of public spending on education: An empirical comparison of EU countries. *European Journal of Education, 49*(4), 543–557.

Agasisti, T., & Zoido, P. (2018). Comparing the efficiency of schools through international benchmarking: Results from an empirical analysis of OECD PISA 2012 data. *Educational Researcher, 47*(6), 352–362.

Agasisti, T., & Zoido, P. (2019). The efficiency of schools in developing countries, analysed through PISA 2012 data. *Socio-Economic Planning Sciences*. https://doi.org/10.1016/j.seps.2019.05.002. forthcoming.

Ammermüller, A., Heijke, H., & Woessmann, L. (2005). Schooling quality in Eastern Europe: Educational production during transition. *Economics of Education Review, 24*(5), 579–599.

Aparicio, J., Cordero, J. M., & Pastor, J. T. (2017). The determination of the least distance to the strongly efficient frontier in data envelopment analysis oriented models: Modelling and computational aspects. *Omega, 71,* 1–10.

Aparicio, J., Cordero, J. M., González, M., & López-Espin, J. J. (2018). Using non-radial DEA to assess school efficiency in a cross-country perspective: An empirical analysis of OECD countries. *Omega, 79,* 9–20.

Aristovnik, A., & Obadić, A. (2014). Measuring relative efficiency of secondary education in selected EU and OECD countries: The case of Slovenia and Croatia. *Technological and Economic Development of Economy, 20*(3), 419–433.

Badin, L., Daraio, C., & Simar, L. (2010). Optimal bandwidth selection for conditional efficiency measures: A data-driven approach. *European Journal of Operational Research, 201*(2), 633–640.

Badin, L., Daraio, C., & Simar, L. (2012). How to measure the impact of environmental factors in a nonparametric production model? *European Journal of Operational Research, 223,* 818–833.

Badin, L., Daraio, C., & Simar, L. (2019). A bootstrap approach for bandwidth selection in estimating conditional efficiency measures. *European Journal of Operational Research, 277*(2), 784–797.

Bogetoft, P., Heinesen, E., & Tranæs, T. (2015). The efficiency of educational production: A comparison of the Nordic countries with other OECD countries. *Economic Modelling, 50,* 310–321.

Bray, M., & Thomas, R. M. (1995). Levels of comparison in educational studies: Different insights from different literatures and the value of multilevel analyses. *Harvard Educational Review, 65*(3), 472–490.

Cazals, C., Florens, J. P., & Simar, L. (2002). Nonparametric frontier estimation: A robust approach. *Journal of Econometrics, 106,* 1–25.

Cazals, C., Fève, F., Florens, J. P., & Simar, L. (2016). Nonparametric instrumental variables estimation for efficiency frontier. *Journal of Econometrics, 190*(2), 349–359.

Cherchye, L., De Witte, K., Ooghe, E., & Nicaise, I. (2010). Efficiency and equity in private and public education: A nonparametric comparison. *European Journal of Operational Research, 202*(2), 563–573.

Clements, B. (2002). How efficient is education spending in Europe? *European Review of Economics and Finance, 1*(1), 3–26.

Coco, G., & Lagravinese, R. (2014). Cronyism and education performance. *Economic Modelling, 38,* 443–450.

Cordero, J. M., Santín, D., & Simancas, R. (2017). Assessing European primary school performance through a conditional nonparametric model. *Journal of the Operational Research Society, 68*(4), 364–376.

Cordero, J. M., Cristobal, V., & Santín, D. (2018a). Causal inference on education policies: A survey of empirical studies using PISA, TIMSS and PIRLS. *Journal of Economic Surveys, 32*(3), 878–915.

Cordero, J. M., Polo, C., Santín, D., & Simancas, R. (2018b). Efficiency measurement and cross-country differences among schools: A robust conditional nonparametric analysis. *Economic Modelling, 74,* 45–60.

Creemers, B., & Kyriakides, L. (2008). *The dynamics of educational effectiveness: A contribution to policy, practice and theory in contemporary schools*. Abingdon, Oxon: Routledge.

Crespo-Cebada, E., Pedraja-Chaparro, F., & Santín, D. (2014). Does school ownership matter? An unbiased efficiency comparison for regions of Spain. *Journal of Productivity Analysis, 41*(1), 153–172.

Daraio, C., & Simar, L. (2005). Introducing environmental variables in nonparametric frontier models: A probabilistic approach. *Journal of Productivity Analysis, 24*(1), 93–121.

Daraio, C., & Simar, L. (2007a). *Advanced robust and nonparametric methods in efficiency analysis*. Springer, New York: Methodologies and Applications.

Daraio, C., & Simar, L. (2007b). Conditional nonparametric frontier models for convex and non-convex technologies: A unifying approach. *Journal of Productivity Analysis, 28,* 13–32.

Daraio, C., & Simar, L. (2014). Directional distances and their robust versions: Computational and testing issues. *European Journal of Operational Research, 237*(1), 358–369.

Daraio, C., & Simar, L. (2016). Efficiency and benchmarking with directional distances: A data-driven approach. *Journal of the Operational Research Society, 67*(7), 928–944.

Daraio, C., Simar, L., & Wilson, P. W. (2015). Testing the "separability" condition in two-stage nonparametric models of production, LEM Working Paper Series 2015/21.

Daraio, C., Simar, L., & Wilson, P. W. (2018). Central limit theorems for conditional efficiency measures and tests of the 'separability' condition in non-parametric, two-stage models of production. *Econometrics Journal, 21*(2), 170–191.

Daraio, C., Simar, L., & Wilson, P. W. (2019). Fast and efficient computation of directional distance estimators. https://doi.org/10.1007/s10479-019-03163-9. forthcoming.

David, R., Teddlie, C., & Reynolds, D. (2000). *The international handbook of school effectiveness research*. Psychology Press.

De Jorge, J., & Santín, D. (2010). Determinantes de la eficiencia educativa en la Unión Europea. *Hacienda Pública Española, 193,* 131–155.

De Witte, K., & López-Torres, L. (2017). Efficiency in education: A review of literature and a way forward. *Journal of the Operational Research Society, 68*(4), 339–363.

Deutsch, J., Dumas, A., & Siber, J. (2013). Estimating an educational production function for five countries of Latin America on the basis of the PISA data. *Economics of Education Review, 36,* 245–262.

Dufrechou, P. A. (2016). The efficiency of public education spending in Latin America: A comparison to high-income countries. *International Journal of Educational Development, 49,* 188–203.

Färe, R., & Grosskopf, S. (2000). Theory and application of directional distance functions. *Journal of Productivity Analysis, 13*(2), 93–103.

Florens, J., Simar, L., & van Keilegom, I. (2014). Frontier estimation in nonparametric location-scale models. *Journal of Econometrics, 178,* 456–470.

Giambona, F., Vassallo, E., & Vassiliadis, E. (2011). Educational systems efficiency in European Union countries. *Studies in Educational Evaluation, 37*(2), 108–122.

Giménez, V., Prior, D., & Thieme, C. (2007). Technical efficiency, managerial efficiency and objective-setting in the educational system: An international comparison. *Journal of the Operational Research Society, 58*(8), 996–1007.

Giménez, V., Thieme, C., Prior, D., & Tortosa-Ausina, E. (2017). An international comparison of educational systems: A temporal analysis in presence of bad outputs. *Journal of Productivity Analysis, 47*(1), 83–101.

Gustafsson, J. E. (2008). Effects of international comparative studies on educational quality on the quality of educational research. *European Educational Research Journal, 7*(1), 1–17.

Hanushek, E. A. (1979). Conceptual and empirical issues in the estimation of educational production functions. *Journal of Human Resources, 14,* 351–388.

Hanushek, E. A. (2003). The failure of input-based schooling policies. *The Economic Journal, 113*(485), 64–98.
Hanushek, E. A., & Kimko, D. D. (2000). Schooling, labor-force quality, and the growth of nations. *American Economic Review, 90*(5), 1184–1208.
Hanushek, E. A., & Woessmann, L. (2014). Institutional structures of the education system and student achievement: A review of cross-country economic research. In R. Strietholt, W. Bos, J. E. Gustafsson, & M. Rosen (Eds.), *Educational policy evaluation through international comparative assessments* (pp. 145–176). Waxmann Verlag.
Henry, K. L. (2007). Who's skipping school: Characteristics of truants in 8th and 10th grade. *The Journal of School Health, 77,* 29–35.
Jimerson, S. R. (2001). Meta-analysis of grade retention research: Implications for practice in the 21st century. *School Psychology Review, 30*(3), 420–437.
Jeong, S., Park, B., & Simar, L. (2010). Nonparametric conditional efficiency measures: Asymptotic properties. *Annals of Operations Research, 173,* 105–122.
Johnes, J. (2015). Operational research in education. *European Journal of Operational Research, 243*(3), 683–696.
Le Donné, N. (2014). European variations in socioeconomic inequalities in students' cognitive achievement: The role of educational policies. *European Sociological Review, 30*(3), 329–343.
Levin, H. (1974). Measuring the efficiency in educational production. *Public Finance Quarterly, 2,* 3–24.
Levin, H. M. (2012). More than just test scores. *Prospects, 42*(3), 269–284.
Mastromarco, C., & Simar, L. (2017). Cross-section dependence and latent heterogeneity to evaluate the impact of human capital on country performance. Discussion Paper UCL-Université Catholique de Louvain, 2017/30.
Mastromarco, C., & Simar, L. (2018). Globalization and productivity: A robust nonparametric world frontier analysis. *Economic Modelling, 69,* 134–149.
Mislevy, R. J., Beaton, A. E., Kaplan, B., & Sheehan, K. M. (1992). Estimating population characteristics from sparse matrix samples of item responses. *Journal of Educational Measurement, 29*(2), 133–161.
OECD. (2009). *PISA data analysis manual, SPSS* (2nd ed.). PISA: OECD Publishing, Paris.
OECD. (2016). *PISA 2015 Technical Report.* PISA: OECD Publishing, Paris.
OECD. (2017). *PISA 2015 assessment and analytical framework: Science, reading, mathematic, financial literacy and collaborative problem solving* (revised ed.). Paris: PISA, OECD Publishing.
O'Donnell, C., Rao, D., & Battese, G. (2008). Metafrontier frameworks for the study of firm-level efficiencies and technology ratios. *Empirical Economics, 37*(2), 231–255.
Rasch, G. (1960/1980). *Probabilistic models for some intelligence and attainment tests. Danish Institute for Educational Research* (Expanded edition 1980). Copenhagen: The University of Chicago Press.
Simar, L., Vanhems, A., & Van Keilegom, I. (2016). Unobserved heterogeneity and endogeneity in nonparametric frontier estimation. *Journal of Econometrics, 190*(2), 360–373.
Simar, L., & Vanhems, A. (2012). Probabilistic characterization of directional distances and their robust versions. *Journal of Econometrics, 166*(2), 342–354.
Simar, L., & Wilson, P. W. (2007). Estimation and inference in two-stage, semi-parametric models of production processes. *Journal of Econometrics, 136*(1), 31–64.
Simar, L., & Wilson, P. W. (2011). Two-stage DEA: Caveat emptor. *Journal of Productivity Analysis, 36*(2), 205.
Sutherland, D., Price, R., & Gonand, F. (2009). Improving public spending efficiency in primary and secondary education. *OECD Journal: Economic Studies, 2009*(1), 1–30.
Tauchmann, H. (2012). Partial frontier efficiency analysis. *Stata Journal, 12*(3), 461–478.
Thieme, C., Giménez, V., & Prior, D. (2012). A comparative analysis of the efficiency of national education systems. *Asia Pacific Education Review, 13*(1), 1–15.

Thieme, C., Prior, D., & Tortosa-Ausina, E. (2013). A multilevel decomposition of school performance using robust nonparametric frontier techniques. *Economics of Education Review, 32*, 104–121.

Todd, P. E., & Wolpin, K. I. (2003). On the specification and estimation of the production function for cognitive achievement. *The Economic Journal, 113*(485), 3–33.

Verhoeven, M., Gunnarsson, V., & Carcillo, S. (2007). *Education and health in G7 countries: Achieving better outcomes with less spending* (No. 2007-2263). International Monetary Fund.

Von Davier, M., & Sinharay, S. (2013). Analytics in international large-scale assessments: Item response theory and population models. In L. Rutkowski, M. Von Davier, & D. Rutkowski (Eds.), *Handbook of international large-scale assessment: Background, technical issues, and methods of data analysis* (pp. 155–174). London: CRS Press.

Willms, J. D., & Smith, T. (2005). *A manual for conducting analyses with data from TIMSS and PISA*. Report prepared for UNESCO Institute for Statistics.

Woessmann, L. (2003). School resources, educational institutions and student performance: The international evidence. *Oxford Bulletin of Economics and Statistics, 65*(2), 117–170.

Worthington, A. C. (2001). An empirical survey of frontier efficiency measurement techniques in education. *Education Economics, 9*(3), 245–268.

Wu, M. (2005). The role of plausible values in large-scale surveys. *Studies in Educational Evaluation, 31*(2–3), 114–128.

Wu, M. (2010). Measurement, sampling, and equating errors in large-scale assessments. *Educational Measurement: Issues and Practice, 29*(4), 15–27.

Chapter 10
A DEA Analysis in Latin American Ports: Measuring the Performance of Guayaquil Contecon Port

Emilio J. Morales-Núñez, Xavier R. Seminario-Vergara, Sonia Valeria Avilés-Sacoto, and Galo Eduardo Mosquera-Recalde

Abstract In this globalized era, the port sector has been a major influence in a country's economic growth. Ports have become one of the main funnels to enhance competitiveness in emerging markets of Latin America. Therefore, it is relevant to carry out an analysis of their performance. A good approach to measure performance is DEA, a mathematical tool that handles a benchmark analysis by an evaluation of multiple factors that describes the nature of an entity. The research herein aims to evaluate and compare the performance of the Ecuadorian Guayaquil Contecon Port in comparison with 14 major ports in Latin American and the Caribbean by using DEA. As a result of the study, the efficiency scores of the ports are analyzed to propose best practices to improve the performance of Guayaquil Contecon Port.

Keywords DEA · Ports · Efficiency · TEU · Best practices · Guayaquil Contecon Port

---

E. J. Morales-Núñez · X. R. Seminario-Vergara
Universidad San Francisco de Quito, USFQ. Diego de Robles entre Francisco de Orellana y Pampite, Cumbaya Campus Zip Code: 17-12 841, Quito, Ecuador
e-mail: emoralesn@estud.usfq.edu.ec

X. R. Seminario-Vergara
e-mail: xseminario@estud.usfq.edu.ec

S. V. Avilés-Sacoto (✉) · G. E. Mosquera-Recalde
Industrial Engineering Department, Institute of Innovation in Logistics, SCM—CATENA, Universidad San Francisco de Quito (USFQ). Diego de Robles entre Francisco de Orellana y Pampite, Cumbaya Campus Zip Code: 17-12 841, Quito, Ecuador
e-mail: svaviless@usfq.edu.ec

G. E. Mosquera-Recalde
e-mail: gemosquera@usfq.edu.ec

V. Charles et al. (eds.), *Data Science and Productivity Analytics*,
International Series in Operations Research & Management Science 290,
https://doi.org/10.1007/978-3-030-43384-0_10

10.1 Introduction

Nowadays, globalization has greatly influenced the world trading systems, and therefore the volume of international trades between countries has increased. Among all the transportation methods, used in trade activities, the seaborne freight is the cheapest transportation method, reducing the cost of making the trade of goods and natural resources between countries (Munisamy and Jun 2013). The importance of every seaport relies on the impact that it creates among the different economic activities in a country and the cities where they are located. Hence, ports play a connection role between sea and land transportation (Dwarakish and Salim 2015). Ports can be considered as "funnels" to economic development because they are catalysts and they encourage specific economic sectors (Rodrigue et al. 2017). Port cities are heavily influenced by the economic benefits, both direct and indirect, that being a port city implies. One of the direct benefits is the support in the supply chain in a city, like the import of raw materials for the manufacturing process and thus, contributing to the gross added value of local products and the reduction in the transportation costs. The indirect benefits are related to the port investment, which leads to increased economic activity and expands the market opportunity of national and international firms, as well as facilitates the access to foreign markets and get cheaper goods for the city and country (Rodrigue et al. 2017).

The port sector in a country plays a crucial role in its economy, which is particularly true when a country develops its port as a regional hub. This is the case of Guayaquil Contecon port in Ecuador located at Guayaquil city. Guayaquil city is the biggest city in Ecuador and the second most important city in the country. Guayaquil Port is the main entrance for containerized goods and products in Ecuador, handling up to 84% of the TEU's moved in the country. Guayaquil Contecon Port (GCP), is the most important port in Ecuador and the seventh ranked port in Latin America with around 2,06 million TEU's moved in 2018 (ECLAC 2019). A TEU is the acronym of the English 20-foot Equivalent Unit and it is a global transportation unit for the capacity of shipping containers (Roa et al. 2013). Ecuador is a developing country that should take advantage of GCP, and therefore, the following questions arise: What are the main differences between Guayaquil Port and other Latin American ports? What Guayaquil Port should do to reach a better efficiency? Which are the worldwide best practices in port management? Does GCP handle best practices? A benchmarking study of GCP and other Latin American ports will be carried out, with the aim to get performance measures and propose, if necessary, best practices for GCP.

Efficiency evaluations in ports have gained importance in the past years. Several studies, such as those done by Cullinane et al. (2004), Cullinane et al. (2005), Cullinane and Wang (2006), Tongzon (2001), Tongzon and Heng (2005), Barros (2003, 2006), Wanke and Barros (2016), Panayides et al. (2009), Pires (2016), Munisamy and Jun (2013), Itoh (2001), Lozano et al. (2010) demonstrates that Data Envelopment Analysis (DEA) constitutes a good approach to evaluate the performance of seaports. DEA is a mathematical benchmarking technique used to compare Decision-Making Units (DMUs), by using multiple variables in order to obtain a relative efficiency.

These variables should reflect a causal relation of the activities of the DMU. Then, the efficiency score obtained from the analysis is between 0 to 1, where 1 represents the most efficient and 0 the less efficient. Those DMU's with a score of 1 creates an efficient frontier.

The study presented here evaluates the top 15 ranked Latin American ports, including Guayaquil Contecon Port, considering the amount of TEUs moved by using DEA modeling to determine the efficiency of each port. The results obtained would help to understand how Guayaquil Contecon Port is performing and if necessary, propose best practices that could improve its development.

10.2 Literature Review

10.2.1 Port Cities

The book *Ports Economics* by Talley (2018) details properly what a port is, the various types of ports, the types of material handled, the difference between the materials, and the importance of the port as a transportation network. "A port (or seaport) is a place at which the transfer of cargo and passengers to and from vessels to waterways and shores occur" (Talley 2018). There are different types of ports, depending on the handling material. There are the cargo ports, passenger ports and, cargo/passenger ports. The cargo ports are the more common type of port, including general and bulk cargo. General cargo includes all dry non-bulk such as packaged cargo or goods of uniform size and weight. The general bulk is either a break-bulk cargo, which is cargo packaged in pallets, in wire or rope or a neo-bulk cargo, which includes automobiles, steel, lumber, or container (Talley 2018). Bulk Cargo is every dry and liquid cargo that is neither package nonuniform (Talley 2018). A port is considered a node in a transportation network, where cargo and/or passenger are moved into or out from a city/country connecting the city to the world (Talley 2018). However, all the types of cargos are mostly measured in TEUs, which is the worldwide measure used for a container or bulk movement within a port.

Port management has become a complex entity that affects directly various locations and functions of the mainland authorities due to the interaction between their maritime functions and foreland trade relationships that are directly related to the land-based urban system relationships (Ducruet and Lee 2006). To understand the port–city relation, Ducruet and Lee (2006) proposed a relationship matrix (Fig. 10.1) based on the concepts of the centrality of a city and the intermediacy level.

Figure 10.1 includes two main diagonals, from which, the first diagonal (top-left to bottom right) shows the balanced progression between the port and the city, from small ports and villages to port metropolis such as New York, Tokyo, and Hong Kong (Ducruet and Lee 2006). The other diagonal (Bottom-left to top-right) shows the unbalanced result of growth for city or a port, leaving behind the port, or city development, respectively. All the elements presented in this matrix represent port

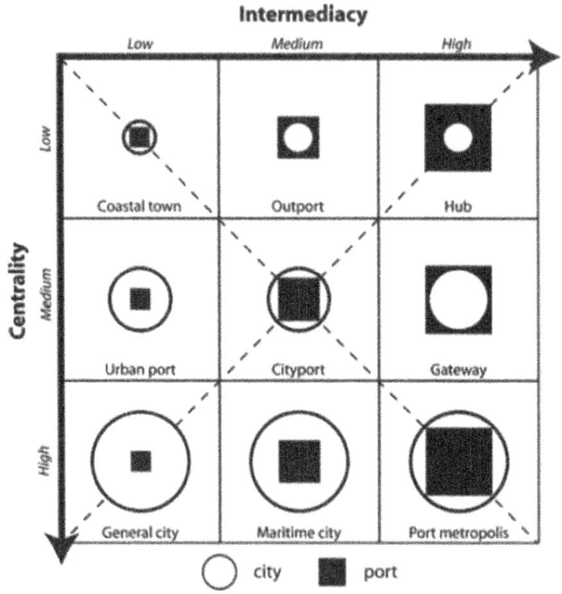

Fig. 10.1 Matrix of port city relation (Ducruet and Lee 2006)

cities, but the name assigned depends on the relation and influenced from the city to the port or vice versa.

10.2.2 Data Envelopment Analysis (DEA)

10.2.2.1 DEA and the CCR Model

DEA is a nonparametric technique based on linear programming to measure the relative performance of different units or entities, where the presence of multiple inputs and outputs makes comparison (Boussofiane et al. 1991). DMUs or Decision-Making Units are the entities of comparison, from which a performance/efficiency measures are obtained; those results are scalar measures (Charnes et al. 1978). The basic model states that with *n* decision-making units (DMU) the efficiency obtained for any DMU is obtained as the maximum of a ratio of weighted outputs to weighted inputs subject to the condition that the similar ratios for every DMU be less than or equal to unity (Charnes et al. 1978). This model is represented mathematically as follows:

$$Max\, G_0 = \frac{\sum_{r=1}^{s} u_r y_{ro}}{\sum_{i=1}^{m} v_i x_{io}} \tag{10.1}$$

Subject to:

$$\frac{\sum_{r=1}^{s} u_r y_{rj}}{\sum_{i=1}^{m} v_i x_{ij}} \leq 1; j = 1, \ldots, n \tag{10.2}$$

$$u_r, v_i \geq 0; \tag{10.3}$$

$$r = 1, \ldots, s; \quad i = 1, \ldots, m.$$

where G_0 characterizes the efficiency score for every DMUj. The value for y_{rj} represents the value of the various outputs r for each DMUj used. At the same time, x_{ij} are the amount of inputs i to each DMUj. DEA uses weighted inputs and outputs so the model can calculate efficiency of every DMUj. Finally, n is equal to the number of units, s is equal to the number of outputs, m number of inputs. This model represents a fractional linear model that needs to be converted into a linear form. Several authors, such as (Munisamy and Jun 2013), (Cullinane et al. 2004), (Pires 2016) agree that model (10.1) should be transformed into a linear form. Charnes, Cooper, and Rhodes transformed the fractional model (Charnes et al. 1978) as follows:

$$Min\ G_0 = \sum_{i=1}^{m} v_i x_{io} \tag{10.4}$$

Subject to

$$\sum_{i=1}^{m} v_i x_{ij} - \sum_{r=1}^{s} u_r y_{rj} \geq 0; j = 1, \ldots, n \tag{10.5}$$

$$\sum_{r=1}^{s} u_r y_{ro} = 1 \tag{10.6}$$

$$u_r, v_i \geq 0 \tag{10.7}$$

$$r = 1, \ldots, s; \quad i = 1, \ldots, m.$$

If a DMU where $G_i = 1$, it can be considered as an efficient DMU, and it is located at the efficient frontier. If $G_i < 1$ the DMU is considered as inefficient, and the difference between an inefficient DMU and the efficient frontier is known as the slack.

The CCR model, introduced by Charnes, Cooper, and Rhodes in 1978, measures the relative efficiency of an entity often referred to as a Decision-Making Unit (DMU) (Leung et al. 2016). The CCR model assumes constants returns to scale so that all observed production combinations can be scaled up or down proportionally (Cullinane et al. 2004).

DEA models can be oriented in two ways, namely input-oriented and output-oriented. The input-oriented model objective is to minimize the inputs while producing at least the given output levels (Cooper et al. 2007). The models presented previously referred to the input-oriented model (10.4), The output-oriented model attempts to maximize outputs while using no more than the observed amount of any input (Cooper et al. 2007). The model is presented next (Munisamy and Jun 2013).

$$Min\, G_o = \sum_{s=1}^{m} v_i x_{io} \tag{10.8}$$

Subject to:

$$-\sum_{i=1}^{m} v_i x_{ij} + \sum_{r=1}^{s} u_r y_{rj} \leq 0; j = 1, \ldots, n \tag{10.9}$$

$$\sum_{r=1}^{s} u_r y_{ro} = 1 \tag{10.10}$$

$$u_r, v_i \geq 0 \tag{10.11}$$

$$r = 1, \ldots, s; \quad i = 1, \ldots, m.$$

Figure 10.2 details both the efficient frontiers for the input and output orientation. The difference between the efficient frontier and the inefficient DMU is known as the "input-slack or output-slack".

The slacks in DEA models represent the difference needed, input-wise or output-wise, for a variable to achieve efficiency. Charnes, Cooper, and Rhodes in 1978 develop a second DEA model that includes the slack variables by using the dual

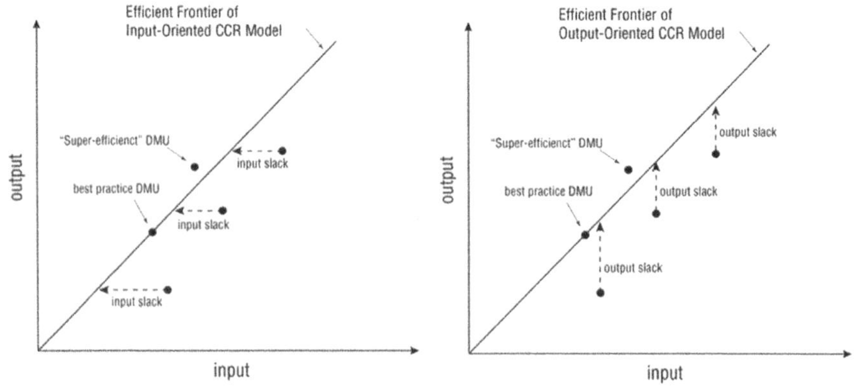

Fig. 10.2 CCR efficient frontier for input-oriented, output-oriented models (Leung et al. 2016)

Table 10.1 DEA models for slack calculation (Cooper et al. 2007)

Model (10.12)	Model (10.13)
Input-oriented	Output-oriented
$\min \theta - \varepsilon\left(\sum_{i=1}^{m} s_i^- + \sum_{r=1}^{s} s_r^+\right)$	$\min \theta - \varepsilon\left(\sum_{i=1}^{m} s_i^- + \sum_{r=1}^{s} s_r^+\right)$
Subject to	Subject to
$\sum_{j=1}^{n} \lambda_j x_{ij} + s_i^- = \theta x_{i0};\ i \in I$	$\sum_{j=1}^{n} \lambda_j x_{ij} + s_i^- = x_{i0};\ i = 1, 2, \ldots, m$
$\sum_{j=1}^{n} \lambda_j x_{ij} + s_i^- = x_{i0};\ i \notin \mathrm{I}$	$\sum_{j=1}^{n} \lambda_j y_{rj} - s_i^+ = \theta y_{i0};\ r \in O$
$\sum_{j=1}^{n} \lambda_j y_{rj} - s_i^+ = y_{r0};\ r = 1, 2, \ldots, s$	$\sum_{j=1}^{n} \lambda_j y_{rj} - s_i^+ = y_{r0};\ r \notin \mathrm{O}$
$\lambda \geq 0; j = 1, 2, \ldots, n$	$\lambda \geq 0; j = 1, 2, \ldots, n$

problem. The slack variables evaluate the degree of possibilities in order to become DEA efficient and represent the values of excessive inputs or output shortfalls (Itoh 2002).

Table 10.1 shows the slack model for the input-oriented (Model (10.12)) and output-oriented (Model (10.13)) where s_r^+, s_i^-, represent the output shortfalls and the input excesses, respectively, (Cooper et al. 2007). The models are the result of a two-phase linear programming, using the dual problem to discover the input/output slacks with the objective of finding a solution that maximize s_r^+, s_r^- without altering the optimal θ^* or φ^*. An optimal solution is defined when $s_r^+ = s_r^{+*}$; $s_i^- = s_i^{-*}$; $\lambda = \lambda^*$ then the solution is called a max-slack solution and if $s_r^+ = 0$; $s_r^- = 0$ and the solution is the called optimal then it is called zero-slack (Cooper et al. 2007).

10.2.2.2 Banker, Charnes, and Cooper Model (BCC)

In 1984, Banker, Charnes, and Cooper proposed a new approach to the previous CCR-DEA models, also known as the Variable Return to Scale (VRS). This model considers the increasing return to scale (IRS), the constant return to scale (CRS), and the decreasing return to scale (DRS) in the efficient frontier while the CCR model only considers the (CRS). The BCC model presents a linear-concave characteristic in the efficient frontier graphic. Both frontiers are shown in Fig. 10.3.

The Variable Return to Scale (VRS) makes the efficient frontier changes from a straight line in the CCR model to a convex line. The BCC model also has an input-oriented version and an output-oriented version. Both satisfy the same logic explained in the CCR model. The BCC, output-oriented, model is presented as follow:

$$Max\ \varphi \tag{10.14}$$

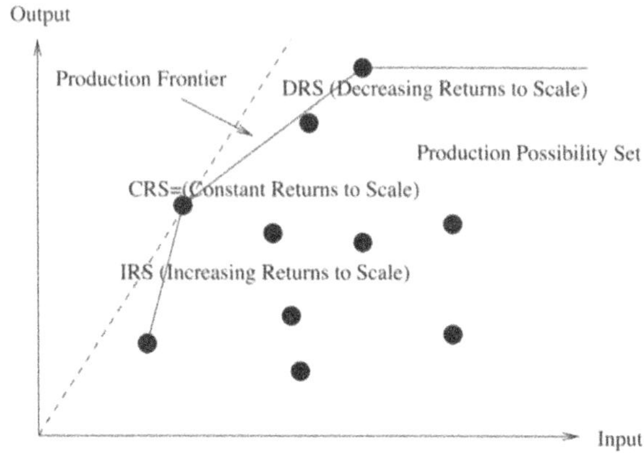

Fig. 10.3 BCC efficient frontier model (Cooper et al. 2007)

Subject To:

$$\sum_{j=1}^{n} \lambda_j x_{ij} = x_{io};\ i = 1, \ldots, m$$

$$\sum_{j=1}^{n} \lambda_j y_{rj} = \varphi y_{ro};\quad r \in O$$

$$\varphi \geq 0; \sum \lambda = 1$$

The main difference between CCR and BCC is the associate restriction regarding to λ, from $\lambda >= 0$ to $\sum \lambda = 1$. In model (10.14), the efficiency scores obtained for φ are $\varphi \geq 1$, where $\varphi = 1$ is considered as efficient and, when $\varphi > 1$ is considered as inefficient. In order to obtain scale efficiency from 0 to 1, the φ score obtained needs to be transformed. The new efficiency score is $\theta = 1/\varphi$, giving as a result a value between 0 and 1. The scores obtained using the BCC model are known as "pure technical efficiency scores" because the model eliminates the "scale part" of the analysis (Jemric and Vujcic 2002).

10.2.3 DEA Studies in Seaports

The number of studies regarding seaport analysis using DEA has increased in a significant matter over the years. Roll and Hayuth started this topic in (1993) using DEA to analyze and compare port performance between 20 different DMUs (Roll and Hayuth 1993). From there, many studies of port efficiency have been done, i.e., the study of 53 ports between Asia and the United States (Cullinane et al. 2005), the study of the north east's ports of Brazil in the year 2010 analyzing 15 ports (Cazuza de Sousa Junior 2010), and later upgraded by a new study by Wanke and Barros in

2016 analyzing 27 ports of the coast of Brazil (Wanke and Barros 2016). Moreover, a study of 30 Latin American ports and comparing their efficiency over the years 2000 to 2008 using DEA by Munisamy and Jun (2013), or a study of the efficiency of 11 Korean ports using DEA by Ro-kyung Park and Prabir De in 2004 (Park and De Prabir 2004), and many others. A crucial part of the study is the variable selection. The studies mentioned previously and others have used different variables. Table 10.2 details the variables that each author used for their studies.

10.2.4 Smart Port Cities and Best Practices

"Smart Cities are smart once their inhabitants are supported in their everyday lives in such a way that they don't even realize how it's done."—Ralph Grothmann, Siemens Corporate Technology (Rockel 2017). Smart port cities do not only involve renovation in the structure of the port city or installing more equipment or having the best technology, but also includes a change of the culture and the mind of the habitants of the port cities.

Port Cities are well known for their potential for growth because they are critical for commerce. They constitute a keystone for economic growth and play a significant role in the goods movement supply chain (EPA 2018). Every port city in a country has an important influence on the economy. For example, in the South African economy, the overall impact effect of the port sector per unit shortage on all other products was found to be 1.1705. This means that one-unit shortage in the port sector would have incurred a 17% loss to the entire economy in 2002 (Chang et al. 2014). Moreover, Guayaquil Contecon Port accumulated a utility of $45 million from 2016 to 2017 (Logística 2017). Ports traditionally serve as economic catalysts for surrounding cities (Zhao et al. 2017). This shows how important a port can be for a country and its economy, and it is important to analyze it and have the best port quality possible so that a country has economic growth.

Jens Meier, Hamburg Port Authority board chairman, states that is no good to continue building roads, canals or railroads to go wherever people want, instead, a smart port city project should focus on improving the existing routes and the quality of port services (Rockel 2017). The way of improving a port and becoming the level of a smart port city is by the adoption of best practices. A best practice is a technique that, through studies or experiences, can produce better results. There is a wide range of best practices. For example, there are policy best practices, process best practices, information best practices, technology best practices, and organizational best practices (Axson 2010). A brief description of the best practices in ports will be detailed forward in the study.

1. Best Practice in Ballast Waters

Every ship has tanks at the bottom of the ship called ballast tanks, where water is stored to maintain the ship's stability and balance while sailing. This retained water is called ballast water, and it is constantly retrieved and released. The International

Table 10.2 DEA studies in Seaport

Author	Year	Paper title	DEA model	Outputs	Inputs
Culliname, Kevin	2005	The Application of Mathematical Programming Approaches to Estimating Container Port Production Efficiency	CCR and BCC	Throughput (TEU)	Quay length (m)
					Terminal area (ha)
					Quayside gantries (no.)
					Yard gantries (no.)
					Straddle carriers (no.)
Park, Ro-Kyung and De, Prabir	2004	An Alternative Approach to Efficiency Measurement of Seaports	CCR and BCC	Cargo throughput (ton)	Berth capacity (no. of ships)
				No. of ships call	Cargo handling capacity (million tons)
				Customer satisfaction (score)	Revenue (billion)
Wanke, P and Barros, C	2016	New Evidence of the determinants of efficiency at Brazilian Ports: a bootstrapped DEA analysis	CCR, BCC & SESE	Solid bulk loading hours (per year)	Quay length (m)
				Container loading hours (per year)	Maximal quay depth (m)
				Solid Bulk throughput (tons/year)	Number of berths
				Container throughput (containers/year) Solid bulk frequency (ships/year)	Warehousing area (sq. m)Yard area (sq. m)
				Container frequency (shipments/year)	Channel width (m)
					Channel depth (m)
Tongzon, Jose and Heng, Wu	2005	Port privatization, efficiency and competitiveness: Some empirical evidence from container ports (terminals)	CCR	Throughput (TEU)	Quay length (m) Terminal surface (ha)
					Terminal surface (ha)
					Number of container quay cranes
					Port size (ha)

(continued)

Table 10.2 (continued)

Author	Year	Paper title	DEA model	Outputs	Inputs
Roll, Y and Hayuth, Y.	2006	Port performance comparison applying data envelopment analysis (DEA)	BCC	Cargo throughput (ton)	Manpower
				Level of service	Capital invested
				Users satisfaction	Cargo uniformity
				Ship calls	
Munisamy, S and Jun, O.	2013	Efficiency of Latin American Container Seaports using DEA	CCR and BCC	Throughput (TEU)	Berth length (m)
					Terminal area (sq. m)
					Quay equipment
					Yard gantries (no.)
					Yard equipment
					Number of forklifts and yard tractors
Barros, Carlos	2006	A Benchmark Analysis of Italian Seaports Using Data Envelopment Analysis	CCR and BCC	Liquid Bulk	Personnel
				Dry Bulk	Capital invested
				Number of ships arrival and departures	Operational costs
				Passengers	
				Containers with TEU	
				Containers without TEU	
				Sales	
Lozano, S., Villa, G. and Canca, D.	2010	Application of centralized DEA approach to capital budgeting in Spanish ports	CCR and BCC	Port traffic (tons)	Land area (sq. m)
				TEUs	Quay length (m)
				Ship calls	Number of cranes
					Number of tugs

Convention for the Control and Management of Ships' Ballast Water and Sediments focus on minimize the risk of invasive introductions of damaging aquatic species and pathogens and subsequent dangers to the environment and human health (IMO 2017). Under the IMO convention, ships need to have ballast water management plans to minimize the insertion of pathogens and marine species. When loading ballast water, the ship should maintain practices that minimize the uptake of invasive. For example, ships should not load in very shallow waters or in the dark when bottom organisms move upwards (Bacchioni and Ramus 2008). Ships should avoid the unnecessary release of ballast water and the tanks should be cleaned regularly. Every port needs to have a facility strictly made for ballast water reception for ships to discharge water into this location (IMO 2017).

2. Best Practices in Land Operations

If there is a port construction and construction operations inside the port, due to a land improvement, a machine installment, an expansion or any other type of construction, shoreline, and land impacts should be assessed, so that there is minimal impact on the land and water, especially those areas of high biodiversity or with endangered species (IFC-WBG 2017). Green belts and open areas around the port should be increased for the conservation of water, energy, and absorbing air and water pollution, lessening noise diffusion, lessening erosion, and producing an aesthetic background (Gupta et al. 2005). Best practices in this area require a strong commitment to risk prevention about the environment. All pipes, valves, and hoses that are used most have periodic quality inspections so that they don't have any holes, cracks or perforations. Also, all industrial safety and security occupational rules must be applied, like giving always personal protection equipment to all workers, have always fire extinguisher equipment around the installations, all toxic or chemical material must be well signalized. Finally, another practice is that all periodic cleaning of the fueling area and all process of refueling any equipment in water must be done with natural light, from 6:30 am to 6:00 pm maximum, as if any unfortunate case of spills or accidents, instant active measures can be applied more effectively in daylight (Contecon 2012).

3. Best Practices in Dredging

Dredging refers to the removal of sediments and soils from the bottom of lakes, rivers, harbors, and other water bodies. As sedimentation is a normal natural process that gradually fills channels and harbors, it is necessary to make a routine of dredging to avoid problems in channels when ships arrive at docks. Dredging process should be carried out only when it is required, i.e., new infrastructure is needed, a creation and maintenance of safe navigation channels, and environmental purposes. The most appropriate dredging method should be used to minimize sediment suspension and destruction of habitat (Bacchioni and Ramus 2008).

4. Best Practices with Dust management

When talking about dust it refers to any material that can be easily spread in the air and cause pollution and contamination. It is common that ships transport bulk cargo that consists basically in granular material is responsible for dust generation. Dust must be controlled at all stages of the process with a docking ship, when loading and unloading materials into or out of a ship, and even when storing in port. Best practices in this subject to prevent dust dispersion includes installation of dust suppression mechanisms, including water spray and covered storage areas; Telescoping chutes best control dust when loading, along with continuous screw conveyor for unloading; Freefall of material should be avoided, and cargo that contains dry material piled up should remain low heights; When material isn't being moved or handled, it should always be covered (IFC-WBG 2017). A way of controlling dust is adapting the facility so that is designed to have a set of double doors for the trucks to enter. This system requires that when one door opens, another door closes. This guarantee that one door is always closed preventing exposure to wind and the spreading of dust (Bacchioni and Ramus 2008).

5. Best Practices in Emissions

Seaports are major contributors to emissions of various types of substances, such as nitric oxide, sulfur monoxide, and other air toxics (EPA 2018). Because of this, good practice should include the tracking of emissions by the creation of an emissions inventory, which could help to assess and notice the impacts of the port activities and provide a baseline for developing emission mitigation strategies for each specific emission and control the performance over time. An emissions inventory is necessary for port authorities to understand and quantify the air quality impacts of every port operations.

6. Best Practices with Spills

Spills are one of the most dangerous forms of contamination and pollution to water and oceans. Whether it is oil, garbage or petroleum, it causes huge toxic damage to water quality and living species. The IMO Manual on Oil Pollution states that spill prevention, control, and countermeasure plan is crucial that every port should have (IMO 2004). A good prevention plan includes identifying areas sensitive to spills, detailing responsibilities for managing spills, preparing specialized spill response equipment such as containment booms, and training all personnel so is capable and qualify in any spill response measures (Bacchioni and Ramus 2008).

10.3 Methodology

10.3.1 How to Approach DEA

A DEA study includes several steps. Kevin Cullinane and Teng-Fei Wang (2006) provide a guide regarding the DEA model application. Figure 10.4 details the several steps that this guide includes.

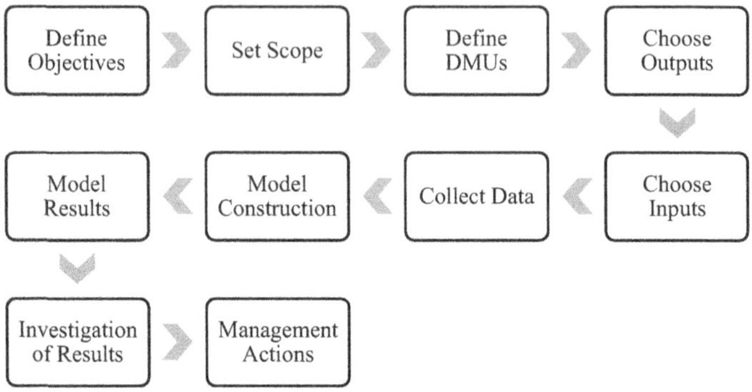

Fig. 10.4 DEA methodology steps

1. **Define Objective and Goals**: The first step for using DEA is to define and identify the objectives and goals of the study. Why this study is important? What is the purpose of the study? What does the researcher want to obtain or prove through the study? This step is very important because a DEA application project can become very overwhelming.
2. **Set Scope**: The second step is to set the scope of the study. The second phase of any project according to project management theory is the planning, which includes time scope, a resources utilization plan, budget planning, and the quality and the quantity of the work (Kerzner 2017).
3. **Define DMUs**: The third step is to choose the DMUs of the study. The DMUs are all the units that the researcher wants to evaluate. The number of DMUs, a researcher must consider some requirements that exist for a DEA study. This number of DMUs for an upper limit there are no restrictions, however, for lower limits, there exist rules to comply. This is because with too few DMUs to compare, all the DMUs may result in optimal, which would never be the goal for a study. In literature, there are many perspectives of many authors of these minimum requirements of DMUs that must be used for a DEA study. Joe Zhu and Wade Cook in their book of Modeling Data Irregularities in DEA (Zhu and Wade 2007) made a great literature review on all studies that shows the minimum number of DMUs that should be used. Among this literature review, many authors appear such as Boussofiane et al. in 1991 establishing that the minimum number of DMUs should be at least the multiplication of the number of inputs and the number of outputs; Golany and Roll in 1989 explained a rule of thumb that this minimum number is twice the sum of the number of inputs and outputs; Friedman and Sinuary-Stern in 1998 stated that the number of DMUs should be at least three times the sum of inputs and outputs. Cooper et al. in 2000 recommends that the minimum number should be the maximum between the number of Friedman and Sinuary-Stern and the result from the multiplication between the number of inputs and outputs (Cullinane and Wang 2006) as shown in the formula above:

$$Min\, number\ of\ DMU = Max\begin{bmatrix} 3 * (inputs + outputs), \\ inputs * outputs \end{bmatrix}$$

4. **Choose output variables**: The next step is to define the output variables. This step is extremely important because a badly defined output variable will draw inconclusive or erroneous conclusions on the study. The output variable must agree with the objectives of the study, and that truly represents the situation of the DMUs
5. **Choose input variables**: This fifth step is equally important as the previous step. The inputs are the variables that every DMU in the study has as resources and that are going to be compared between each other for every DMU. These inputs must agree with the output variable and with the objective of the study. Moreover, this input variable must be related in some way that explains the output chosen.
6. **Collect the data**: This sixth step is crucial for the study, as data is very important for constructing and running the model. DEA model can't be performed if a data value for a given input or output is null. DEA can't handle missing values. If this situation happens, the researcher or experimenter has three choices: First, replace the input or output variable with another variable that still follows the objectives of the study and that have available data to collect. A second option is to simply eliminate the variable from the model. Finally, a third choice is to eliminate that DMU. This choice should be the last option as a DEA model work better with greater numbers of DMUs because it has more units to compare, and it is relatively easier to replace an input or output variable than a DMU.
7. **Model construction**: This step is one of the more technical parts. Resolving the model by hand is practically impossible, as there are too many variables to solve. If n DMUs are going to be analyzed, then there will be $(n + 1) * n$ variables that need to be found. For this, softwares like AMPL, Python, and Matlab are recommended to solve the DEA model.
8. **Model results**: In this step, the experimenter retrieves the results from the model construction programming. The results would give a phi value of efficiency and n number of lambdas, being n the number of DMUs in the study. This phi value could be 1 if DMU is efficient, or greater than 1 of the DMU is not efficient—from the output orientation perspective. For an input-oriented point of view, the efficiency value is theta instead of phi. When the score obtained is equal to 1, the DMU assign to that theta is efficient, but if the score is less than 1 the DMU is considered not efficient.
9. **Investigation of results**: After getting the model results, the step that follows for every experimenter should be an in-depth analysis of the results. Not only observe those DMUs that are efficient, but also investigate and question why the DMUs are efficient or why not. Results should adjust to reality.
10. **Management actions**: In this final phase, all the objectives and goals of the study are resolved. Why this study was made? What does the experimenter want to get and do with the results? This is where commonly best practices enter a study. If a DMU is not efficient, what needs to be done to make it efficient? What does a resulting efficient DMU do differently than an inefficient DMU that it

could apply and start performing? For this, best practices from those efficient DMUS should be analyzed. In this way, the efficiency can be explained by tangible practices, and then propose how to implement them for an inefficient DMU to make them improve. Those practices should be realistic improvement proposals.

10.4 Case Study: Guayaquil Contecon Port

10.4.1 Define the Objective: Case Study

Guayaquil Contecon Port (GCP) is the most important port in Ecuador, which handles up to 84% of all the containerized and non-containerized goods that enter the country (El Telégrafo 2013). Being the most important port in Ecuador, GCP is an important contributor to business development in the country. Additionally, GCP is not only facilitating business development within the country, but the port is also competing against other ports of the region to be more profitable. To become more attractive to international corporations, GCP needs to be efficient so it is able to offer lower costs for the investors to used GCP. The efficiency and performance would cause costs reduction in operations that would improve exportation and importation rate, which benefits the local business (Rodrigue et al. 2017). For these reasons the study is going to use DEA to benchmark GCP against some of the best Latin American ports to understand how efficient GCP is against the other ports.

10.4.2 Define the Scope

The scope of this study is to compare the efficiency of the Ecuadorian Guayaquil Contecon Port against some of the top-ranked ports in Latin America and the Caribbean continent.

10.4.3 Define the Units: Port Selection

To select the ports in the study to make the benchmarking analysis using DEA, the ranking of the 20 top ports of Latin America and the Caribbean in 2018 was used. It is important to mention that because of data availability and access to information during the literature review and data collection phase, only 15 ports of this ranking were chosen, including the port of Guayaquil. It was important that each port selected had available information to research for each variable selected, because if a piece of data of any variable of any port were missing, the results of the study would

Table 10.3 List of selected ports

Country	Port	ID
Ecuador	Guayaquil	1
Panama	Balboa	2
	Colon	3
Chile	Valparaíso	4
	San Antonio	5
Argentina	Buenos Aires	6
Uruguay	Montevideo	7
Peru	Callao	8
Colombia	Buenaventura	9
	Cartagena	10
Costa Rica	Limon-Moin	11
Jamaica	Kingston	12
Mexico	Manzanillo	13
	Veracruz	14
Brazil	Tecon Santos	15

be compromised and as so, its validity. Each of these ports is going to be a DMU for the study, so DEA will work with 15 different DMUs to compare and analyze, fulfilling the requirement of a minimum number of DMUs posed by Golany and Roll in 1989 (Zhu and Wade 2007). These ports were from different countries around Latin America and the Caribbean and are shown in Table 10.3.

A specific ID was added to each of these ports in no specific order to be shown in a world map in Fig. 10.5.

10.4.4 Selection of Output and Input Variables

One of the most important parts of the study is the variable selection. For the DEA model to work properly when evaluating the efficiency of Guayaquil Port and the rest of the selected ports of Latin America, the variables selected must apply for all the DMUs. There are two different types of variables, the input variables, and the output variables. The input and output variables should reflect the actual objectives and represent the process of container port production as accurately as possible (Lu and Wang 2017).

In this study, the chosen objective is maximizing the performance of each port by maximizing the outputs and keeping the inputs constant. For this, a clear stated variable as an output must be chosen to explain the behavior of the port, a variable that truly represents the actual state and condition of the port. This variable must also be universal, meaning that it should apply for every port. Among the literature in

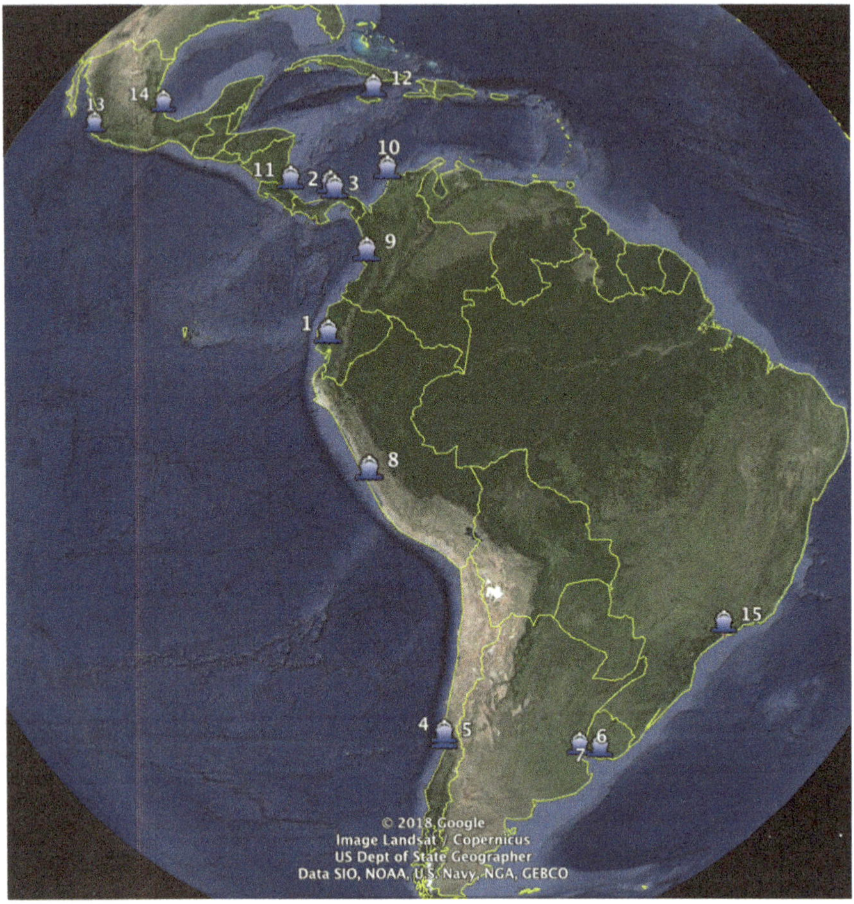

Fig. 10.5 Location of each of the 15 ports in the study (Done by the authors)

studies of ports, several researchers including Cullinane (2004) from the University of Gothenburg, state that the ideal variable for outputs in studying ports is the number of TEUs moved per year. It is useful to use moved TEUs as a measure of the efficiency of ports because it shows how much the port produces, is the output product, and the final goal of every port. Moreover, international commissions such as the Economic Commission of Latin America and the Caribbean (ECLAC 2018) updates a ranking of Latin America Ports every year based on the container throughput meaning the TEUs moved per year. For this, the first output variable is the container throughput measured in TEUs per year.

After defining this first output, another thing to consider is that it is not the same moving a thousand TEUs while having a hundred quayside cranes, then moving the same amount of TEUs while having only ten quayside cranes. There is a difference because the quayside crane productivity in both cases will be distinct if the same number of TEUs are moved, the fewer the cranes the more productive they are. This idea is followed by a FAL bulletin made by CEPAL about the productivity of the assets at container terminals in Latin America and the Caribbean between 2005 and 2013. It defines the quays productivity of the ports by dividing the TEUs moved by the number of quays a port have (Doerr 2014). Extending this idea, two more outputs where defined: First is the quayside cranes productivity, and second the port area productivity. The area productivity follows the idea that is not the same to move a thousand TEUs in a small port of 100,000 square meters, than moving the same amount of TEUs while being a larger port of 1,000,000 square meters. Moreover, FAL Bulletin also defines area productivity as a proper measure of defining overall port productivity. For this reason, these two new outputs where defined each one measured by TEUs/quayside cranes, and TEUs/sq meter, respectively. However, to obtain the values of these variables, two new variables were added.

1. The first variable is the Total Port Area, which relates to the total area that belongs to the port. This variable is important to determine the capacity of the port itself. A port too small will never be able to satisfy a big demand for containers. This variable was measure in square meters.
2. The second variable is the number of quayside cranes, which is the number of cranes that the port uses to move the containers from the ships to the port. This variable is important due to its crucial part in the container unloading process.

Once the data of these two variables were obtained for each port, the number of TEUs moved, data from the first output, had to be divided by the number of quayside cranes and port area for every port. For example, the TEU throughput for Guayaquil Contecon Port is about 3 million, and Guayaquil has 25 cranes and a total area of 1,8 million square meters. The quayside crane productivity and area productivity would be calculated in the following way:

$$Quayside\ Crane\ Productivity = \frac{Throughput\ of\ TEUs}{Number\ of\ Quayside\ Cranes}$$

$$Quayside\ Crane\ Productivity = \frac{3,000,000}{25} = 120,000\ ,\ TEUs/crane$$

$$Area\ Productivity = \frac{Throughput\ of\ TEUs}{Total\ Port\ Area}$$

$$Area\ Productivity = \frac{3,000,000}{1,800,000} = 1.67\ TEUs/sqmeter$$

In this way, the output productivity variables could be obtained for every one of the 15 ports in the study.

For the input variables, all the variables should be available for every port in the study. These variables are as follows:

Table 10.4 Selected input and output variables for study

Variable	Type (Input/out)	Reference
TEU's moved	Output	Cullinane et al. (2004), Wanke and Barros (2016), Munisamy and Jun (2013)
Area productivity	Output	FAL Bulletin by CEPAL (2014)
Quayside cranes productivity	Output	FAL Bulletin by CEPAL (2014)
Number of berth	Input	Wanke and Barros (2016), Tongzon and Heng (2005)
Total berth length	Input	Wanke and Barros (2016), Lozano et al. (2010), Munisamy and Jun (2013)
Container storage area	Input	Lonza and Marolda (2016)

1. First, the container storage area, which is the total area that the ports give for container storage. This is important because it represents the total capacity of the port to store containers. This variable was measured in square meters.
2. Second, the number of berths, which are the total number of berths of quays that the port has for container reception. This variable is important because it is directly proportional with the container capacity reception.
3. Third, the total berth length, which is the total length of the sum of all the berths measured in meters. The total berth length is of great importance because with a longer berth, the greater the capacity of the port for reception of shipments. This variable was measured in meters.

All chosen variables for the study are shown in Table 10.4, each one with the respective author that has used it before in a previous DEA study.

10.4.4.1 Google Earth Mapping System

Total Port Area and Container Storage Area (CSA) are variables considered by many researchers to be one of the most important variables in seaports efficiency calculations. Although some information regarding these variables can be obtained from port authorities' web pages, the terminal area and container storage area requires special treatment. For them, the areas of the ports must be mapped of all terminals and container storages using the Google Earth Pro mapping system. It is important to consider that some images from Google Earth Pro are not updated because some countries can report that this can affect their national security (Dunn 2018) However, it was decided to use the images available from the software for the analysis.

In order to understand what a port terminal is, it is necessary to consider both, the area containing port operations and the administrative offices (Dunn 2018). Figure 10.6 shows the Buenaventura port in Colombia (referred to as DMU 9 in Table 10.3) colored in blue which determines the considered terminal area.

For the container storage area, the same procedure was used as for Total Port Area (Google Earth Pro Mapping system). In this case, the areas where containers could

Fig. 10.6 Buenaventura port in Colombia (Done by the authors)

Fig. 10.7 Container storage area from Guayaquil Contecon Port (Done by the author)

be stored were analyzed. These areas include warehouse or open areas, and they have been colored in yellow, as shown in Fig. 10.7.

10.4.5 Collect Data

One of the most important parts of the study was to obtain all the data so that the model proposed could be run. For this, an intense research for every port was made until all the data was collected. The main sources for gathering all the information

Table 10.5 Collected Data for every port

Port	Outputs			Inputs		
	$Y_{1j}\left(TEU/_{year}\right)$	$Y_{2j}\left(TEU/_{crane}\right)$	$Y_{3j}\left(TEU/_{m^2}\right)$	$X_{1j}(m^2)$	X_{2j}	$X_{3j}(m)$
Guayaquil	2,064,281	233,949	1.65	273,270	10	1,625
Balboa	2,520,587	119,465	1.64	400,000	7	1,710
Colon	4,324,478	299,324	5.24	278,000	6	1,258
Valparaiso	903,296	119,304	2.15	358,848	8	1,611
San Antonio	1,660,832	162,111	0.91	290,000	9	800
Buenos Aires	1,797,955	91,810	0.71	378,390	26	3,000
Montevideo	797,880	134,204	0.85	238,264	9	638
Callao	2,340,657	225,022	1.49	264,561	7	560
Buenaventura	1,369,139	65,714	0.62	190,000	13	2,029
Cartagena	2,862,787	95,643	3.04	900,000	8	1,700
Limon-moin	1,187,760	199,938	1.86	400,000	9	1,545
Kingston	1,833,053	82,105	1.53	526,875	9	2,310
Manzanillo	3,078,505	117,932	6.48	259,423	14	2,164
Veracruz	1,176,253	124,145	0.66	327,492	11	2,935
Tecon Santos	3,836,487	255,585	1.23	980,000	4	1,290

was the port authority's website for every port. In some cases, other literature about ports like international rankings, papers of study, and official government websites were needed to collect all the data needed for every port. Table 10.5 details the information collected from the ports.

10.4.6 Model Construction: Model Selection and Model Proposal

10.4.6.1 DEA Model Selection

Two traditional DEA models, namely the CCR model and the BCC model, have been widely used in seaports efficiency analysis. Table 10.2 shows a variety of examples of DEA models applied in seaports, all in different scenarios. Some of those studies prioritize the inputs, and some of the outputs, nonetheless, every variable that is chosen has a direct relationship with the objective established by the authors.

The main objective is to maximize the amount of TEU's moved, by Guayaquil Contecon Port (GCP) without changing any important infrastructure from a cost perspective. The reason behind this objective is to understand Guayaquil's port current situation from an efficiency perspective. The DEA model to be used in this study is the BCC model with an output orientation in order to maximize the outputs of the port while maintaining the actual level of inputs. The BCC model or Variable

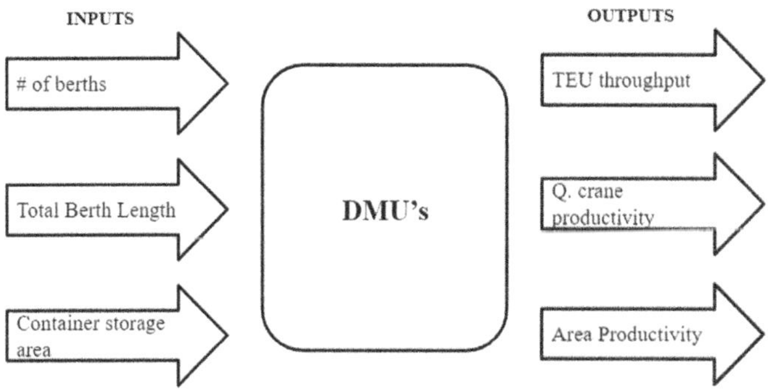

Fig. 10.8 Model proposed for the case study

Return to Scale (VRS) model allows to determine a convex efficient frontier allowing the variables to have positive or negative effects. A Variable Return to Scale helps understand the slack differences between the efficient DMU's and inefficient DMU's allowing to analyze a reason why Guayaquil Contecon Port (GCP) is an efficient port in comparison to other ports in Latin America. The best practices in ports and management tendencies could be identified after the results obtained in the study.

10.4.6.2 Model Proposal

The previous subsections were used to show the critical components for the model construction. Figure 10.8 shows the model proposed for this study.

The inputs value is assigned to a X_{ij} and Y_{rj} for the output values. The DEA-BCC is going to be used because it allows for a more realistic assumption of the Variable Return to Scale. The model proposed is based on the principles of the model (10.14) shown in Sect. 10.2. The written model is shown next where it includes the sets, the objective function, restriction, the inputs, and the outputs expressions

Sets

$$Set\ I = 1, 2, 3\ for\ inputs$$

$$Set\ J = 1, 2, \ldots, 15\ for\ DMUs$$

$$Set\ R = 1, 2, 3\ for\ outputs$$

Inputs

$$X_{1j} = Container\ Storage\ Area\ for\ DMU_j$$

$$X_{2j} = \#\ of\ Berths\ for\ DMU_j$$

$$X_{3j} = Total\ Berth\ Length\ for\ DMU_j$$

Output

$$Y_{1j} = TEUs\ moved\ for\ DMU_j$$

$$Y_{2j} = Crane\ productivity\ for\ DMU_j$$

$$Y_{3j} = Area\ productivity\ for\ DMU_j$$

Variables

$$\varphi = Efficiency\ score\ for\ a\ DMU_j$$

$$\lambda_J = Multiplicative\ factor\ for\ the\ \varphi_j\ need\ to\ be\ efficient$$

Objective Function

$$Max\,\varphi \tag{10.15}$$

Subject to:

$$\sum_{j=1}^{J} \lambda_j * X_{ij} \le X_{ik};\ i = 1, \ldots, I\&k = 1, \ldots, J \tag{10.16}$$

$$\varphi * y_{rk} - \sum_{j=1}^{J} \lambda_j * Y_{rj} \le o;\ i = 1, \ldots, I\&k = 1, \ldots, J \tag{10.17}$$

$$\sum_{j=1}^{J} \lambda_j = 1;\ VRS \tag{10.18}$$

And

$$\lambda_j \ge 0;\ \varphi \ge 0 \tag{10.19}$$

Equation (10.15) represents the objective function for the model which tries to maximize the efficiency *phi,* where *phi* represents the efficiency score for every DMU*j*. The objective function determines the efficiency for one DMU*j* at a time, in an iterative manner, by comparing each input (X_{ij}) and output (Y_{rj}) between each other. Equation (10.16) is the first constraint of the model which compares the inputs

(X_{ij}) of every port, against the input of one port at a time. The λ_J variable is a factor value calculated as the factor the input X_{ij} needs to be efficient, input-wise. Equation (10.17) is the second constraint in the model compares the output level of every port (Y_{rj}) against an output of a specific DMU*j*. The λ_J factor is also calculated. In this manner for every data of efficiency obtained, 15 λ_J would be obtained. Equation (10.18) is the VRS (DEA-BCC) model restriction that states that the sum of each λ_J must be equal to 1. Finally, Eq. (10.19) is the non-negativity constraints for the λ_J and φ variables.

10.4.7 Run the Model: Computer-Based Model Application

In order to solve the case study, a computer-based method is required to get optimal results. For this application, the linear programming maximization model was implemented in AMPL system. AMPL is an algebraic programming language that is used to solve large scale programming design. The AMPL language tries to make the model generation easier and less error-prone (Fourer et al. 2003). In addition, Excel sheets were used to prepare the data for the AMPL system and to obtain a better result display. The AMPL software can display the results in a .txt files allowing the file copied to an Excel sheet for a more organized result display.

10.5 Results

The following results are based on the Guayaquil Contecon Port case study. It is important to mention that the set of steps mentioned in Sect. 10.3 can be replicated for other port studies or any other performance study.

10.5.1 Management Investigation of Results: Obtained Results

DEA performs a comparative efficiency model that benchmarks a set of Decision-Making Units, and its results obtained only are applicable within the group of DMU's analyzed. The results for this case study consider only the input and output factor used in the mathematical model, model (10.15 to 10.19), and the ports selected as DMU's. The efficiency score obtained, *phi*, using the BCC model are presented as $\varphi \geq 1$, being $\varphi = 1$ an efficient DMU and $\varphi > 1$ 1 is considered inefficient. The results obtained by the computer-based model (AMPL) are shown in Table 10.6.

Table 10.6 Results obtained for each DMU

Port	DMU	phi
Guayaquil	1	1.000
Balboa	2	1.000
Colon	3	1.000
Valparaiso	4	1.358
San Antonio	5	1.263
Buenos Aires	6	1.418
Montevideo	7	2.198
Callao	8	1.000
Buenaventura	9	1.533
Cartagena	10	1.000
Limon-Moin	11	1.059
Kingston	12	1.352
Manzanillo	13	1.000
Veracruz	14	2.017
Tecon Santos	15	1.000

Expressing the data graphically, in Fig. 10.9, it denotes the ports that have the highest efficiency score obtained, highlighting that Guayaquil Contecon Port as a member of the efficient ports in Latin America.

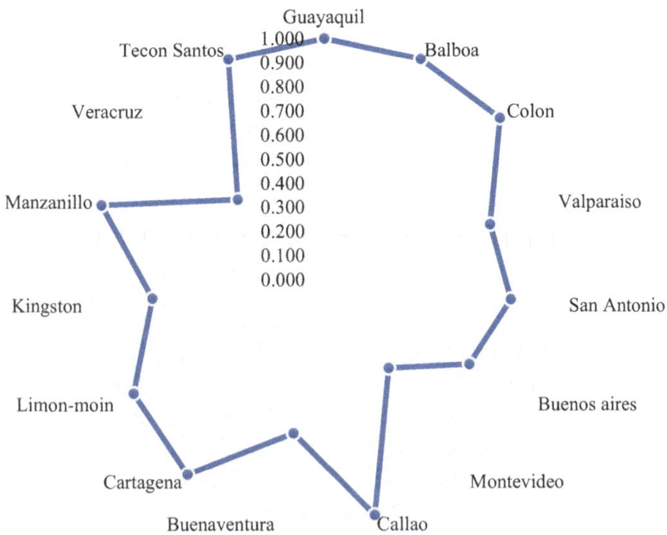

Fig. 10.9 Efficiency scores obtained

For the case study, the input and output factors considered gave a result of seven ports (Guayaquil, PPC Balboa, Puerto Colon, Callao, Manzanillo, Cartagena, Tecon Santos) with an efficiency score of $phi = 1$. The remaining eight ports (Valparaiso, San Antonio, Buenos Aires, Montevideo, Buenaventura, Limon-Moin, Kingston, Veracruz) are considered inefficient with a score of $phi > 1$. When performing DEA, the scores obtained could be a starting point for potential improvements to become more efficient. The inefficient ports can introduce any improvement method, such as best practices, technologies, among others, and observe the influence of the changes made periodically using DEA. Munisamy (2013) performed a similar study in Latin American ports with an 8-year window analysis, between 2000 and 2008 to observe the evolution in the performance in Latin American ports. The same type of study could be used to understand the changes and the impact, positive or negative, of any improvement introduced to an inefficient DMU.

10.5.2 Management Actions: Best Practices

From the results obtained after running the model, Guayaquil Contecon Port is efficient. However, measuring a port is not only based on the structural variables like cranes, area, and berths, but also the practices, the activities and the operations that are performed in it. Therefore, the best practices of the port must be analyzed in addition to the DEA results. The authors presume that all ports similarly to Guayaquil Contecon Port, have proper best practices that are applied apart from the structural variables, i.e., Guayaquil Contecon has a well-stated set of security norms and protocols, and risk prevention plan that will be analyzed in this section. For Guayaquil Contecon Port, all this information is available on the Contecon website as well as the intern regulations of the port.

In the best practices in land operation, Guayaquil has much norms about it. First, the warehouses in the port have cleaning and order maintenance regularly. Moreover, all chemicals that are stored must be well and visually labeled so that there are no confusions and errors while manipulating this type of materials. Additionally, to the warehouses, all rain gutters, fueling stations, floors, kitchen, berths, and equipment must be regularly cleaned and inspected. This includes all the cranes, forklifts, trucks and all vehicles that are used for containers flow. However, when there are constructions inside the port, all debris and dust must be cleaned by a third-party company. The maintenance also includes the regular inspection of the sediments at the bottom of the berths (Contecon 2012). If it is necessary, a cleaning procedure applies, which fulfill the best practice in dredging.

For the hazardous wastes like fluorescent lamps, empty chemical containers, paint residues, solvents, fuel, adsorbent material, waste inks, used oils, and grease lubricants among others, they are stored in a special clean area above the floor until proper disposal that is made by a certified environmental manager.

Moreover, all operators must have gloves, dust masks, aprons and any personal protection equipment that is required for their labor. This includes that the installations hold fire extinguishers, first aid kits, and all safety and security equipment. It is important to emphasize that there are training for all workers before they start working in the port. These training include proper management of hazardous wastes, best environmental practices, handling of fire extinguishers and first aid training, and even training in the unfortunate case of a spill (Contecon 2012), which can be very dangerous to the environment.

The port also measures the quality of the air and the water, so that it is sure that no environmental harms are done. The quality of the air is measured every 6 months based on some established indicators of quality. In the same manner, the water is measured monthly also basing on other indicators la temperature, pH level, oxygen level, among others (Contecon 2012). All of this is to ensure that the port is not affecting the environment and if so, take proper actions to rectify the damage.

To help minimize the spreading in the dust, the port applies a norm stating that not any type of work can be performed in the areas that are designated for the loading and unloading of material. This practice is not focused only on controlling the spreading of dust, but to the exposition of gases, radiations, and noises (Contecon n.a.).

It is noticeable with all this information about Guayaquil Contecon Port that it has a strong base of best practices that also support its efficient DEA result. GCP not only worries about the environment but also about its workers. All practices are executed to guarantee the correct performance of all port activities. These regulations, norms, and practices applied by Guayaquil Contecon Port are helping to improve its performance.

10.6 Conclusions and Recommendations

Data Envelopment Analysis can be performed in multiple industries as a benchmarking tool to compare efficiency between different organizations. The set of steps proposed in this study was developed to be an easier and simpler way to realize performance analysis. These steps would allow anyone to perform a DEA study in ports and in other productive industries, facilitating researchers to be more efficient in this type of study.

This study uses Data Envelopment Analysis to compare the performance between 15 Latin American container seaports, focusing on comparing them to Guayaquil Contecon Port (GCP). The results obtained provide a technical idea of Guayaquil Contecon Port performance against the other ports.

The DEA analysis shows that multiple ports in Latin America are considered VRS-efficient, including Guayaquil Contecon Port (GCP), which stated that the main port in Ecuador is an efficient and productive entity. Moreover, having an efficient port in a developing country would boost the economic activities on it, which benefits and increase the country's Gross Domestic Product (GDP). Guayaquil Contecon Port efficiency score is not surprising because while being a relatively small port, it does have a great throughput of TEUs. This paper shows the data for an annual throughput of TEUs for the ports based on the CEPAL ranking in 2018, however, this ranking is also calculated every year. The CEPAL ranking for 2017 displays that Guayaquil Contecon Port had an annual throughput of 1,871,591 TEUs (ECLAC 2018). This shows how Guayaquil Port has increased in one year its annual throughput in almost 200,000 TEUs. This data clearly shows that Guayaquil Contecon Port is moving more TEUs, with the same inputs, which is the technical definition of efficiency. It is important to notice that there are administrative and operational aspects in the port that the study is not considering that could affect the scores obtained. This study only focuses on the technical and structural aspects of the ports. That is why it is also important to separately consider the best practices in operation and administrative that the GCP applies to their functional system as explained in the previous section. On the other hand, inefficient ports could apply some of the mentioned best practices to improve their efficiency to become more competitive by comparing what efficient ports do that could be applied to their reality.

For further research, it would be worthy to handle a DEA study with time window analysis to compare DMU's efficiency score overtime. This approach would allow to understand in more detail, how Latin American ports are evolving into more efficient entities and the main reasons. Another consideration to improve this research is to change the scope of the study and include other important ports of the world to understand the current situation of Latin American ports compared to these ports.

References

Axson, D. (2010). *Best practices in planning and performance management.* New Jersey: Wiley.

Bacchioni, A., & Ramus, J. (2008, May). *Best practices in port management: An assessment of two ports.* Retrieved from https://dukespace.lib.duke.edu/dspace/bitstream/handle/10161/490/MP_arb34_b_200805.pdf?seque

Barros, C. (2003). Incentive regulation and efficiency of portuguese port authorities. *Marit Econ Logist, 5*, 55–69. https://doi.org/10.1057/palgrave.mel.9100060.

Barros, C. A. (2006). Benchmark analysis of italian seaports using data envelopment analysis. *Maritime Economics & Logistics, 8*, 347–365. https://doi.org/10.1057/palgrave.mel.9100163.

Boussofiane, A., Dyson, R. G., & Thanassoulis, E. (1991). Applied data envelopment analysis. *European Journal of Operational Research, 52*(1), 1–15. https://doi.org/10.1016/0377-2217(91)90331-O.

Cazuza de Sousa Junior, J. N. (2010). *AVALIAÇÃO DA EFICIÊNCIA DOS PORTOS UTILIZANDO ANÁLISE ENVOLTÓRIA DE DADOS : ESTUDO DE CASO DOS PORTOS DA REGIÃO NORDESTE DO BRASIL José Nauri Cazuza de Sousa Júnior. 3*(2), 74–91.

Chang, Y. T., Shin, S. H., & Lee, P. T. W. (2014). Economic impact of port sectors on South African economy: An input-output analysis. *Transport Policy, 35,* 333–340. https://doi.org/10.1016/j.tranpol.2014.04.006.

Charnes, A., Cooper, W. W., & Rhodes, E. (1978). Measuring the efficiency of decision making units. *European Journal of Operational Research*, (2), 429–444. Retrieved from https://doi.org/10.1016/0377-2217(78)90138-8

Contecon. (2012). *Plan de manejo ambiental.* (16), 59.

Cooper, W. W., Seiford, L., & Tone, K. (2007). *Data envelopement analysis* (2nd ed.). Springer.

Cullinane, K., & Wang, T. F. (2006). Chapter 23 Data envelopment analysis (DEA) and improving container port efficiency. *Research in Transportation Economics*, *17*(06), 517–566. https://doi.org/10.1016/S0739-8859(06)17023-7

Cullinane, K., Song, D., & Wang, T. (2005). The Application of mathematical programming approahces to estimating container port production efficiency. *Journal of Productivity Analysis*, *24*, 73–92. https://doi.org/10.1007/s11123-005-3041-9

Cullinane, K., Song, D., Ji, P., & Wang, T. (2004). An application of DEA windows analysis to container port production efficiency. *The Review of Network Economics, 3*(2), 184–206.

Doerr, O. (2014). *Asset productivity at container terminals in Latin America and the Caribbean: 2005–2013* (p. 13).

Ducruet, C., & Lee, S. W. (2006). Frontline soldiers of globalisation: Port-city evolution and regional competition. *GeoJournal, 67*(2), 107–122. https://doi.org/10.1007/s10708-006-9037-9.

Dunn, J. (2018). *Lecture about Ports infrastructure*.

Dwarakish, G. S., & Salim, A. M. (2015). Review on the role of ports in the development of a nation. *Aquatic Procedia*, *4*(Icwrcoe), 295–301. https://doi.org/10.1016/j.aqpro.2015.02.040

ECLAC. (2018). Ports Ranking. The top 20 in Latin America and the Carribbean in 2017. Retrieved from https://www.cepal.org/en/infographics/ports-ranking-top-20-latin-america-and-caribbean-2017

ECLAC. (2019). Ports Ranking. The top 20 in Latin America and the Carribbean in 2018. Retrieved from https://www.cepal.org/en/infographics/ports-activity-2018-top-20-ports-latin-america-and-caribbean

El Telégrafo. (2013, August 9). *83,95% de la carga pasa por Guayaquil*. Retrieved from https://www.eltelegrafo.com.ec/noticias/economia/4/8395-de-la-carga-pasa-por-guayaquil

EPA. (2018). Best Practices for Port Operations.

Fourer, R., Gay, D. M., & Kernighan, B. W. (2003). *AMPL—A Modeling Language for Mathematical Programming* (2nd ed.) (pp. 519–554). https://doi.org/10.1287/mnsc.36.5.519

Gupta, A. K., Gupta, S. K., & Patil, R. S. (2005). Environmental management plan for port and harbour projects. *Clean Technologies and Environmental Policy, 7*(2), 133–141. https://doi.org/10.1007/s10098-004-0266-7.

IFC-WBG. (2017). *Environmental, health and safety guidelines for ports, harbours and terminals*.

IMO. (2017). International Convention for the Control and Management of Ships' Ballast Water and Sediments (BWM). Retrieved from International Maritime Organization website: http://www.imo.org/en/About/Conventions/ListOfConventions/Pages/International-Convention-for-the-Control-and-Management-of-Ships'-Ballast-Water-and-Sediments-(BWM).aspx

Itoh, H. (2002). Efficiency changes at major container ports in Japan: A window application of data envelopment analysis. *Review of Urban and Regional Development Studies, 14*(2), 133–152. https://doi.org/10.1111/1467-940X.00052.

Jemric, I., & Vujcic, B. (2002). Efficiency of banks in Croatia: A DEA approach. *Comparative Economic Studies, 44*(2–3), 169–193. https://doi.org/10.1057/ces.2002.13.

Kerzner, H. (2017). *Project management: A systems approach to planning, scheduling, and controlling* (12th ed.). Wiley.

Leung, A., Burke, M. I., & Yen, B. T. H. (2016, December). *Oil vulnerability of Australian capital cities: A pilot study using data envelopment analysis (DEA) for vulnerability benchmarking*. Retrieved from http://www.atrf.info

Logística, Z. (2017). El puerto de Guayaquil: Una joya para la economía del Ecuador. Retrieved from https://www.zonalogistica.com/el-puerto-de-guayaquil-una-joya-para-la-economia-del-ecuador/

Lonza, L., Maroida, M. (2016). Ports as drivers of urban and regional growth *Transportation Research Procedia,* 14, 2507–2516. https://doi.org/10.1016/j.trpro.2016.05.327

Lozano, S., Villa, G., & Canca, D. (2010). Application of centralised DEA approach to capital budgeting in Spanish ports. *Computers & Industrial Engineering, 60*(3), 455–465. https://doi.org/10.1016/j.cie.2010.07.029.

Lu, B., & Wang, S. (2017). *Container port production and management*. https://doi.org/10.1007/978-981-10-2428-3

Munisamy, S., & Jun, O. B. (2013, February 17). Efficiency of Latin American Container Seaports using DEA. In: *Proceedings of 3rd Asia-Pacific Business Research Conference*.

Panayides, P. M., Maxoulis, C. N., Wang, T-F., & Ng, K. Y. A. (2009). A critical analysis of DEA applications to seaport economic efficiency measurement. *Transport Reviews, 29*(2), 183–206. https://doi.org/10.1080/01441650802260354

Park, R. K., & De Prabir, P. (2004). An alternative approach to efficiency measurement of seaports. *Maritime Economics and Logistics, 6*(1), 53–69. https://doi.org/10.1057/palgrave.mel.9100094.

Pires, G. C. (2016). *ESTUDO DA EFICIÊNCIA DE TERMINAIS DE CONTÊINERES USANDO O MÉTODO DA ANÁLISE ENVOLTÓRIA DE DADOS (DEA)*. UNIVERSIDADE FEDERAL DE SANTA CATARINA.

Roa, I., Peña, Y., Amante, B., & Goretti, M. (2013). Ports: Definition and study of types, sizes and business models. *Journal of Industrial Engineering and Management, 6*(4), 1055–1064. https://doi.org/10.3926/jiem.770.

Rockel, M. (2017). What it takes to make a port city smart. *DotMagazine*, p. 2.

Rodrigue, J.-P., Comtois, C., & Slack, B. (2017). *The geography of transport systems* (4th ed.). New York: Routledge, Taylor & Francis Group.

Roll, Y., & Hayuth, Y. (1993). Port performance comparison applying data envelopment analysis (DEA). *Maritime Policy and Management, 20*(2), 153–161. https://doi.org/10.1080/03088839300000025.

Talley, W. K. (2018). Port economics. In *Port economics*. https://doi.org/10.4324/9781315667720

Tongzon, J. (2001). Efficiency Measurement of Selected Australian and other International ports using data envelopment analysis. *Transportation Research Part A: Policy and Practice, 35*, 107–122. https://doi.org/10.1016/S0965-8564(99)0049-X

Tongzon, J., Heng, W. (2005) Port privatization, efficiency and competitiveness: Some empirical evidence from container ports (terminals). *Transportation Research Part A: Policy and Practice, 39*(5), 405–424, ScholarBank@NUS Repository. https://doi.org/10.1016/j.tra.2005.02.001

Wanke, P., & Barros, C. P. (2016). New evidence on the determinants of efficiency at Brazilian ports: A bootstrapped DEA analysis. *International Journal of Shipping and Transport Logistics, 8*(3), 250. https://doi.org/10.1504/ijstl.2016.076240.

Zhao, Q., Xu, H., Wall, R. S., & Stavropoulos, S. (2017). Building a bridge between port and city: Improving the urban competitiveness of port cities. *Journal of Transport Geography, 59*, 120–133. https://doi.org/10.1016/j.jtrangeo.2017.01.014.

Zhu, J., & Wade, C. (2007). *Modeling data irregulatities and structural complexities in data envelopment analysis*. New York: Springer Science.

Chapter 11
Effects of Locus of Control on Bank's Policy—A Case Study of a Chinese State-Owned Bank

Cong Xu, Guo-liang Yang, Jian-bo Yang, Yu-wang Chen, and Hua-ying Zhu

Abstract This paper investigates how Locus of Control (*LOC*) will impact the bank's policies through a case study of a Chinese state-owned bank. At the end of 2008, the investigated bank implemented a personal business-preferred policy. We established two Data Envelopment Analysis (*DEA*) models to test the impacts of *LOC* on the implementation of the policy. The results show that internal-controlled branches tend to be more sensitive to the bank's policy. When it is a positive policy, the internal-controlled branches tend to improve more than the external-controlled branches, while the regression of internal branches is also more significant when it is a negative policy. Location and managers' personalities are identified as the two direct reasons that cause LOC effects. Several suggestions are also provided in this paper to alleviate the negative effects of LOC.

Keywords Bank policy · Data envelopment analysis · Locus of control

---

C. Xu · J. Yang · Y. Chen · H. Zhu
Alliance Manchester Business School, Manchester University, Manchester M15 6PB, UK
e-mail: samson88992211@gmail.com

J. Yang
e-mail: jian-bo.yang@manchester.ac.uk

Y. Chen
e-mail: Yu-wang.Chen@manchester.ac.uk

H. Zhu
e-mail: december.zhy@gmail.com

G. Yang (✉)
Institutes of Science and Development, Chinese Academy of Sciences, Beijing 100190, China
e-mail: glyang@casipm.ac.cn

V. Charles et al. (eds.), *Data Science and Productivity Analytics*,
International Series in Operations Research & Management Science 290,
https://doi.org/10.1007/978-3-030-43384-0_11

11.1 Introduction

The definition of Locus of Control (*LOC*) was first proposed by Rotter (1966), and it soon became a popular topic in a broad field. According to the definition of *LOC*, work behaviors can be explained by whether employees believe that their working outcomes are controlled internally or externally. The internal-controlled employees believe that their outcomes are more related to their personal characters, like skills and efforts. On the contrary, external-controlled employees believe that their outcomes depend more on the external factors like environments or luck which are out of their control. Employees under different types of *LOC* could react differently facing the same policy. In this paper, we investigate how *LOC* influences the bank's policy through a case study of a Chinese state-owned bank.

After China joined the World Trade Organization (*WTO*), Chinese state-owned banks have to face competitions from numerous foreign banks and joint-stock banks with high-quality assets and advanced management systems. In order to maintain the dominant position in the Chinese financial market, Chinese state-owned banks have to adjust their strategies. According to served customer types, banks in China generally classify their business into two categories. Company business serves companies (private, joint-stock, and state-owned), government organizations, and social groups, while personal business specialize in personal customers. In the first decade of this century, it is commonly agreed that personal business is the Chinese bank's foundation. Before 2010, most banks' personal business accounted for the majority of their total business. As a result, decision-makers believe that the key to saving state-owned banks is to save their personal business. At the end of 2008, the investigated bank published a personal business preferred policy, known as the "Personal Business Oriented Policy" or "*PBP*". The policy includes two aspects. One is that a personal business team is organized in each branch. Customer officers, especially outstanding personal customers, are reallocated to these teams to specialize in personal business. The other one is to significantly increase the performance assessment weights for personal business.

For internal-controlled employees, positive and negative reinforcements lead to greater increments and decrements in verbalized expectancies (Rotter et al. 1961). The publication of *PBP* is obviously a positive reinforcement for personal business, while it is also a negative reinforcement for company business. Obviously, the implementation of *PBP* will lead all branches to rebalance the positions of personal business and company business, but the degrees of changes are expected to be different. On one hand, internal-controlled branches' personal business is expected to be significantly improved, while their company business is expected to regress seriously. On the other, although the external-controlled branches will also adjust their strategies, the degrees of improvements and regressions are expected to be not as much as the internal-controlled branches. In order to test the effects of *LOC*, we apply Data Envelopment Analysis (*DEA*) in this paper.

DEA is a linear programming based nonparametric approach proposed by Charnes et al. (1978). By referencing the performances of Decision-Making Units (*DMU*s),

DEA identifies a Performance Possibility Set (*PPS*) or possibility region based on the assumption that any linear combination of *DMU*s is considered to be possible or that the performances inside PPS are regarded as achievable. The set of the best performances of PPS is called "efficient frontier", and the *DMU*s with best performances are defined as efficient *DMU*s. The *DMU*s dominated by efficient frontier are defined as inefficient *DMU*s. By calculating the *DEA* Malmquist index and its decompositions, we can observe the developments of both efficient frontier and inefficient *DMU*s over time. One difficulty of *LOC* research is to judge whether an employee is internal-controlled or external-controlled (Hersch and Scheibe 1967; Allen et al. 1974; Broedling, 1975). *DEA* could be used as a new method to help identify different types of *LOC*. One significant character of *LOC* is that employees with outstanding performances tend to be internal-controlled, while the backward ones tend to be more external-controlled. Therefore, it is reasonable to indicate that the efficient branches are majorly internal-controlled, while the inefficient branches are more external-controlled. By observing the relative locations of *DEA* frontier and inefficient *DMU*s before and after *PBP*, we can indirectly indicate whether internal- and external-controlled branches react differently to the policy.

Two *DEA* models are established in this paper to test *LOC* effects on personal business and company business, respectively. The results are significant for both models. For personal business model, the efficient branches improve dramatically, while the impacts on inefficient branches are not as significant as efficient branches. The decompositions of MI show that the distance between efficient frontier and inefficient branches is increased. For company business model, the efficient branches for company business are seriously depressed. Although the development of inefficient branches also slows down to some extent, the negative effect is not as serious as efficient branches. The MI compositions show that the distance between efficient frontier and inefficient branches is significantly decreased after *PBP*.

The contributions of this paper can be summarized into the following points. We proposed a *DEA* model that could be used to test whether employees under different types of *LOC* will react differently with a positive (or negative) policy. We also applied this model to a Chinese state-owned bank to test whether *LOC* will influence bank's policy. The results show that internal-controlled branches are significantly more sensitive to the new published policy. Some further considerations and suggestions are also provided for bank's future policy design.

The remainder of this paper is organized as follows. Section 11.2 reviews the past relative researches. After that, the investigated bank's background and the *PBP* policies are introduced in the Sect. 11.3. The corresponding hypotheses are also proposed in the same section. Section 11.4 explains the methodology, and the testing results are displayed in Sect. 11.5. Section 11.6 further investigates the potential reasons for the LOC effects, some suggestions are also proposed in the same section for bank's future policy design. Conclusions are given in the final section.

11.2 Previous Research

The concept of *LOC* was first proposed by Rotter (1966). After that, a large number of following researches have proved that internal-controlled employees tend to have several advantages over external-controlled employees. For example, internal-controlled employees are easier to feel satisfied with their jobs (Mitchell et al. 1975), more likely to be considerate of subordinates (Pryer and Distefano 1971), tend not to burn out (Glogow 1986), and follow a more strategic style of executive action (Miller et al. 1982). A direct implication of these studies is that employees with advanced performances are likely to be more internal-controlled, while employees with poor performances tend to be more external-controlled. A large number of solid evidence have proved this implication in practical cases (Hersch and Scheibe 1967; Allen et al. 1974; Broedling 1975; Majumder et al. 1977; Anderson 1977; Anderson and Schneier 1978; Pittman and Pittman 1979; Carden et al. 2004). Although a number of past researchers have found that *LOC* has effects on banks operations (Chiu et al. 2005; Katsaros and Nicolaidis 2012; Lee 2013; Katsaros et al. 2014; Ruiz-Palomino and Bañón-Gomis 2015; Karkoulian et al. 2016), none of them focus on how *LOC* will impact banks' policies.

The idea of *DEA* was first proposed by Charnes et al. (1978). Since then, it has been proved that *DEA* is a powerful tool in many areas, for example, academy institutes (Yang et al. 2014a, 2014b), hospital (Aksezer 2016; Portela et al. 2016), environment (Sarkis and Cordeiro 2001; Sueyoshi and Goto 2011), etc. Using DEA to analyze bank branches' performances is not a new topic. Paradi et al. (2011) evaluated the efficiency of 816 branches of a Canadian bank from the aspects of productivity, intermediary business, and profitability through three *DEA* models. Golany and Storbeck (1999) classified inputs as nondiscretionary inputs and discretionary inputs, classified outputs as short-term outputs and long-term outputs, and built-up an input-oriented *BCC* model to assess the efficiency of 200 bank branches. Due to the different requirements of the managers from different levels in hierarchy, Paradi and Schaffnit (2004) established two input-oriented models for branch managers and senior managers, respectively, to analyze the efficiency of a Canadian bank's 90 branches.

Some fairly recent studies have applied *DEA* to examine the policy impacts on the efficiency of financial institutions. Casu and Girardone (2010) tested the effects of EU deregulation policies on financial integration and convergence with *DEA*. Azad et al. (2014) assessed the influence of Financial Services Agency's recent policy changes on the efficiency and returns-to-scale of Japanese financial institutions. Masum et al. (2016) collected the panel data of 48 banks from Bangladesh from 2004 to 2013 to analyze the direct impacts of human resource management practices on traditional bank performance measurement. Fethi et al. (2011) investigated the liberalization impacts on the efficiency of Egyptian banks with different ownerships and sizes via Malmquist index. Rajput and Gupta (2011) established *DEA* models to validate the improvement of foreign banks under the financial liberalization policy in India from

2005 to 2010. Similar researches can also be found in the papers authored by Fujii et al. (2014), Sri and Shieh (2014), and Piatti and Cincinelli (2015).

We believe that our research will fill in the research gap by proposing a *DEA* model that could be used to test whether employees under different types of *LOC* will react differently with a positive (or negative) policy. Our research will provide a new area for *DEA* applications.

11.3 Background and Theory

11.3.1 Bank Background

In this paper, we have selected a Chinese state-owned bank as our research target. The selected bank has a history for over 60 years, and its 24,000 branches cover all major cities in China. The bank adopts a strict four-hierarchy management system, including the head office in Beijing, city head offices, branches, and secondary branches.

(a) Beijing head office

Beijing head office is on the top level of bank's management hierarchy. Its responsibilities include general strategy (regulation) planning, deliberation and adjustment, city head office supervision and assessment, city branch managers' personnel assignment, etc.

(b) City head offices

According to the specific market environment, city branches tailor the strategies planned by Beijing head office into detailed strategies or regulations for their branches and secondary branches. Besides, city head offices also take the responsibility of branch supervision and assessment, etc.

(c) Branches

Unlike the head office in Beijing and city head offices, branches do not have the authority to make or adjust strategies or regulations. The responsibilities for branches are to manage and organize secondary branches' daily operations under the regulations planned by upper head offices. Normally, a branch supervises 6 to 12 secondary branches. A branch has the authority to open a new secondary branch in a suitable location as well as to close secondary branches with poor performances.

(d) Secondary branches

Secondary branch is the basic production unit for a bank, and it is the only level with operation function in the four-hierarchy structure. It is responsible for bank's daily operations, for example, transactions, deposits, loans, selling financial and insurance productions, etc.

According to customer types, the bank classifies over 100 types of financial productions and services into two major categories: personal business and company business. Personal business represents the business that serves employee customers, for example, personal deposits, personal loans (mainly property mortgage), payment settlements, electronic bank, investments and financing products, and so on. Company business serves enterprise and institution customers and it involves company deposits, company loans, fund business, bill business, payment settlements, surety business, and so on.

11.3.2 Policy Changes from 2006 to 2015

PBP is implemented by a city head office of the bank at the end of 2008, and it is effective to its 13 branches and 112 Secondary branches in the city. From 2003 to 2008, steadily increasing income enriches the capital amount possessed by personal customers. At the same time, the immigration waves brought numerous personal investors with a considerate amount of capital to the city. Therefore, bank managers had a very high expectation on the development of personal business. In order to absorb the new entering personal capitals, the bank implemented the *PBP* policy at the end of 2008. *PBP* included the following two specific aspects. First, the bank significantly increased the performance assessment weights for personal business and decreased the weights for company business. Second, a new management position named "personal business director" was allocated to each lattice. Most of the new directors were selected from outstanding customer officers with rich experience of developing personal business, and they were given the authority to establish their own groups with other customer officers to explore the potential personal business market.

11.3.3 LOC *Effects on* PBP *Policy*

Individuals' expectations for future reinforcements are systematically affected by whether they see a task as being determined or controlled by chance, random or other factors beyond their control, or see the reinforcement in the situation as an outcome of their own characteristics (Rotter et al. 1961). For bank branches, *PBP* actually contains two aspects of information. On one hand, the implementation of *PBP* is a positive reinforcement for personal business. Specialized personal business teams significantly reinforce branches' business capability. Besides, the increased performance assessment weights also motivate branches to pay more efforts on extending personal business. On the other, *PBP*'s negative effects on branches' company business cannot be ignored as well. A number of company business officers are relocated to the newly formulated personal business specialized teams, which seriously weakens branches' company business capability. Besides, the decreased performance assessment weights also discourage branches' motivations for company business.

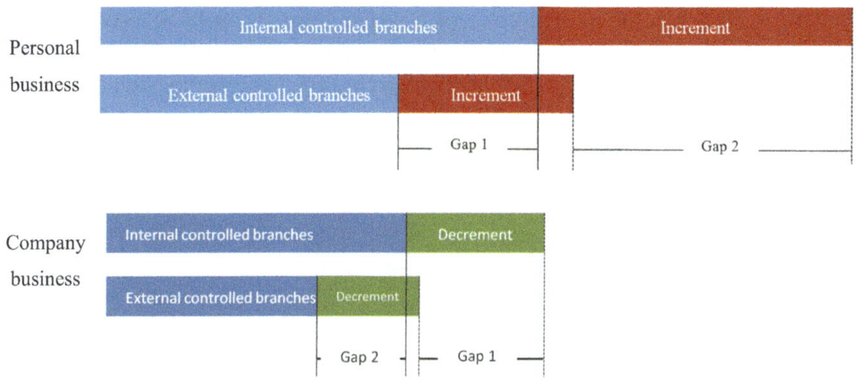

Fig. 11.1 *LOC* effects on branches' performances

Although all branches are expected to be influenced by *PBP*, we argue that the increment (or decrement) degrees are going to be different for internal-controlled branches and external-controlled branches.

As a result, the gaps between advanced branches' performances and backward branches' performances are expected to be broadened. As a result, the gap between advanced branches' performances and backward branches' performances is expected to be narrowed (Fig. 11.1).

The *LOC* effects on *PBP* can be summarized into the following two hypotheses:

H1: For personal business, the implementation of *PBP* will broaden the gap between the advanced branches and backward branches.

H2: For company business, the implementation of *PBP* will narrow down the gap between the advanced branches and backward branches.

11.4 Methodology

11.4.1 DEA *and Efficiency Evaluation*

DEA, proposed by Charnes et al. (1978), is a nonparametric efficiency evaluation method. One significant advantage of *DEA* is that no prior knowledge is required in terms of the weights of inputs and outputs criteria. The core idea of *DEA* is to formulate a Production Possibility Set (*PPS*) via referencing the existing *DMU*s' performances. Suppose there are n *DMU*s, and they all consume m inputs $X = (x_1, x_2, \ldots, x_m)$ and produce s outputs $Y = (y_1, y_2, \ldots, y_s)$. The PPS is defined by:

$$PPS = \{(X, Y)|XcanproduceY\} \tag{1}$$

Return To Scale (RTS) is a concept adopted from the field of economics, it describes the changes in the scale of inputs and outputs (Yang et al. 2016). Adopting different assumptions of RTS will lead to different PPS. For example, the *DEA* models under Constant Return to Scale (*CRS*) assumption, also known as *CCR* model, accept that outputs will change by the same proportion as the changes of inputs or assumptions. The corresponding PPS is defined by Charnes et al. (1978):

$$PPS(X, Y) = \Big\{(X, Y)|X \geq \sum\nolimits_{j=1}^{n} \lambda_j X_j, \\ Y \leq \sum\nolimits_{j=1}^{n} \lambda_j Y_j, , \lambda_j \geq 0, j = 1, \ldots, n\Big\} \tag{2}$$

Banker et al. (1984) proposed the *DEA* model under the assumption of Variable Return to Scale (*VRS*) known as the *BCC* model. Under *VRS* the PPS is defined as:

$$PPS(X, Y) = \Big\{(X, Y)|X \geq \sum\nolimits_{j=1}^{n} \lambda_j X_j, Y \leq \sum\nolimits_{j=1}^{n} \lambda_j Y_j, \\ \sum\nolimits_{j=1}^{n} \lambda_j, = 1,\ \lambda_j \geq 0, j = 1, \ldots, n\Big\} \tag{3}$$

Definition 1 A *DMU* is to be rated as fully (100%) **efficient** on the basis of available evidence if and only if the performances of other *DMU*s does not show that some of its inputs or outputs can be improved without worsening some of its other inputs or outputs (Cooper et al. 2011). For a characterized PPS, the set of efficient performances are defined as **efficient frontier**. In economic field, if there is a unique output, efficient frontier is also called **production function**.

No matter under *CRS* or *VRS* assumption, *DEA* model can construct a piecewise linear efficient frontier. A *DMU*'s efficiency can be evaluated by measuring the distance between its position and the corresponding efficient frontier. If the assessed *DMU* is on the efficient frontier, it is efficient. Otherwise, it is inefficient. Suppose we investigate the minimum inputs without sacrificing *DMU*'s any outputs and using radial measurement, the *DEA* input-oriented model under *CRS* is displayed as follows (Charnes et al. 1978):

$$Min_{\lambda_j} \theta_{j_0} \tag{4}$$

$$s.t. \sum_{j=1}^{n} \lambda_j x_{ij} \leq \theta_{j_0} x_{ij_0} i = 1, \ldots, m;$$

$$y_{rj_0} - \sum_{j=1}^{n} \lambda_j y_{rj} \leq 0 r = 1, \ldots, s$$

$$\lambda_j \geq 0 \text{ for all } j.$$

Suppose we accept the *VRS* assumption, the *DEA BCC* model is displayed as follows (Banker et al. 1984):

$$Min_{\lambda_j}\theta_{j_0} \tag{5}$$

$$s.t.\sum_{j=1}^{n}\lambda_j x_{ij} \leq \theta_{j_0}x_{ij_0}i = 1, \ldots, m;$$

$$y_{rj_0} - \sum_{j=1}^{n}\lambda_j y_{rj} \leq 0r = 1, \ldots, s$$

$$\sum_{j=1}^{n}\lambda_j = 1$$

$$\lambda_j \geq 0\,\text{for all}\,j,$$

where y_{rj} denotes the amount of "rth output" produced by DMU_j and x_{ij} represents the amount of "ith input" consumed by DMU_j. λ_j are composite variables. θ_{j_0} is the efficient score with the measurement scale from 0 to 1. If θ_{j_0} is equal to 1, the assessed *DMU* is efficient. Otherwise, it is inefficient. It should be noticed that the only difference between *CCR* and *BCC* model is that *BCC* model has an extra constraint $\sum_{j=1}^{n}\lambda_j = 1$.

The idea of *DEA* can also be illustrated graphically (Fig. 11.2). Suppose there are six *DMU*s: A, B, C, D, E, and F. They all consume the two inputs and produce one output. Efficient *DMU*s A, B, C, D, and E together formulate a piecewise linear efficient frontier. F, as an inefficient *DMU*, its efficient score can be obtained by calculating OG'/OG.

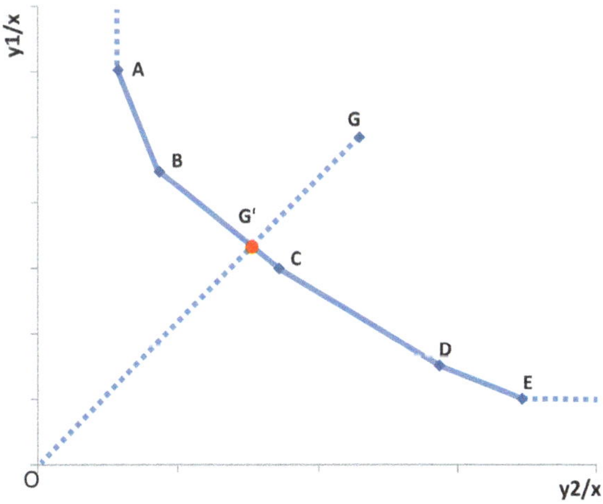

Fig. 11.2 Illustration of *DEA* concept

DMU's inefficiency can be decomposed into two sources: from *DMU*'s inefficient operation and disadvantageous environment that a *DMU* operates. *DMU*'s operation efficiency, also named as pure technical efficiency, can be identified by calculating *DMU*'s efficient score under *VRS* using *DEA BCC* model. *DMU*'s technical efficiency can be evaluated by *DMU*'s efficient score under *CRS* using *DEA CCR* model. We have the relationship between *DMU*s' technical efficiency, scale efficiency and pure technical efficiency (Cooper et al. 2007):

$$\begin{aligned} \textit{Technical Efficiency } (TE) &= \textit{Pure Technical Efficiency } (PTE) \\ &\quad \times \textit{Scale Efficiency } (SE) \end{aligned} \tag{6}$$

11.4.2 Malmquist Index and Its Decompositions

When measuring the productivity change of a *DMU* between two different periods, Malmquist index is a usual measurement method for *DEA* models. Malmquist index is defined by *DMU*'s two types of productivity changes: *Catch-up effect* and *Frontier shift*. *Catch-up* effect refers to a *DMU*'s efficiency changes over these two periods, while frontier shift describes the development of efficient frontier. Denote $(X, Y)^t$ as the PPS in period t. Let Y_j^t and X_j^t be the amount of outputs and inputs that DMU_j performs, respectively, in period t. The *CCR* PPS in period t is defined by:

$$(X, Y)^t = \{(X, Y)|X \geq \sum_{j=1}^{n} \lambda_j X_j^t, Y \leq \sum_{j=1}^{n} \lambda_j Y_j^t, \lambda_j \geq 0, j = 1, \ldots, n\} \tag{7}$$

and the *BCC* PPS in period t is defined by:

$$\begin{aligned} (X, Y)^t &= \left\{ (X, Y)|X \geq \sum_{j=1}^{n} \lambda_j X_j^t, \; Y \leq \sum_{j=1}^{n} \lambda_j Y_j^t, \right. \\ &\left. = \sum_{j=1}^{n} \lambda_j \; 1, \; \lambda_j \geq 0, \; j = 1, \ldots, n \right\} \end{aligned} \tag{8}$$

When evaluation the productivity of DMU_0 from period 1 to period 2, the catch-up effect can be calculated by:

$$\text{Catch up effect} = \frac{\text{Efficiency of}(X_0, Y_0)^2 \text{with the respect to period 2frontier}}{\text{Efficiency of}(X_0, Y_0)^1 \text{with the respect to period 1frontier}} \tag{9}$$

and the frontier-shift can is defined by:

$$\varphi_1 = \frac{\text{Efficiency of}(X_0, Y_0)^1 \text{with the respect to period 1frontier}}{\text{Efficiency of}(X_0, Y_0)^1 \text{with the respect to period 2frontier}} \quad (10)$$

$$\varphi_2 = \frac{\text{Efficiency of}(X_0, Y_0)^2 \text{with the respect to period 1frontier}}{\text{Efficiency of}(X_0, Y_0)^2 \text{with the respect to period 2frontier}} \quad (11)$$

$$\text{Frontier shift} = \sqrt{\varphi_1 \times \varphi_2} \quad (12)$$

Catch-up > 1 indicates that DMU_0's technical efficiency improves from period 1 to period 2. On the contrary, catch-up < 1 and catch-up = 1 represent no change and regress in efficiency, respectively. Frontier-shift > 1 indicates that DMU_0's frontier technology improves from period 1 to period 2, while frontier-shift = 1 and frontier-shift < 1 represent no change and regress in efficiency, respectively.

Malmquist index summarized these two types of changes, which is defined as:

$$Malmquistindex = (Catchup) \times (Frontier - shift). \quad (13)$$

Ray and Desli (1997) proposed a decomposition based on *BCC* model as the benchmark to measure technical change over multiple periods by the ratio of *VRS* distance functions.

11.4.3 Inputs and Outputs Identification

From the Scopus database, we identified 60 papers about bank branches' efficiency published from 1988 to 2015, and we summarize the inputs and outputs criteria in Table 11.1. On input side, labor and labor-related indexes are the most frequently used input criteria, and they can be found in over 87% of the related papers. Specifically, most researchers directly used employee number as an input (Camanho and Dyson 2005; Cook and Zhu 2008; Tsolas 2010; LaPlante and Paradi 2015), and some also involved employee quality, for example, employee score(Khalili-Damghani et al. 2015) and employee training time(Avkiran and McCrystal 2012). Cost is another popular input adopted by 45% of the past researches. The types of costs are various, for example, operation expenditure (Athanassopoulos and Giokas 2000), interest cost (Rezaei Taziani et al. 2009; Yang and Liu 2012), and non-labor cost (Giokas 2008). 16 papers involved the number of facilities used by branches as an input, like number of computer terminals (Zenios et al. 1999; Athanassopoulos and Giokas 2000; Amirteimoori and Nashtaei 2006; Hasannezhad and Hosseini 2011); and number of ATM machines (Golany and Storbeck 1999; Liu and Tsai 2012). Other papers considered some relatively less popular inputs. LaPlante and Paradi (2015) considered the average household income and the population of a branch belonged region as an input criterion to represent branches' location, while Tsolas (2010) simply used branches' rental cost. Besides the objective criteria, Hasannezhad and Hosseini (2011) also involved managers' experience as an input.

Table 11.1 Inputs and outputs in literatures

	Adopted times	Percentage (%)
Inputs		
Labor/labor cost	52	87
Cost	27	45
Facilities	16	27
Capital/assets	12	20
Space/area	11	18
Deposits	6	10
Location	5	8
Outputs		
Loan	34	57
Deposit	30	50
Transactions	18	30
Accounts/cards	17	28
Non-interest income	12	20
Revenue	11	18
Insurance business	8	13

In past papers, the selection of outputs is basically based on specific research targets. Their targets can be categorized into three general types: assessing branches' productivities (Deville 2009; Meepadung et al. 2009; Deville et al. 2014), assessing branches' profitability (Lang and Welzel 1999; Lu et al. 2014), and assessing branches' services (Camanho and Dyson 2005). Correspondingly, researches for different targets will choose different output bundles. For example, when decision-makers want to evaluate branches' performances in terms of their productivity, the most popular outputs include loan/mortgage (used by 34 papers over 60), deposit (30 over 60), and insurance business (8 over 60). When decision-makers want to assess branches' profitability, branches' revenue (11 over 60) and non-interest income (12 over 60) are the common outputs. Only a few studies consider bank branches as a service organization unit, and they would like to use transactions (18 over 60) and card users/accounts (17 over 60) as their outputs.

Branches' productivity data set used by this research is collected from the bank's accounting department central database established in 2006. Before 2006, city head office actually knew very little about the detail performances of its branches. Over 100 types of financial productions or services were provided by branches', but at the end of every season branches only reported head office several summary indexes, like profit and overheads. Because there was not a database to store data, some branches even did not have the electronic version of their detail performance data. Insufficient performance information makes the head office very difficult to supervise its branches. Corresponding to this requirement, at the beginning of 2006, the bank established its first information platform. Every operation terminal is connected to this information platform via bank's local area network. The system then uploads every transaction's data to the central database which is supervised by city head office's accounting department.

When the bank assesses branches' performance according to customer types, four criteria are usually used to assess bank's personal business and company business. They are "personal deposit", "personal loan", "company deposit", and "company loan". For example, the bank set up the criterion "personal deposit" to represent branches' deposit volume and financial productions with similar characteristics for personal customers. Similar financial productions or services include investment and financing products, electronic bank business, etc. The volumes of these productions and services are transferred into personal deposits with equal economic value added for the bank. The transfer ratio is settled by city head office and could be adjusted every season according to the market and the bank's strategies. Although at the end of 2013, the bank involved a new criterion "intermediary business volume" to present bank's non-deposit and non-loan business, we still decide to use the four criteria structure to collect branches' performances data in order to maintain the consistency.

On the input side, the data are collected from the network construction department. When a branch decides to open a new lattice, it needs to get the approval from network construction department. Therefore, this department stores detailed information on the established date and closing date for each secondary branch. So the number of secondary branches supervised by each branch for each year can be easily obtained out of it. For the investigated bank, the investments of secondary branches basically share a uniform scale in terms of human resources and fixed asset investments. On one hand, according to different functions, employees' positions are categorized into two types: management positions and operation positions. Management positions are further classified into a three-level hierarchy: P1, P2, P3 (from low to high), and similarly operation positions have a hierarchy: L1, L2, L3 (from low to high). For management positions, every secondary branch has one P3 position (president), one P2 position (vice-president), and two P1 positions (Chef of operation). For operation positions, every secondary branch has five L1 positions, three L2 positions, and two L3 positions. It should be noticed that there might be some limited differences between the actual staff number and this scale. On the other, the fixed assets allocated to each branch are similar as well. The shared fixed asset like printers, ATM machines, water dispensers, common furniture, etc., are the same from one secondary branch to the other. The other personal office assets, like operation terminals, computers, desk and chairs, office supplies, etc., are allocated to each employee staff. As what has been discussed above, the number of staff follows the same scale for all secondary branches, so the personal office assets are quite similar as well. The only cost difference between secondary branches is from the rent. Some secondary branches are located in the city center while some secondary branches are located in the integration area of urban and rural area. The different locations lead to a gap in operation cost. Unfortunately, the rental cost is not available for this research. However, it will not significantly influence the final results. It is because rental cost only takes about 10% of secondary branches' overall cost. About 90% of the costs

Table 11.2 Inputs and outputs bundles

	Outputs				Input
	Personal deposit	Personal loan	Company deposit	Company loan	No. of secondary branches
Personal business model	×	×			×
Company business model			×	×	×

are actually quite the same from one secondary branch to the other. Since the cost for each secondary branch is quite similar to each other, the secondary branch numbers can basically represent the resources used by branches.

According to the cooperation agreement with the bank, we are able to access to all productivity data in terms of personal deposit, personal loan, company deposit, and company loan for all 13 branches from 2006 to 2011. Besides, we are also able to collect the data regarding the number of secondary branches that supervised by branches during this period.

This study builds up two *DEA* models with different outputs bundles to investigate the *PBP*'s effects on branches' personal business and company business, respectively. The inputs and outputs criteria bundles selected are list in Table 11.2. For company business model, company deposits and company loans are used as the output criteria. Personal deposits and personal loans are selected as the inputs indexes for the personal business model. The number of supervised secondary branches is used as the only common input index for both models.

11.4.4 Hypothesis Testing

Efficiency is the key concept of *DEA* and it measures *DMU*s performances by measuring the distance between assessed *DMU*s and reference *DMU*s. Catch-up effect is one of the decompositions of MI, and it measures how efficiency changes over time. Efficiency and efficiency change can be applied here to test H1 and H2. In particular, we expect that *PBP* has a significant influence on both personal business and company business models.

For personal business, we expect that the distances between efficient branches and inefficient branches will be broadened after the bank applies *PBP* because the increment for efficient branches is expected to be more than inefficient branches. The broadened distance will lead to a significant increase in branches' catch-up effect. In Fig. 11.3, A1, B1, C1, D1, and E1 are efficient *DMU*s, and they together formulate an efficient frontier in period 1. The efficiency of the assessed *DMU* G1

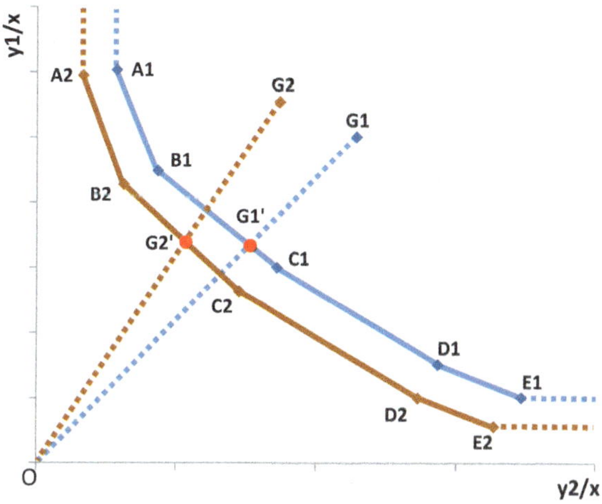

Fig. 11.3 Illustration of efficiency change

is measured by $\frac{OG1'}{OG1}$. In period 2, the efficient frontier in period 1 develops a new frontier formulated by A2, B2, C2, D2, and E2. The assessed *DMU*'s performance in period 2 is recorded to be G2. Then the efficiency of assessed *DMU* is measured by $\frac{OG2'}{OG2}$. If $\frac{OG1'}{OG1} < \frac{OG2'}{OG2}$ (or efficiency change > 1), then the efficiency of assessed *DMU* is improved, which indicates that the distance between assessed *DMU* and efficient frontier is narrowed. On the contrary, if $\frac{OG1'}{OG1} > \frac{OG2'}{OG2}$ (or efficiency change < 1), then the efficiency of assessed *DMU* is decreased, which indicates that the distance between assessed *DMU* and efficient frontier is broadened. For personal business model, we expect that branches' average efficiency is decreased (or average catch-up effect > 1) because the distances between advanced branches and backward branches are expected to be broadened. For personal business, we expect that the distances between efficient branches and inefficient branches will be narrowed after the bank applies *PBP* because the decrements for efficient branches are expected to be more than inefficient branches. Correspondingly, the narrowed distances will lead to a significant increase in bank's catch-up effect.

When testing H1 and H2, we are particularly interested in the change of efficiency change corresponding to the implementation of *PBP*. Therefore, a further regression analysis is applied to test the relationship between the implementation of *PBP* and the catch-up effects for both personal business model and company business model. The key specification is the following:

$$Efficiencychange = \beta_0 + \alpha_0 * t_{PBP} + \varepsilon,$$

where t_{PBP} is a time dummy variable for whether the efficiency change is before (0) or after (1) *PBP*. For personal business model, we expect that the implementation of *PBP* will broaden the gap between efficient and inefficient branches. Therefore, we expect the coefficient α_0 to be a negative coefficient. For company business model, we expect that *PBP* will narrow the distance between efficient branches and inefficient branches, so we expect that the coefficient α_0 to be a positive coefficient.

11.5 Results

Because *PBP* was published at the end of 2008, so we collect branches' performance data during the period from 2006 to 2011 to observe *PBP*'s impacts. Table 11.3 displays the descriptive statistics of the collected data. During this period, branches' average personal business volume is 7307.53 million RMB, where personal deposit takes 5081.014 million RMB and personal loan takes 2226.517 million RMB. Branches' average company business volume is 12129.82 million RMB, where the company deposit takes 6706.861 million RMB and company loan takes 5422.96 million RMB.

The results for *DEA* personal business model are displayed in Table 11.4. We are particularly interested in the average efficiency score and efficiency change (or catch-up effects) because the efficiency score measures the relative position between advanced branches and backward branches and the catch-up effect reflects the change of efficiency. Figures 11.4 and 11.5 illustrate branches' average efficiency score and

Table 11.3 Descriptive statistics

Variable	Obs	Mean	SD	Min	Max
Personal deposit	78	5081.014	1745.318	2341.07	9501.327
Personal loan	78	2226.517	1618.743	56.89	6876.315
Company deposit	78	6706.861	5683.171	2005.16	28870.47
Company loan	78	5422.96	4750.541	714.30	26601.46
Branch number	78	8.68	1.25	6.00	11.00

Table 11.4 MI and its decompositions for personal business model

Year	Average efficiency score	Efficiency changes (*CCR*)	Frontier shift	Pure technical efficiency change (*BCC*)	Scale efficiency changes	Total factor productivity changes
2006	0.751	–	–	–	–	–
2007	0.747	0.971	1.18	0.979	0.993	1.147
2008	0.765	1.064	1.095	1.026	1.037	1.165
2009	0.727	0.929	1.587	0.966	0.962	1.474
2010	0.710	0.966	1.295	0.963	1.003	1.251
2011	0.714	1.013	1.07	0.994	1.019	1.083

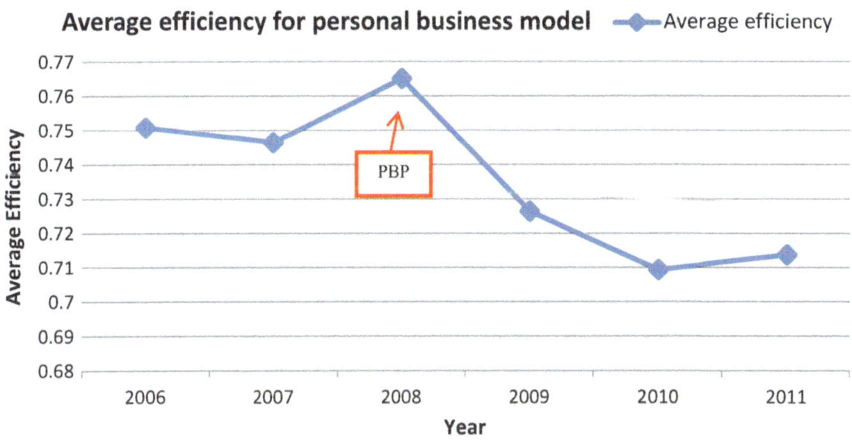

Fig. 11.4 Average efficiency for personal business model

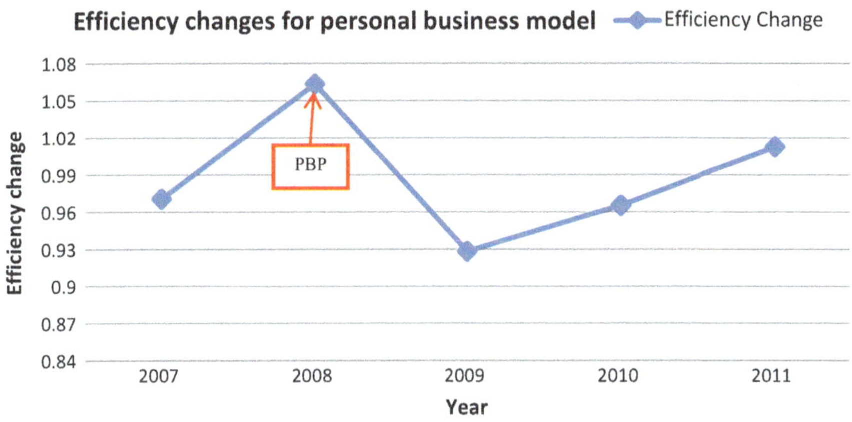

Fig. 11.5 Efficiency changes for personal business model

efficiency change, respectively, during this period. In 2006 and 2007, branches' efficiency score remains at around 0.75. From 2007 to 2008, the average efficiency score slightly increased to 0.76 with a catch-up effect of 1.06. After the implementation of *PBP* at the end of 2008, the average efficiency score dramatically dropped to around 0.73 with a catch-up effect of 0.93. It indicates that the implementation of *PBP* reversed the previous increasing trend and significantly broadened the gap between advanced branches and backward branches. After that, the influence of *PBP* lasted until the end of 2011 when the average efficiency score reached 0.71. From 2008 to 2010, the implementation of *PBP* caused a 6.5% decrease in average efficiency. Figure 11.6 displays the kernel density distributions for branches' efficient scores before (from 2006 to 2008) and after (from 2009 to 2011) *PBP*. The efficient scores of

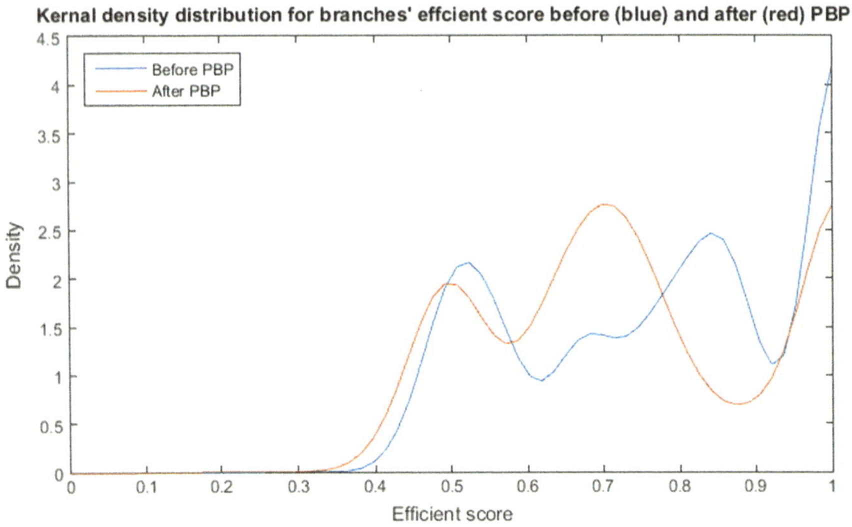

Fig. 11.6 Kernel density distribution (personal business model)

inefficient branches are significantly decreased. All of the results above have proved that the implementation of *PBP* significantly broadened the gaps between advanced branches and backward branches.

Table 11.5 displays the *DEA* results for the company business model. Like personal business, we are particularly interested in branches' average efficiency score and the catch-up effects before and after *PBP*'s implements. Figures 11.7 and 11.8 illustrate the average efficiency score and the average catch-up effects from 2008 to 2011, respectively. In 2006, the average efficiency score stood at 0.5. From 2006 to 2008, the average efficiency score experienced a dramatically decrease from 0.5 to 0.35, which means for a couple of years before 2008 the average distance between efficient branches and inefficient branches was increasing. Besides, the catch-up effect is

Table 11.5 MI and its decompositions for company business model

Year	Average efficiency	Efficiency changes (*CCR*)	Frontier shift	Pure technical efficiency change (*BCC*)	Scale efficiency changes	Total factor productivity changes
2006	0.501	–	–	–	–	–
2007	0.478	0.943	1.355	0.993	0.950	1.277
2008	0.351	0.668	1.459	1.062	0.629	0.974
2009	0.306	0.876	1.391	1.001	0.875	1.218
2010	0.352	1.215	0.965	1.012	1.200	1.173
2011	0.387	1.105	1.129	1.053	1.049	1.248

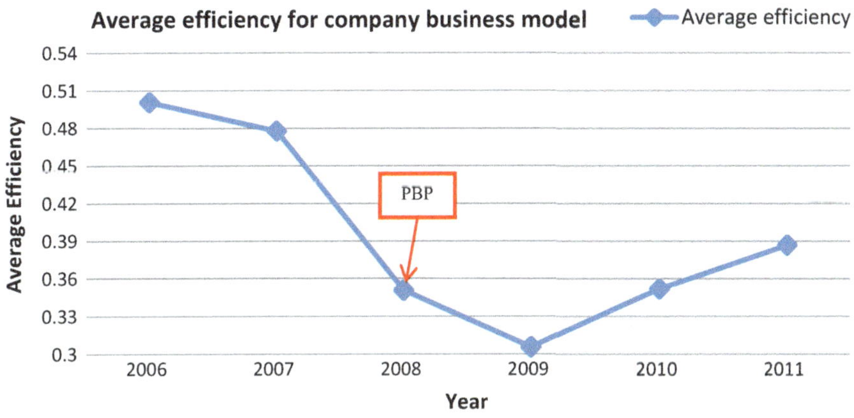

Fig. 11.7 Average efficiency for company business model

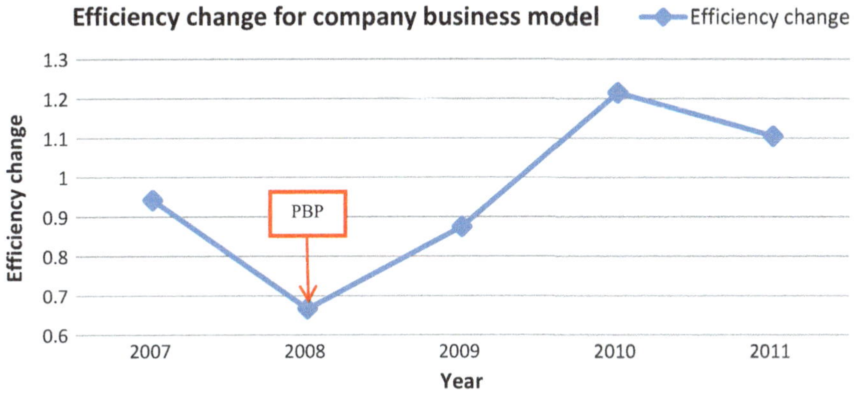

Fig. 11.8 Efficiency change for company business model

decreased from 0.95 in 2007 to 0.68 in 2008, which represents the increasing speed was increasing as well. After the implementation of *PBP*, the gap's increasing speed is significantly slowed down. The catch-up effects increased from 0.68 in 2008 to around 0.9 in 2009. In the following years, *PBP*'s effects continuously increased the catch-up effects which reached 1.2 and 1.1, respectively, in 2010 and 2011. The corresponding average efficiency score revised the decreasing trend in 2009 and increased to 0.35 in 2010 and 0.39 in 2011. In summary, the implementation of *PBP* slowed down the increasing gap between efficient branches and inefficient branches in 2009 and narrowed down the gap in 2010 and 2011. All of the results above agree with our expectations and provide solid evidence for H2.

The regression results further illustrate the relationship between *PBP* and branches' catch-up effects (Table 11.6). On one hand, *PBP* has an obvious negative effect on the branches' catch-up effect. The coefficient is –0.051. On the other,

Table 11.6 Regression results

	Efficiency change for personal business model	Efficiency change for company business model
After *PBP* or before *PBP* 1 for after, 0 for before	−0.051[a]	0.174[a]

[a]Significant at 1% level

PBP has a significant effect on the branches' catch-up effect. The coefficient is 0.174. Both results are significant in 1% level.

11.6 Further Discussions and Implications

11.6.1 Potential Reasons for LOC

In our following research, we investigated the potential reasons to explain the fact that branches are diversified into internal-controlled branches and external-controlled branches, through interviewing the managers of branches. We found that branches' locations and branch managers' personalities are the two commonest factors that contribute to the LOC effects.

Branch's location is found to make a significant contribution to branch's control system. On one hand, the efficient branches which are majorly internal-controlled for both personal business model and company business model are found to be located in relatively densely populated regions, where the branches can easily access a large amount of potential personal business customers. When the bank implements PBP, the efficient branches will reallocate their most qualified customer officers to develop personal business with little hesitation because they have the confidence that the extra personal business volumes they are going to get will definitely cover the volumes they are going to lose in company business. As a result, efficient branches' personal business volumes are significantly improved. However, because the internal-controlled branches reallocate their most qualified customer officers to develop personal business, branches' company business volume is seriously frustrated. On the other, employee reallocation is not such an easy decision for inefficient branches. It is because the inefficient branches in the personal business model are usually located in the sparsely populated regions, where the inefficient branches doubt the potential of developing personal business. It is unlikely for the inefficient branches to allocate their most professional customer officers to the newly formulated personal business team. Therefore, the increasing degrees for their personal business volumes and the decreasing degrees for their company business volumes are not as significant as efficient branches.

Besides, we found that branch managers' personalities could be another reason that contributes to branches' diversely controlled attitudes. During the interviewing process, we find out that some external-controlled branches' personalities tend to be conservative. They prefer to maintain what they have achieved now and worry that if they change the current employees' allocations and business strategies they will lose what they have got. The conservative attitudes lead to the fact that when receiving positive or negative policies the external managers would not make a significant change in their existing business strategies. As a result, their branches are not sensitive to bank's new policies.

11.6.2 Results Implements and Suggestions for Future Policy Design

The hypothesis tests H1 and H2 have proved that LOC has significant effects on branches' sensitivity in terms of bank's policy. A direct implementation of this conclusion is that bank's policies may not have a significant effect on external-controlled branches as they do on internal-controlled branches, which could have a negative effect on the implementation of bank's strategies. Considering the potential negative effect of LOC, we propose several suggestions for bank's future policy design.

It will be helpful for the decision-makers to communicate with branch managers before and after they design a specific policy. If there exist external-controlled branches, bank managers could find out the underlying reasons in time through communications. Besides, involving the concept of equivalent competitions and opportunities in bank's co-operated culture development could help to turn external-controlled branch managers into internal-controlled.

If a large number of branches have been found to be external-controlled, bank managers should think twice whether the implement policies are suitable for all branches. For instance, this case, although bank's decision-makers believe that it is the priority for the bank to develop personal business, the potential for some branches to develop personal business is limited due to location reasons. For these branches with poor locations, the personal business preferred policy is rather a burden than positive reinforcement. Personalized policies could be an idea to alleviate the negative effect brought by LOC. Under personalized policies, branches with different characters could be encouraged to develop the business types that they have advantages in. For example, the branches located in sparsely populated region could be encouraged to develop company business like settlement business and guarantee business which relies less on location. Several personalized target setting models could be utilized to assist bank managers to design personalized policies (Yang et al. 2009; Yang et al. 2012; Yang and Xu 2014).

11.7 Conclusions

The results of both *DEA* models provide valid evidences to H1 and H2, which indicates that *LOC* has a significant impact on the bank's policy effects. When there is positive reinforcement for the environment, for example, bank published a personal business preferred policy, advanced branches which are majorly internal-controlled will improve their performances more significantly than the backward branches which are majorly external-controlled. On the contrary, when there is a negative reinforcement for the environment, internal-controlled branches' performances will be more seriously frustrated. Comparing to internal-controlled branches, external-controlled branches act more conservatively. The scale of both improving and frustrating is not as significant as internal-controlled branches. The psychological difference leads to the fact that positive reinforcements tend to increase the gaps between two types of branches while negative reinforcements will narrow the gaps. In the following research, we found that location and branch managers' personalities could be two reasons to explain why LOC will have a significant effect on bank's policy. One potential negative effect of LOC is that bank's policy tends to have a less significant effect on backward branches which are majorly external-controlled. We suggest that bank managers should increase communication with branches before and after the policy design. Besides, it could be helpful to emphasize the concept of equivalent competitions and opportunities in bank's culture development. If a large number of branch managers have been found to be external-controlled because of objective reasons that cannot be solved, we suggest that the bank could consider applying personalized policies instead of common policies.

References

Aksezer, Ç. S. (2016). A nonparametric approach for optimal reliability allocation in health services. *International Journal of Quality and Reliability Management*, 33, 284–294.

Allen, G. J., Giat, C., & Cherney, R. J. (1974). Locus of control, test anxiety, and student performance in a personalized instruction course. *Journal of Educational Psychology, 66,* 968–973.

Amirteimoori, A., & Nashtaei, R. A. (2006). The role of time in multi-component efficiency analysis: An application. *Applied Mathematics and Computation, 177,* 11–17.

Anderson, C. R. (1977). Locus of control, coping behaviors, and performance in a stress setting: a longitudinal study. *Journal of Applied Psychology, 62,* 446–451.

Anderson, C. R., & Schneier, C. E. (1978). Locus of control, leader behavior and leader performance among management students. *Academy of Management Journal, 21,* 690–698.

Athanassopoulos, A. D., & Giokas, D. (2000). The use of data envelopment analysis in banking institutions: Evidence from the Commercial Bank of Greece. *Interfaces, 30,* 81–95.

Avkiran, N. K., & McCrystal, A. (2012). Sensitivity analysis of network DEA: NSBM versus NRAM. *Applied Mathematics and Computation, 218,* 11226–11239.

Azad, A. S. M. S., Yasushi, S., Fang, V., & Ahsan, A. (2014). Impact of policy changes on the efficiency and returns-to-scale of Japanese financial institutions: An evaluation. *Research in International Business and Finance, 32,* 159–171.

Banker, R. D., Charnes, A., & Cooper, W. W. (1984). Some models for estimating technical and scale inefficiencies in data envelopment analysis. *Management Science, 30,* 1078–1092.

Broedling, L. A. (1975). Relationship of internal-external control to work motivation and performance in an expectancy model. *Journal of Applied Psychology, 60,* 65.

Camanho, A. S., & Dyson, R. G. (2005). Cost efficiency, production and value-added models in the analysis of bank branch performance. *Journal of the Operational Research Society, 56,* 483–494.

Carden, R., Bryant, C., & Moss, R. (2004). Locus of control, test anxiety, academic procrastination, and achievement among college students. *Psychological Reports, 95,* 581–582.

Casu, B., & Girardone, C. (2010). Integration and efficiency convergence in EU banking markets. *Omega, 38,* 260–267.

Charnes, A., Cooper, W. W., & Rhodes, E. (1978). Measuring the efficiency of decision making units. *European Journal of Operational Research, 2,* 429–444.

Chiu, C. K., Lin, C. P., Tsai, Y. H., & Hsiao, C. Y. (2005). Modeling turnover intentions and their antecedents using the locus of control as a moderator: A case of customer service employees. *Human Resource Development Quarterly, 16,* 481–499.

Cook, W. D., & Zhu, J. (2008). CAR-DEA: context-dependent assurance regions in DEA. *Operations Research, 56,* 69–78.

Cooper, W. W., Seiford, L. M., & Tone, K. (2007). *Data envelopment analysis a comprehensive text with models, applications, references and DEA-solver software* (2nd ed.) (p. 490). Berlin: Springer, ISBN, 387452818.

Cooper, W. W., Seiford, L. M., & Zhu, J. (2011). *Handbook on data envelopment analysis.* Berlin: Springer.

Deville, A. (2009). Branch banking network assessment using DEA: A benchmarking analysis-A note. *Management Accounting Research, 20,* 252–261.

Deville, A., Ferrier, G. D., & Leleu, H. (2014). Measuring the performance of hierarchical organizations: An application to bank efficiency at the regional and branch levels. *Management Accounting Research, 25,* 30–44.

Fethi, M. D., Shaban, M., & Weyman-Jones, T. (2011). Liberalisation, privatisation and the productivity of Egyptian banks: A non-parametric approach. *Service Industries Journal, 31,* 1143–1163.

Fujii, H., Managi, S., & Matousek, R. (2014). Indian bank efficiency and productivity changes with undesirable outputs: A disaggregated approach. *Journal of Banking & Finance, 38,* 41–50.

Giokas, D. I. (2008). Assessing the efficiency in operations of a large Greek bank branch network adopting different economic behaviors. *Economic Modelling, 25,* 559–574.

Glogow, E. (1986). Research note: Burnout and locus of control. *Public Personnel Management, 15,* 79–83.

Golany, B., & Storbeck, J. E. (1999). A data envelopment analysis of the operational efficiency of bank branches. *Interfaces, 29,* 14–26.

Hasannezhad, M., & Hosseini, S. E. (2011). A comparative study on performance measurement of decision-making units: A case study in Iranian tejarat banks. *Advances in Operations Research.*

Hersch, P. D., & Scheibe, K. E. (1967). Reliability and validity of internal-external control as a personality dimension. *Journal of Consulting Psychology, 31,* 609.

Karkoulian, S., Srour, J., & Sinan, T. (2016). A gender perspective on work-life balance, perceived stress, and locus of control. *Journal of Business Research.*

Katsaros, K. K., & Nicolaidis, C. S. (2012). Personal traits, emotions, and attitudes in the workplace: Their effect on managers' tolerance of ambiguity. *Psychologist Manager Journal, 15,* 37–55.

Katsaros, K. K., Tsirikas, A. N., & Nicolaidis, C. S. (2014). Managers' workplace attitudes, tolerance of ambiguity and firm performance: The case of Greek banking industry. *Management Research Review, 37,* 442–465.

Khalili-Damghani, K., Taghavi-Fard, M., & Karbaschi, K. (2015). A hybrid approach based on multi-criteria satisfaction analysis (MUSA) and a network data envelopment analysis (NDEA) to evaluate efficiency of customer services in bank branches. *Industrial Engineering and Management Systems, 14,* 347–371.

Lang, G., & Welzel, P. (1999). Mergers among german cooperative banks: A panel-based stochastic frontier analysis. *Small Business Economics, 13,* 273–286.
Laplante, A. E., & Paradi, J. C. (2015). Evaluation of bank branch growth potential using data envelopment analysis. *Omega, 52,* 23–41.
Lee, H. W. (2013). Locus of control, socialization, and organizational identification. *Management Decision, 51*, 1047–1055.
Liu, F. H. F., & Tsai, L. C. (2012). Reallocating multiple inputs and outputs of units to improve overall performance. *Applied Mathematics and Computation, 219,* 237–247.
Lu, S. L., Lee, K. J., & Huang, Y. F. (2014). An investigation of the performances of regional centres and traditional branches: evidence from Taiwanese banks. *Applied Financial Economics, 24,* 639–648.
Majumder, R. K., Macdonald, A., & Greever, K. B. (1977). A study of rehabilitation counselors: Locus of control and attitudes toward the poor. *Journal of Counseling Psychology, 24,* 137.
Masum, A. K. M., Azad, M. A. K., & Beh, L. S. (2016). The role of human resource management practices in bank performance. *Total Quality Management and Business Excellence, 27,* 382–397.
Meepadung, N., Tang, J. C. S., & Khang, D. B. (2009). IT-based banking services: Evaluating operating and profit efficiency at bank branches. *Journal of High Technology Management Research, 20,* 145–152.
Miller, D., de Vries, M. F. K., & Toulouse, J.-M. (1982). Top executive locus of control and its relationship to strategy-making, structure, and environment. *Academy of Management Journal, 25,* 237–253.
Mitchell, T. R., Smyser, C. M., & Weed, S. E. (1975). Locus of control: Supervision and work satisfaction. *Academy of Management Journal, 18,* 623–631.
Paradi, J. C., Rouatt, S., & Zhu, H. (2011). Two-stage evaluation of bank branch efficiency using data envelopment analysis. *Omega, 39,* 99–109.
Paradi, J. C., & Schaffnit, C. (2004). Commercial branch performance evaluation and results communication in a Canadian bank—a DEA application. *European Journal of Operational Research, 156,* 719–735.
Piatti, D., & Cincinelli, P. (2015). Measuring social efficiency: The case of Italian mutual banks. *Academy of Accounting and Financial Studies Journal, 19,* 205–224.
Pittman, N. L., & Pittman, T. S. (1979). Effects of amount of helplessness training and internal-external locus of control on mood and performance. *Journal of Personality and Social Psychology, 37,* 39–47.
Portela, M. C. A. S., Camanho, A. S., Almeida, D. Q., Lopes, L., Silva, S. N., & Castro, R. (2016). Benchmarking hospitals through a web based platform. *Benchmarking, 23,* 722–739.
Pryer, M. W., & Distefano JR, M. (1971). Perceptions of leadership behavior, job satisfaction, and internal-external control across three nursing levels. *Nursing research*, *20*, 534–536.
Rajput, N., & Gupta, M. (2011). Assessing the efficiency of foreign banks in Indian context. *Banks and Bank Systems, 6,* 99–106.
Ray, S. C., & Desli, E. (1997). Productivity growth, technical progress, and efficiency change in industrialized countries: comment. *The American Economic Review, 87,* 1033–1039.
Rezaei Taziani, T., Sanei, M., Jahanshahloo, G. R., Jablonsky, J., & Mozaffari, M. R. (2009). Ranking bank branches with interval data by IAHP. *International Journal of Mathematical Analysis, 3,* 971–983.
Rotter, J. B. (1966). Generalized expectancies for internal versus external control of reinforcement. *Psychological monographs*, *80*, 1–28.
Rotter, J. B., Liverant, S., & Crowne, D. P. (1961). The growth and extinction of expectancies in chance controlled and skilled tasks. *The Journal of Psychology*, *52*, 161–177.
Ruiz-Palomino, P., & Bañón-Gomis, A. (2015). The negative impact of chameleon-inducing personalities on employees' ethical work intentions: The mediating role of Machiavellianism. *European Management Journal.*

Sarkis, J., & Cordeiro, J. J. (2001). An empirical evaluation of environmental efficiencies and firm performance: Pollution prevention versus end-of-pipe practice. *European Journal of Operational Research, 135,* 102–113.

Sri, G., & Shieh, C. J. (2014). Application of data envelopment analysis to operating performance evaluation of financial system. *Anthropologist, 17,* 831–836.

Sueyoshi, T., & Goto, M. (2011). DEA approach for unified efficiency measurement: Assessment of Japanese fossil fuel power generation. *Energy Economics, 33,* 292–303.

Tsolas, I. E. (2010). Modeling bank branch profitability and effectiveness by means of DEA. *International Journal of Productivity and Performance Management, 59,* 432–451.

Yang, C., & Liu, H. M. (2012). Managerial efficiency in Taiwan bank branches: A network DEA. *Economic Modelling, 29,* 450–461.

Yang, G., Rousseau, R., Yang, L., & Liu, W. (2014a). A study on directional returns to scale. *Journal of Informetrics, 8,* 628–641.

Yang, G., Ahlgren, P., Yang, L., Rousseau, R., & Ding, J. (2016). Using multi-level frontiers in DEA models to grade countries/territories. *Journal of Informetrics, 10,* 238–253.

Yang, G., Yang, L., Liu, W., Li, X., & Fan, C. (2014b). Directional returns to scale of biological institutes in Chinese Academy of Sciences.

Yang, J. B., Wong, B. Y. H., Xu, D. L., & Stewart, T. J. (2009). Integrating DEA-oriented performance assessment and target setting using interactive MOLP methods. *European Journal of Operational Research, 195,* 205–222.

Yang, J. B., & Xu, D. L. (2014). Interactive minimax optimisation for integrated performance analysis and resource planning. *Computers & Operations Research, 46,* 78–90.

Yang, J. B., Xu, D. L., & Yang, S. (2012). Integrated efficiency and trade-off analyses using a DEA-oriented interactive minimax reference point approach. *Computers & Operations Research, 39,* 1062–1073.

Zenios, C. V., Zenios, S. A., Agathocleous, K., & Soteriou, A. C. (1999). Benchmarks of the efficiency of bank branches. *Interfaces, 29,* 37–51.

Chapter 12
A Data Scientific Approach to Measure Hospital Productivity

Babak Daneshvar Rouyendegh (B. Erdebilli), Asil Oztekin, Joseph Ekong, and Ali Dag

Abstract This study is aimed at developing a holistic data analytic approach to measure and improve hospital productivity. It is achieved by proposing a fuzzy logic-based multi-criteria decision-making model so as to enhance business performance. Data Envelopment Analysis is utilized to analyze the productivity and then it is hybridized with the Fuzzy Analytic Hierarchy Process to formulate the decision-making model. The simultaneous hybrid use of these two methods is utilized to compile a ranked list of multiple proxies containing diverse input and output variables which occur in two stages. This hybrid methodology presents uniqueness in that it helps make the most suitable decision with the consideration of the weights determined by the data from the hybrid model.

Keywords Data Envelopment Analysis (DEA) · Data science · Analytics · Analytic Hierarchy Process (AHP) · Fuzzy logic · Hospital efficiency

12.1 Motivation

Health care companies encounter new obstacles each day. New regulations are imposed, advanced technologies are introduced, and new organizations are developed on a regular basis as a result of new public policies being adopted with regard to industry standards. Health care managers need to have the ability to respond to

B. Daneshvar Rouyendegh (B. Erdebilli)
Department of Industrial Engineering, Ankara Yıldırım Beyazıt University, 06010 Ankara, Turkey

A. Oztekin (✉)
Department of Operations & Information Systems, University of Massachusetts Lowell, One University Ave., Lowell, MA 01854, USA
e-mail: Asil_Oztekin@uml.edu
URL: https://www.uml.edu/MSB/faculty/Oztekin-Asil.aspx

J. Ekong
Department of Technology, Ohio Northern University, Ada, OH 45810, USA

A. Dag
Department of Business Intelligence & Analytics, Creighton University, Omaha, NE 68178, USA

V. Charles et al. (eds.), *Data Science and Productivity Analytics*,
International Series in Operations Research & Management Science 290,
https://doi.org/10.1007/978-3-030-43384-0_12

such challenges using sound performance evaluation and decision-making models, thus taking as much guesswork and human error out of the equation as possible. Performance evaluation in the health care sector is essential for hospitals to properly compete in order to determine their shortages with respect to rival companies based on the determined inputs and outputs. To this end, there are a number of serious tasks that need to be noticed and fulfilled for a successful implementation by managers and administrators so as to gain competitive advantage. These include utilizing the maximum capacity of the hospital, achieving staff efficiency by considering all of the relevant criteria when comparing oneself to rival hospitals, detecting strong and weak points within the organization, having the ability to see the "big picture" from outside of the company, evaluating all aspects of performance, and effectively responding and making decisions in accordance with the observations. Falling in rhythm with the tasks mentioned earlier, efficiency is a premier element of management and decision-making among all establishments and sectors because it explicitly and directly affects the financial outcome of the business and the edge gained by the competition. Accurate and instantaneous decisiveness is essential in health care management, especially considering the spiking of health care expenses as well as escalating the number of individuals insured through government-funded insurance plans. To ensure pragmatic measures in terms of productivity in this sector, the element of efficiency must be frequently measured in a valid and accurate manner (Barnum et al. 2011).

Data Envelopment Analysis (DEA), pioneered by Charnes, provides a data-focused approach to Decision-Making Units (DMU)s for decision-making by focusing on the measurement of vital and relevant inputs and outputs (Charnes et al. 1978; Cook et al. 2010, 2014). Charnes' method is grounded in linear programming (LP) and the capability to evaluate the decision unit's analogous fashion. However, the method does have complications when quantifying different scales, measuring multiple scales, and correlating entries and outputs that are measured in disparate units. The Multi-Criteria Decision-Making model (MCDM) is a procedural method to resolve intricate engineering dilemmas. Until 1988, the data from the MCDM and the DEA were kept separated. In 1988, Gonlay hybridized interactive and multiple-objective LP with DEA. The literature available from the MCDM does not propose an all-inclusive grading as a primary goal. It does dissert the use of preferential data to further distill the inequitable power of the DEA model. This approach enables users of the model to appropriate varying levels of significance to each input and output in the model, thereby influencing the model's outcome. This could also be considered the Achilles' heel of this process since biased knowledge from the individuals using this model is required and the inputted data does not carry a concrete weight. As shown by Golany (1988), Kornbluth (1991), Golany and Roll (1994), and Halme et al. (1999), each factor inputted into the DEA model would be put through an allotment of favored input/output marks or hypothetical DMU's. It has been shown that the application of preferred data through weighted parameter values can result in a complete DMU ordering (Adler et al. 2002).

DEA, a non-confining method, is a surrogate for multi-variation mathematical methods when it is used to quantify information with several inputs and outputs.

DEA offers researchers a broad spectrum of occasions to use the technique, and unlike the multi-variate statistical methods, DEA does not require any assumptions and it also includes accommodations to insert new constraints to models according to the investigator's parameters (Rouyendegh and Erol 2010). DEA separates units into two classes: efficient and inefficient. This classification is based on a dual arrangement of several outputs that provide a positive influence on the overall assessment. The initial DEA excludes the facilitation of fully-ranked studies. It merely provides a dual grouping of efficient and inefficient units but does not subsequently position the units by weighted value. According to this model, all units classified as efficient are considered to be the same positive equivalence. DEA segregates the units into two groupings: efficient and less-efficient, which are based on a binary output that positively subsidizes the overall quantification process. The original DEA simply facilitates the categorization of units into efficient and less-efficient dichotomic categories without ranking them. All of the units classified as efficient are technically good in the Pareto sense (Ganley and Cubbin 1992; Zhu 2015).

On the other hand, Saaty's AHP (Saaty 1980) has become a preferred technique in the models based on the preferential relationship in the regular multi-criteria decision-making (MCDM) routine. In an MCDM technique, the decision-maker (DM) first executes comparisons by pair. Then, the pair-wise comparison matrix and the eigenvector are determined to specify the importance of each factor in the decision. The weighted rank guides the decision-maker thought process of picking the more favorable choice. In this study, a hybridization method that combines fuzzy AHP and DEA is studied and suggested to help circumvent the unique incompetencies of each method, and it is applied to choose the premier health care facility among the seven locations as an illustrative case study.

12.2 Literature Review

DEA is a non-parametric LP model that requires no suppositions to be made about the operational form of the construction. Over one thousand articles and studies have been composed on the subject of DEA's, showing multiple examples and continued expansion of the method. In the most basic problem of this model, where the unit has one input and one output, the efficiency of the unit is ranked on the ratio of output to input. For cases with more than one input and one output factor, the DEA model is the only solution. Several outputs and inputs can be added into the efficiency measurement, where the sum of the outputs is divided by the sum of the inputs (Friedman and Sinuany-Stern 1997).

Tanbour (1997) used two input factors (labor and beds) and four output factors (three surgical processes and one doctor's appointment) to approximate the expansion in productivity of a surgical specialty procedure in Sweden while covering it with the maximum amount of waiting time. It applied a DEA-based Malmquist model and retrieved data for six years, starting from 1988 to 1993. It concluded that progressions in efficiency occurred mainly because of technological advancements rather than progressions in relative efficiency. Surprisingly, no DEA-based Malmquist procedure research has conducted a national assessment on the productivity of nonprofit medical facilities. Another body of research uses the first query that quantifies the efficiency of US NPO hospitals in a DEA-based Malmquist model (Tanbour 1997). Linna (1998) measured hospital financial efficiency and productivity in Finland from 1988 to 1994 by applying parametric section models, multiple DEA methods, and the Malmquist Productivity index. To generate output variables, that study combined all the emergency room check-ins, the patients that returned for follow-up appointments, the DRG-weighted number of the sum of all new and returning admissions, all the beds being used, all the patients residing in the building, the number of weeks for nurses' on-site training, the total quantity of impact-weighted, and peer-reviewed scientific studies. It totaled the net operating expenses, total figure of beds, median hourly salary for all employees, annual government health care price index, and instructor status. For input factors, Linna used all of the above in addition to the readmission rate for repeat patients. It concluded that cost efficiency and technological advancements annually contributed a 3–5% growth rate increase in productivity (Linna 1998).

Gjuffrida (1999) utilized a DEA-based Malmquist model to ascertain the efficiency of primary care over a five-year period from 1991 to 1995 in England. That study used a total of 12 factors (two input variables and ten output variables) and concluded that technical progression and scale efficiency added only minor advancements to the overall amount of productivity. It did also conclude that technological progression did not significantly increase productivity (Gjuffrida 1999). Hollingsworth et al. (1999) reviewed ninety-one studies of DEA applications on health care and primary care facilities. These studies measured the efficiency of health districts, hospitals, nursing homes, and primary health care facilities (Hollingsworth et al. 1999). Bhat et al. (2001) conducted a work by analyzing the efficiency of district-level government hospitals in Gujarat, comparing the relative efficiency among the hospitals using the DEA approach. They revealed that efficiency variations are more significant within district-level government hospitals than within the government fund hospitals and that the overall efficiency levels of the government fund facilities are relatively higher than the district-level government hospitals (Bhat et al. 2001).

Ozgen and Ozcan (2004) studied the productivity levels of 140 independently run dialysis services in the United States from 1994 to 2000. They also used a DEA-based Malmquist index to measure efficiency and assimilated various output factors containing: outpatient dialysis, dialysis training, and home dialysis treatments; labor input variables: doctors, medical staff, and equipment; and several financial input variables: medicine, pharmaceuticals, medical supplies, laboratory supplies, maintenance supplies and labor, administrative expenses, and other general expenditures. They discovered that independent dialysis services did not positively progress in productivity, but they did improve practical efficiency (Ozgen and Ozcan 2004). Du et al. (2014) conducted a case study over 119 general acute care hospitals in Pennsylvania. They used a DEA-based super-efficiency model. This model took both the quantity and the quality of the output into account. In addition to the conventional choice of input and output variables, they include the survival rate as a quality measure of health outcome in the set of output variables (Du et al. 2014). Kontodimopoulos and Niakas (2006) conducted a study over 73 dialysis facilities in Greece over a twelve-year timeframe, 1993–2004. They also used a DEA-based Malmquist model and incorporated the nurses and dialysis machines as input factors and the dialysis appointments as the output factors. Despite the fact that the researchers did not uncover a solid conclusion because of uncovering corrupt business practices, they were able to ascertain that productivity and efficiency would increase by as much as 5% on a yearly basis and the technical efficiency changed by 30% after the introduction of new technology (Ozgen and Ozcan 2004).

The remaining of this study is arranged as follows: Sect. 12.3 offers materials and methods—primarily Fuzzy Set Theory (FST), Data Envelopment Analysis (DEA), and Analytic Hierarchy Process (AHP). Section 12.4 introduces the DEA-FAHP and DEA-AHP, and Sect. 12.5 illustrates the hybrid model via a case study. The conclusion of the research is given in the final section, Sect. 12.6.

12.3 Materials and Methods

12.3.1 Fuzzy Set Theory

Zadeh proposed the Fuzzy Set Theory (FST) as a mathematical model to deal with imprecision and uncertainty. This theory has greatly impacted the mathematical field by allowing operators, programmers, and users to be applied in the fuzzy domain, and also by creating a method to mathematically be able to represent vague data. A fuzzy number (FN) is a category of units with a continuation of levels of membership. This style of model is noted by a membership function, which appoints a membership grade to each object between the range of zero and one (Kahraman et al. 2003; Kahraman et al. 2002).

A tilde "~" is placed above any symbol representing an FSN. A Triangular Fuzzy Number (TFN) $\tilde{M}$ can be depicted as in Fig. 12.1. A TFN is represented as (l, m, u). The parameters l, m, and u, respectively, signify the most diminutive value possible, the most promising value, and the greatest potential value that defines a fuzzy dilemma.

Fig. 12.1 Triangular fuzzy number depiction

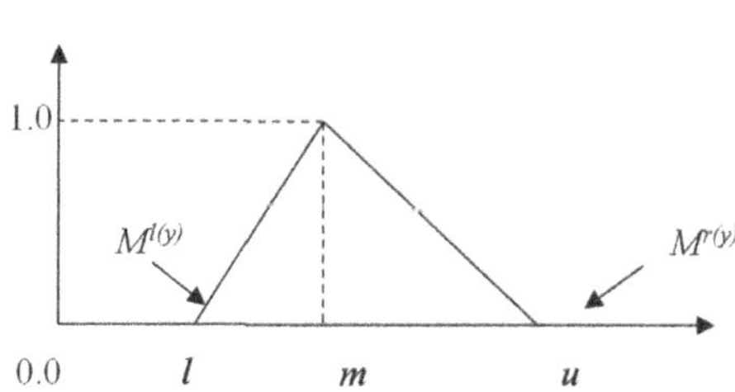

Each TFN has linear exemplifications, such that its membership function can be portrayed as in Eq. (12.1)

$$\mu\left(\frac{x}{\tilde{M}}\right) = \begin{cases} 0, & x < l, \\ (x-l)/(m-l), & l \le x \le m, \\ (u-x)/(u-m), & m \le x \le u, \\ 0, & x > u. \end{cases} \tag{12.1}$$

The α-cut of A of X as in Eq. (12.2)

$$[A]^{\alpha} = \begin{Bmatrix} t \in X\ A(t) \ge \alpha & if\,\alpha > 0 \\ Closure\ of\ the\ support\ of\ A\ if\,\alpha = 0 \end{Bmatrix} \tag{12.2}$$

If $[A]^{\alpha}$ is a convex (subset of X $\forall\alpha \in [0, 1]$), then fuzzy set A of X is named convex (Zadeh 1965).

A fuzzy number can always be presented by its relating left and right depiction of each degree of membership as in Eq. (12.3):

$$\tilde{M} = \left(M^{l(y)}, M^{r(y)}\right) = (L + (m-l)y,\ u + (m-u)y),\ y \in [0, 1] \tag{12.3}$$

where $l(y)$ and $r(y)$ signify the left side depiction and the right side depiction of a fuzzy set number (FSN), respectively. Multiple ordering models for FSNs have been expanded upon in the text. These models may provide different ordering outcomes, and most of them are monotonous in graphic manipulation demanding multi-layered mathematical calculations (Kahraman et al. 2002).

While there are different operations available with TFNs, only the vital processes used in this research are defined. If we show two positive TFNs to be (l_1, m_1, u_1) and (l_2, m_2, u_2), then Eqs. (12.4–12.6) hold true:

$$(l_1, m_1, u_1) + (l_2, m_2, u_2) = (l_1 + l_2, m_1 + m_2, u_1 + u_2) \tag{12.4}$$

$$(l_1, m_1, u_1) * (l_2, m_2, u_2) = (l_1 * l_2, m_1 * m_2, u_1 * u_2) \tag{12.5}$$

$$(l_1, m_1, u_1) * k = (l_1 * k, m_1 * k, u_1 * k), \text{ where } k > 0. \tag{12.6}$$

12.3.2 Basic Concepts of DEA

DEA is a common method that has been widely used for classifying and grading the decision-making units. Researchers use DEA to evaluate DMU's across a wide range of professional fields: education, health care, agriculture, banking, market research, and military (for a bibliography, see Emrouznejad 2001) (Corredoira et al. 2011). The DEA model removes the necessity of certain ascertains and parameters required by traditional efficiency measurement models. In its basic form, the DEA model which was originally developed by Farrell in 1957 and expanded to the CCR Model by Charnes, Cooper, and Rhodes, utilizes an oriented radial quantity of efficiency, which marks a spot on the borderline with the identical combination of inputs (input orientation) or outputs (output orientation) of the chosen unit (Kontodimopoulos and Niakas 2005).

Researchers use DEA to measure the immediate efficiency in a group of DMUs with similar input and output quantities. *The relative efficiency of a DMU is defined as the ratio of the sum of its weighted outputs, to the sum of its weighted inputs.* This process exists to mark the units that are below the efficiency standard, and set progression goals for them on the basis of the operations of the units defined as "efficient". Pareto optimality serves as the base of DEA (Charnes et al. 1978). Efficiency in a DMU is attained when there is no other single or combination of DMUs that can produce a minimum of the same outputs while removing one fewer of any input (Emrouznejad 2001).

The model computes the comparative ratio of outputs to inputs for each unit, with the score expressed as 0–1 or 0–100%. A DMU with a score of less than 100% is ranked as inefficient compared to the other units. The DEA is used to identify areas of efficiency and areas in need of improvement and is growing in popularity in its use among management. DEA has been initially used to investigate the relative efficiency of nonprofit organizations, but now the use of the model has spread to other industries such as hospitals, educational institutions, banks, and network industries, among many others. In the first stage of the model, a frontier based estimation on the input or output orientation is utilized by DEA to assess efficiency. Then, each DMU is assigned an efficiency score by comparing the output and input ratio of the DMU on the efficient frontier. A mathematical computation for each DMU's technical efficiency is presented as follows (Lavado and Cabanda 2009):

Symbols used

e_k: *Efficiency score for DMU,*
y_{rj}: *Amount of input r for DMU* $_j$,
x_{ij}: *Amount of input i for DMU* $_j$,
u_r: *Weight attached to output r,*
v_i: *Weight attached to input i,*
n: *Number of DMUs,*
t: *Number of outputs,*
m: *Number of inputs*

$$e_k = \max \sum_{r=1}^{t} u_r y_{rj}$$

$$Subject\ to$$

$$\sum_{i=1}^{m} v_i x_{rj} = 1 \tag{12.7}$$

$$\sum_{r=1}^{t} u_r y_{rj} - \sum_{i=1}^{m} v_i x_{rj} \leq 0 \tag{12.8}$$

$$u_r \geq 0, r = 1, \ldots, t$$

$$v_i \geq 0, i = 1, \ldots, m$$

$$u_r \geq 0, r = 1, \ldots, t$$
$$v_i \geq 0, i = 1, \ldots, m$$

12.3.3 Basic Concepts of AHP

AHP is a generalized philosophy of measurement. The scale shows relative priorities on an absolute plane from both discrete and continuous dual assessments in multi-level hierarchical structures. These assessments can be obtained from actual measurements, or alternatively they may be derived from a fundamental scale that shows the relative power of penchants and impressions. The AHP model has a specific caveat with departure from reliability and the measurement of this departure, and with reliance within and among the clusters of elements of its nucleus. The most common applications of AHP are observed in multi-criteria thinking and decision-making in logistics, resource allocation, and conflict resolution. When in general form, the AHP model consists of a nonlinear frame for fleshing out the aspects of both deductive and inductive thinking without the use of syllogism. That use is made achievable by using multiple factors at the same time, thus permitting the decision to conclusively resolve in a synthesis (Saaty and Vargas 2006).

With respect to decision layers, the AHP decision model pathway utilizes a one-way hierarchical relation. The middle level(s) constitutes the hierarchical pathway, while the bottom level is composed of the decision alternatives, as illustrated in Fig. 12.2 (Wang et al. 2008). Pair-wise comparisons measured on a 1–9 scale, as demonstrated in Table 12.1, are used to determine priorities within the hierarchy at every level based on the structured model provided by the AHP method.

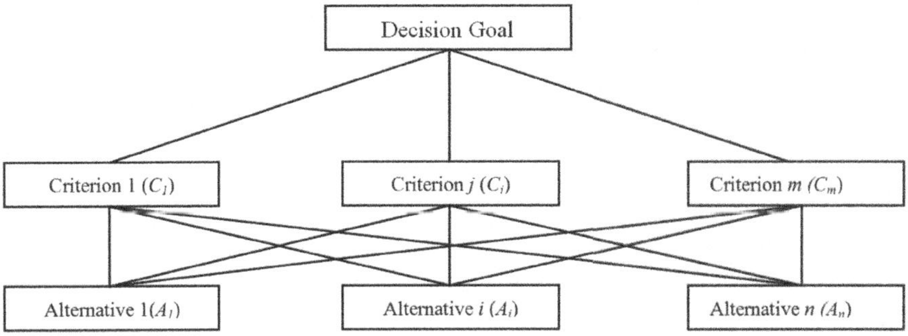

Fig. 12.2 A typical three-level MCDM problem hierarchy

Table 12.1 The 1–9 fundamental scale of absolute numbers for AHP

Importance intensity	Definition	Explanation
1	Equal importance	Multiple behaviors contribute equally to the outcome
3	Moderate importance of one over another	Circumstances and experience moderately prefer one over another
5	Strong importance of one over another	Circumstances and experience strongly favor one over another
7	Very strong importance of one over another	Dominate activity that is strongly favored and its supremacy is demonstrated in examples
9	Extreme importance of one over another	Importance of one behavior affirmed on the highest possible level
2, 4, 6, 8	Intermediate values	Representation of compromise of values listed above

12.4 Hybrid DEA-FAHP and DEA-AHP Models

12.4.1 DEA-FAHP

The DEA model is based on two sets of several outputs adding positively to the complete assessment and primarily deals with categorizing units into two groups, efficient and less-efficient (Ganley and Cubbin 1992). The first DEA model created does not facilitate full ranking, it simply gives the categorization of the two groups: efficient and inefficient. No rankings are provided, making all units equal in the Pareto sense (Rouyendegh and Erol 2010; Sinuany-Stern et al. 2000; Rouyendegh 2011). However, AHP, the pair-wise comparison matrix data are based on the subjective decision-makers' preferences while AHP-DEA builds an objective matrix. The AHP-DEA model, the MCDM are taken into account through DEA while the ranking is performed by AHP, thus the model does not suffer from limitations of either model. What follows is a fuzzy logic enhanced AHP evaluation along with DEA via the linguistic terms. DEA-FAHP method essentially integrates two popular methods. The steps of the method are as follows:

Step 1: Calculate the decision matrix from DEA method $e_{k,k'}$. With m alternatives and n criteria, $e_{k,k'}$ can be represented as in Eq. (12.9) through Eq. (12.10) (Rouyendegh and Erol 2010; Sinuany-Stern et al. 2000; Rouyendegh 2011):

$$e_{k,k'} = max \sum_{r=1}^{t} u_r y_{rk} \tag{12.9}$$

Subject to

$$Subjectto$$

$$\sum_{i=1}^{m} v_i x_{rk} = 1 \tag{12.10}$$

$$\sum_{r=1}^{s} u_r y_{rk} - \sum_{i=1}^{m} v_i x_{rk} \leq 0 \tag{12.11}$$

$$\sum_{r=1}^{s} u_r y_{rk'} - \sum_{i=1}^{m} v_i x_{rk'} \leq 0 \tag{12.12}$$

$$u_r \geq 0, r = 1, \ldots, t$$

$$v_i \geq 0, i = 1, \ldots, m$$

The solution to this problem gives $e_{k,k'}$ elements values as well as the binary compared E matrix ($k' = 1, \ldots, n, k = 1, \ldots, n$ and $k \neq k'$).

Step 2: Arrange the Triangular Fuzzy Numbers (TFNs). Each decision-maker makes a dual comparison of the decision criteria and qualities and assigns each one a relative score. The fuzzy transformation scale is as shown in Table 12.2 (Mikhailov 2003).

$$\breve{G}_1 = (l_1, m_1, u_1) \tag{12.13}$$

Step 3: Calculate the fuzzy extent value. With acknowledgement of ith the object is computed as follows:

Table 12.2 The 1–9 fuzzy transformation scale

Importance	Triangular fuzzy scale	Importance	Triangular fuzzy scale
1	(1, 1, 1)	1/1	(1/1, 1/1, 1/1)
2	(1.6, 2.0, 2.4)	1/2	(1/2.4, 1/2.0, 1/1.6)
3	(2.4, 3.0, 3.6)	1/3	(1/3.6, 1/3.0, 1/2.4)
5	(4.0, 5.0, 6.0)	1/5	(1/6.0, 1/5.0, 1/4.0)
7	(5.6, 7.0, 8.4)	1/7	(1/8.4, 1/7.0, 1/5.6)
9	(7.2, 9.0, 10.8)	1/9	(1/10.8, 1/9.0, 1/7.2)

$$s_i = \sum_{j=1}^{m} H_{gi}^{j} \times \left[\sum_{i=1}^{n} \sum_{j=1}^{m} H_{gj}^{j} \right]^{-1} \tag{12.14}$$

where

$$\sum_{j=1}^{n} H_{gi}^{j} = i \left(\sum_{j=1}^{m} l_i \sum_{j=1}^{m} m_i \sum_{j=1}^{m} u_i \right) \tag{12.15}$$

The inverse of the vector is computed as in Eqs. (12.16), (12.17):

$$\sum_{i=1}^{n} \sum_{j=1}^{m} H_{gi}^{j} = \left(\sum_{j=1}^{n} l_i \sum_{j=1}^{n} m_i \sum_{j=1}^{n} u_i \right) \tag{12.16}$$

$$\left[\sum_{i=1}^{n} \sum_{j=1}^{m} H_{gi}^{j} \right]^{-1} = \left(\frac{1}{\sum_{i=1}^{n} u_i}, \frac{1}{\sum_{i=1}^{n} m_i}, \frac{1}{\sum_{i=1}^{n} l}\right) \tag{12.17}$$

Step 4: The degree of possibility of $H_2 = (l_2, m_2, u_2) \geq H_1 = (l_1, m_1, u_1)$ is defined as in Eq. (12.18):

$$V(H_2 \geq H_1) = {}_{y}^{sup} \geq x\left[\min\left(\mu_{H_1}(x), \mu_{H_2}(y)\right)\right] \tag{12.18}$$

The corresponding membership functions are defined by Eq. (12.19):

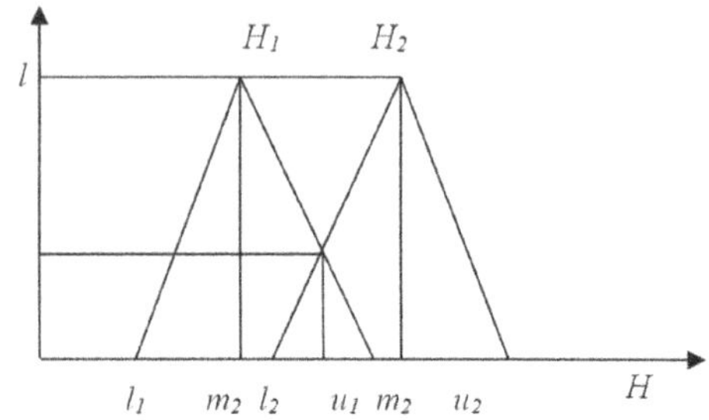

Fig. 12.3 The intersection between H_1 and H_2

$$V(H_2 \geq H_1) = hgt(H_1 \cap H_2) = \mu_{H_2}(d) = \begin{cases} 1 & if\ m_1 \geq m_2 \\ 0 & if\ l_1 \geq u_2 \\ \frac{l_1 - u_2}{(m_2 - u_2) - (m_1 - u_1)} & otherwise \end{cases} \tag{12.19}$$

The intersection of H_1 and H_2 can be shown as in Fig. 12.3. To compare H_1 and H_2, both values of $v(H_2 \geq H_1) and v(H_1 \geq H_2))$ are needed.

Step 5: A degree of possibility for a convex fuzzy number greater than k convex fuzzy number $M_i (i = 1, 2,\ldots, k)$ may be described as in Eq. (12.20):

$$\begin{aligned} v(H \geq H_1), H_1, H_2, \ldots., H_k) &= v[(H \geq H_1) \text{ and, } (H \geq H_2), \ldots, (H \geq H_k)] \\ &= \min v(H \geq H_i), \quad i = 1, 2, 3, \ldots, k \end{aligned} \tag{12.20}$$

where

$$d'(A_i) = \min v(s_i \geq s_k)$$

For $k = 1, 2,\ldots, n$; a weight vector is expressed.

$$W' = (d'(A_1), d'(A_2), \ldots, d'(A_n))^T$$

Step 6: The normalized weight vectors can be defined via Eq. (12.21)

$$W = (d(A_1), d(A_2), \ldots, d(A_n))^T \tag{12.21}$$

12.4.2 DEA-AHP

The procedure for the DEA-AHP method can be summarized as in the following steps:

Step 1: Calculate the decision matrix from DEA method ($e_{k,k'}$). With m alternatives and n criteria, $e_{k,k'}$ is as follows (Rouyendegh and Erol 2010; Sinuany-Stern et al. 2000; Rouyendegh 2011):

$$e_{k,k'} = \max \sum_{r=1}^{t} u_r y_{rk} \tag{12.22}$$

$$Subjectto$$

$$\sum_{i=1}^{m} v_i x_{rk} = 1 \tag{12.23}$$

$$\sum_{r=1}^{s} u_r y_{rk} - \sum_{i=1}^{m} v_i x_{rk} \leq 0 \tag{12.24}$$

$$\sum_{r=1}^{s} u_r y_{rk'} - \sum_{i=1}^{m} v_i x_{rk'} \leq 0 \tag{12.25}$$

$$u_r \geq 0, r = 1, \ldots, t$$

$$v_i \geq 0, i = 1, \ldots, m$$

The solution of this problem yields the values for $e_{k,k'}$ elements as well as the binary E comparison matrix ($k' = 1, \ldots, n$, $k = 1, \ldots, n$ and $k \neq k'$).

Step 2: Calculate: The pair-wise comparative matrix components are derived from Eq. (12.26):

$$a_{k,k'} = \frac{e_{k,k'}}{e_{k',k}} \tag{12.26}$$

Step 3: Each component derived from the second step is divided by the column's total value. The matrix obtained here is a normalized matrix as shown in Eq. (12.27)

$$a'_{k,k\prime} = \frac{a_{k,k'}}{\sum_{k=1}^{n} a_{k,k'}} \tag{12.27}$$

Step 4: The column vector elements are computed via the collection over the rows as in Eq. (12.28)

$$a''_{k,k\prime} = \sum_{k=1}^{n} a'_{k,k'} \tag{12.28}$$

Step 5: Normalize the column vector via Eq. (12.29)

$$a'''_{k,k'} = \frac{a''_k}{\sum_{k=1}^{n} a''_k} \tag{12.29}$$

12.5 Case Study Results and Discussion

The aim of this research is prioritizing the decision-making models used in health care organizations to improve performance and productivity. This study has two stages. In the first stage, DEA-FAHP is deployed for prioritizing the alternatives, and in the second stage, DEA-AHP is used to compare the results against FAHP.

In this section, the description of the DEA-FAHP application is explained using an actual dataset obtained through the selection of seven hospitals in Turkey. For the hybrid model, two input variables and three output variables are used in the study which can be listed as the following:

Inputs: Number of Doctors, Number of beds
Outputs: Bed Ratio, Total Surgery Operations, Number of Patients, depicted in Fig. 12.4.

Step 1: The first process using the DEA was solved using the LINDO program as demonstrated in Tables 12.3 and 12.4 in which the crisp and fuzzy evaluation matrices for seven hospitals are presented.

Step 2:
See Table 12.4.

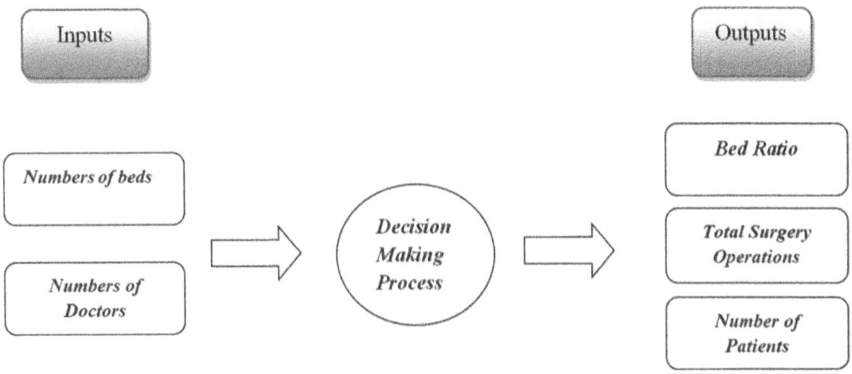

Fig. 12.4 Decision-making process with two input and three output variables

Table 12.3 Evaluation matrix for seven hospitals

	Hospital 1	Hospital 2	Hospital 3	Hospital 4	Hospital 5	Hospital 6	Hospital 7
Hospital 1	0.005	0.143	0.139	0.143	0.448	0.143	0.683
Hospital 2	0.005	0.143	0.144	0.143	0.092	0.143	0.053
Hospital 3	0.005	0.143	0.144	0.143	0.092	0.143	0.053
Hospital 4	0.441	0.143	0.144	0.143	0.092	0.143	0.053
Hospital 5	0.005	0.143	0.144	0.143	0.092	0.143	0.053
Hospital 6	0.532	0.143	0.144	0.143	0.092	0.143	0.053
Hospital 7	0.005	0.143	0.144	0.143	0.092	0.143	0.053

Table 12.4 Fuzzy evaluation matrix of seven hospitals

	Hospital 1	Hospital 2	Hospital 3	Hospital 4	Hospital 5	Hospital 6	Hospital 7
Hospital 1	(0.004, 0.005, 0.006)	(0.11, 0.14, 0.17)	(0.01, 0.14, 0.17)	(0.11, 0.14, 0.17)	(0.36, 0.45, 0.54)	(0.11, 0.14, 0.17)	(0.55, 0.68, 0.82)
Hospital 2	(0.004, 0.005, 0.006)	(0.11, 0.14, 0.17)	(0.12, 0.14, 0.17)	(0.11, 0.14, 0.17)	(0.07, 0.09, 0.11)	(0.11, 0.14, 0.17)	(0.04, 0.05, 0.06)
Hospital 3	(0.004, 0.005, 0.006)	(0.11, 0.14, 0.17)	(0.12, 0.14, 0.17)	(0.11, 0.14, 0.17)	(0.07, 0.09, 0.11)	(0.11, 0.14, 0.17)	(0.04, 0.05, 0.06)
Hospital 4	(0.353, 0.441, 0.530)	(0.11, 0.14, 0.17)	(0.12, 0.14, 0.17)	(0.11, 0.14, 0.17)	(0.07, 0.09, 0.11)	(0.11, 0.14, 0.17)	(0.04, 0.05, 0.06)
Hospital 5	(0.004, 0.005, 0.006)	(0.11, 0.14, 0.17)	(0.12, 0.14, 0.17)	(0.11, 0.14, 0.17)	(0.07, 0.09, 0.11)	(0.11, 0.14, 0.17)	(0.04, 0.05, 0.06)
Hospital 6	(0.426, 0.532, 0.638)	(0.11, 0.14, 0.17)	(0.12, 0.14, 0.17)	(0.11, 0.14, 0.17)	(0.07, 0.09, 0.11)	(0.11, 0.14, 0.17)	(0.04, 0.05, 0.06)
Hospital 7	(0.004, 0.005, 0.006)	(0.11, 0.14, 0.17)	(0.12, 0.14, 0.17)	(0.11, 0.14, 0.17)	(0.07, 0.09, 0.11)	(0.11, 0.14, 0.17)	(0.04, 0.05, 0.06)

Step 3:

$$\sum_{i=1}^{7}\sum_{j=1}^{7} H_{gi}^{j} = (5.466,\ 7.005,\ 8.417)$$

$$\left[\sum_{i=1}^{7}\sum_{j=1}^{7} H_{gi}^{j}\right]^{-1} = \left(\frac{1}{8.417},\ \frac{1}{5.466},\ \frac{1}{7.005}\right) = (0.118,\ 0.143,\ 0.183)$$

$$H_1 = (1.257,\ 1.704,\ 2.047) \times (0.118,\ 0.143,\ 0.183) = (0.148, 0.244, 0.357)$$

$$H_2 = (0.573, 0.723, 0.869) \times (0.118, 0.143, 0.183) = (0.068, 0.103, 0.159)$$

$$H_3 = (0.573, 0.723, 0.869) \times (0.118, 0.143, 0.183) = (0.068, 0.103, 0.159)$$

$$H_4 = (0.922, 1.159, 1.393) \times (0.118, 0.143, 0.183) = (0.109, 0.166, 0.255)$$

$$H_5 = (0.573, 0.723, 0.869) \times (0.118, 0.143, 0.183) = (0.068, 0.103, 0.159)$$

$$H_6 = (0.995, 1.250, 1.501) \times (0.118, 0.143, 0.183) = (0.117, 0.179, 0.275)$$

$$H_7 = (0.573, 0.723, 0.869) \times (0.118, 0.143, 0.183) = (0.068, 0.103, 0.159)$$

Step 4:

$$V(H_1 \geq H_2) = 1$$

$$V(H_1 \geq H_3) = 1$$

$$V(H_1 \geq H_4) = 1$$

$$V(H_1 \geq H_5) = 1$$

$$V(H_1 \geq H_6) = 1$$

$$V(H_1 \geq H_7) = 1$$

$$V(H_2 \geq H_1) = \frac{0.148 - 0.159}{(0.103 - 0.159) - (0.244 - 0.148)} = 0.072$$

$$V(H_2 \geq H_3) = 1$$

$$V(H_2 \geq H_4) = 0.442$$

$$V(H_2 \geq H_5) = 1$$

$$V(H_2 \geq H_6) - 0.36$$

$$V(H_2 \geq H_7) = 1$$

$$V(H_3 \geq H_1) = \frac{0.148 - 0.159}{(0.103 - 0.159) - (0.244 - 0.148)} = 0.072$$

$$V(H_3 \geq H_3) = 1$$

$$V(H_3 \geq H_4) = 0.442$$

$$V(H_3 \geq H_5) = 1$$

$$V(H_3 \geq H_6) = 0.36$$

$$V(H_3 \geq H_7) = 1$$

$$V(H_4 \geq H_1) = 0.58$$

$$V(H_4 \geq H_2) = 1$$

$$V(H_4 \geq H_3) = 1$$

$$V(H_4 \geq H_5) = 1$$

$$V(H_4 \geq H_6) = 0.91$$

$$V(H_4 \geq H_7) = 1$$

$$V(H_5 \geq H_1) = 0.072$$

$$V(H_5 \geq H_2) = 1$$

$$V(H_5 \geq H_3) = 1$$

$$V(H_5 \geq H_4) = 0.442$$

$$V(H_5 \geq H_6) = 0.35$$

$$V(H_5 \geq H_7) = 1$$

$$V(H_6 \geq H_1) = 0.66$$

$$V(H_6 \geq H_2) = 1$$

$$V(H_6 \geq H_3) = 1$$

$$V(H_6 \geq H_4) = 0.442$$

$$V(H_6 \geq H_5) = 0.35$$

$$V(H_6 \geq H_7) = 1$$

$$V(H_7 \geq H_1) = 0.072$$

$$V(H_7 \geq H_2) = 1$$

$$V(H_7 \geq H_3) = 1$$

$$V(H_7 \geq H_4) = 0.442$$

$$V(H_7 \geq H_5) = 1$$

$$V(H_7 \geq H_6) = 0.35$$

Step 5:

$$\mathrm{V(H_1 \geq H_2, H_3, H_4, H_5, H_6, H_7) = minV(H_1 \geq H_2, H_3, H_4, H_5, H_6, H_7) = 1}$$

$$\mathrm{V(H_2 \geq H_1, H_3, H_4, H_5, H_6, H_7) = 0.072}$$

$$\mathrm{V(H_3 \geq H_1, H_3, H_4, H_5, H_6, H_7) = 0.072}$$

$$V(H_4 \geq H_1, H_2, H_3, H_5, H_6, H_7) = 0.058$$

$$V(H_5 \geq H_1, H_2, H_3, H_4, , H_6, H_7) = 0.072$$

$$V(H_6 > H_1, H_2, H_3, H_4, H_5, H_7) = 0.066$$

$$V(H_7 \geq H_1, H_2, H_3, H_4, H_5, H_6) = 0.072$$

Step 6:

$$W' = (1, 0.072, 0.072, 0.58, 0.072, 0.66, 0.072)$$

$$W = (0.40, 0.028, 0.028, 0.229, 0.028, 0.26, 0.028)$$

$$W' = (1, 0.072, 0.072, 0.58, 0.072, 0.66, 0.072)$$

$$W = (0, 40, 0.028, 0.028, 0.229, 0.028, 0.26, 0.028)$$

As the next step, DEA-FAHP results are compared against that of DEA-AHP. Steps of DEA-AHP as explained in Sect. 4.2 are followed to prioritize the hospitals and they are ranked as W = (0.243;0.10;0.10;0.165;0.10;0.178;0.10) from hospital one to hospital seven, respectively.

As shown in Table 12.5, two methods—DEA-FAHP and DEA-AHP—reveal the same result in terms of the ranking of the analyzed hospitals as though the weights are different for the two models. The comparative outcomes can be tabulated as in Table 12.5.

For every test facility, the DEA-FAHP and DEA-AHP obtain the ratio of the matching efficient frontier of the efficacy that is attained by the facility. A larger score refers to a better result. The results demonstrate that with regard to the chosen inputs

Table 12.5 Comparison table for different techniques

Hospitals	DEA-FAHP	Rank	DEA-AHP
H_1	0.40	1	0.243
H_2	0.028	4	0.10
H_3	0.028	4	0.10
H_4	0.229	3	0.165
H_5	0.028	4	0.10
H_6	0.26	2	0.178
H_7	0.028	4	0.10

and outputs the efficiency score of Hospital 1 has the highest score (DEA-FAHP: 0.40, DEA-AHP: 0.243) due to having the best efficiency and hence performance. Hospitals 2, 3, 5, and 7 have the lowest scores out of the seven hospitals (DEA-FAHP: 0.028, DEA-AHP: 0.10).

12.6 Conclusions

This current research utilizes the DEA-FAHP and DEA-AHP methods to scrutinize certain aspects in the health care system with a specific focus on hospitals as exemplified in Turkey. With health care competition being constantly on the rise, the availability of a hybrid model for hospitals to use is very critical in order to ascertain the discrepancies in their productivity and efficiency in regards to that of their competitors. The DEA and Fuzzy AHP present unique power if collectively used to reveal potential outcome improvements, but both systems have limitations that omit a complete picture to be gained if utilized alone. The hybridization of these models constructs the DEA-FAHP model that would integrate the necessary pieces of the two models by circumventing the drawbacks of each model.

The integration of two models has some advantages. DEA-AHP and DEA-FAHP methodologies are simple enough, easy to use, applicable to any number of decision DMUs, and particularly useful and effective to complex MCDM problems with a large number of decision alternatives, where pair-wise comparisons are certainly impossible to be made. Consequently, we have devised an effective template for the classification and grading of multi-variate DMUs to evaluate the efficiency of public health care institutions in Turkey. In this study, a dual-stage methodology is postulated where the binary evaluation of the results derived from the model which is founded on DEA for the first stage. Second stage of the study consists of completely grading the proxies established on the findings acquired from the first stage. The benefit of the DEA-FAHP evaluation methodology is that the FAHP dual evaluations have been acquired mathematically from different information (input/output) by processing pair-wise DEA sequences. This creates results based on hardcore, objective, unbiased evidence with no subjective evaluations creeping into the process. Conversely, the ambiguous human decision-making procedure typically incorporates fuzziness and vagueness and the FAHP is designed to overcome that caveat of the process.

In conclusion, the suggested DEA-FAHP hybrid model can be flexibly applied throughout all business industries, not merely the health care sector. In this study, the example model of seven hospitals has been applied using determined inputs and outputs so as to determine the rank of efficiencies of the hospitals. For our sample of the seven piloted hospitals in Turkey, the same results were observed in a comparison of DEA-FAHP and DEA-AHP methods. Finally, the DEA-FAHP model has the competency to assess similar forms of circumstances where vagueness exists in MCDM issues such as project selection, supplier selection, etc. Removing the uncertainty and vagueness associated with large-scale, important decisions can be eliminated with this hybrid model. Thus, large organizations can make effective decisions with confidence knowing that they are essentially making a sound and valid decision.

References

Adler, N., Friedman, L., & Sinuany-Stern, Z. (2002). Review of ranking methods in the data envelopment analysis context. *European Journal of Operational Research, 140*(2), 249–265.

Barnum, D. T., Walton, S. M., Shields, K. L., & Schumock, G. T. (2011). Measuring hospital efficiency with data envelopment analysis: Nonsubstitutable vs. substitutable inputs and outputs. *Journal of Medical Systems, 35,* 1393–1401.

Bhat, R., Verma, B. B., & Reuben, E. (2001). Hospital efficiency: An empirical analysis of district hospitals and grant in aid hospitals in Gujarat. *Journal of Health Management, 3*(2), 167–197.

Charnes, A., Cooper, W. W., & Rhodes, E. (1978). Measuring the efficiency of decision making units. *European Journal of Operational Research, 2*(6), 429–444.

Cook, W. D., Liang, L., & Zhu, J. (2010). Measuring performance of two-stage network structures by DEA: A review and future perspective. *Omega, 38,* 423–430.

Cook, W. D., Tone, K., & Zhu, J. (2014). Data envelopment analysis: Prior to choosing a model. *Omega, 44,* 1–4.

Corredoira, R. A., Chilingerian, J. A., & Kimberly, J. R. (2011). Analyzing performance in addiction treatment: An application of data envelopment analysis to the state of Maryland system. *Journal of Substance Abuse Treatment, 41,* 1–13.

Du, J., Wang, J., Chen, Y., Chou, S.-Y., & Zhu, J. (2014). Incorporating health outcomes in Pennsylvania hospital efficiency: An additive super-efficiency DEA approach. *Annals of Operations Research, 221*(1), 161–172.

Emrouznejad, A. (2001). DEA HomePage. Warwick Business School Coventry CV4 7AL, UK, 1995–2001. http://www.deazone.com/books/DEA-socioEcoPlanning.pdf.

Friedman, L., & Sinuany-Stern, Z. (1997). Scaling units via the canonical correlation analysis in the DEA context. *European Journal of Operational Research, 100*(3), 629–637.

Ganley, J. A., & Cubbin, J. S. (1992). Public sector efficiency measurement: Applications of data envelopment analysis. Elsevier Science Publishers.

Gjuffrida, A. (1999). Productivity and efficiency changes in primary care: A Malmquist index approach. *Health Care Management Science, 2,* 11–26.

Golany, B. (1988). An interactive MOLP procedure for the extension of data envelopment analysis to effectiveness analysis. *Journal of the Operational Research Society, 39*(8), 725–734.

Golany, B., & Roll, Y. A. (1994). Incorporating standards via data envelopment analysis. In A. Charnes, W. W. Cooper, A. Y. Lewin, & L. M. Seiford (Ed.), Data *Envelopment analysis: Theory, methodology and applications*. Norwell, Mass, USA: Kluwer Academic Publishers.

Halme, M., Joro, T., Korhonen, P., Salo, S., & Wallenius, J. (1999). A value efficiency approach to incorporating preference information in data envelopment analysis. *Management Science, 45*(1), 103–115.

Hollingsworth, B., Dawson, P. J., & Maniadaleis, N. (1999). Efficiency measurements of health care: A review of non-parametric methods and applications. *Health Care Management Science, 2*(3), 161–172.

Kahraman, C., Ruan, D, & Ethem, T. (2002). Capital budgeting techniques using discounted fuzzy versus probabilistic cash flows. *Information Sciences, 42,* 57–76.

Kahraman, C., Ruan, D., & Dogan, I. (2003). Fuzzy group decision-making for facility location selection. *Information Sciences, 157,* 135–215.

Kontodimopoulos, N., & Niakas, D. (2005). Efficiency measurement of hem dialysis units in Greece with data envelopment analysis. *Health Policy, 71,* 195–204.
Kontodimopoulos, N., & Niakas, D. (2006). A 12-year analysis of Malmquist total factor productivity in dialysis facilities. *Journal of Medical Systems, 30,* 333–342.
Kornbluth, J. S. H. (1991). Analysing policy effectiveness using cone restricted data envelopment analysis. *Journal of the Operational Research Society, 42*(12), 1097–1104.
Lavado, R. F., & Cabanda, E. C. (2009). The efficiency of health and education expenditures in the Philippines. *Central European Journal of Operations Research, 17,* 275–291.
Linna, M. (1998). Measuring hospital cost efficiency with panel data models. *Econometrics and Economics, 7,* 415–427.
Mikhailov, L. (2003). Deriving priorities from fuzzy pair wise comparison judgments. *Fuzzy Sets and Systems, 134,* 365–385.
Ozgen, H., & Ozcan, Y. (2004). Longitudinal analysis of efficiency in multiple output dialyses. *Health Care Management Sciences, 7,* 253–261.
Rouyendegh, B. D. (2011). The DEA and intuitionistic fuzzy TOPSIS approach to departments' performances: A pilot study. *Journal of Applied Mathematics.* https://doi.org/10.1155/2011/712194.
Rouyendegh, B. D., & Erol, S. (2010). The DEA–FUZZY ANP department ranking model applied in Iran Amirkabir University. *Acta Polytechnica Hungarica, 7,* 103–114.
Saaty, T. L. (1980). *The analytic hierarchy process.* New York: McGraw-Hill.
Saaty, T. L., & Vargas, L. G. (2006). Decision making with the analytic network process. Spring Science, LLC, pp 1–23.
Sinuany-Stern, Z., Mehrez, A., & Hadad, Y. (2000). An AHP/DEA methodology for ranking decision making units. *International Transactions in Operational Research, 7,* 109–124.
Tanbour, M. (1997). The impact of health care policy initiatives on productivity. *Health Economics, 6,* 57–70.
Wang, Y. M., Liu, J., & Elhag, T. M. S. (2008). An integrated AHP-DEA methodology for bridge risk assessment. *Computers & Industrial Engineering, 54*(3), 1–13.
Zadeh, L. A. (1965). Fuzzy sets. *Information and Control, 8*(3), 338–353.
Zhu, J. (2015). DEA based benchmarking models. *International Series in Operations Research & Management Science, 221,* 291–308.

Chapter 13
Environmental Application of Carbon Abatement Allocation by Data Envelopment Analysis

Anyu Yu, Simon Rudkin, and Jianxin You

Abstract China's commitment to significantly reducing carbon emissions faces the twin challenges of focusing on costly reduction efforts, whilst preserving the rapid growth that has defined the country's recent past. However, little work has been able to meaningfully reflect the collaborative way in which provinces are assigned targets on a subnational regional basis. Suggesting a meta-frontier allocation approach by using data envelopment analysis (DEA), this chapter introduces the potential collaboration between heterogeneous industrial units to the modelling framework. Our theoretical work exposits the roles collectives of industrial decision making units may play in optimizing against multiple target functions, doing so whilst recognizing the two objectives of income maximization and pollution abatement cost minimization. Considering the period 2012–2014, we illustrate clearly how China's three regional collaborations interact with the stated aims of national policy. Developed eastern China may take on greater abatement tasks in the short term, thus freeing central and western China to pursue the economic growth which will then support later abatement. Policymakers are thus given a tool through which an extra layer of implementation can be evaluated between the national allocation and setting targets for regional individual decision making units. China's case perfectly exemplifies the conflicts which must be accounted for if the most economical and efficient outcomes are to be achieved.

This chapter is based upon article in the Applied Energy, Yu A, You J, Rudkin S, Zhang H, Industrial carbon abatement allocations and regional collaboration: Re-evaluating China through a modified data envelopment analysis, 2019, 233–234,232–243, with permission form Elsevier.

A. Yu (✉)
International Business School, Zhejiang Gongshang University, Hangzhou, China
e-mail: yuanyu1990yy@163.com

A. Yu · J. You
School of Economics and Management, Tongji University, Shanghai, China
e-mail: yjx2256@vip.sina.com

S. Rudkin
School of Management, Swansea University, Swansea, UK
e-mail: simonrudkin@outlook.com

V. Charles et al. (eds.), *Data Science and Productivity Analytics*,
International Series in Operations Research & Management Science 290,
https://doi.org/10.1007/978-3-030-43384-0_13

Keywords Data envelopment analysis · Carbon allocation · Carbon abatement cost · Regional collaboration

13.1 Introduction

China's emergence as the "global factory" sits firmly at odds with the governmental desire to be a world leader in tackling climate change (Hilton and Kerr 2017). At the Copenhagen climate change summit of 2009 the Chinese government committed to reduce carbon intensity, carbon dioxide (CO_2) emissions per unit gross domestic product; by 45% compared to its 2005 value. Carbon emission abatement is of major national importance and involves collaboration between the central government and the regional governments where polluting factories are located. China's approach thus far has been to identify targets for provinces (regions) and then to construct carbon trading markets to perfect the allocation at the firm level (Wang et al. 2016). Pilot markets for carbon emissions trading have been launched in seven Chinese provinces or municipalities in 2003, that is, in Shenzhen, Beijing, Tianjin, Shanghai, Guangdong, Hubei, and Chongqing (Wang et al. 2016).

China's central role in global carbon dioxide abatement is the focus of a plethora of academic and policy works approaching the issue from multiple disciplines. In literature, regional carbon emissions cannot be reduced independently, but must be with the consideration of all related production factors (e.g., resource input, economic output, and energy consumption) (Wu et al. 2016a; Wang et al. 2016; Zhou et al. 2015). Thus, carbon abatement tasks should be allocated (called carbon allocation for short) in a comprehensive way considering all the related production factors. DEA, as an important non-parametric method for measuring carbon allocation in a total-factor evaluation is a natural choice for adoption here.

The allocation estimation of resources is central to DEA techniques (Feng et al. 2015). DEA methods can be used to allocate the resources of input, output, or both (Beasley 2003). Two primary types of allocations exist within DEA: fixed cost allocation and resource allocation and, representing fixed costs allocated to each decision unit or the process of allocating the resource. Each is based on the DEA efficiency results from the allocation (Du et al. 2014). In DEA, the fixed cost is regarded as a complement of inputs, or outputs, in allocation, Further the fixed cost forms a single type of input measure in efficiency evaluation. Meanwhile, resource allocation is assumed to optimize inputs and outputs simultaneously, the selection being subject to corresponding limitations on resources or production possibilities. DEA also offers another mechanism called centralized allocation, which aims to allocate resources by a centralized decision maker controlling overall units (Fang 2013). The centralized allocation model is derived from Lozano and Villa (2004). There are three objectives for centralized allocation: maximizing desirable outputs, minimizing undesirable outputs, and minimizing inputs (Lozano et al. 2009). An important feature of the centralized allocation is that the optimized target is to consider the overall benefits for all the decision makers (Fang 2013; Lozano and Villa 2004), but ignores the benefits of individuals (Feng et al. 2015).

DEA allocation approaches have been the basis of many studies of carbon emission allocations. Gomes and Lins (2008) proposed a zero-sum gains DEA model to allocate CO_2 emission permits amongst countries. Lozano et al. (2009) provided three levels of centralized models to consider the allocation of emission permits. The application of centralized DEA allocation models to carbon allocations is proposed by Feng et al. (2015) and Wu et al. (2016b). Sun et al. (2014) analysed variations of the mechanisms to allocate permits amongst a group of manufacturing companies. Zhou et al. (2014) introduced spatial, temporal, and a joint spatial-temporal allocation strategies for controlling CO_2 emissions at the provincial level in China. Crucially however, the carbon abatement costs of those allocations were omitted, offering us the opportunity to improve thereupon. Unlike the aforementioned studies, this chapter uses an improved DEA approach to allocate carbon abatement tasks.

A major contribution of this chapter is to propose a meta-frontier DEA allocation model, and this lays in its investigation of potential regional collaborations on carbon abatement allocations. This allocation model can ensure the DEA measure considering the existence of heterogeneous technologies for different groups of allocated decision making units (DMUs). The innovation of the DEA model presented here nests a suitable extension for dealing with the large amount of DMUs present in a big data context. Regional collaborations in carbon allocation reduce the heterogeneity of conflicting desires but there remains a tension with the centralized allocation scheme. However many DMUs are included the centralized allocation suffers implementation difficulties arising from the inconsistency between the interests of individuals and the overall economy (Feng et al. 2015). Inherent interest conflicts exist between coalitions and the national interest. Coalitions here stand for the groups homogenous within the group and heterogeneous across groups with heterogeneous technologies.

Regional collaboration activities for carbon abatement have huge potential. In the extant literature, regional collaborations are being applied in increasing numbers. These works focus on the issues of political challenges relating to energy transfer, climate change mitigation, resource sharing, and energy security (Uddin and Taplin 2015; Huda and Mcdonald 2016; Srivastava and Misra 2007). This collaboration can be an attempt for the meta-frontier allocation modelling. The collaborations form the basis of working together on joint interests and are a platform for agreed abatement task allocations. Provinces are more likely to accept costs if they see that their neighbours are also taking their perceived fair share as well. An existing case is the joint air pollution control amongst Beijing-Tianjin-Hebei and surrounding regions decided by China's National Development and Reform Commission and their relative ministries (Zhang et al. 2014).

As another contribution, carbon emission abatement in production generates corresponding costs. China's industrial production systems have significant disparities of resources, economic well-being, and technological capabilities for carbon abatement. In this circumstance, carbon abatement costs, such as the costs of industrial structure modification, costs of modifying the structure of energy consumption, or costs of technological updates, may vary wildly amongst different regions (Wang et al. 2016; Cui et al. 2014). Wang et al. (2016) identified that larger carbon abatement tasks would be allocated to regions with the lowest cost; be that either the financial or opportunity cost associated with reducing production. This can then accelerate the forming of regional collaboration, and regions with different allocated tasks would become more incentivized to further reduce their abatement task in collaboration with regions with lower abatement costs. Thus, the carbon emission abatement cost can also be treated as an incentive for regional collaboration. Failure to account for this within the allocation process is liable to bring unreasonable and inequitable outcomes.

Existing DEA carbon allocation estimations primarily aim to either maximize the potential gross domestic product (GDP) gains, or minimize the carbon emissions solely. Costs generated from the abatement processes are always ignored. A notable exception is Wu et al. (2016a), which estimated carbon allocations considering corresponding costs. However, Wu et al. (2016a) incorporated the simplifying assumption that the price of the carbon emission allowance is equal to the cost of allocation. Such an assumption would mean a single value for the whole of China, which is unrepresentative.

A modified DEA model is proposed to allocate carbon emission abatement tasks for regional industrial systems. The aforementioned carbon allocation model is extended to incorporate regional collaborations and detect their impacts. To achieve this purpose, a meta-frontier DEA approach for carbon abatement allocations is proposed. Moreover, we treat the minimized total carbon abatement costs as a part of the allocation target to obtain effective allocation results and explore potential regional collaborations; an estimation thereof is also derived.

We make two key contributions to the literature. Methodologically, we adjust the basic allocation model into an improved meta-frontier form to analyse the impact of technological heterogeneity for different DMU groups. Empirically, we extend the proposed DEA allocation method to focus on the potential for regional collaboration and the evaluation thereof, an area typically ignored by the literature on regional allocations. Moreover, our modified DEA offers an improved carbon abatement allocation estimation considering carbon abatement costs at the regional level, thus providing a more robust starting point for analyzing regional collaboration. Through this study, we are able to make stronger recommendations for the regional allocation of carbon abatement tasks.

The remainder of the chapter is organized as follows. First, we outline our DEA approach, introducing the modifications for carbon abatement allocations. Second, we then present the Chinese data and our results on potential allocations. Finally, we draw conclusions and provide signposts for future carbon abatement and broader environmental policies.

13.2 DEA Meta-Frontier Methodology for Carbon Abatement Allocation

In this section, we aim to illustrate the extended meta-frontier DEA allocation model. Meta-frontier DEA model is first proposed in O'Donnell et al. (2008) to evaluate the performance efficiency considering the existence of technology heterogeneity. This method extension can help to solve the resource allocation issue for DMUs with heterogeneous technologies, and the heterogeneity is identified as meta-technology frontiers in DEA. Heterogeneous technologies are also considered in the DEA allocation model. Centralized fixed cost and resource allocation models are first introduced in Ding et al. (2018) and adopted in the empirical analysis in He et al. (2018). By detecting the potential information on heterogeneous technologies, meta-frontier DEA allocation model is suitable for dealing with the evaluation with a large amount of DMUs. Consequently, it can be regarded as an important method extension under the big data context.

13.2.1 A Meta-Frontier DEA Approach for the Centralized Allocations

We first propose a basic centralized allocation model. Here the proposed centralized model setting is similar to Feng et al. (2015). DEA allocation analysis has three traits: (1) The performance evaluation is formed by all the DMUs and output targets may not be achievable in the short term; (2) after the certain amounts of permits or resources are allocated, there must be changes in DMU production (Wu et al. 2016b); (3) the ex-ante planning is adopted in DEA allocation, and the allocation results are used to forecast the performance of resource utilization in the next period (Feng et al. 2015). However such adopted practice does not preclude the analysis of alternatives.

Assume there are n DMUs participated in the allocation mechanism, denoted by decision making unit j (i.e., DMU j, $j = 1, 2, \ldots, n$). Each DMU uses h inputs x to produce r outputs y. s is the allocated resource or cost, v is the number of resource s available to the system, and Δs is the abatement amount. In all that follows Δs is the carbon abatement task. $\hat{y}$ here denotes the expected output in the allocation. g is the total allocated resource amount for all the DMUs. Notably, this model aims to maximize the potential output by conducting the resource allocation. The intensity of production within DMU$_i$ is given as λ_i. It is noteworthy that, model (13.1) is a constant returns to scale (CRS) model, which can capture the overall technical efficiency (pure technical efficiency and scale efficiency) of the evaluated DMU, and can satisfy all relevant production technologies (Zhou and Ang 2008). Thus we proceed with CRS to maintain a simplicity of exposition. Resulting is (13.1).

$$\text{Min} \sum_{i=1}^{n} \hat{y}_i$$
$$\text{s.t.} \sum_{j=1}^{n} \lambda_j x_{hj} \le x_{hi},\ \ h = 1, 2, \ldots H,$$
$$\sum_{j=1}^{n} \lambda_j y_{rj} \ge \hat{y}_{ri},\ \ r = 1, 2, \ldots R,$$
$$\sum_{j=1}^{n} \lambda_j s_{vj} = s_{vi} - \Delta s_{vi},\ \ v = 1, 2, \ldots V,$$
$$\sum_{j=1}^{n} \Delta s_{vj} = g_v,$$
$$\hat{y}_{ri}, \lambda_j \ge 0,\ \ j = 1, 2, \ldots, n. \tag{13.1}$$

Model (13.1) is extended into a meta-frontier allocation model as follows. Based on O'Donnell et al. (2008), the properties of the meta-frontier DEA model are derived. T^p and T^{meta} are used to represent the production technology for the group frontier and the meta-frontier. If $(x_h, y_r, s_v) \in T$, then $(x_h, y_r, s_v) \in T^{meta}$; $T^{meta} = \{T^1, T^2, \ldots, T^p\}$, $p = 1, 2, \ldots, P$. *D(p)* denotes the subset of observed DMU belonging to the specific group frontier *p*. This setting can express the existence, the heterogeneous technology, and the DMUs within the groups of different production technologies are projected to the corresponding sub-frontier. In conventional allocation, the resource should be allocated to make DMUs that can have higher performance efficiency. If the technological heterogeneity is not considered, it is not fair for DMUs within low production technology levels to be asked to improve their performance efficiency in allocation. Thus, the meta-frontier allocation model is proposed. This model setting is also discussed in Ding et al. (2018), but in a multiplier DEA model. Considering the technology heterogeneity, the meta-frontier model is suitable for the allocation attempt for measuring a large amount of DMUs.

$$\text{Min} \sum_{i=1}^{n} \hat{y}_i$$
$$\text{s.t.} \sum_{j \in D(p)} \lambda_{ij} x_{hj} \le x_{hi},\ i = 1, 2, \ldots, n,\ p = 1, 2, \ldots, P, \forall i \in D(p), h = 1, 2, \ldots H,$$
$$\sum_{j \in D(p)} \lambda_{ij} y_{rj} \ge \hat{y}_{ri},\ i = 1, 2, \ldots, n,\ p = 1, 2, \ldots, P, \forall i \in D(p), r = 1, 2, \ldots R,$$
$$\sum_{j \in D(p)} \lambda_{ij} s_{vj} = s_{vi} - \Delta s_{vi},\ i = 1, 2, \ldots, n,\ p = 1, 2, \ldots, P, \forall i \in D(p), v = 1, 2, \ldots V,$$
$$\sum_{j \in D(p)} \Delta s_{vj} = g_v,\ p = 1, 2, \ldots, P, \forall j \in D(p),$$
$$\hat{y}_{ri}, \lambda_{ij} \ge 0,\ j = 1, 2, \ldots, n. \tag{13.2}$$

Furthermore, we introduce time variation into the proposed allocation model. Here, we introduce the panel data which cross successive time periods t, and $t = 1, 2, \ldots T$. The allocation result then can be optimized for a multi-period observation as model (13.3). The technological heterogeneity and time variation can help disclose the information hiding in the collected data. Maintaining some tractability, we assume that the production technology keeps constant across multi-periods in the observations, and the technological heterogeneity is derived from the difference amongst DMUs.

$$
\begin{aligned}
&\text{Min} \sum_{t=1}^{T} \sum_{i=1}^{n} \hat{y}_{ri}^{t} \\
&\text{s.t.} \sum_{j \in D(p)} \lambda_{ij}^{t} x_{hj}^{t} \le x_{hi}^{t}, i = 1, 2, \ldots, n, \ p = 1, 2, \ldots, P, \forall i \in D(p), h = 1, 2, \ldots H, t = 1, 2, \ldots T, \\
&\sum_{j \in D(p)} \lambda_{ij}^{t} y_{rj}^{t} \ge \hat{y}_{ri}^{t}, \ i = 1, 2, \ldots, n, p = 1, 2, \ldots, P, \forall i \in D(p), r = 1, 2, \ldots R, \\
&\sum_{j \in D(p)} \lambda_{ij}^{t} s_{vj}^{t} = s_{vi}^{t} - \Delta s_{vi}^{t}, \ i = 1, 2, \ldots, n, \ p = 1, 2, \ldots, P, \forall i \in D(p), v = 1, 2, \ldots V, \\
&\sum_{j \in D(p)} \Delta s_{vj}^{t} = g_{v}, \ p = 1, 2, \ldots, P, \forall j \in D(p) \\
&\hat{y}_{ri}^{t}, \lambda_{ij}^{t} \ge 0, \ j = 1, 2, \ldots, n.
\end{aligned} \tag{13.3}
$$

13.2.2 A Modified DEA Approach for Carbon Abatement Allocations

The allocation goal for China's central government is to achieve the given CO_2 abatement task, with minimized carbon abatement costs, and maximized regional economic output from industrial production. This chapter considers allocation process affected by national collaboration at first, termed national allocation for short. This indicates that the industrial production system of each region has a higher level of involvement in the handling of its emission abatement target. Each region is treated as a co-operator in the national platform of carbon abatement allocation, the central government then setting levels. The national allocation also assumes that each region uses all efforts to reduce its allocated carbon abatement cost to minimize the total national abatement costs. This could be regarded as a centralized resource allocation problem (Lozano and Villa 2004) and derives from Wang et al. (2016) and Wu et al. (2016a) amongst others.

In this application assume that there are n independent evaluation regions. In common with past work, in the process of regional industrial production each region employs labour, capital, and energy as inputs and produces both desirable and undesirable outputs. We consider labour (l), capital stock (k), and energy consumption (e) as the three inputs. Gross Domestic Product (GDP) (y) and CO_2 emissions (c) play the role of desirable and undesirable outputs, respectively. Following Wang et al. (2012), this chapter considers that CO_2 emissions are subjected by an equal constraint, which indicates a null-joint relationship between carbon emission and GDP (Sueyoshi and Goto 2012). The relationship means the joint-production process between GDP and CO_2 emissions.

In model (13.4), $\hat{y}_i$ indicates the maximized post allocation GDP output for region i, and the target function means the maximized GDP output for all the regions. Notably, $\hat{y}_i$ is affected by all the factors in the carbon allocation, including the expenditure of carbon abatement. Δc_i denotes the abatement task of CO_2 for region i, and is decided by the central government. b is the total future CO_2 abatement task and thus the constraint $\sum_{i=1}^{n} \Delta c_i = b$ means that the sum of CO_2 abatement tasks of all the regions should be equal to the national total. An upper limit on the size of the task to be given to any region is set at c_i^u .

We assume that each region participates in the carbon abatement processes. Thus $0 \le \Delta c_i \le c_i^u$ implies that the most any region can be asked to do is eliminate its current output c_i and the lower limit any potential ask implies that the region does nothing. All the other variables and constraints have identical meanings to those in models (13.2) and (13.3). We aim to obtain the optimal solutions of λ_{ij}^*, Δc_i^*, and $\hat{y}_i^*$ by solving model (13.4). To this end, $(c_i - \Delta c_i^*)$ denotes the optimal allocated carbon emission quota for region i in next period at the current production level. As in Wang et al. (2016) and Wu et al. (2016b), $\hat{y}_i^*$ here represents the maximized GDP. Moreover, the optimal GDP output and carbon emission in model (13.4) are replaced by $\hat{y}_i$ and $c_i - \Delta c_i$. These are the optimized GDP output and remaining carbon emission, in the allocation process, but are not obtained by the same proposition as the weak disposability. This indicates that the potential technology improvement could be assumed in the regional carbon abatement to reduce the loss of GDP output.

$$
\begin{aligned}
&\text{Max} \sum_{i=1}^{n} \hat{y}_i \\
&\text{s.t.} \quad \sum_{j=1}^{n} \lambda_{ij} e_j \le e_i, i = 1, 2, \ldots, n, \\
&\qquad \sum_{j=1}^{n} \lambda_{ij} k_j \le k_i, i = 1, 2, \ldots, n, \\
&\qquad \sum_{j=1}^{n} \lambda_{ij} l_j \le l_i, i = 1, 2, \ldots, n, \\
&\qquad \sum_{j=1}^{n} \lambda_{ij} y_j \ge \hat{y}_i, i = 1, 2, \ldots, n,
\end{aligned}
$$

$$
\begin{aligned}
&\sum_{j=1}^{n} \lambda_{ij} c_j = c_i - \Delta c_i, \ \ i = 1, 2, \ldots, n, \\
&\sum_{i=1}^{n} \Delta c_i = b, \\
&0 \le \Delta c_i \le c_i^u, \\
&\hat{y}_i, \lambda_j \ge 0, \ j = 1, 2, \ldots, n.
\end{aligned} \tag{13.4}
$$

Furthermore, the model is also extended to model (13.5) based on the multi-period allocation process ($t = 1, 2, \ldots T$) for panel data. This model assumes the allocation process is conducted in a multi-period observation, and the carbon abatement is allocated with the consideration of the total abatement target for the whole period.

$$
\begin{aligned}
\text{Max} &\sum_{i=1}^{n} \hat{y}_i^t \\
\text{s.t.} \ &\sum_{j \in D(p)} \lambda_{ij}^t e_j^t \le e_i^t, i = 1, 2, \ldots, n, \ p = 1, 2, \ldots, P, \forall i \in D(p), \\
&\sum_{j \in D(p)} \lambda_{ij}^t k_j^t \le k_i^t, i = 1, 2, \ldots, n, \ p = 1, 2, \ldots, P, \forall i \in D(p) \\
&\sum_{j \in D(p)} \lambda_{ij}^t l_j^t \le l_i^t, i = 1, 2, \ldots, n, \ p = 1, 2, \ldots, P, \forall i \in D(p) \\
&\sum_{j \in D(p)} \lambda_{ij}^t y_j^t \ge \hat{y}_i^t, i = 1, 2, \ldots, n, \ p = 1, 2, \ldots, P, \forall i \in D(p) \\
&\sum_{j \in D(p)} \lambda_{ij}^t c_j^t = c_i^t - \Delta c_i^t, i = 1, 2, \ldots, n, \ p = 1, 2, \ldots, P, \forall i \in D(p) \\
&\sum_{j \in D(p)} \Delta c_j^t = b, \ p = 1, 2, \ldots, P, \forall j \in D(p), \\
&0 \le \Delta c_i^t \le c_i^{tu}, \\
&\hat{y}_i^t, \lambda_{ij}^t \ge 0, \ j = 1, 2, \ldots, n, t = 1, 2, \ldots T.
\end{aligned} \tag{13.5}
$$

Furthermore, carbon abatement costs for the observed DMUs are also introduced in the meta-frontier model. The carbon abatement cost p_{ci} is regarded as an exogenous variable for the following models. Current studies of centralized carbon allocation, to the best of our knowledge, always focus on the target of maximized economic output or minimized carbon emission solely (Feng et al. 2015; Wu et al. 2016b). This chapter proposes a DEA allocation model, which aims to obtain the maximized GDP outputs and minimized carbon abatement costs simultaneously after the carbon abatement tasks are allocated.

$$
\begin{aligned}
\text{Min}\ & \frac{\sum_{i=1}^{n} p_{ci} \cdot \Delta c_i}{\sum_{i=1}^{n} \hat{y}_i} \\
\text{s.t.}\ & \sum_{j=1}^{n} \lambda_{ij} e_j \le e_i, i = 1, 2, \ldots, n, \\
& \sum_{j=1}^{n} \lambda_{ij} k_j \le k_i, i = 1, 2, \ldots, n, \\
& \sum_{j=1}^{n} \lambda_{ij} l_j \le l_i, i = 1, 2, \ldots, n, \\
& \sum_{j=1}^{n} \lambda_{ij} y_j \ge \hat{y}_i, i = 1, 2, \ldots, n, \\
& \sum_{j=1}^{n} \lambda_{ij} c_j = c_i - \Delta c_i, i = 1, 2, \ldots n, \\
& \sum_{i=1}^{n} \Delta c_i = b, \\
& 0 \le \Delta c_i \le c_i^u, \\
& \hat{y}_i, \lambda_j \ge 0, j = 1, 2, \ldots, n.
\end{aligned} \tag{13.6}
$$

Though carbon emission abatement is environmentally desirable potential costs must be considered and hence the discussion is extended to the treatment of the issue of income. Our model delivers on both objectives, incorporating regional collaboration for the first time.

Definition 13.1 The economical level of total national carbon abatement is defined as:

$$
EL = \frac{\sum_{i=1}^{n} \hat{y}_i}{\sum_{i=1}^{n} p_{ci} \cdot \Delta c_i}, i = 1, 2, \ldots, n \tag{13.7}
$$

The economical level is the reciprocal of the national target function, which is the ratio of maximized industrial GDP output to total carbon abatement costs for the country. Hence this ratio can measure if the national carbon abatement allocation cost is economical for its economic gain. The economical level is adopted to compare the national economic performance derived from different allocation plans.

Model (13.6) is non-linear, but it can be transformed into the linear model (13.8). In model (13.8), $\eta_{ij} = z\lambda_{ij}$ denotes the transformed intensity variable corresponding to λ_{ij} . $\hat{Y}_i = z\hat{y}_i$ and $\Delta C_i = z\Delta c_i$ represent the transformed values for $\hat{y}_i$ and Δc_i , respectively. By solving model (13.8), we can obtain the optimal values of $\hat{Y}_i^*$, ΔC_i^* , η^* and z^* . Based on these results, regional optimal outputs of industrial GDP and carbon abatement tasks are acquired, $\hat{y}_i^* = \hat{Y}_i^*/z^*$, $\Delta c_i^* = \Delta C_i^*/z^*$, respectively.

$$
\begin{aligned}
& \text{Min} \sum_{i=1}^{n} p_{ci} \cdot \Delta C_i \\
\text{s.t.} \quad & \sum_{i=1}^{n} \hat{Y}_i = 1, i = 1, 2, \ldots, n, \\
& \sum_{j=1}^{n} \eta_{ij} e_j \leq z e_i, i = 1, 2, \ldots, n, \\
& \sum_{j=1}^{n} \eta_{ij} k_j \leq z k_i, i = 1, 2, \ldots, n, \\
& \sum_{j=1}^{n} \eta_{ij} l_j \leq z l_i, i = 1, 2, \ldots, n, \\
& \sum_{j=1}^{n} \eta_{ij} y_j \geq \hat{Y}_i, i = 1, 2, \ldots, n, \\
& \sum_{j=1}^{n} \eta_{ij} c_j = z c_i - \Delta C_i, i = 1, 2, \ldots, n, \\
& \sum_{i=1}^{n} \Delta C_i = z b, \\
& 0 \leq \Delta C_i \leq z c_i^u, \\
& \hat{Y}_i, \eta_j \geq 0, j = 1, 2, \ldots, n.
\end{aligned}
\tag{13.8}
$$

Furthermore, the allocation model with the multi-period observations can be illustrated as model (13.9).

$$
\begin{aligned}
& \text{Min} \sum_{i=1}^{n} p_{ci}^t \cdot \Delta C_i^t \\
\text{s.t.} \quad & \sum_{i=1}^{n} \hat{Y}_i^t = 1, i = 1, 2, \ldots, n, \\
& \sum_{j=1}^{n} \eta_{ij}^t e_j^t \leq z e_i^t, i = 1, 2, \ldots, n,
\end{aligned}
$$

$$
\begin{aligned}
&\sum_{j=1}^{n} \eta_{ij}^{t} k_j^{t} \le z k_i^{t}, i = 1, 2, \ldots, n, \\
&\sum_{j=1}^{n} \eta_{ij}^{t} l_j^{t} \le z l_i^{t}, i = 1, 2, \ldots, n, \\
&\sum_{j=1}^{n} \eta_{ij}^{t} y_j^{t} \ \ge \hat{Y}_i^{t}, i = 1, 2, \ldots, n, \\
&\sum_{j=1}^{n} \eta_{ij}^{t} c_j^{t} = z c_i^{t} - \Delta C_i^{t}, i = 1, 2, \ldots, n, \\
&\sum_{i=1}^{n} \Delta C_i^{t} = zb, \\
&0 \le \Delta C_i^{t} \le z c_i^{tu}, \\
&\hat{Y}_i^{t}, \eta_{ij}^{t} \ge 0, j = 1, 2, \ldots, n, t = 1, 2, \ldots T.
\end{aligned} \tag{13.9}
$$

In practice, the national collaboration may be hard to achieve as it requires the maximum possible levels of trust and involvement for the collaborators. Thus we propose a second scenario that a region can achieve limited collaborations with its neighbours to pursue the most economical carbon allocation for that grouping. In this circumstance, regions may jointly plan their allocated carbon reduction amounts and reassess their practical carbon emission abatement tasks. Collaborations here are assumed as regional groups with technology heterogeneity: this is modelled by the proposed meta-frontier allocation model. The corresponding DEA allocation model is as follows:

$$
\text{Min} \frac{\sum_{i=1}^{n} p_{ci} \cdot \Delta c_i}{\sum_{i=1}^{n} \hat{y}_i}
$$

$$
\begin{aligned}
\text{s.t.} \quad &\sum_{j \in D(p)} \lambda_{ij} e_j \le e_i, i = 1, 2, \ldots, n, p = 1, 2, \ldots, P, \forall i \in D(p) \\
&\sum_{j \in D(p)} \lambda_{ij} k_j \le k_i, i = 1, 2, \ldots, n, p = 1, 2, \ldots, P, \forall i \in D(p) \\
&\sum_{j \in D(p)} \lambda_{ij} l_j \le l_i, i = 1, 2, \ldots, n, p = 1, 2, \ldots, P, \forall i \in D(p) \\
&\sum_{j \in D(p)} \lambda_{ij} y_j \ge \hat{y}_i, i = 1, 2, \ldots, n, p = 1, 2, \ldots, P, \forall i \in D(p)
\end{aligned}
$$

$$
\begin{aligned}
&\sum_{j \in D(p)} \lambda_{ij} c_j = c_i - \Delta c_i, i = 1, 2, \ldots, n, p = 1, 2, \ldots, P, \forall i \in D(p) \\
&\sum_{j \in D(p)} \Delta c_j = b, p = 1, 2, \ldots, P, \forall j \in D(p), \\
&0 \le \Delta c_i \le c_i^u, \\
&\hat{y}_i, \lambda_j \ge 0, j = 1, 2, \ldots, n.
\end{aligned}
\tag{13.10}
$$

Once again the model is non-linear and so model (13.10) is transformed into the linear model (13.11). *D(p)* denotes the subset of observed DMU belonging to the regional coalition *p*. A coalition of size *Q* is made up of multiple DMU_q ($q = 1, 2, \ldots, Q$). All the other variables and constraints in model (13.10) and (13.11) have the similar interpretations as those in models (13.6) and (13.8), respectively. The main difference between the models (13.8) and (13.11) is the participators of allocation. In scenario 1, model (13.8) aims to reach the national optimal carbon abatement cost by CO_2 abatement allocation at the national level. In scenario 2, model (13.11) aims to reach the national allocation target by allocating carbon abatement tasks across the local collaboration regions. DMUs only aim to obtain the optimal allocation solution to their own *D(p)*.

$$
\begin{aligned}
&\text{Min} \sum_{i=1}^{n} p_{ic} \cdot \Delta C_i \\
\text{s.t.} \quad &\sum_{i=1}^{n} \hat{Y}_i = 1, i = 1, 2, \ldots, n, \\
&\sum_{j \in D(p)} \eta_{ij} e_j \le z e_i, i = 1, 2, \ldots, n, \ \ p = 1, 2, \ldots, P, \forall i \in D(p) \\
&\sum_{j \in D(p)} \eta_{ij} k_j \le z k_i, i = 1, 2, \ldots, n, \ \ p = 1, 2, \ldots, P, \forall i \in D(p) \\
&\sum_{j \in D(p)} \eta_{ij} l_j \le z l_i, i = 1, 2, \ldots, n, \ \ p = 1, 2, \ldots, P, \forall i \in D(p) \\
&\sum_{j \in D(p)} \eta_{ij} y_j \ge \hat{Y}_i, i = 1, 2, \ldots, n, \ \ p = 1, 2, \ldots, P, \forall i \in D(p) \\
&\sum_{j \in D(p)} \eta_{ij} c_j = z c_i - \Delta C_i, i = 1, 2, \ldots, n, \ \ p = 1, 2, \ldots, P, \forall i \in D(p) \\
&\sum_{j \in D(p)} \Delta C_j = z b, p = 1, 2, \ldots, P, \forall j \in D(p), \\
&0 \le \Delta C_i \le z c_i^u, \\
&\hat{Y}_i^t, \eta_{ij}^t \ge 0, j = 1, 2, \ldots, n, t = 1, 2, \ldots T.
\end{aligned}
\tag{13.11}
$$

In model (13.11), the regional collaboration coalition is viewed as treating all participators as having the same observation set *D(p)*. Participators of different coalitions are evaluated in different frontiers within the DEA allocation model (13.11). Models (13.10) and (13.11) are meta-frontier DEA allocation models. We assume that all members of the coalition share the same best practice through the available inputs and outputs. This constraint can help to identify the impacts of technology heterogeneity in national allocation, for example, geographic disparities of economic fundamentals and technological levels. Furthermore, this allocation process may be labelled as a resource-pooling-only game of lower level collaboration, that is, one modification of the Linear Transformation of Products (LTP) games, proposed by Timmer et al. (2000) and extended by Lozano (2013) in a DEA form. In this case, regions jointly plan and allocation of pooled available resources (CO_2 abatement tasks) in their own coalition.

Similarly, model (13.11) is also extended to deal with the panel data for a multi-period measure as model (13.12).

$$
\begin{aligned}
&\text{Min} \sum_{i=1}^{n} p_{ci}^{t} \cdot \Delta C_i^t \\
\text{s.t.}\quad &\sum_{i=1}^{n} \hat{Y}_i^t = 1, i = 1, 2, \ldots, n, \\
&\sum_{j \in D(p)} \eta_{ij}^t e_j^t \le z e_i^t, i = 1, 2, \ldots, n, p = 1, 2, \ldots, P, \forall i \in D(p) \\
&\sum_{j \in D(p)} \eta_{ij}^t k_j^t \le z k_i^t, i = 1, 2, \ldots, n, p = 1, 2, \ldots, P, \forall i \in D(p) \\
&\sum_{j \in D(p)} \eta_{ij}^t l_j^t \le z l_i^t, i = 1, 2, \ldots, n, p = 1, 2, \ldots, P, \forall i \in D(p) \\
&\sum_{j \in D(p)} \eta_{ij}^t y_j^t \ge \hat{Y}_i^t, i = 1, 2, \ldots, n, p = 1, 2, \ldots, P, \forall i \in D(p) \\
&\sum_{j \in D(p)} \eta_{ij}^t c_j^t = z c_i^t - \Delta C_i^t, i = 1, 2, \ldots, n, p = 1, 2, \ldots, P, \forall i \in D(p) \\
&\sum_{j \in D(p)} \Delta C_j^t = z b, p = 1, 2, \ldots, P, \forall j \in D(p), \\
&0 \le \Delta C_i^t \le z c_i^{ut}, \\
&\hat{Y}_i^t, \eta_{ij}^t \ge 0, j = 1, 2, \ldots, n, t = 1, 2, \ldots T.
\end{aligned}
\tag{13.12}
$$

Proposition 13.1 *National collaboration produces an outcome which weakly dominates other independent regional allocations.*

Proof It is readily apparent that the optimal solution of carbon abatement task allocation in model (13.12) is a feasible solution to that in model (13.9). Thus the target function value obtained from model (13.12) is less than or equal to that from model (13.9).

Property 13.1 *The optimal value of model* (13.11) *is convex with respect to the carbon abatement amount* ΔC_i. *The target function* $f(\Delta C_i)$ *is convex with respect to* ΔC_i.

Proof For any carbon allocation task ΔC_i which satisfies the constraint $0 \le \Delta C_i \le zc_i^u$, we can obtain the optimal values of $\hat{Y}_i^*$, ΔC_i^* , η_j^* and z^* by solving model (13.11). Here we assume that, ($\hat{Y}_{i1}$, η_{j1} , z_1) and ($\hat{Y}_{i2}$, η_{j2}, z_2) are the corresponding optimal solutions for the optimal ΔC_{i1} and $\Delta C_{i2}(\Delta C_{i1}, \Delta C_{i2} \in [0, zc_i^u])$ As in Feng et al. (2015), a feasible solution by a linear combination is constructed, that is, $\omega\hat{Y}_{i1} + (1-\omega)\hat{Y}_{i2}$, $\omega\Delta C_{i1} + (1-\omega)\Delta C_{i2}$ and $\omega z_1 + (1-\omega)z_2$, $0 \le \omega\Delta C_{i1} + (1-\omega)\Delta C_{i2} \le zc_i^u$, $0 \le \omega \le 1$, $j = 1, 2, \ldots, n$. Model (13.11) is solved with the constructed linear combination solution. The optimal result of the objective target should be less than or equal to $\omega f_1 + (1-\omega) f_2$, that is, $f(\omega\Delta C_{i1} + (1-\omega)\Delta C_{i2}) \le \omega f_1 + (1-\omega) f_2$. Thus, the target function $f(\Delta C_i)$ is convex with respect to ΔC_i.

Property 13.1 indicates that there exists an optimal carbon abatement task in the national abatement allocation process. The convexity means that the optimal allocated carbon abatement task is a balanced result, which is more economical than other allocated carbon allocation tasks. To achieve the most economical carbon abatement process, the central government would have the motivation to adjust the carbon abatement allocation continuously. We believe that the optimal carbon abatement task is then accurately obtained by the proposed DEA approach of this chapter.

13.2.3 The Carbon Abatement Costs Measure

The abatement costs are considered in the aforementioned carbon allocation. However, acquiring the actual abatement costs for pollutants is hard, and hence the shadow prices of pollutants are commonly accepted as proxies. Estimation of these shadow prices can be done in a number of ways, but with the DEA approach nesting their evaluation, this performance analysis is treated as an accepted approach. Wang and Wei (2014) and Wang and He (2017) are amongst the recent examples of works seeking to estimate the prices of carbon abatement tasks for China's different industrial sectors. However, to the best of our knowledge, studies of carbon allocations based on abatement costs are still scarce. This chapter speaks to that gap, first estimating the allocation of carbon abatement tasks with the presence of corresponding carbon abatement costs by employing a DEA model as following model (13.13).

When evaluating a representative decision making unit *i*, DMU_i we are interested in the energy efficiency θ_{ei} and the carbon efficiency θ_{ci}. Efficiency values are all in the range of [0, 1]. When evaluating DMU i's environmental efficiency score, the target function contains mixed effects of energy consumption and carbon emission. This model measures the environmental performance efficiency in relation to input and output simultaneously. ρ_e and ρ_c mean the weight of the energy efficiency θ_{ei} and the carbon efficiency θ_{ci} in target function.

$$
\begin{aligned}
& \min \rho_e \theta_{ei} + \rho_c \theta_{ci} \\
\text{s.t.} \quad & \sum_{j=1}^{n} \lambda_j e_j \le \theta_{ei} e_i, \\
& \sum_{j=1}^{n} \lambda_j k_j \le k_i, \\
& \sum_{j=1}^{n} \lambda_j l_j \le l_i, \\
& \sum_{j=1}^{n} \lambda_j y_j \ge y_i, \\
& \sum_{j=1}^{n} \lambda_j c_j = \theta_{ci} c_i, \\
& \lambda_j \ge 0, \; j = 1, 2, \ldots, n.
\end{aligned} \tag{13.13}
$$

Data on the abatement costs of pollutants is difficult to obtain so, as previously motivated, we use the shadow prices of carbon emissions to represent the real ones. This chapter aims to use the DEA method to estimate the shadow price of carbon emission, and this is also adopted in Wang and Wei (2014). Then the shadow price and carbon abatement allocation could be measured with similar model settings; and forbidding any presenting of production functions to avoid the inherent misspecification risk present in the parametric method (Choi et al. 2012). We firstly present the dual programming of model (13.14):

$$
\begin{aligned}
& \max(-k_i w_k - l_i w_l + y_i w_y) \\
\text{s.t.} \quad & e_i w_e = \rho_e, \\
& c_i w_c = \rho_c, \\
& y_j w_y - e_j w_e - k_j w_k - l_j w_l - c_j w_c \le 0, \; j = 1, 2 \ldots, n, \\
& w_e \ge 0, w_k \ge 0, w_l \ge 0, w_y \ge 0, w_c \; is \; free.
\end{aligned} \tag{13.14}
$$

In model (13.14), w_e, w_k, w_l, w_y, w_c are dual variables corresponding to the constraints of energy, capital, labour, GDP, and carbon emissions, respectively. The target function is the efficiency of DMU$_i$. As Wang et al. (2015) and Wang and He (2017) we assume that the absolute shadow price of our marketable desirable output

(GDP) is equal to its market price. The shadow prices of carbon emission with respect to the desirable output are transformed as:

$$p_{ci} = p_{yi}\frac{w_c}{w_y} = \frac{w_c}{w_y} * 1\,\text{CNY}, \tag{13.15}$$

Here p_{ci} and p_{yi} are the relative shadow prices of carbon emission and GDP for region i, respectively. These shadow prices reflect the trade-off between desirable and undesirable outputs (Wang et al. 2015). The shadow price of CO_2 denotes the marginal rate of transformation between CO_2 and GDP, which could be regarded as being a price proxy of carbon abatement cost for China's regions. For instance, the shadow price can be derived from the technology expenditure and production reduction loss for carbon abatement in practice.

The implementation procedures of estimating allocated carbon abatement considering regional collaborations are summarized as follows:

Step 1: Estimate the marginal abatement costs (called MACs for short) for each region by Eq. (13.15) based on models (13.13) and (13.14).
Step 2: Evaluate the allocated carbon abatement amount ($\Delta c_j, j = 1, 2, \ldots n$) of each region by model (13.9) based on the obtained MAC_j from step 1.
Step 3: Re-evaluate the carbon abatement allocation considering possible regional collaborations by using model (13.12).
Step 4: Calculate the maximized industrial GDP ($\hat{y}_i$) and economical levels considering regional collaborations by model (13.12). Compare the allocated results from different regional collaborations.

13.3 A Case Study of China's Regional Industrial Systems

13.3.1 Data

Our data contains 30 provinces, autonomous regions, and municipalities in the mainland of China. As with most empirical studies of China a lack of data availability for Tibet leads to its exclusion from the modelling process. These regions can be grouped into three major areas, that is, the eastern, central, and western areas (Hu and Wang 2006). Regional groupings can be seen in Table 13.2. The eastern area has the best level of economic development in China, its GDP output contributed 55.34% of Chinese total GDP in 2014 (National Bureau of Statistics of China, 2015). The central area is regarded as the agricultural base for the country, whilst the western area has the lowest population density and the lowest level of economic development in China. It is thus highly reasonable to presume heterogeneity in regions.

We focus on China's regional carbon allocations during the period 2012–2014. Labour, capital stock, and energy consumption are the three inputs, industrial added value is the desirable output, and CO_2 denotes the undesirable output. Capital stock and industrial GDP are all expressed at 2012 prices for consistency. In the industrial production process, SO_2 emission, soot emission, dust emission, and NO_x emission can also be regarded as undesirable outputs. Whilst SO_2, NO_x, and other emissions may have their own abatement processes, that is, reduced by technical investments by government such as installing scrubbers and dust collection (Wang et al. 2016). Compared with other emissions, the CO_2 emissions abatement is more directly affected by fossil energies consumption and therefore the generated CO_2 levels may be directly related to industrial production. Thus, this study only uses the CO_2 emission as the undesirable output.

DEA models require the number of evaluated DMUs to be more than three times the total number of inputs and outputs to maintain validity (Friedman and Sinuany-Stern 1998). When permitting collaborations this may not be true of our modelling and hence we use multiple years of panel data to maintain sufficient quantity for robust inference. As three years is a short period, it is reasonable to assume that no significant technical changes occur in the period (Charnes et al. 1994; Halkos and Tzeremes 2009). The three-year panel data for our DEA ensures the collaborating region has the least sample amount of DMUs (27), which is greater than three times of total number of inputs and outputs (total five inputs and outputs are used in the model). By adopting models (13.9) and (13.12), the panel data are used in our multi-period allocation. Notably, as an application, our initial model focuses on carbon emissions and the consumption of energy resources with an equal target weight setting by solving model (13.13). The equal weights mean that the targets of energy-saving and carbon abatement are treated as equally important. This equality mirrors the approach in Wang et al. (2012) and Wu et al. (2016b).

Data on labour and capital of industrial production systems are derived from the Industrial Statistical Yearbook of China issued in each of 2013, 2014, and 2015. The industrial added value is collected from the Statistical Yearbook of China over the same time frame. Data on energy consumption is obtained from Energy Statistical Yearbook of China during the same period. Regional CO_2 emissions are not available in existing data sources but following Li et al. (2012), they can be estimated by multiplying the amounts of combined energy consumptions with their corresponding carbon emission coefficients. The carbon emission coefficients are obtained from the Core Writing Team et al. (2007). Table 13.1 shows the descriptive statistics for the data of all the variables in China during 2012–2014.

13.3.2 Efficiency and Carbon Abatement Costs

Based on the data for 2012 to 2014, we estimate the annual average environmental efficiencies and corresponding annual average shadow prices following models (13.14) and (13.15), respectively. The arithmetic average results for three years are shown in Table 13.2. There are five regions with efficiencies which are higher than 0.85: Beijing, Tianjin, Guangdong, Inner Mongolia, and Chongqing. These regions

Table 13.1 Descriptive statistics (2012–2014)

Indicators		Unit	Max	Min	Average	Std. Dev
Input	Energy	10^4 tons[a]	19392.8	839.0	6864.8	4469.0
	Labour	10^4 people	1470.5	11.7	324.7	337.9
	Capital stock	10^9 CNY	4087.3	71.9	1102.1	852.8
Desirable output	Industrial GDP	10^9 CNY	2859.6	45.9	845.6	703.9
Undesirable output	CO_2 emission	10^4 tons	72313.2	1833.9	22767.6	16413.4

[a]*Note* The unit refers to standard coal equivalent

have better performance in energy consumption and carbon emission than other regions in China. There also exist efficiencies in 14 regions which are less than the average efficiency (0.55). This indicates that these regions have not performed well in energy consumption and carbon emission abatement. Interestingly, it is observed that there exist no efficient regions in Table 13.2. This phenomenon arises because efficiencies for some regions for one specific year might be efficient (i.e., efficiency is equal to 1), but not for other years. For example, Beijing and Tianjin are efficient in 2014, but not efficient during 2012–2013. Employing annual average DEA efficiencies with three-year data reduces the gap of regional efficiencies, and helps avoid the time disturbance on efficiencies in a short period.

Table 13.2 presents remarkable spatial disparities. The eastern area has the highest average efficiency result amongst the three areas, 0.64. The central area sits just below this at 0.60 and is higher than the western area, 0.37. Amongst the regions, Ningxia province has the lowest, just 0.11. A strong correlation between economic development and energy efficiency is suggested. Higher GDP regions might have more possibilities to invest in eliminating heavy pollutant industries and to adopt advanced production technologies; both may increase regional environmental efficiencies.

Similarly, Table 13.2 shows remarkable geographic disparities are also seen in the carbon MACs of each region in Table 13.2. The MACs denote the opportunity costs for carbon abatement tasks converted into CNY values of GDP, and are measured in CNY per ton. The average MAC is 1861.56 CNY per ton, which tells us that 1861.56 CNY must be spent to reduce carbon emission by one ton. Average MAC values for the east, central, and west areas are 1957.3, 1679.97, and 1946.31 CNY. Interestingly, Pearson's correlation coefficient of efficiencies and MACs is −0.9520 for only the central regions, and is significant at the 1% level. This indicates that regions with higher efficiencies would have lower MACs only for central cases. For example, Inner Mongolia has the highest efficiency of 0.9479, but the lowest MAC of 81.25 CNY per ton.

By contrast Beijing and Guangdong, both economically developed regions, have much higher MACs than average, 3565.50 and 2100.61, respectively. This phenomenon might be partially explained by the gap between CO_2 emissions and industrial GDP in the provinces. The industrial carbon intensities (i.e., the ratio of CO_2

Table 13.2 Efficiency results and corresponding carbon abatement costs

	Region	Efficiency	MAC
Eastern area	Beijing	0.8635	3565.50
	Tianjin	0.8772	867.68
	Hebei	0.2648	1240.08
	Liaoning	0.3746	1489.38
	Shanghai	0.6968	2651.56
	Jiangsu	0.5897	2726.34
	Zhejiang	0.7267	1988.96
	Fujian	0.7487	1483.71
	Shandong	0.5159	2063.21
	Guangdong	0.9371	2100.61
	Hainan	0.4284	1353.22
Central area	Shanxi	0.1997	4444.68
	Inner Mongolia	0.9479	81.25
	Jilin	0.5594	1359.78
	Heilongjiang	0.6917	1360.91
	Anhui	0.4527	2222.34
	Jiangxi	0.5467	1558.63
	Henan	0.5746	2285.12
	Hubei	0.4809	2312.28
	Hunan	0.8378	566.41
	Guangxi	0.7504	608.30
Western area	Chongqing	0.8542	815.03
	Sichuan	0.4633	1670.93
	Guizhou	0.2210	4030.62
	Yunnan	0.3758	1373.27
	Shaanxi	0.6108	1363.01
	Gansu	0.2191	1279.99
	Qinghai	0.2377	1238.95
	Ningxia	0.1126	4378.38
	Xinjiang	0.1980	1366.59

Note Region in this chapter is used to define the provinces and hence we refer to the three groupings of provinces as "areas"

emission and industrial GDP) for Beijing and Guangdong are 1.3223 and 1.2779, respectively, which are less than those of other regions. It is not economical for regions with the lowest industrial carbon intensities to reduce their corresponding MACs by technological improvements or industrial structure transformations.

Comparing MACs in this chapter with ones in existing relative studies, we discover that our average MAC is larger than the ones in Wang and Wei (2014), whose average

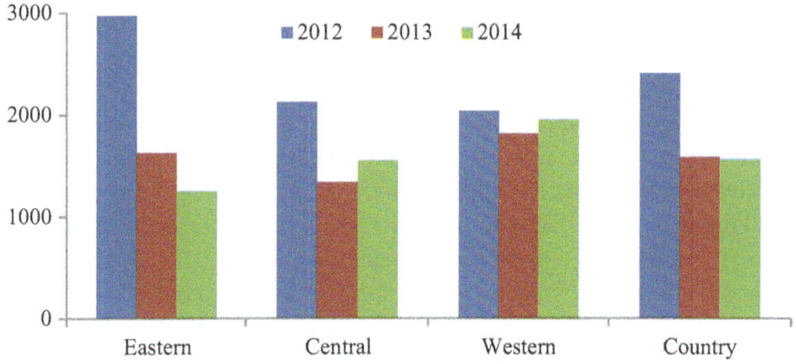

Fig. 13.1 Average MACs of three areas and the country during 2012–2014

industrial MAC is 45.81 USD per ton for China's major cities during 2006–2010. However, our result is as similar to the average MAC estimated in Zhou et al. (2015). Differences in MACs might be caused by different technical efficiencies, including differential impacts from primitive technology, operational scale, industrial structure, and the variation of the data on efficiency evaluation (Wang et al. 2015; Ha et al. 2008).

Figure 13.1 further illustrates the dynamic changing trend of the MACs. During 2012–2014, China's average MACs decreased from 2418.17 to 1572.73 CNY per ton. Only a slight decrease is shown during 2013–2014. Moreover, our three areas MACs also show decreasing trends from 2012 to 2013. During 2013–2014 MACs fall in the east, but rise in the less developed central and western areas. Over the three years the eastern area has decreased its MAC by 57.69%, whilst central and western areas have reduced their MACs by 26.87% and 3.89%, respectively. By 2014 the west had the highest MAC, having been the lowest in 2012. The rationality behind these changes is that national level emission abatement policies and regulations have been widely advocated by the regions, especially eastern China. These policies provide incentives for regions to reduce their carbon abatement costs. The eastern area significantly reduced its MAC to enable it to perform more carbon abatement tasks effectively. China's eastern region has the strongest economic and technological foundations amongst the three areas and therefore has the ability to do the most. However, for central and western areas there still exists scope to improve their MACs in the future, but the present focus is on wealth generation.

13.3.3 Allocating CO_2 Emissions Abatement

To effectively reduce China's large carbon emissions, the total carbon abatement task should be reasonably allocated to each region, paying attention to their environmental performance. This chapter uses the proposed approach of model (13.12) to allocate carbon abatement tasks amongst China's regions. Regions may prefer to work collectively to allocate within their areas. In calibrating model (13.9), we draw upon policy announcements relevant to the period.

The "13th–five–year working scheme of controlling greenhouse gases" sets the carbon intensity reduction target as 22% for China's industrial sector. Assuming that China's GDP growth rate keeps constant in the next period for the allocation, we set the annual total CO_2 abatement ratio at 4.4%. This value is taken from "China's low carbon energy saving and emission abatement plan during 2014–2015 issued by the State Council". Although this may be a simplification in times of slowing economic growth in China, it is still reasonable given policy efforts to restore the growth path under its specific target. Efforts to achieve growth should avoid becoming distracted by non-carbon-abatement issues. The next three-year abatement task is obtained as 26.40% (i.e., the sum of 4.40, 8.80, and 13.20% of carbon abatement tasks) of the CO_2 abatement amount in 2014. Then we set the three-year national total CO_2 abatement amount b as 186601×10^4 tons. To set the target CO_2 abatement task for each collaboration area, we divided the national total CO_2 abatement amount b according to the proportion of the total CO_2 emission amount coming from that collaboration. To avoid the total carbon abatement task being allocated to a small group of regions, we assume that each region could not reduce more than 30% of its current carbon emissions due to the limitations of current production scale and technology ($c_i^u = 0.3c_i$). The 30% abatement upper limit follows Wu et al. (2016b). b and c_i^u can be easily adjusted to represent different scenarios.

We continue to propose three regions for collaboration, which is the treatment afforded by most Chinese policy. The eastern, central, and western areas discussed above thus form our regions for the purpose of the analysis that follows. By solving models (13.9) and (13.12), detailed allocation results affected by national collaboration and intra-area collaborations (i.e., collaborations within eastern, central, or western areas, respectively) during 2012–2014 are outlined. We illustrate these in Figs. 13.2 and 13.3. For carbon abatement tasks, the comparison of the national allocation and intra-area allocations for each region is shown in Fig. 13.2 (unit: 10^4 tons), and the comparison of maximized industrial GDP output (i.e., $\hat{y}_i$ in model (13.9)) of each region is shown in Fig. 13.3 (Unit: 10^9 CNY).

Figure 13.2 shows that in the national allocation, 17 regions should decrease their carbon emissions, and 13 regions may keep their carbon emissions constant (i.e., each carbon abatement is equal to zero). The rationality of regions keeping their carbon emission constant can be attributed to two aspects: (1) the relative carbon abatement costs are too high to reduce their carbon emissions, for example, MACs of Beijing, Shanghai, Shanxi, Anhui, Henan, Hubei, Guizhou, and Ningxia are all higher than 2000 CNY per ton, much higher than the average MAC for the country. (2) They have lower environmental efficiency performances. The average efficiency of regions without any more allocated CO_2 abatement (0.47) is lower than the total average efficiency of 0.55. To achieve the national target, it is more economical for these regions to increase economic outputs than to reduce carbon emissions. Carbon allocation tasks are affected by the mixed impacts of their carbon abatement costs and environmental efficiencies.

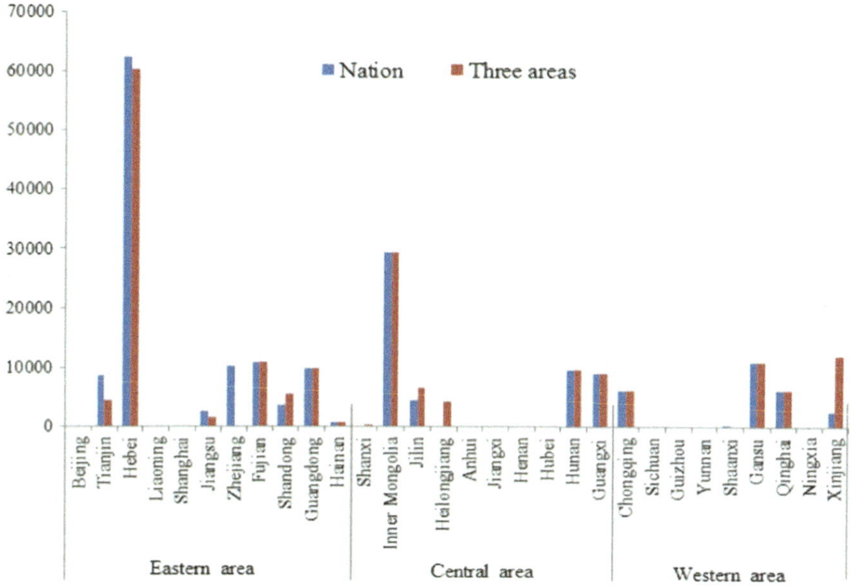

Fig. 13.2 Result comparison of allocated carbon abatement tasks

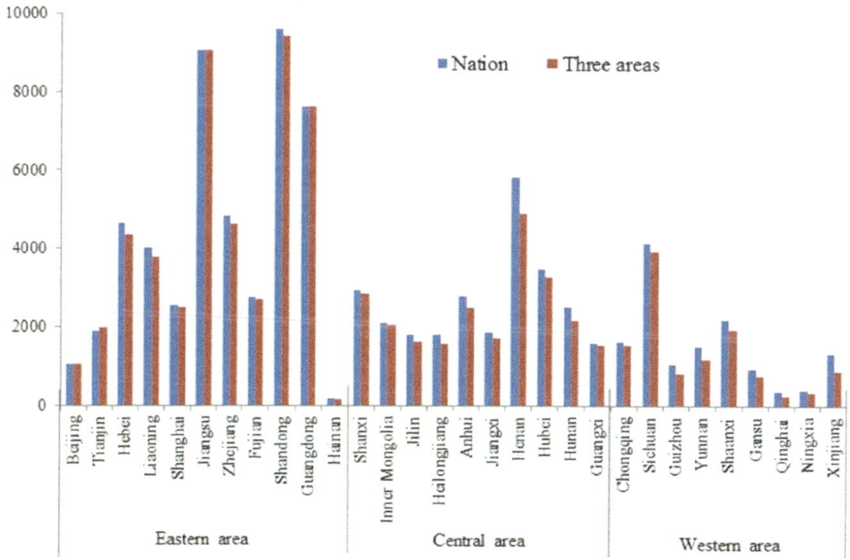

Fig. 13.3 Result comparison of allocated maximized industrial GDP

Four regions, Hebei, Inner Mongolia, Gansu, and Qinghai should, due to their low MACs, undertake the upper-level carbon abatement task (i.e., reduce 30% of c_i). Interestingly Inner Mongolia also has the highest efficiency and the largest carbon abatement proportion. Even though Inner Mongolia has the highest efficiency of carbon emission abatement, its MAC is still the lowest (81.25 CNY per ton). Thus output in Inner Mongolia might be sacrificed to reduce more carbon emissions to achieve the highest economical level for the country.

Total carbon abatements for eastern, central, and western areas are 92963, 58750, and 34888 $\times 10^4$ tons, respectively. Ratios of carbon abatement tasks to carbon emissions (called carbon abatement ratio for short) for eastern, central, and western areas are 10.6453%, 8.1265%, and 6.6568%, respectively. Here, the eastern area has the highest abatement ratio and the western area has the lowest. As such, it is reasonable to allocate a greater proportion of the abatement to the east.

The explanation behind the areal diversity is that, the eastern area has the strongest economic foundation and highest technology level in China. It is easier for more economically developed regions to adjust their industrial structure or energy structure or adopt advanced technologies to reduce their carbon emissions. Whilst underdeveloped areas, such as the western area, face greater policy pressure to enact economic growth. Thus the carbon emission task would be reduced to enable the pursuit of economic growth. A similar rank amongst areas is also achieved by Wu et al. (2016b). Our approach has stronger motivation provided by our consideration of MACs.

Comparing allocated results affected by national collaboration with those affected by intra-area collaborations, the carbon allocation tasks of 10 regions have been adjusted. In the eastern intra-area collaboration, Tianjin, Hebei, Jiangsu, and Zhejiang should be allocated a lower carbon abatement task, whilst Shandong should increase its efforts. In the central intra-area collaboration, Shanxi, Jilin, and Heilongjiang should increase their carbon abatement tasks. In the western intra-area collaboration, Shaanxi could slightly reduce its corresponding carbon abatement task, and Xinjiang should increase its contribution. Based on the comparison of allocation results, we conclude that regional collaborations would result in increased carbon abatement tasks for central regions, but mixed effects would exist in eastern and western regions. In both collaborations, the eastern area has the greatest total GDP output, followed by the central area and then the western area. This matches the ranking currently observed.

To confirm the effectiveness of the proposed allocation model, we further calculate the allocation results estimated by the conventional DEA allocation model. The conventional DEA allocation model for carbon allocation treats the maximized national total GDP as the target function, but other settings are identical to those in model (13.9). Necessarily, only the comparison of national allocation results between the conventional and proposed models is illustrated in Fig. 13.4.

The results indicate that: (1) with the consideration of MACs, the eastern area could reduce its carbon abatement task (i.e., 16915 $\times 10^4$ tons) and central and western areas would have more carbon abatement tasks (i.e., 7879 and 9036 tons, respectively). (2) Considering the existence of MACs, the potential maximized GDPs for

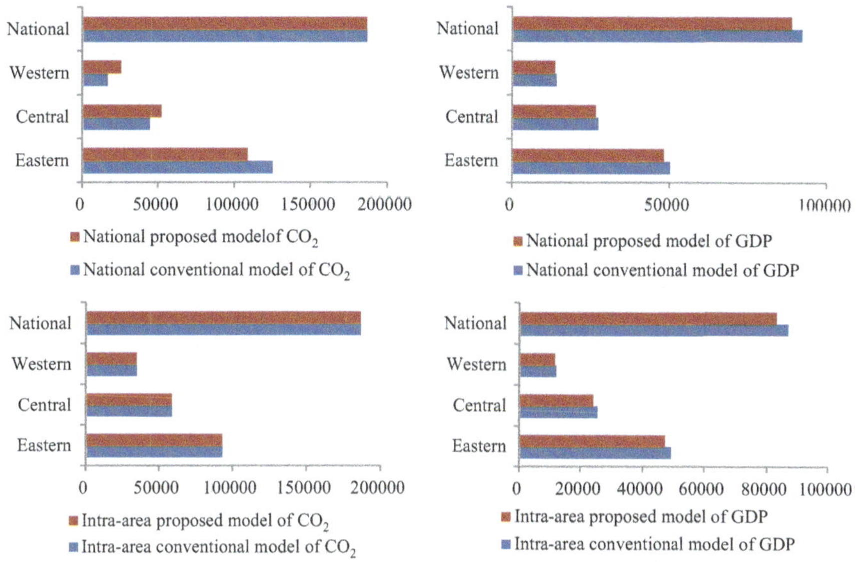

Fig. 13.4 Comparisons of allocation results between the conventional model and proposed model

all the areas of China would be significantly decreased, for example, the GDP reduction is 3149×10^9 CNY for the whole country. The above results indicate that the proposed DEA allocation could effectively estimate the carbon allocations affected by corresponding MACs. The ignorance of MACs would result in overestimation of maximized GDP and changes of allocation results.

13.3.4 *Regional Collaboration and Carbon Allocations*

To further illustrate the effects of regional collaborations, we present the carbon abatement task allocations for the three areas under a series of different groupings in Table 13.3. We compare the grand coalition of the three separate areas, and the

Table 13.3 Results of allocated carbon abatement amounts affected by regional collaborations

Area	Allocated carbon abatement tasks					
	{E, C, W}	{E}, {C}, {W}	{E, C}, {W}	{E, W}, {C}	{E}, {C, W}	Shapley V
E	108671	92963	99379	104839	92963	92963
C	52427	58750	52334	58750	56888	58750
W	25503	34888	34888	23012	36750	34888
China	186601	186601	186601	186601	186601	186601

three combinations that see two areas paired together. E, C, and W denote eastern, central, and western areas specifically, respectively, and braces, { }, denote collaboration. From Table 13.3, according to the proportions of carbon emission to national carbon emissions of the three areas during the period 2012–2014, the allocated carbon dioxide abatement tasks of eastern, central, and western areas are 92963, 56888, and 36750 × 10^4 tonnes, respectively. Considering the eastern area, we can see it receives its highest allocation of tasks when in the grand coalition (108761 × 10^4 tons) and its lowest when it acts alone (92963 × 10^4 tons). By contrast the other two regions receive their smallest allocation when they are acting in collaboration with the eastern area. Comparing with the Shapley (1953) allocations confirms that these are equal to the values for each region when acting alone. This certifies the robustness of the DEA method.

We also consider the impact of coalitions on GDP in Table 13.4. A national coalition maximizes GDP for each region whilst acting alone delivers the lowest. Finally, Table 13.5 considers the economical level for each regional collaboration combination. For the whole nation and central area, the full national coalition is the most economical. Meanwhile, the eastern area would do well to act alone, and the western area achieves its highest economical level when collaborating with the east. From these results, we conclude that the national level coalition achieves the nationally most economical results and maximized GDP output. This is consistent with Proposition 13.1. In collaborations, developing regions have a more economical carbon abatement process than the already developed east; the task may thus be transferred across to the more economically developed east. The western area would lobby for this, as it performs best in collaboration with the east.

Table 13.4 Results of maximized industrial GDP affected by regional collaborations

Area	Maximized industrial GDP				
	{E, C, W}	{E}, {C}, {W}	{E, C}, {W}	{E, W}, {C}	{E}, {C, W}
E	48241	47213	48090	47663	47213
C	26714	24204	26500	24204	25036
W	13631	11600	11600	12546	13059
China	88586	83017	86190	84413	85308

Table 13.5 Economical levels affected by regional collaborations

Area	Allocated economical levels				
	{E, C, W}	{E}, {C}, {W}	{E, C}, {W}	{E, W}, {C}	{E}, {C, W}
E	30.67	35.97	33.90	31.15	35.97
C	136.99	86.21	136.99	86.21	98.04
W	45.66	27.25	27.25	47.62	28.90
China	42.92	40.98	42.37	40.65	42.19

13.3.5 Further Discussion

Based on the aforementioned analysis, some conclusions and implications are derived for the benefit of carbon abatement policy and the practical implementation thereof. Our conclusions demonstrate the key tension between regional objectives and the nationally efficient allocation. For example the east may seek to limit its allocation of carbon abatement tasks by avoiding collaboration, but this would be to the detriment of the other areas and the whole country; as these areas would prefer the national collaboration. A practical explanation is that, as mentioned above, intuitively, the economically developed area has more potential to reduce carbon emissions. Thus, the eastern area should accept more of the carbon allocation burden in the national allocation vision. However, such compliance from the east is against its economic increment target; a collaboration struggle appears in east China.

That the eastern economically developed regions have limited carbon emission quotas compared with other areas is also indirectly proved by Zhang et al. (2014) with a collaboration measure by the Shapley value method. Different from the existing literature, one merit of our collaborative DEA allocation is to provide exact allocated results with all possible regional collaborations. Consequently, a novel vision of carbon allocation, considering the collaborative relationship between allocated objects in the DEA method, may be achieved.

Policy should encourage collaboration at the national level (e.g., joint carbon abatement or the integrated carbon trading market amongst regions) to allow the central and western areas to concentrate on upping their industrial output whilst the east shoulders the burden of abatement. Such a move could facilitate the transfer of technology and energy-intensive industries westward and hence create a greater future economical carbon abatement process in central and western provinces. Considering the struggle against collaboration implied for the eastern area, the appropriate compensation for carbon reduction (e.g., low-carbon subsidies for industrial sectors or firms) should be enacted by the central government to facilitate the potential regional collaborations on carbon reduction. Given the better economic development recognized in the eastern area, energy-saving and carbon-free technological updating could be advocated to reduce the extra carbon abatement task (Jia et al. 2018).

Necessarily, the DEA approach is linked to historic data, but it is an effective means for identifying the carbon abatement task allocations based on corresponding costs. Our results have permitted the consideration of regional collaborations in China, highlighting tensions and delivering the case for collaboration. For China, the national level solution is optimal.

13.4 Conclusions

This chapter proposes a meta-frontier DEA allocation model to reflect the potential technology heterogeneity of DMUs. Introducing this modified allocation approach permitted the study to present an analysis of the division of carbon abatement tasks considering corresponding regional level collaboration and the dual optimization of carbon reduction and output maximization. The proposed DEA allocation model can be expected to function as a suitable selection in future applications in the big data context. As decision making units, and the number of observations on the activities they take part in, increase so it becomes more important to have this robustness.

For industrialized and industrializing nations alike, the challenge of controlling carbon emissions is a pressing one, and the lessons from China should resonate. Our work is, like existing studies, a closed system which focuses entirely on the industrial sector. We have demonstrated there exist remarkable geographic disparities in environmental efficiencies and carbon abatement costs which previous DEA works have struggled to internalize. Regional collaborations can help influence members' abatement tasks, and our framework gives, for the first time, clear insight into how.

Our most important findings, however, concern the roles regional collaborations might play at the national level. A clear case is made for greater allocations for eastern China, where high economic development and lower abatement costs mean that greater efforts can be accommodated. Allocating more to a block like this gives the two less developed areas, especially western China, the chance to develop economically such that they two might take on greater tasks in the future. Consequently greater discrepancies between regions emerge on the abatement task allocated, but wealth differentials narrow in the long run; both processes embed regional identity and facilitate the continuance of such coalitions.

Some policy suggestions are that, the consideration of regional collaborations against the grand coalition of all provinces demonstrates that the latter delivers the most efficient outcome, but it is inherently unstable due to the optimality of other contributions for individual regions. China's most developed eastern area has the most to gain from allocating independently, whilst the less-developed central and western areas wish to join with the east. Lessons in promoting collaboration are clear and policy should seek to ensure that this is done. Policymakers should consider our findings carefully and ensure that the conflicts of carbon abatement task allocation are resolved.

However, some limitations in discussion also exist. We limited our collaborations to the most common Chinese relationships. There is, however, no reason why longer term coalitions between geographically disparate provinces should not engage. For example, the third-highest MAC region, Shanghai may work with the third lowest, Guangxi. Whilst the economic motivations for such a relationship are clear, the lack of geographical connectivity is likely to raise questions about the costs of working together. Here we argue that maintaining a sector focus is pertinent to the current economic make-up of the constituent provinces and positing coalitions with existing infrastructure remains most realistic now and into the future.

For China, the national allocation should be adopted, a result that extends globally from the theoretical work. China's need to achieve over the three study years is very clear for its international position and domestic environment, but other nations face similar dilemmas and the DEA modelling process constructed above should resonate in their decision-making. We have simplifications informed by policy but these may be readily adjusted to other settings and signposts for wider adoption.

Furthermore, considering that the areal collaboration in China is still currently difficult to realize, the collaborative DEA model could be more meaningful if adopted in a small-scale regional analysis, for example, the collaboration amongst provinces or cities. The novel vision of this study could also help the central government to reasonably decompose the national carbon abatement target for local governments and facilitate regional collaboration in the future. In addition, the proposed meta-frontier allocation method can also be adopted in other applications with related collaborations, such as the resource allocation for companies or institutions. Notably, the technology heterogeneity identified in this study is assumed derived from the geographic agglomeration and this can be modified based on other perspectives in future applications.

DEA has an important role to play in addressing pressing environmental issues in an efficient and transparent way and the modifications we make in this chapter will aid that process. For all concerned, the options, and consequences, of costly improvement allocations are clear and must be heeded by all collaboration efforts.

References

Beasley, J. E. (2003). Allocating fixed costs and resources via data envelopment analysis. *European Journal of Operational Research, 147*(1), 198–216.

Charnes, A., Cooper, W. W., Lewin, A. Y., & Seifor, L. M. (1994). *Data envelopment analysis: Theory, methodology, and application*. Norwell: Kluwer Academic Publishers.

Choi, Y., Zhang, N., & Zhou, P. (2012). Efficiency and abatement costs of energy-related CO_2 emissions in China: A slacks-based efficiency measure. *Applied Energy, 98,* 198–208.

Core Writing Team, Pachauri, R. K., & Reisinger, A. (2007). *Intergovernmental Panel on Climate Change (IPCC). IPCC fourth assessment report: Mitigation of climate change 2007.* http://www.ipcc.ch/publications_and_data/publications_ipcc_fourth_assessment_report_synthesis_report.htm. Last accessed 14 March 17.

Cui, L., Fan, Y., Zhu, L., & Bi, Q. (2014). How will the emissions trading scheme save cost for achieving China's 2020 carbon intensity reduction target? *Applied Energy, 136,* 1043–1052.

Ding, T., Chen, Y., Wu, H., & Wei, Y. (2018). Centralized fixed cost and resource allocation considering technology heterogeneity: a DEA approach. *Annals of Operations Research, 268,* 497–511.

Du, J., Cook, W. D., Liang, L., & Zhu, J. (2014). Fixed cost and resource allocation based on DEA cross-efficiency. *European Journal of Operational Research, 235*(1), 206–214.

Fang, L. (2013). A generalized DEA model for centralized resource allocation. *European Journal of Operational Research, 228*(2), 405–412.

Feng, C., Chu, F., Ding, J., Bi, G., & Liang, L. (2015). Carbon Emissions Abatement (CEA) allocation and compensation schemes based on DEA. *Omega, 53,* 78–89.

Friedman, L., & Sinuany-Stern, Z. (1998). Combining ranking scales and selecting variables in the DEA context: The case of industrial branches. *Computers & Operations Research, 25*(9), 781–791.

Gomes, E. G., & Lins, M. E. (2008). Modelling undesirable outputs with zero sum gains data envelopment analysis models. *Journal of the Operational Research Society, 59*(5), 616–623.

Ha, N. V., Kant, S., & Maclaren, V. (2008). Shadow prices of environmental outputs and production efficiency of household-level paper recycling units in Vietnam. *Ecological Economics, 65*(1), 98–110.

Halkos, G. E., & Tzeremes, N. G. (2009). Exploring the existence of Kuznets curve in countries' environmental efficiency using DEA window analysis. *Ecological Economics, 68,* 2168–2176.

He, W., Yang, Y., Wang, Z., & Zhu, J. (2018). Estimation and allocation of cost savings from collaborative CO_2 abatement in China. *Energy Economics, 72,* 62–74.

Hilton, I., & Kerr, O. (2017). The paris agreement: China's 'New Normal' role in international climate negotiations. *Climate Policy, 17*(1), 48–58.

Hu, J. L., & Wang, S. C. (2006). Total-factor energy efficiency of regions in China. *Energy Policy, 34,* 3206–3217.

Huda, M. S., & McDonald, M. (2016). Regional cooperation on energy in South Asia: Unraveling the political challenges in implementing transnational pipelines and electricity grids. *Energy Policy, 98,* 73–83.

Jia, P., Li, K., & Shao, S. (2018). Choice of technological change for China's low-carbon development: Evidence from three urban agglomerations. *Journal of Environmental Management, 206,* 1308–1319.

Li, H., Mu, H., Zhang, M., & Gui, S. (2012). Analysis of regional difference on impact factors of China's energy-Related CO_2 emissions. *Energy, 39,* 319–326.

Lozano, S., & Villa, G. (2004). Centralized resource allocation using data envelopment analysis. *Journal of Productivity Analysis, 22*(1), 143–161.

Lozano, S., Villa, G., & Brännlund, R. (2009). Centralised reallocation of emission permits using DEA. *European Journal of Operational Research, 193*(3), 752–760.

Lozano, S. (2013). DEA production games. *European Journal of Operational Research, 231,* 405–413.

National Bureau of Statistics of China (NBSC). (2015). *China statistical year book.* Beijing, China: China Statistics Pres.

O'Donnell, C. J., Rao, D. S. P., & Battese, G. E. (2008). Metafrontier frameworks for the study of firm-level efficiencies and technology ratios. *Empirical Economics, 34*(2), 231–255.

Shapley, L. S. (1953). A value for N person games. *Annals Mathametical Studies, 28,* 307–317.

Srivastava, L., & Misra, N. (2007). Promoting regional energy co-operation in South Asia. *Energy Policy, 35,* 360–3368.

Sueyoshi, T., & Goto, M. (2012). Weak and strong disposability vs. natural and managerial disposability in DEA environmental assessment: Comparison between Japanese electric power industry and manufacturing industries. *Energy Economics, 34,* 686–699.

Sun, J., Wu, J., Liang, L., Zhong, R. Y., & Huang, G. Q. (2014). Allocation of emission permits using DEA: centralised and individual points of view. *International Journal of Production Research, 52*(2), 419–435.

Timmer, J., Borm, P., & Suijs, J. (2000). Linear transformation of products: games and economies. *Journal Optimizing Theory and Applications, 105*(3), 677–706.

Uddin, N., & Taplin, R. (2015). Regional cooperation in widening energy access and also mitigating climate change: Current programs and future potential. *Global Environment Change, 35,* 497–504.

Wang, Y. S., Bian, Y. W., & Xu, H. (2015). Water use efficiency and related pollutants' abatement costs of regional industrial systems in China: a slacks-based measure approach. *Journal Cleaner Production, 101,* 301–310.

Wang, Z., & He, W. (2017). CO_2 emissions efficiency and marginal abatement costs of the regional transportation sectors in China. *Transportation Research D, 50,* 83–97.

Wang, K., Wei, Y. M., & Zhang, X. (2012). A comparative analysis of China's regional energy and emission performance: Which is the better way to deal with undesirable outputs? *Energy Policy, 46,* 574–584.

Wang, K., & Wei, Y. (2014). China's regional industrial energy efficiency and carbon emissions abatement costs. *Applied Energy, 130,* 617–631.

Wang, K., Wei, Y., & Huang, Z. (2016). Potential gains from carbon emissions trading in China: A DEA based estimation on abatement cost savings. *Omega, 63,* 48–59.

Wu, J., Zhu, Q., Chu, J. An, Q., & Liang, L. (2016a). A DEA-based approach for allocation of emission reduction tasks. *International Journal of Production Research*, *54*(20), 5990–6007.

Wu, J., Zhu, Q., & Liang, L. (2016b). CO_2 emissions and energy intensity reduction allocation over provincial industrial sectors in China. *Applies Energy*, *166*, 282–291.

Zhang, Y. J., Wang, A. D., & Da, Y. B. (2014). Regional allocation of carbon emission quotas in China: Evidence from the Shapley value method. *Energy Policy, 74,* 454–464.

Zhou, P., & Ang, B. W. (2008). Linear programming models for measuring economy-wide energy efficiency performance. *Energy Policy, 36,* 2911–2916.

Zhou, P., Sun, Z. R., & Zhou, D. Q. (2014). Optimal path for controlling CO_2 emissions in China: a perspective of efficiency analysis. *Energy Economics, 45,* 99–110.

Zhou, X., Fan, L. W., & Zhou, P. (2015). Marginal CO_2 abatement costs: Findings from alternative shadow price estimates for Shanghai industrial sectors. *Energy Policy, 77,* 109–117.

Chapter 14
Pension Funds and Mutual Funds Performance Measurement with a New DEA (MV-DEA) Model Allowing for Missing Variables

Maryam Badrizadeh and Joseph C. Paradi

14.1 Introduction[1]

One of the assumptions in Data Envelopment Analysis (DEA) is that the active work units (Decision Making Units "DMU") under study are operating under the same "culture". Culture means a wider definition than usual and, in fact, it requires that the DMUs operate under technical, environmental, legal, geographical, and other issues. Thus, the DMUs have to have the same input and output measures and the influence of outside factors must be the same.

It happens that there does not exist a model that can appropriately consider some aspects that are different in DMU's environment (culture). This research has created a DEA model, Mixed Variable DEA (MV-DEA), which can accept "similar" but not too many differences in the various DMU groups (here Pension and Mutual Funds). In this case, the two investment funds are quite similar but government rules are different.

While there are several methods in academia as well in practitioners' hands, and other such organizations, DEA has proven to be a very flexible and useful tool for the past 40+ years. Financial services have been, and still are, prime domain applications. DEA has not been utilized for evaluating such firms that operate under different (but not radically different) environments together. Therefore, the "big picture" of various DEA analyses for different financial investment vehicles can be quite similar. For instance, mutual funds and pension funds were examined separately using conventional DEA approaches. From an overall view, there are no fundamental differences

[1] Part of this chapter is based upon our other work: Paradi et al. (2018).

M. Badrizadeh (✉) · J. C. Paradi
C/O Joseph C. Paradi, University of Toronto, The Center for Management of Technology and Entrepreneurship, 200 College St, Toronto, ON M5S3E5, Canada
e-mail: maryam.badrizadeh@mail.utoronto.ca

V. Charles et al. (eds.), *Data Science and Productivity Analytics*,
International Series in Operations Research & Management Science 290,
https://doi.org/10.1007/978-3-030-43384-0_14

between these studies, however, on closer examination, there are significant differences between the two fund types in their investment strategy, tax status, reporting requirements, and other measures. Hence, by evaluating them relative to each other, the importance of their differences can be illustrated, yet the results can be readily compared.

As this work focuses on Mutual Funds and Pension funds with an emphasis on the latter, past work is briefly reviewed here.

DEA is considered as one of the most useful techniques for managers to measure the efficiency of mutual funds they manage. For instance, Murthi et al. (1997) compared the results of DEA approaches with traditional indices of performance. McMullen and Strong (1998) assessed 135 common stock mutual funds by using DEA. Morey and Morey (1999) examined the DEA approaches on a sample of 26 mutual funds. Choi and Murthi (2001) analyzed the benchmark problem and controls for economies of scale in administering mutual funds and proposed an alternative approach for it. Wilkens and Zhu (2001) developed a methodology that a piecewise linear efficient frontier considering various factors identifies benchmarks for underperforming funds. Basso and Funari (2005) evaluated the Italian mutual funds market by using DEA approaches. Malhotra et al. (2007) examined the U.S. mutual funds industry from 1998 to 2003 and Premachandra et al. (2012) assessed the relative performance of 66 U.S. mutual fund families from 1993 to 2008 by introducing a new two-stage DEA model.

DEA has been used to evaluate pension fund performance by several researchers over the past 20 years or so. Barrientos and Boussofiane (2005) studied the efficiency of pension fund managers in Chile by using DEA for the period of 1982–1999. Barros and Garcia (2006) evaluated Portuguese pension funds' performance from 1994 to 2003 by using different DEA models such as CCR, BCC, and others. Garcia (2010) analyzed changes in the productivity of Portuguese pension funds management institutions from 1994 to 2007 by using DEA and the Malmquist index. Sathye (2011) estimated the production efficiency of pension funds in Australia for the years 2005–2009 by using CCR and BCC models. Galagedera and Watson (2015) assessed pension funds in Australia by using DEA for the year 2012. Zamuee (2015) evaluated Namibian pension funds by using a CCR model for years 2010–2014.

The previous studies on pension funds using DEA predominantly focused on comparing different models instead of having a clear methodology and framework. Moreover, DEA requires a sufficient sample size to allow for good separation and discrimination among DMUs. Most of these studies had very few DMUs considering the number of inputs and outputs which decreases the accuracy of their results and they should be considered with caution.

There are government regulations for pension funds that impact the managers' control (there are regulations for mutual funds also, but some are different) over their investment strategies, hence, differentiate them from others that do not have such restrictions. For instance, the acceptable range for contribution amounts and benefit payments, the main variables in the pension funds industry, are prescribed in law. Also, one of the important issues to consider is the pension funds' ability to meet their financial obligations to its members; those which do are "fully funded" while those which do not or have deficits, are "underfunded" (industry people often call this being "underwater"). Therefore, fully funded plans are referenced only to other such plans while the underfunded plans are references to underfunded plans as well as fully funded plans. None of the previous papers had considered these important characteristics unique to pension funds.

The objective of this research is to introduce a new MV-DEA model that provides an approach to deal with moderately different cultures and rules, but in the same industry, to be evaluated together while the main characteristics of each are retained. Canadian private pension funds, regulated by the Federal Government of Canada, that are required to submit their annual financial statements to the Office of the Superintendent of Financial Institutions Canada (OSFI), and Canadian open-ended mutual funds were studied. This study makes a novel contribution to both the methodology and the application of DEA. The results of the new MV-DEA model were compared to traditional DEA models and it was shown that the MV-DEA model provided more realistic results in our study.

The rest of the research is organized as follows: The methodology is explained in Sect. 14.2. The application to data is provided in Sect. 14.3 and the experimental results are discussed in Sect. 14.4. The main conclusions are presented in Sect. 14.5.

14.2 Methodology

First, the main characteristics of pension funds and mutual funds were investigated. Then, the available data was evaluated and cleaned before using in the model. Finally, the new DEA model was developed.

One of the important considerations in pension funds is government regulations which are in place to protect the retirement income of the participants. Pension laws and regulations shape the unique legal investment environment in which pension funds operate. In general, regulations can be categorized into two types. One deals with plan administration. There are numerous rules and these change from time-to-time and from one situation to another based on age, actuarial statistics, conditions on funds transfer to spouse/common-law partner after death, etc. The second type deals with asset allocation. As a result, if regulations impact asset allocation, then the managers' performance can be assessed in this dimension. Consequently, the standard deviation of returns was estimated based on asset allocation and asset categories as some does/does not require certain reportable data to the appropriate regulator. At the same time, some non-discretionary variables, such as amounts and benefit payments were designated as such due to the fact that they were not under managements' control—a situation not applicable to Mutual Funds.

One of the DEA requirements is that DMUs have to be from the same "cultural environment" meaning, from the same industry (Dyson et al. 2001). Here, a new DEA model was developed that creates an opportunity for analysis of two cases with similar, but not identical "culture"—in this case, Pension Funds (PF) and Mutual Funds (MF). This new model was designated as the MV-DEA model and it allows for the comparison of two (or possibly more) cases of financial entities by accounting for their differences while retaining their characteristics.

The authors' interest here is to discover how effectively pension funds and mutual funds perform not just individually, but when evaluated together. The goal of the MV-DEA model is, as some of the variables are different or even non-existent in one or the other, to bridge the two different approaches. In the MV-DEA model, an analyst may designate a unit as a pension fund and the constraints for non-discretionary cases were considered only. Other variables and constraints were not used. However, for mutual funds, the constraints for discretionary cases were considered only. Therefore, in the MV-DEA model, the standard radial VRS model is run for DMUs (mutual funds) while the non-discretionary radial VRS model (Non-Dis-VRS) is used for some of the DMUs (pension funds).

The VRS model was proposed by Banker, Charnes, and Cooper in 1984 (Banker et al. 1984). The VRS frontier does not pass through the origin as the Charnes, Cooper, and Rhodes model (CCR) does. This frontier is comprised of the best-performing DMUs and envelops the inefficient DMUs. Up to this point, all the variables are controlled at the discretion of the fund managers. However, sometimes variables are not subject to management decisions and should be considered as non-discretionary variables. Consequently, such variables are removed from the objective function of the linear program but are included in the constraints to assure their influence and their values remain constant while the discretionary variables are optimized (Banker and Morey 1986).

Let's consider DMUs (i = 1, 2, …, n), inputs (j = 1, …, m), and outputs (r = 1, …, s). The "D" and "ND" refer to the "Discretionary" and "Non-Discretionary" input and output. The mathematical model for the multiplier form is as follows:

$$
\begin{array}{ll}
\text{Objective 1} & Min\ z = \sum_{j\in D} v_j x_{jo} + \sum_{j\in ND} v_j x_{jo} - \sum_{r\in ND} u_r y_{ro} - v_o \\
\text{Objective 2} & Min\ z = \sum_{j=1}^{m} v_j x_{jo} - v_o \\
\text{Constraint 1} & \sum_{j\in D} v_j x_{ji} + \sum_{j\in ND} v_j x_{ji} - \sum_{r\in ND} u_r y_{ri} - \sum_{r\in D} u_r y_{ri} - v_o \geq 0 \\
\text{Constraint 2} & \sum_{r\in D} u_r y_{ro} = 1 \\
\text{Constraint 3} & v_j \geq \varepsilon\ ,\ \ j \in D \\
\text{Constraint 4} & v_j \geq 0\ ,\ \ j \in ND \\
\text{Constraint 5} & u_r \geq \varepsilon\ ,\ \ r \in D \\
\text{Constraint 6} & u_r \geq 0\ ,\ \ r \in ND \\
\text{Constraint 7} & \sum_{j=1}^{m} v_j x_{ji} - \sum_{r=1}^{s} u_r y_{ri} - v_o \geq 0 \\
\text{Constraint 8} & \sum_{r=1}^{s} u_r y_{ro} = 1 \\
\text{Constraint 9} & v_j \geq 0\ ,\ \ j = 1, \ldots, m \\
\text{Constraint 10} & u_r \geq 0\ ,\ \ r = 1, \ldots, s
\end{array}
\tag{14.1}
$$

Therefore, in the MV-DEA model, for pension funds, the Non-Dis-VRS model (objective 1 and constraints 1–6 of the multiplier form) is used while for mutual funds the VRS model (objective 2 and constraints 7–10 of the multiplier form) is used.

The envelopment form of the model is as follows:

$$
\begin{array}{ll}
\text{Objective 3} & Max\ \varphi + \varepsilon\,(\sum_{j\in D} s_j + \sum_{r\in D} t_r) \\
\text{Objective 4} & Max\ \varphi + \varepsilon\,(\sum_{j=1}^{m} s_j + \sum_{r=1}^{s} t_r) \\
\text{Constraint 11} & \sum_{i=1}^{n} \lambda_i y_{ri} = t_r + \varphi\, y_{rio}\,, \quad r \in D \\
\text{Constraint 12} & \sum_{i=1}^{n} \lambda_i y_{ri} = t_r + y_{rio}\,, \quad r \in ND \\
\text{Constraint 13} & \sum_{i=1}^{n} \lambda_i x_{ji} = -s_j + x_{jio}\,, \quad j = 1, \dots, m \\
\text{Constraint 14} & \sum_{i=1}^{n} \lambda_i = 1 \\
\text{Constraint 15} & \sum_{i=1}^{n} \lambda_i y_{ri} = t_r + \varphi\, y_{rio}\,, \quad r = 1, \dots, s
\end{array}
\tag{14.2}
$$

For mutual fund DMUs, the VRS model with objective 4 and constraints 13–15 were used while for pension fund DMUs the Non-Dis-VRS with objective 3 and constraints 11–14 were used.

To further clarify, Eq. (14.3) is used for pension funds and Eq. (14.4) is used for mutual funds in the multiplier form of the MV-DEA model.

Non-Dis-VRS multiplier form for only pension funds' DMUs in the data set

$$
\begin{array}{ll}
Min\ z = \sum_{j\in D} v_j x_{jo} + \sum_{j\in ND} v_j x_{jo} - \sum_{r\in ND} u_r y_{ro} - v_o & \\
\text{Subject to:} & \sum_{j\in D} v_j x_{ji} + \sum_{j\in ND} v_j x_{ji} - \sum_{r\in ND} u_r y_{ri} - \sum_{r\in D} u_r y_{ri} - v_o \geq 0 \\
& \sum_{r\in D} u_r y_{ro} = 1 \\
& v_j \geq \varepsilon\,, \quad j \in D \\
& v_j \geq 0\,, \quad j \in ND \\
& u_r \geq \varepsilon\,, \quad r \in D \\
& u_r \geq 0\,, \quad r \in ND
\end{array}
\tag{14.3}
$$

VRS multiplier form for only mutual funds' DMUs in the data set

$$
\begin{array}{ll}
Min\ z = \sum_{j=1}^{m} v_j x_{jo} - v_o & \\
\text{Subject to:} & \sum_{j=1}^{m} v_j x_{ji} - \sum_{r=1}^{s} u_r y_{ri} - v_o \geq 0 \\
& \sum_{r=1}^{s} u_r y_{ro} = 1 \\
& v_j \geq 0\,, \quad j = 1, \dots, m \\
& u_r \qquad r \qquad s
\end{array}
\tag{14.4}
$$

Also, in the envelopment form of the MV-DEA model, Eq. (14.5) is used for pension funds and Eq. (14.6) is used for mutual funds in the same dataset.

Non-Dis-VRS envelopment form for only pension funds' DMUs in the data set

$$
\begin{aligned}
&Max\ \varphi + \varepsilon \left(\textstyle\sum_{j \in D} s_j + \sum_{r \in D} t_r\right) \\
&\text{Subject to:} \quad \textstyle\sum_{i=1}^{n} \lambda_i y_{ri} = t_r + \varphi\, y_{rio}\,, \quad r \in D \\
&\qquad \textstyle\sum_{i=1}^{n} \lambda_i y_{ri} = t_r + y_{rio}\,, \quad r \in ND \\
&\qquad \textstyle\sum_{i=1}^{n} \lambda_i x_{ji} = -s_j + x_{jio}\,, \quad j = 1, \ldots, m \\
&\qquad \textstyle\sum_{i=1}^{n} \lambda_i = 1
\end{aligned}
\tag{14.5}
$$

VRS envelopment form for only mutual funds' DMUs in the data set

$$
\begin{aligned}
&Max\ \varphi + \varepsilon \left(\textstyle\sum_{j=1}^{m} s_j + \sum_{r=1}^{s} t_r\right) \\
&\text{Subject to:} \quad \textstyle\sum_{i=1}^{n} \lambda_i x_{ji} = -s_j + x_{jio}\,, \quad j = 1, \ldots, m \\
&\qquad \textstyle\sum_{i=1}^{n} \lambda_i = 1 \\
&\qquad \textstyle\sum_{i=1}^{n} \lambda_i y_{ri} = t_r + \varphi\, y_{rio}\,, \quad r = 1, \ldots, s
\end{aligned}
\tag{14.6}
$$

Moreover, in order to be able to correctly evaluate the value of pension funds, the analyst must consider their status (whether they are fully funded or underfunded). But, in the mutual funds' world, the concepts of fully funded/underfunded are not relevant. Therefore, the categorical DMUs are also considered when developing MV-DEA model and include the funds' status. As a result, fully funded plans are referenced only to their own group while the underfunded plans are referenced to both.

All pension funds were active plans (meaning that they were fully operational) in this research. Those which were underfunded were under the scrutiny of the regulator, OSFI. They were obliged to bring their funds to fully funded status within 5–10 years. If the fund managers do not meet these time targets, their plans may be terminated by the regulator. However, this still enabled the authors to utilize these funds in the study, even to find them on the DEA frontier—but very few such instances occurred. A detailed applicability of the MV-DEA model is explained in the following sections.

14.3 Application to Data

14.3.1 Data Collection

Each Country where formal pension plans exist, naturally, there are rules on how these operate. In Canada, most Plans are either a defined benefit plan (DB) or defined contribution (DC) plan. These plans offer their employees a fundamentally different payout process. The DC plan version starts with a fixed payment per period when a member retires, but the amount will change with inflation and other reasons. The investment risks are with the contributor and they can gain or lose value based on their fund managers' skills and market conditions. This is the type that most private-sector plans are favoring now in Canada. The DB type plans are different from the DC plans in a very important aspect as these plans guarantee the pension payout amounts regardless of market conditions or fund manager success in investing the funds. Typically, these plans are offered by Government and Institutional organizations, such as universities and unionized firms. But, of course, there are exceptions to these practices. There are other types of plans, often hybrid, such as the combination pension plans (Combo) which incorporate some positive elements from both the DB and DC plans. In these cases, a pension is promised (i.e., the defined benefit concept) and accordingly, from time-to-time, employers are able to use some pension surplus (i.e., the defined contribution concept) to fund their defined benefit plan's current service costs. Only these three types of private pension plans were studied.

Open-ended mutual funds do not have restrictions on the number of shares they issue. Also, the shares are generally purchased directly from the fund rather than from the existing shareholders and can be issued and redeemed at any time. Close-ended mutual funds have a fixed number of shares. New shares cannot be created by managers to meet investors' demands and unlike open-ended funds; the shares can usually be traded between investors. In this study, only open-ended mutual funds were considered. Variables were chosen based on the available data for private pension funds and appropriate adjustments, such as considering their tax status, were taken into account to provide a defensible comparison between the two types of funds.

Variable selection was a difficult and well-considered effort. Several methods were employed to assure that such selection would apply to all funds under study. Hence, such selection approaches as sensitivity analysis, statistical tests, and outliers without a valid reason to be such were removed. After applying these approaches, there were 173 federally regulated Canadian pension plans which encompassed some, 90 DB plans (52%), 37 DC plans (21.4%) and 46 Combo plans (26.6%), and 61 Canadian open-ended mutual funds.

14.3.2 Variables

All Canadian pension plans are required to submit their annual financial statements to OSFI, the Regulator. However, for some of the variables such as benefit payments and management fees, only DB plans and Combo plans have to report; this is optional for DC plans as they vary over time. The financial reporting requirements for DC plans are straightforward since there are no future obligations for them to report on. As a result, financial reporting for DB and Combo plans is much more complicated because the employer must estimate the present value of future obligations to its employees. Consequently, the available data for DC plans is fewer than DB and Combo plans. This study was carried out in two parts. Based on available data, DB plans and Combo plans were combined while DC plans were processed on their own.

Table 14.1 Inputs and outputs for DB plans and Combo plans

Inputs	Outputs
• Investment expenses • Management fees • Contribution amounts • Standard deviation of returns	• Net investment income • Benefit payments

Table 14.2 Inputs and outputs for DC plans

Inputs	Outputs
• Investment expenses • Contribution amounts • Standard deviation of returns	• Net investment income

14.3.2.1 DB and Combo Plans

In building the DEA model, inputs were selected for Combo and DB plans as investment expenses, professional fees, and contribution amounts, plus the standard deviation of returns. The outputs were benefit payments and investment income; all variables are shown in Canadian Dollars (CAD). In Table 14.1, variables for DB and Combo plans are shown and reproduced with permission from Paradi et al. (2018).

Also, the Wilcoxon rank-sum test was applied to assure statistically that DB plans and Combo plans have the same population and can be considered in one group.

14.3.2.2 DC Plans

DC plan managers do not need to report on management fees and benefit payments with OSFI, hence these fields are left blank in their financial reports and this differentiates DC plans' variables from DB and Combo plans. The variables for DC plans are shown in Table 14.2 and reproduced with permission from Paradi et al. (2018).

14.3.2.3 Mutual Funds

Mutual funds were treated in the same manner as pension funds. Inputs are management fees, investment expenses, and Funds' sales values, plus the standard deviation of returns calculated in the same consistent process as for the pension funds. Outputs were redemptions plus dividends and net investment income (after-tax). All payments are in CAD.

14.3.3 Model

Clearly, both types of funds were designed to produce maximum returns in an output-oriented model. To make it simple and easy to understand, the VRS model was chosen for the new MV-DEA model. Additionally, a VRS model could be modified and expanded simply without introducing much complexity. Moreover, the results were easy to understand and not likely to be easily misinterpreted. The other feature of this new model is that it is easy to adopt to any DEA model as long as the objectives are clear for such projects.

The MV-DEA model was considered to be useful in various situations to better understand its usability. As the two fund types were processed together, the outcomes could be compared from the new MV-DEA model. This task was approached by using all the variables for both types as discretional variables in a VRS output-oriented model. Next, the Non-Dis-VRS (output-oriented) fit the authors' needs as there are some non-discretional variables for both fund types. Then, the third combination was using the MV-DEA output-oriented model where for the pension funds the relevant measures were used as non-discretional, while for mutual funds these were discretional.

Then, the efficient DMUs were selected from models for both fund types. Now this combined efficient DMU data was run and the results were analyzed.

14.4 Results and Discussion

14.4.1 All DMUs Under Consideration

The total number of DMUs for each type of plans and mutual funds is as follows:

Defined Benefit Plans (DB): 90
Defined Contribution Plans (DC): 37
Combination Plans (Combo): 46
Mutual Funds (MF): 61.

The results for the VRS, Non-Dis-VRS, and MV-DEA models for all pension funds and mutual funds are represented in Tables 14.3, 14.4, and 14.5 and reproduced with permission from Badrizadeh and Paradi (2017).

In Tables 14.3 and 14.4, the funds' status (fully funded/underfunded) for DB and Combo plans are considered as categorical variables in the DEA models. Fully funded plans are referenced only to fully funded plans while the underfunded plans are referenced to both fully and underfunded plans. In Table 14.5, since funds' status is not an issue for DC plans the categorical DMUs are not included.

In Table 14.3, the average efficiency scores for the new MV-DEA model (Non-Dis-VRS for PFs and VRS for MFs at the same time) are in-between the average efficiency scores for the VRS and the Non-Dis-VRS models. To highlight this, the

Table 14.3 Considering all DB and MFs for VRS, Non-Dis-VRS, and MV-DEA models

Plans	VRS_O_CAT[a]	Non-Dis-VRS_O_CAT[b]	MV-DEA_O_CAT[c]
DB	#DMUs: 90 #Efficient: 33 Average: 0.606 Max: 1 Min: 0.106 Ave of lowest quartile (22 DMUs): 0.173	#DMUs: 90 #Efficient: 33 Average: 0.576 Max: 1 Min: 0.106 Ave of lowest quartile (22 DMUs): 0.171	N/A
DB and MF	#DMUs: 151 #Efficient: 45 (DB:33, MF:12) Average: 0.581 Max: 1 Min: 0.073 Ave of lowest quartile (38 DMUs): 0.162	#DMUs: 151 #Efficient: 45 (DB:33, MF:12) Average: 0.497 Max: 1 Min: 0.001 Ave of lowest quartile (38 DMUs): 0.113	#DMUs: 151 #Efficient: 45 (DB:33, MF:12) Average: 0.564 Max: 1 Min: 0.073 Ave of lowest quartile (38 DMUs): 0.161

[a]VRS_O_CAT: VRS model for all DMUs (PFs and MFs), Output-Oriented, Categorical DMUs (fully funded and underfunded)
[b]Non-Dis-VRS_O_CAT: Non-Discretionary VRS model for all DMUs (PFs and MFs), Output-Oriented, Categorical DMUs (fully funded and underfunded)
[c]MV-DEA_O_CAT: Mixed Variable DEA model with Non-Discretionary VRS model for pension plans and at the same time VRS model for mutual funds (NEW DEA Model), Output-Oriented, Categorical DMUs (fully funded and underfunded)

Table 14.4 Considering all DB, Combo, and MFs for VRS, Non-Dis-VRS, and MV-DEA models

Plans	VRS_O_CAT	Non-Dis-VRS_O_CAT	MV-DEA_O_CAT
DB and Combo	#DMUs: 136 #Efficient: 39 (DB: 34, Combo: 5) Average: 0.537 Max: 1 Min: 0.068 Ave of lowest quartile (34 DMUs): 0.156	#DMUs: 136 #Efficient: 39 (DB: 34, Combo: 5) Average: 0.508 Max: 1 Min: 0.064 Ave of lowest quartile (34 DMUs): 0.148	N/A
DB and Combo and MF	#DMUs: 197 #Efficient: 50 (DB: 33, Combo: 5, MF: 12) Average: 0.539 Max: 1 Min: 0.068 Ave of lowest quartile (49 DMUs): 0.151	#DMUs: 197 #Efficient: 50 (DB: 33, Combo: 5, MF: 12) Average: 0.467 Max: 1 Min: 0.001 Ave of lowest quartile (49 DMUs): 0.113	#DMUs: 197 #Efficient: 50 (DB: 33, Combo: 5, MF: 12) Average: 0.519 Max: 1 Min: 0.064 Ave of lowest quartile (49 DMUs): 0.146

Table 14.5 Considering all DC and MFs for VRS, Non-Dis-VRS, and MV-DEA models

Plans	VRS_O[a]	Non-Dis-VRS_O	MV-DEA_O
DC	#DMUs: 37 #Efficient: 13 Average: 0.696 Max: 1 Min: 0.194 Ave of lowest quartile (9 DMUs): 0.329	#DMUs: 37 #Efficient: 13 Average: 0.696 Max: 1 Min: 0.194 Ave of lowest quartile (9 DMUs): 0.329	N/A
DC and MF	#DMUs: 98 #Efficient: 15 (DC: 7, MF: 8) Average: 0.462 Max: 1 Min: 0.001 Ave of lowest quartile (24 DMUs): 0.115	#DMUs: 98 #Efficient: 15 (DC: 7, MF: 8) Average: 0.462 Max: 1 Min: 0.001 Ave of lowest quartile (24 DMUs): 0.115	#DMUs: 98 #Efficient: 15 (DC: 7, MF: 8) Average: 0.462 Max: 1 Min: 0.001 Ave of lowest quartile (24 DMUs): 0.115

[a]DC plans are very similar to the mutual funds and the funds' status (fully funded/underfunded) is not an issue for DC and MFs. Therefore, the categorical DMUs are not considered for DEA models for DC as well as DC and MFs

MV-DEA average score is 0.5642 while for the VRS model it is 0.5811 and for the Non-Dis-VRS model, it is 0.4974.

As shown in Fig. 14.1, the VRS model overestimates the efficiency scores for DB plans and the Non-Dis-VRS model underestimates them for MFs. As presented in Fig. 14.2, in the descending order of efficiency scores for these three models, the MV-DEA model's descending line of scores is in-between the VRS and the Non-Dis-VRS lines. All Figures in this research are reproduced with permission from Badrizadeh and Paradi (2017).

Also, when the efficiency scores for pension funds from the Non-Dis-VRS model and mutual funds from the VRS model are compared to scores from the MV-DEA model, the percentage change is zero for all DMUs in the DB and MFs' datasets. Such change shows the relationship between the new value (MV-DEA) and the old value (VRS model for MF and the Non-Dis-VRS model for PFs) that can be calculated as $[((new\ X - old\ X)/old\ X) \times 100]$. Percentage change can be easily interpreted. Therefore, the result of the percentage change indicates that the MV-DEA model is working as expected.

To sum up the findings, the results show that the average efficiency scores for the new MV-DEA model increased compared to the Non-Dis-VRS model for all DMUs (both pension funds and mutual funds) and decreased compared to the VRS model for the same dataset (without considering non-discretionary variables).

In Table 14.4, the results for the VRS, Non-Dis-VRS, and MV-DEA models for DB, Combo, and MFs are presented.

As shown in Table 14.4 and Figs. 14.3 and 14.4, the average efficiency scores for MV-DEA for DB, Combo, and MFs are in-between the average efficiency scores for the VRS and the Non-Dis-VRS models. Also, when the efficiency scores for the

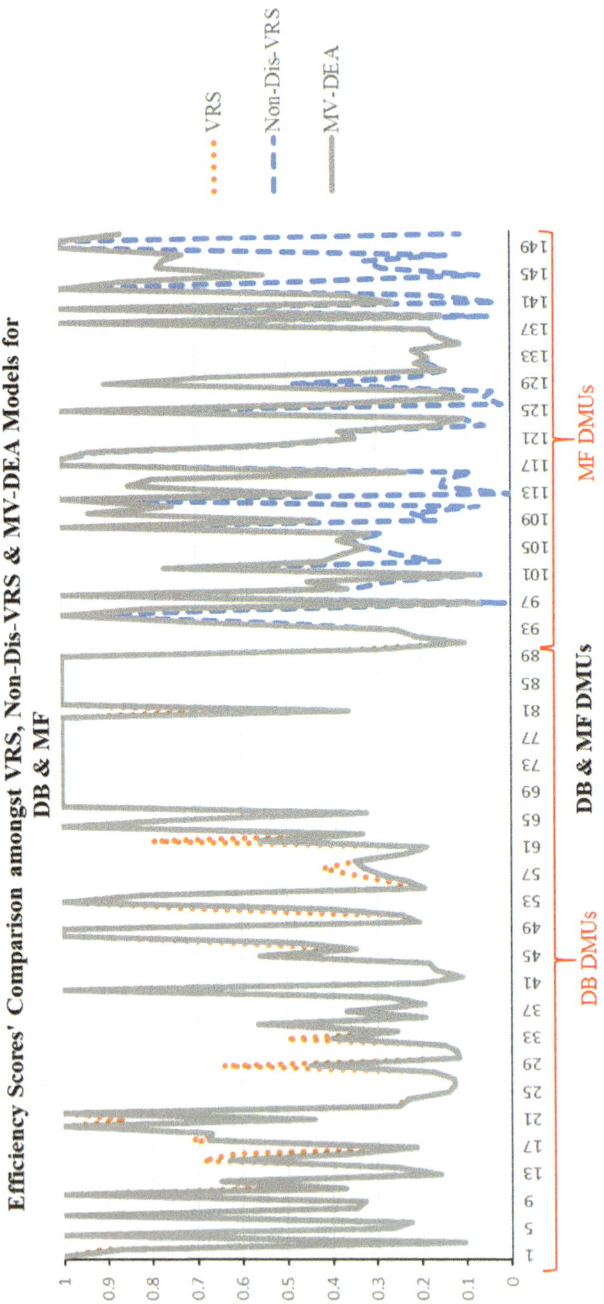

Fig. 14.1 Efficiency scores' comparison among VRS, Non-Dis-VRS, and MV-DEA models for DB and MF

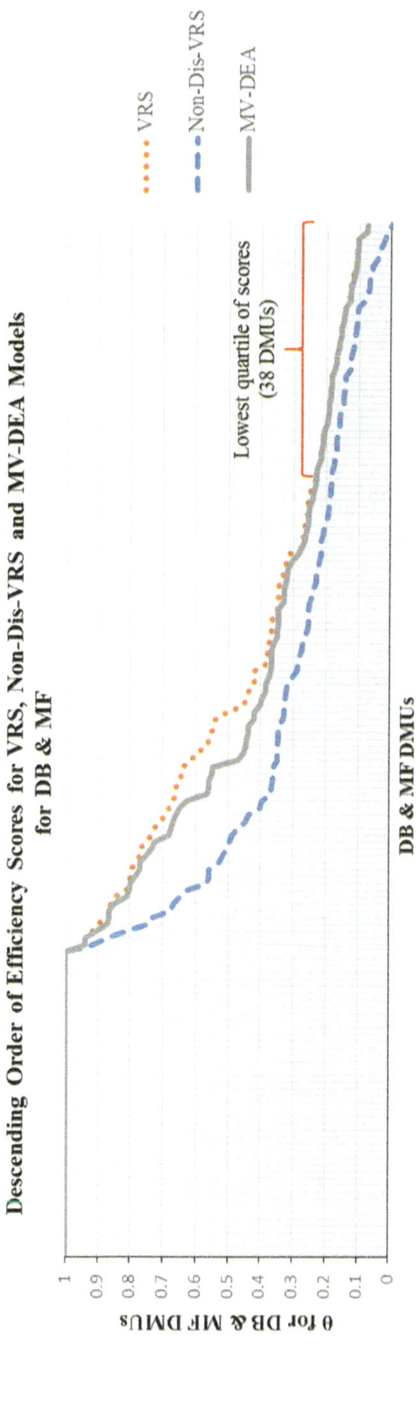

Fig. 14.2 Descending order of efficiency scores for VRS, Non-Dis-VRS, and MV-DEA models for DB and MF

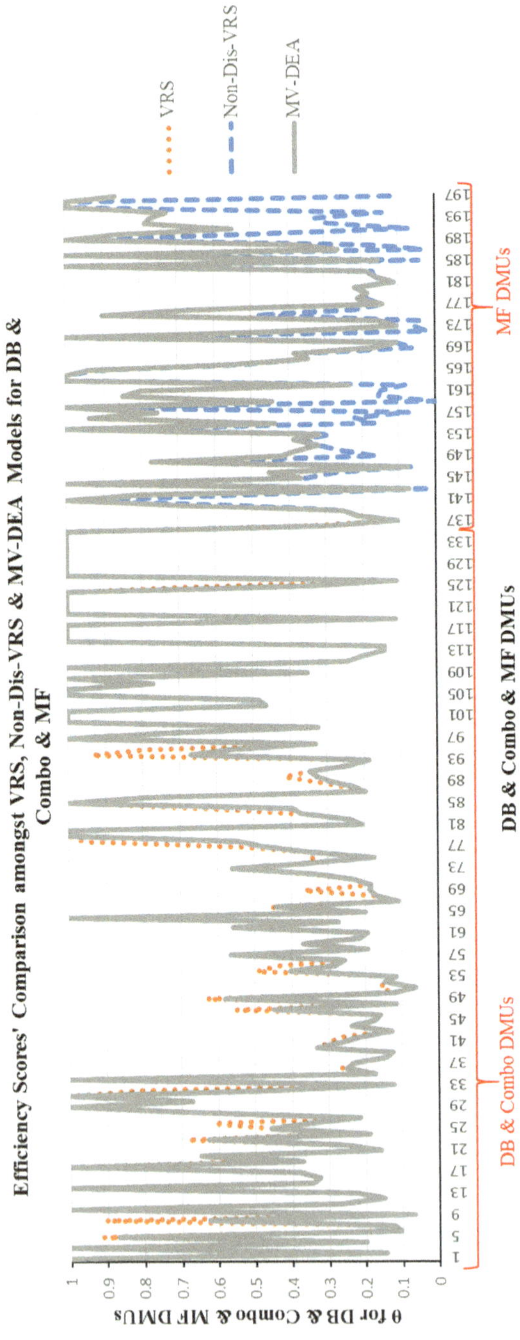

Fig. 14.3 Efficiency scores' comparison among VRS, Non-Dis-VRS, and MV-DEA models for DB, Combo, and MF

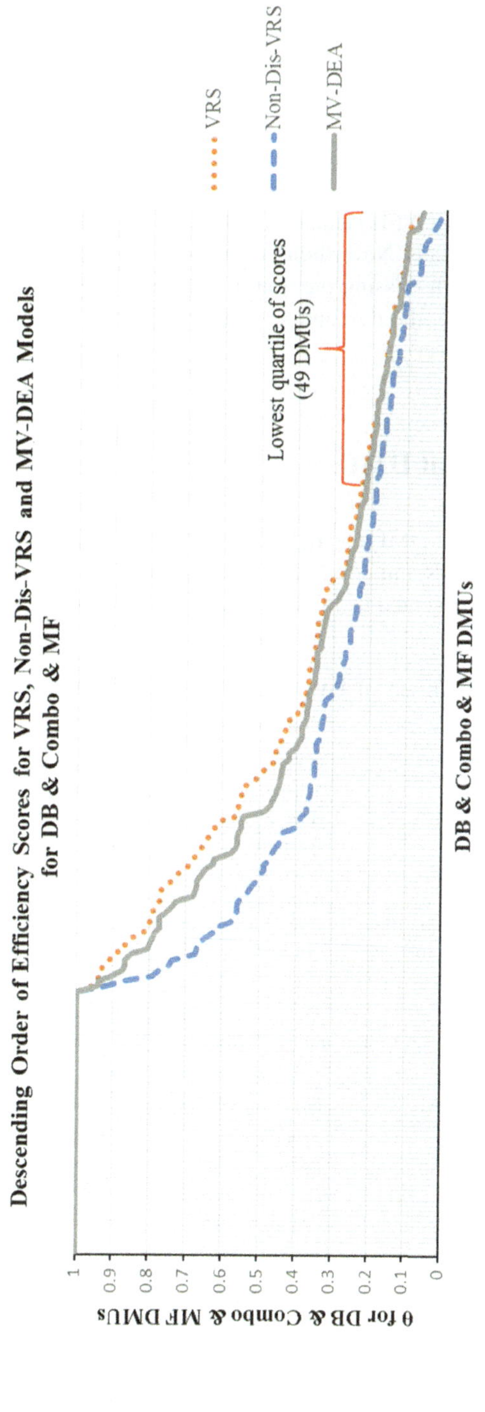

Fig. 14.4 Descending order of efficiency scores for VRS, Non-Dis-VRS, and MV-DEA models for DB, Combo, and MF

DB and Combo plans from the Non-Dis-VRS model and the mutual funds from the VRS model are compared to the efficiency scores from the MV-DEA model, the percentage change is zero for all DMUs in DB, Combo, and MFs. The reason for having low minimum θs is that both fully funded and underfunded plans are included and since most pension plans are underfunded, very low minimum efficiency scores are often found in this industry.

Furthermore, in Table 14.5, all three DEA models yielded the same results; but worthy of note is that DC funds are not required to file records of certain fees (management) and (benefits) to OSFI. Hence, the measure for such plans was different from the DB and Combo plans. Nevertheless, the model was output-oriented and the Non-Dis-VRS model, when missing the non-discretionary output, acts the same as the VRS model. Therefore, the descending efficiency scores' lines for these three models for DC and MFs overlap.

14.5 Mixing Efficient DMUs

In this section, the efficient DMUs were selected from the VRS model for MFs and from Non-Dis-VRS for PFs and combined. Various DEA models were run on this dataset. The results are presented in Tables 14.6, 14.7, and 14.8 and reproduced with

Table 14.6 Mixing efficient DB and MF DMUs for VRS, Non-Dis-VRS, and MV-DEA models

Plans	VRS_O_CAT	Non-Dis-VRS_O_CAT	MV-DEA_O_CAT
DB and MF	#DMUs: 55 #Efficient: 45 (DB: 33, MF: 12) Average: 0.941 Max: 1 Min: 0.155 Ave of lowest quartile (14 DMUs): 0.768	#DMUs: 55 #Efficient: 45 (DB: 33, MF: 12) Average: 0.891 Max: 1 Min: 0.045 Ave of lowest quartile (14 DMUs): 0.572	#DMUs: 55 #Efficient: 45 (DB: 33, MF: 12) Average: 0.941 Max: 1 Min: 0.155 Ave of lowest quartile (14 DMUs): 0.768

Table 14.7 Mixing efficient DB, Combo, and MF DMUs for VRS, Non-Dis-VRS, and MV-DEA models

Plans	VRS_O_CAT	Non-Dis-VRS_O_CAT	MV-DEA_O_CAT
DB and Combo and MF	#DMUs: 75 #Efficient: 50 (PF: 38, MF: 12) Average: 0.858 Max: 1 Min: 0.111 Ave of lowest quartile (19 DMUs): 0.467	#DMUs: 75 #Efficient: 50 (PF: 38, MF: 12) Average: 0.808 Max: 1 Min: 0.045 Ave of lowest quartile (19 DMUs): 0.331	#DMUs: 75 #Efficient: 50 (PF: 38, MF: 12) Average: 0.845 Max: 1 Min: 0.111 Ave of lowest quartile (19 DMUs): 0.432

Table 14.8 Mixing efficient DC and MF DMUs for VRS, Non-Dis-VRS, and MV-DEA models

Plans	VRS_O	Non-Dis-VRS_O	MV-DEA_O
DC and MF	#DMUs: 35 #Efficient: 15 (DC: 7, MF: 8) Average: 0.72 Max: 1 Min: 0.031 Ave of lowest quartile (9 DMUs): 0.276	#DMUs: 35 #Efficient: 15 (DC: 7, MF: 8) Average: 0.72 Max: 1 Min: 0.031 Ave of lowest quartile (9 DMUs): 0.276	#DMUs: 35 #Efficient: 15 (DC: 7, MF: 8) Average: 0.72 Max: 1 Min: 0.031 Ave of lowest quartile (9 DMUs): 0.276

permission from Badrizadeh and Paradi (2017).

The number of DMUs for each fund type are as follows:

DB: Out of 90 DB plans 33 DMUs were efficient;
DC: Out of 37 DC plans 13 DMUs were efficient;
Combo: Out of 46 DMUs 20 DMUs were efficient;
MF: Out of 61 MFs 22 DMUs were efficient.

In Table 14.6 and Figs. 14.5 and 14.6, as the results were combined from various DEA models (Non-Dis-VRS and VRS) for both fund types, it was found that the efficient pension funds performed better than their mutual funds' equivalents. Not surprisingly the frontier was mostly formed from pension fund DMUs.

Also, the results for the VRS and the MV-DEA models are the same since all DB plans become efficient in these two models (VRS is used for DB DMUs in the VRS model and Non-Dis-VRS is used for DB DMUs in the MV-DEA model) and VRS is used in both the VRS and the MV-DEA models for MFs' DMUs. Therefore, the results are the same and the lines overlap.

As presented in Table 14.7 and Figs. 14.7 and 14.8, the efficient DB and Combo plans perform better than the efficient mutual funds. Since most of the pension fund DMUs were efficient (but not all), the outcomes from the VRS, Non-Dis-VRS, and MV-DEA models were not quite the same.

In Table 14.8, as no benefit data existed for DC plans (the non-discretionary output variable in the output-oriented model) the Non-Dis-VRS model acted identically to the VRS model. Finally, the MV-DEA model also produced the same results as the VRS and Non-Dis-VRS models in this case. The minimum score in Table 14.8 was unusually low (0.0306). Although this was an efficient mutual fund in the mutual fund dataset, however, in the combined sample of efficient MFs and efficient DCs, it becomes highly inefficient. The reason is that sometimes efficient DMUs are referenced to themselves and when the frontier is moving (because of mixing efficient DCs and efficient MFs here), the self-defined efficient DMUs will be highly inefficient for the new frontier.

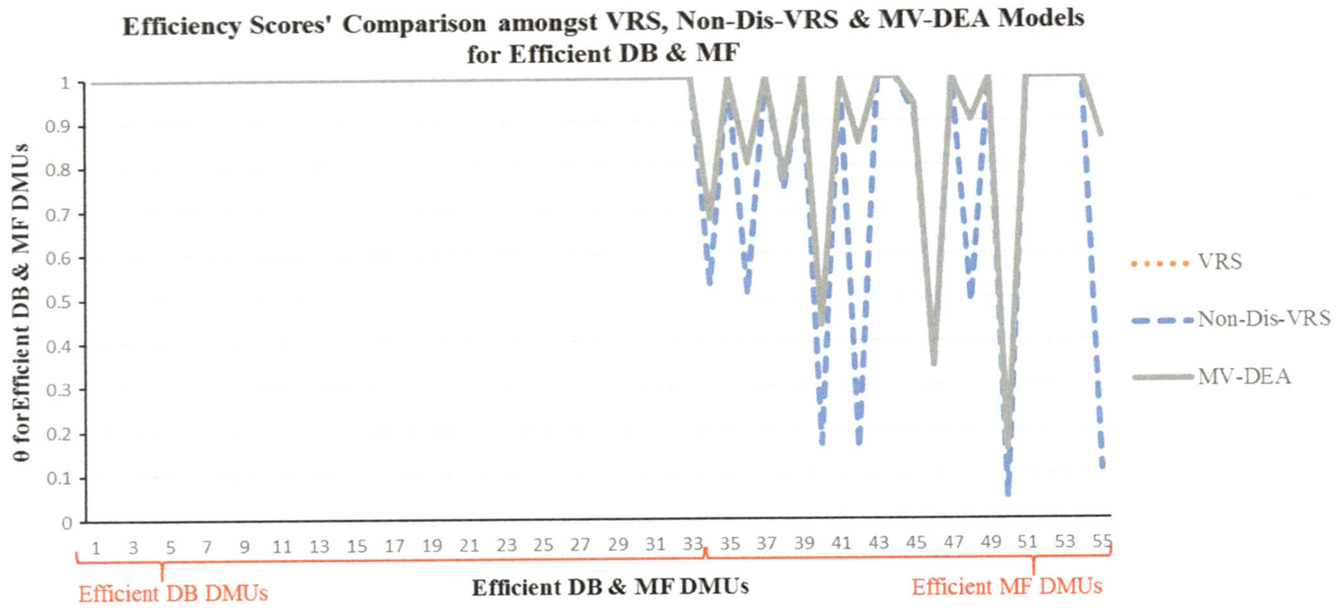

Fig. 14.5 Efficiency scores' comparison among VRS (covered by MV-DEA line), Non-Dis-VRS, and MV-DEA models for efficient DB and MF

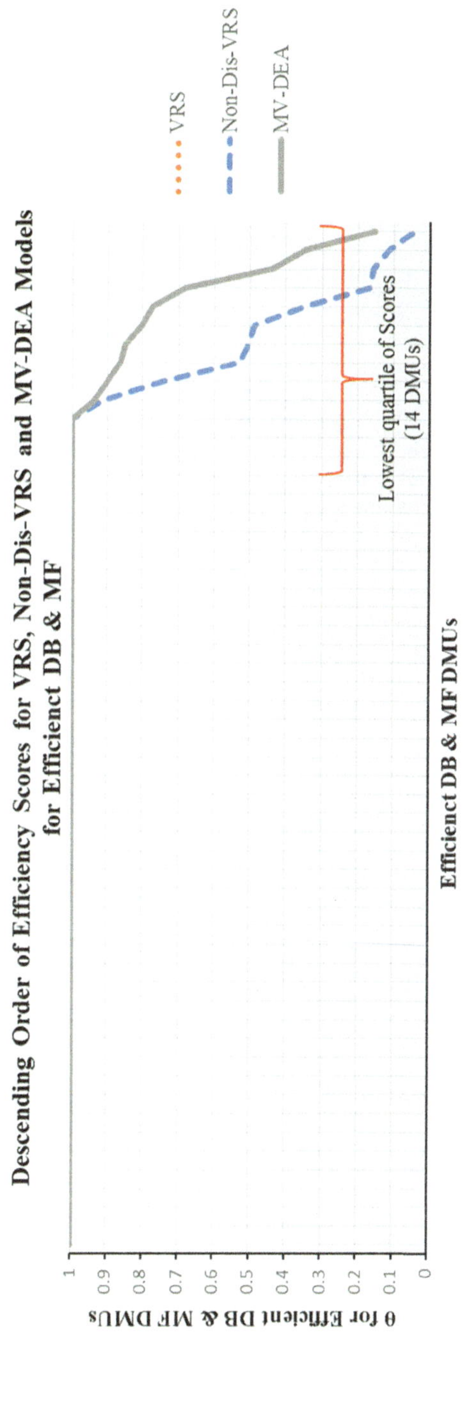

Fig. 14.6 Descending order of efficiency scores for VRS (covered by MV-DEA line), Non-Dis-VRS, and MV-DEA models for efficient DB and MF

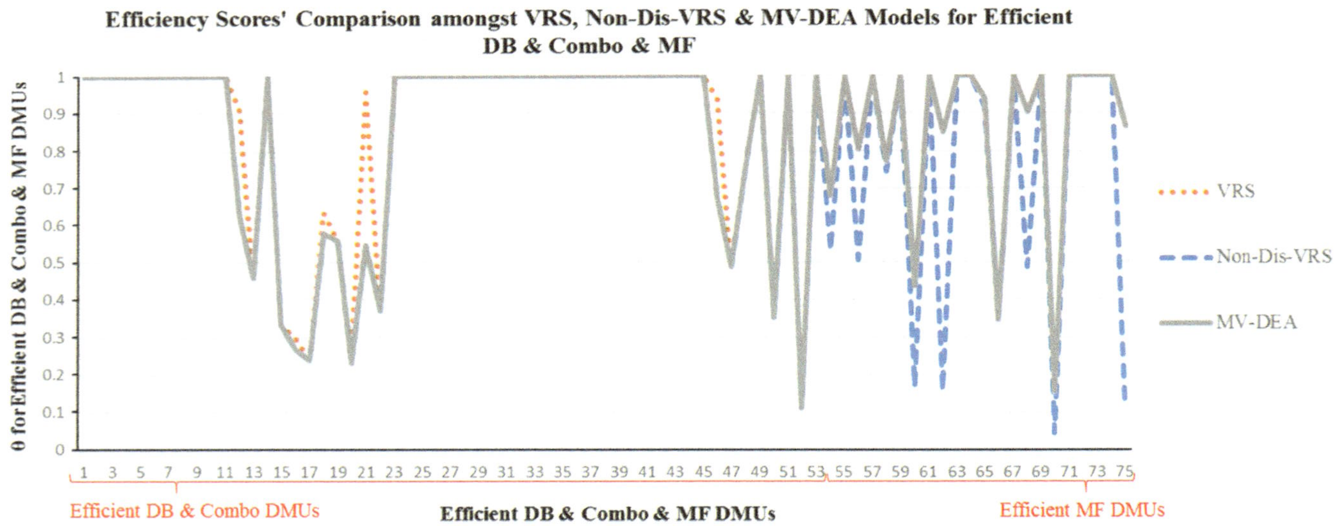

Fig. 14.7 Efficiency scores' comparison among VRS, Non-Dis-VRS, and MV-DEA models for efficient DB, Combo, and MF

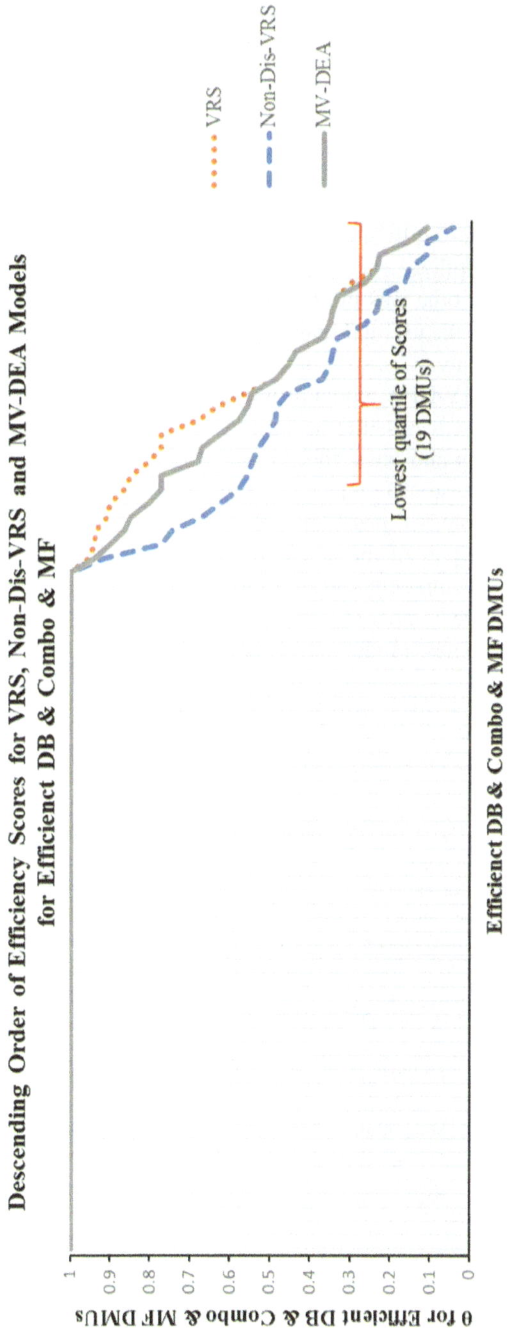

Fig. 14.8 Descending order of efficiency scores for VRS, Non-Dis-VRS, and MV-DEA models for efficient DB, Combo, and MF

14.6 Conclusions

While many researchers had worked on Mutual funds using DEA around the world, Pension funds were not analyzed before this work, at least in North America. This is an odd situation, but probably the result of the paucity of data and the existence of plenty of privacy and confidentiality rules. But even in Canada, where this work was produced, getting reliable data, and particularly complete datasets, while they do exist, they are hard to obtain. The authors were fortunate to have access to good datasets, but did need to carefully obtain and study the many rules that both Pension and Mutual Funds must follow.

The most notable outcome was the development of a new DEA model, namely MV-DEA, which enables the user to break through the "culture" problem; the model developed was able to analyze pension funds and mutual funds separately and combined. Specifically, the MV-DEA model provided a view into the various fund management practices. This was possible because while both fund types are operating within certain regulations and laws, there is sufficient similarity between the two to be able to adjust for these differences. This methodology could be used in other business areas, for example, airlines, shipping, etc.

During the validation part of the research, the MV-DEA model was proven to work well in both its envelopment and multiplier formulations. The Non-Dis-VRS and the VRS models were compared to the new MV-DEA model. The results were quite satisfactory and the new model performed well.

References

Badrizadeh, M., & Paradi, J. C. (2017). *Evaluating pension funds considering invisible variables & bridging pension funds & mutual funds through the development of a new DEA model*. Ph.D. dissertation, University of Toronto.

Banker, R. D., Charnes, A., & Cooper, W. W. (1984). Some models for estimating technical and scale inefficiencies in data envelopment analysis. *Management Science, 30*(9), 1078–1092.

Banker, R. D., & Morey, R. (1986). Efficiency analysis for exogenously fixed inputs and outputs. *Operations Research, 34*(4), 513–521.

Basso, A., & Funari, S. (2005). A generalized performance attribution technique for mutual funds. *Central European Journal of Operations Research, 13*(1), 65–84.

Barrientos, A., & Boussofiane, A. (2005). How efficient are pension fund managers in Chile? *Revista de Economia Contemporânea, 9*(2), 289–311.

Barros, C. P., & Garcia, M. T. M. (2006). Performance evaluation of pension funds management companies with data envelopment analysis. *Risk Management and Insurance Review, 9*(2), 165–188.

Choi, Y. K., & Murthi, B. P. S. (2001). Relative performance evaluation of mutual funds: A non parametric approach. *Journal of Business Finance & Accounting, 28*(7), 853–876.

Dyson, R. G., Allen, R., Camanho, A. S., Podinovski, V. V., Sarrico, C. S., & Shale, E. A. (2001). Pitfalls and protocols in DEA. *European Journal of Operational Research, 132,* 245–259.

Galagedera, D. U. A., & Watson, J. (2015). Benchmarking superannuation funds based on relative performance. *Applied Economics, 47*(28), 2959–2973.

Garcia, M. T. M. (2010). Efficiency evaluation of the portuguese pension funds management companies. *Journal of International Financial Markets, Institutions and Money, 20*(3), 259–266.
Malhotra, D. K., Martin, R., & Russel, P. (2007). Determinants of cost efficiencies in the mutual fund industry. *Review of Financial Economics, 16*(4), 323–334.
McMullan, P., & Strong, R. (1998). Selection of mutual funds using data envelopment analysis. *The Journal of Business and Economic Studies, 4*(1), 1–12.
Morey, M., & Morey, R. (1999). Mutual fund performance appraisals: A multi-horizon perspective with endogenous benchmarking. *Omega, 24*(2), 241–258.
Murthi, B., Choi, Y., & Desai, P. (1997). Efficiency of mutual funds and portfolio performance measurement: A Non-parametric approach. *European Journal of Operational Research, 98*(2), 408–418.
Paradi, J. C., Sherman H. D., & Tam F. K. (2018). *Data envelopment analysis in the financial services industry. A guide for practitioners and analysts working in operations research using DEA international series in operations research & management science*. Springer, Berlin.
Premachandra, I. M., Zhu, J., Watson, J., & Galagedera, D. U. A. (2012). Best performing US mutual fund families from 1993 to 2008: Evidence from a novel two-stage DEA model for efficiency decomposition. *Journal of Banking & Finance, 36*(12), 3302–3317.
Sathye, M. (2011). The impact of financial crisis on the efficiency of superannuation funds. *Journal of Law and Financial Management, 10*(2), 16–27.
Wilkens, K., & Zhu, J. (2001). Portfolio evaluation and benchmark selection: A mathematical programming approach. *The Journal of Alternative Investments, 4*(1), 9–19.
Zamuee, M. R. (2015). Data envelopment analysis to measure efficiency of Namibian pension funds. *American Journal of Marketing Research, 1*(4), 215–221.

Chapter 15
Sharpe Portfolio Using a Cross-Efficiency Evaluation

Mercedes Landete, Juan F. Monge, José L. Ruiz, and José V. Segura

Abstract The Sharpe ratio is a way to compare the excess returns (over the risk-free asset) of portfolios for each unit of volatility that is generated by a portfolio. In this paper, we introduce a robust Sharpe ratio portfolio under the assumption that the risk-free asset is unknown. We propose a robust portfolio that maximizes the Sharpe ratio when the risk-free asset is unknown, but is within a given interval. To compute the best Sharpe ratio portfolio, all the Sharpe ratios for any risk-free asset are considered and compared by using the so-called cross-efficiency evaluation. An explicit expression of the Cross-Efficiency Sharpe Ratio portfolio is presented when short selling is allowed.

Keywords Finance · Portfolio · Minimum-variance portfolio · Cross-efficiency

15.1 Introduction

In 1952 Harry Markowitz made the first contribution to portfolio optimization. In the literature on asset location, there has been significant progress since the seminal work by Markowitz in 1952, Markowitz (1952), who introduced the optimal way of selecting assets when the investor only has information about the expected return and variance for each asset in addition to the correlation between them.

A previous version of this manuscript is available on arXiv.org.

M. Landete · J. F. Monge (✉) · J. L. Ruiz · J. V. Segura
Center of Operations Research, University Miguel Hernandez, Elche, Alicante, Spain
e-mail: monge@umh.es

M. Landete
e-mail: landete@umh.es

J. L. Ruiz
e-mail: jlruiz@umh.es

J. V. Segura
e-mail: jvsh@umh.es

V. Charles et al. (eds.), *Data Science and Productivity Analytics*,
International Series in Operations Research & Management Science 290,
https://doi.org/10.1007/978-3-030-43384-0_15

In 1990, Harry Markowitz, Merton Miller and William Sharpe won the Nobel Prize in Economics for their portfolio optimization theory.

The optimal portfolio obtained by the Markowitz model usually shows high long-term volatility. This feature has motivated a body of research oriented to control the present error in the Markowitz model. Since the variance of the portfolio cannot be considered as an adequate measure of risk, a number of alternative measures have been proposed in the literature in an attempt to quantify the portfolio variance more appropriately (see Markowitz (1959); Jin et al. (2006); Nawrocki (1999) among others). Another way to control the risk in the optimization model is based on setting a minimum threshold for the expected return. Following that approach, several models which incorporate risk measures such as "safety measure", "value at risk", "conditional value at risk", etc., have been proposed in order to control the volatility of the solution. See Artzener et al. (1999); Krokhmal et al. (2002) and references therein.

The incorporation of new restrictions to the problem is also a tool that has been used both to prevent the risk and to incorporate the knowledge of the analyst in search of the best solution. New models have emerged in the last years, which include linear programming models, integer optimization models and stochastic programming models (see Mansini et al. (2014) among others).

Another important feature of the Markowitz model is its myopia about the future scenario of potential returns that will happen. For this reason, producing accurate forecasts in portfolio optimization is of outmost importance. In this sense, forecasting models, factor models in covariance matrix and return estimation, bayesian approaches, or uncertainty estimates (see Ben-Tal and Nemirovski (1999) and references therein) are helpful. The need to improve predictions and consider the present uncertainty in the Markowitz model has motivated the development of what is collectively known as "robust optimization" techniques. Robust methods in mathematical programming were introduced by Bertsimas and Pachamanova (2008) and after studying in a portfolio context by Goldfarb and Iyengar (2003) among others.

There exist several methods in the literature aimed at improving the performance of Markowitz's model, but none of these methods can be considered better than the others. To the authors' knowledge, a systematic comparison of the approaches discussed above has not yet been published. However, in DeMiguel et al. (2009) 14 different models are compared on the basis of a number of datasets with different quality measures. The results obtained show that "*none of the sophisticated models consistently beat the naïve 1/N benchmark*".

Our objective in this paper is to determine the best tangent portfolio, when the free risk rate asset is unknown or the information on this parameter is not deterministic for a long time period. The goal is to find a robust portfolio in the sense of a tangent portfolio better than other tangent portfolios compared with it. To achieve that goal, we use some techniques based on Data Envelopment Analysis (DEA), which provides an analysis of the relative efficiency of the units involved. In the context of portfolio optimization, several authors have used such DEA techniques, specifically the cross-efficiency evaluation (like us here), yet with a different purpose (see, for example, Lim et al. (2014)).

In the next section we present a brief description of the original Markowitz and Sharpe ratio models for portfolio optimization, and discuss some of the features related to the solutions and the efficient frontier that will be needed for the remainder of the paper. In Sect. 15.3 we propose an approach to portfolio optimization based on the cross-efficiency evaluation. In Sect. 15.4 we compare our approach with other classical solutions through the study of two pilot cases. And in the last section we offer a conclusion.

15.2 Overview

In this section, we present a brief description of the Sharpe ratio for asset allocation. The portfolio optimization problem seeks the best allocation of investment among a set of assets. The model introduced by Markowitz provides a portfolio selection as a return-risk bicriteria tradeoff where the variance is minimized under a specified expected return. The mean-variance portfolio optimization model can be formulated as follows:

$$\min \quad \sigma_P^2 = \frac{1}{2} w^T \Sigma w \tag{15.1}$$

$$s.t. \quad w^T \mu = \rho \tag{15.2}$$

$$w^T 1_n = 1. \tag{15.3}$$

The objective function (15.1), σ_P^2, gives the variance of the return $w^T \mu$, where Σ denotes the $n \times n$ variance–covariance matrix of $n-$vector of returns μ, and w is the $n-$vector of portfolio weights invested in each asset. Constraint (15.2) requires that the total return is equal to the minimum rate ρ of return the investor wants. The last constraint (15.3) forces to invest all the money. We denote by 1_n the $n-$dimensional vector of ones. Note that the weight vector w is not required to be non-negative as we want to allow short selling, whose weight of vector w is less than 0.

This model uses the relationship between mean returns and variance of the returns to find a minimum variance point in the feasible region. This minimum variance is a point on the *efficient frontier*, W_ρ. The efficient frontier is the curve that shows all efficient portfolios in a risk-return framework, see Fig. 15.1.

15.2.1 Global Minimum Variance Portfolio

The **Global Minimum Variance** (GMV) portfolio from the Efficient Frontier (W_ρ) is obtained without imposing the expected-return constraint (15.2). The portfolio weights, (w^*_{GMV}), expected return (r^*_{GMV}) and variance (σ^{*2}_{GMV}) are given by

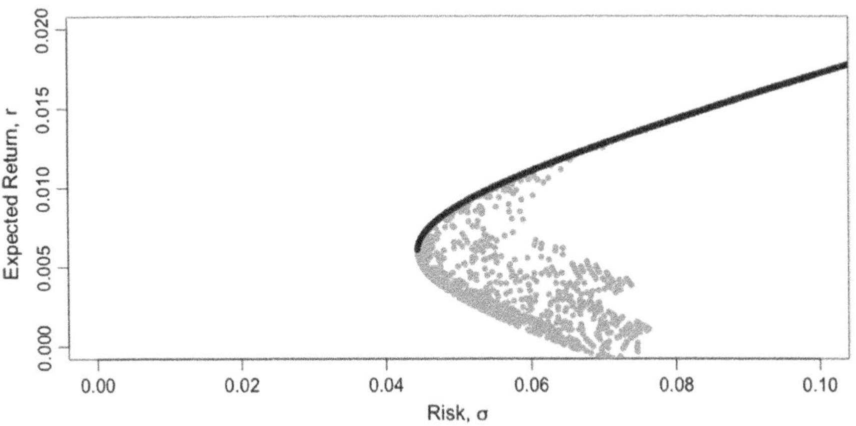

Fig. 15.1 Efficient frontier and cloud of possible portfolios

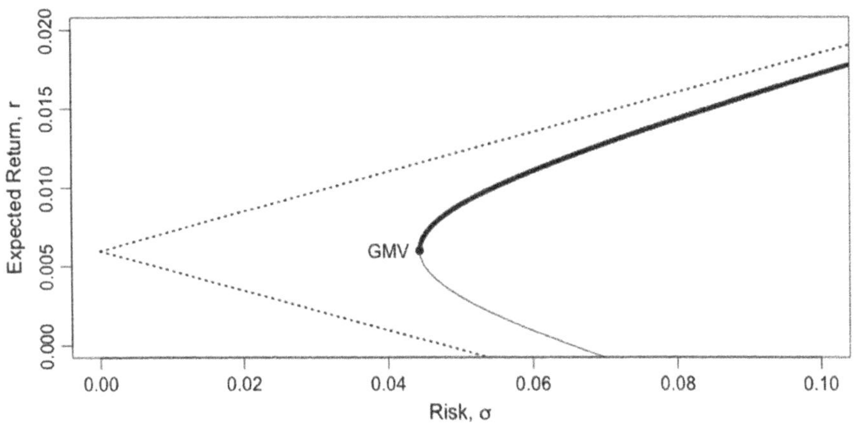

Fig. 15.2 Hyperbola and the assymptotes for mean-variance efficient portfolios

$$w^*_{GMV} = \frac{\Sigma^{-1}1_n}{1_n^T\Sigma^{-1}1_n}, \quad r^*_{GMV} = \frac{1_n^T\Sigma^{-1}\mu}{1_n^T\Sigma^{-1}1_n} \quad \text{and} \quad \sigma^{*2}_{GMV} = \frac{1}{1_n^T\Sigma^{-1}1_n}. \tag{15.4}$$

The hyperbola of the feasible portfolios is enclosed by the asymptotes $r = c/b \pm \sqrt{(ab - c^2)/b}\,\sigma$ with

$$a = \mu^T\Sigma^{-1}\mu, \quad b = 1_n^T\Sigma^{-1}1_n \quad \text{and} \quad c = 1_n^T\Sigma^{-1}\mu. \tag{15.5}$$

The expected return of the global minimum variance portfolio, r_{GMV}, is the apex of the hyperbola. Figure 15.2 represents the hyperbola for the feasible portfolios, the efficient frontier, the global minimum variance portfolio (GMV) and the asymptotes.

15.2.2 Sharpe Ratio

The **Tangent Portfolio** (TP) is the portfolio where the line through the origin is tangent to the efficient frontier W_ρ. This portfolio represents the portfolio with maximum ratio mean/variance.

$$w^*_{TP} = arg \max_w \frac{w^T \mu}{\sqrt{w^T \Sigma w}} \quad s.t. \quad w^T 1_n = 1. \tag{15.6}$$

Another studied portfolio is obtained by maximizing the same ratio when a risk-free asset, r_f, is considered. This portfolio is called the **Maximum Sharpe Ratio** (MSR) portfolio. The Sharpe ratio is the expected excess returns (over the risk-free asset) per unit of risk. Therefore, the Maximum Sharpe Ratio (MSR) portfolio is the solution to the model:

$$w^*_{MSR} = arg \max_w \frac{w^T (\mu - r_f)}{\sqrt{w^T \Sigma w}} \quad s.t. \quad w^T 1_n = 1 \tag{15.7}$$

where r_f denotes the risk-free asset. The allocation w^*_{MSR} is known as market portfolio, M. If the risk-free rate is $r_f = 0$, the market portfolio is identical to the tangent portfolio solution of problem (15.6).

Capital Market Theory asks about the relationship between expected returns and risk for portfolios and free-risky securities.

The solution to (15.7) includes only risky assets. This solution is known as the Market Portfolio (M). A line from the risk-free interest rate through the Market Portfolio (M) is known as the **Capital Market Line** (CML). All the efficient portfolios must lie along this line,

$$CML: \quad E(r) = r_f + \frac{r_M - r_f}{\sigma_M} \sigma$$

where $E(r)$ is the expected portfolio return, r_f the risk-free rate of interest, and r_M, σ_M, respectively, the return and risk of the market portfolio M. All the portfolios on the CML have the same Sharpe ratio. See Fig. 15.3.

The CML summarizes a simple linear relationship between the expected return and the risk of efficient portfolios. Sharpe assumed that the total funds were divided between the market portfolio (M) and security f. The inversion is fully invested here,

$$w_M + w_f = 1.$$

The expected return of the portfolio is

$$E(r_p) = w_f r_f + w_M r_M.$$

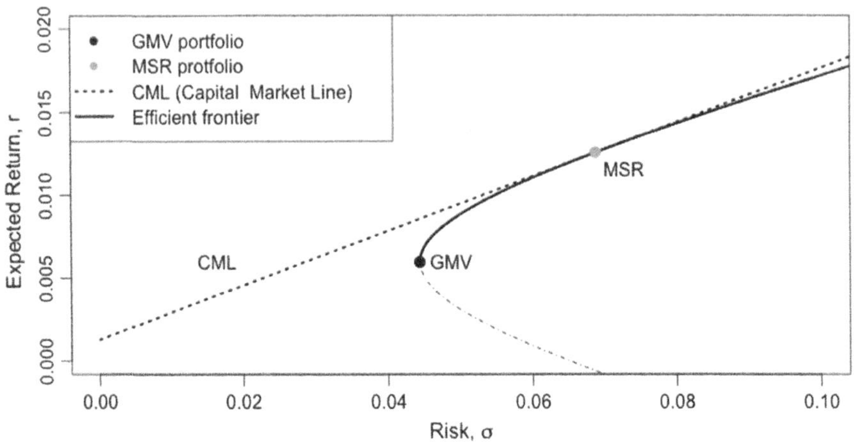

Fig. 15.3 Efficient frontier obtained from four assets. The Global Minimum Variance (GMV) is the portfolio with less risk. The Maximum Sharpe Ratio (MSR) is the tangent portfolio located in the Efficient frontier in the presence of a risk-free asset. The combination of the risk-free asset and the tangency portfolio (MSR) generates the Capital Market Line (CML). CML is the set of non-dominated portfolios when a risk-free asset is present

In order to calculate the optimal *MSR* portfolio of (15.7) we have to maximize (15.7) subject to $w^T 1_n = 1$. In Chapados (2011) we can see how to derive the following expression for the solution of this problem:

$$w^*_{MSR} = \frac{\Sigma^{-1}(\mu - r_f)}{1_n^T \Sigma^{-1}(\mu - r_f)}. \tag{15.8}$$

The risk σ^* and the expected excess returns r^* for the optimal solution to the maximization Sharpe ratio problem with free risk r_f is

$$r^*_{MSR} = w^{*T}_{MSR}\,\mu = \frac{(\mu - r_f)^T \Sigma^{-1} \mu}{1_n^T \Sigma^{-1}(\mu - r_f)} \tag{15.9}$$

$$\sigma^*_{MSR} = \sqrt{w^{*T}_{MSR}\,\Sigma\, w^*_{MSR}} = \frac{\sqrt{(\mu - r_f)^T \Sigma^{-1}(\mu - r_f)}}{1_n^T \Sigma^{-1}(\mu - r_f)}. \tag{15.10}$$

Tangent portfolios are portfolios usually designed for long-term investors. Most investors tend to take on too much risk in good times, and then sell out in bad times. Tangent Portfolios are designed to let investors do well enough in both good and bad times. This lets us reap the long-term benefits of investing in stocks and bonds with a simple, low-maintenance solution.

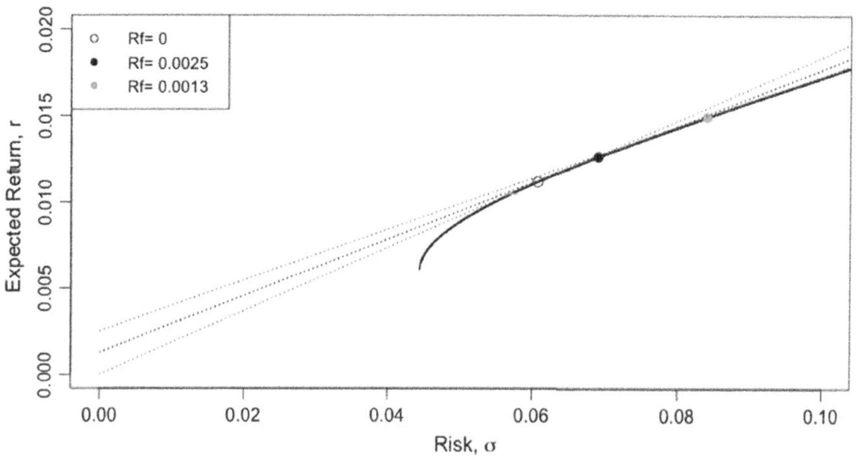

Fig. 15.4 Different CML lines when the risk-free asset is 0, 0.0013 and 0.0025, and their maximum Sharpe ratio portfolios

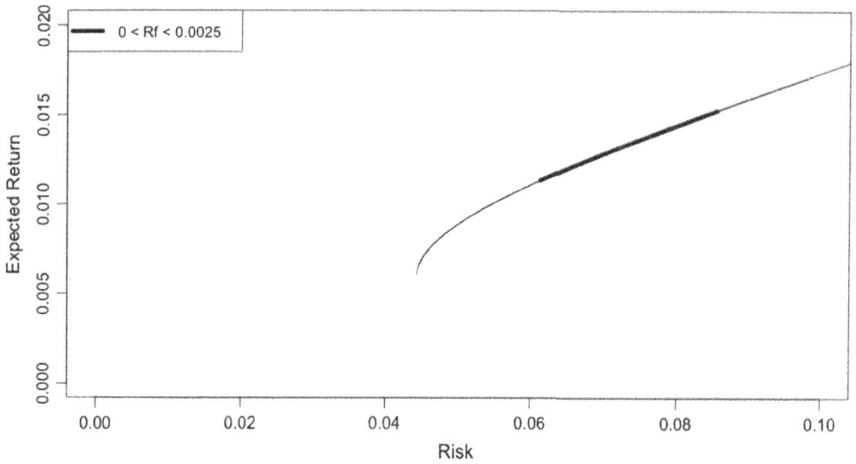

Fig. 15.5 $W_\rho^{r_f}$ set when the risk-free asset is in the interval [0, 0.0025]

Denote by $W_\rho^{r_f}$ the subset of efficient portfolios, $W_\rho^{r_f} \subset W_\rho$, formed for maximum Sharpe ratio portfolios, i.e., tangent portfolios obtained for some value of r_f. Note that, all the tangent portfolios in $W_\rho^{r_f}$ are obtained varying r_f from 0 to the hyperbola apex r^*_{GMV}, i.e., $r_f \in [0, r^*_{GMV}]$. See Figs. 15.4 and 15.5.

15.3 Portfolio Selection Based on a Cross-Efficiency Evaluation

In this section we propose an approach to make the selection of a portfolio within the set $W_\rho^{r_f}$. This approach is inspired by the so-called Data Envelopment Analysis (DEA) and cross-efficiency evaluation methodologies. DEA, as introduced in Charnes et al. (1978), evaluates the relative efficiency of decision-making units ($DMUs$) involved in production processes. For each DMU_0, it provides an efficiency score in the form of a weighted sum of outputs to a weighted sum of inputs. DEA models allow the $DMUs$ total freedom in the choice of the input and output weights. This means that each DMU_0 chooses the weights that show it in its best possible light, with the only condition that the efficiency ratios calculated for the other $DMUs$ with those weights are lower than or equal to a given quantity, usually set at 1. Thus, DMU_0 is rated as efficient if its efficiency score equals 1. Otherwise, it is inefficient, and the lower the efficiency score, the larger its inefficiency. DEA has been successfully used in many real applications to analyze efficiency in areas such as banking, health care, education or agriculture.

Inspired by DEA, we propose to solve the following model for each portfolio (σ_0, r_o) in $W_\rho^{r_f}$

$$\begin{aligned} &\text{Maximize} \frac{u\,(r_0 - u_0)}{v\,\sigma_0} \\ &\text{Subjet to.} \frac{u\,(r - u_0)}{v\,\sigma} \leq 1 \qquad \forall (\sigma, r) \in W_\rho^{r_f} \\ &u, v \geq 0 \\ &u_0 \geq 0. \end{aligned} \tag{15.11}$$

In (15.11), the portfolios in $W_\rho^{r_f}$ would be playing the role of the DMUs, which in this case have one single input (risk, σ) and one single output (return, r). It should be noted that, unlike the problem we address here, in standard DEA we have a finite number of DMUs. Obviously, the optimal value of (15.11) when solved in the evaluation of each portfolio in $W_\rho^{r_f}$, (σ_0, r_o), equals 1, because there exist non-negative weights u^*, v^* and v_0^* such that $u^*(r_0 - u_0^*)/v^*\sigma_0 = 1$, and $u^*(r - u_0^*)/v^*\sigma \leq 1$ for the rest of the portfolios $(\sigma, r) \in W_\rho^{r_f}$. These weights are actually the coefficients of the tangent hyperplane to the curve (efficient frontier) W_ρ at (σ_0, r_0).

Denote in general by $(\sigma^*_{MSR_i}, r^*_{MSR_i}) \in W_\rho^{r_f}$ the MSR portfolio obtained when the risk-free rate r_f is r_f^i. This portfolio maximizes the Sharpe ratio (15.7). Therefore, the optimal solution of (15.11) when $(\sigma^*_{MSR_i}, r^*_{MSR_i})$ is evaluated is

$$u_i^* = \frac{1}{r^*_{MSR_i} - r_f^i}, \quad v_i^* = \frac{1}{\sigma^*_{MSR_i}} \quad \text{and} \quad u_{i0}^* = r_f^i. \tag{15.12}$$

As said before, its optimal value equals 1. Nevertheless, these optimal solutions for the weights allow us to define the cross-efficiency score of a given portfolio obtained with the weights of the others.

Definition 1 Let (u_j^*, v_j^*, u_{j0}^*) be an optimal solution of (15.11) for portfolio $j := (\sigma_{MSR_j}^*, r_{MSR_j}^*)$. The cross-efficiency of a given portfolio $i := ((\sigma_{MSR_i}^*, r_{MSR_i}^*)$ obtained with the weights of portfolio j is defined as follows

$$Ef_i(r_f^j) = \frac{u_j^*(r_{MSR_i}^* - u_{j0}^*)}{v_j^* \sigma_{MSR_i}^*}. \tag{15.13}$$

We can see that (15.13) provides an evaluation of the efficiency of portfolio i from the perspective of portfolio j.

The following proposition holds.

Proposition 1

$$Ef_i(r_f^j) = \frac{(r_{MSR_i}^* - r_f^j)/\sigma_{MSR_i}^*}{(r_{MSR_j}^* - r_f^j)/\sigma_{MSR_j}^*}. \tag{15.14}$$

Proof of Proposition 1.

$$Ef_i(r_f^j) = \frac{u_j^*(r_{MSR_i}^* - u_{j0})}{v_j^* \sigma_{MSR_i}^*} = \frac{\sigma_{MSR_j}^*(r_{MSR_i}^* - r_f^j)}{\sigma_{MSR_i}^*(r_{MSR_j}^* - r_f^j)} = \frac{(r_{MSR_i}^* - r_f^j)/\sigma_{MSR_i}^*}{(r_{MSR_j}^* - r_f^j)/\sigma_{MSR_j}^*}. \qquad \square$$

Equation (15.14) provides a different interpretation of the cross-efficiency scores. Specifically, $Ef_i(r_f^j)$ represents the ratio between the excess return by risk of portfolio i with respect to portfolio j when the risk-free asset is r_j, that is, how bad the Sharpe ratio of portfolio i is compared to the optimal Sharpe ratio of portfolio j.

Cross-efficiency evaluation (Sexton et al. 1986; Doyle and Green 1994) arose as an extension of DEA aimed at ranking DMUs. DEA provides a self-evaluation of DMUs by using input and output weights that are unit-specific, and this makes impossible to derive an ordering. In addition, it is also claimed that cross-efficiency evaluation may help improve discrimination, which is actually the problem we address in the present paper. DEA often rates many DMUs as efficient as a result of the previously mentioned total weight flexibility: DMUs frequently put the weight on a few inputs/outputs and ignore the variables with poor performance by attaching them a zero weight. The basic idea of cross-efficiency evaluation is to assess each unit with the weights of all DMUs instead of its own weights only. Specifically, the cross-efficiency score of a given unit is usually calculated as the average of its cross-efficiencies obtained with the weights profiles provided by all DMUs. Thus, each unit is evaluated with reference to the range of weights chosen by all DMUs, which provides a peer-evaluation of the unit under assessment, as opposed to the conventional DEA self-evaluation. In particular, this makes possible a ranking of the DMUs based on the resulting cross-efficiency scores. Cross-efficiency evaluation has also been widely used to address real-world problems, in particular to deal with issues related to portfolios (see Lim et al. (2014); Galagedera (2013); Pätäri et al. (2012)). Next, we adapt the idea of the standard cross-efficiency evaluation to the problem of portfolio selection we address here. In order to do so, we first define the cross-efficiency score of a given portfolio, which is the measure that will be used for the selection of portfolios among those in $W_\rho^{r_f}$.

15.3.1 Cross-Efficiency Sharpe Ratio Portfolio

In this section we present the average cross-efficiency measure for any portfolio $(\sigma, r) \in W_\rho$ and obtain an expression for the Maximum Cross-Efficiency Sharpe Ratio portfolio (MCESR).

Definition 2 Let r_f be the risk-free rate, which satisfies $r_f \in [r_{\min}, r_{\max}]$, then the average cross-efficiency score (CE_i) of portfolio i, $i = (\sigma^*_{MSR_i}, r^*_{MSR_i}) \in W_\rho$ with $r_i \in [r_{\min}, r_{\max}]$, is given by

$$CE_i = \frac{1}{r_{\max} - r_{\min}} \int_{r_{\min}}^{r_{\max}} Ef_i(r_f)\, \mathrm{d}r_f. \tag{15.15}$$

Note that the expression (15.15) is a natural extension of the cross-efficiency evaluation in DEA for a continuous frontier. Using the expression of $Ef_i(r_f^j)$, see Eq. (15.14), the cross-efficiency CE_i can be written as

$$CE_i = \frac{r^*_{MSR_i}}{\sigma^*_{MSR_i}} I_1 - \frac{1}{\sigma^*_{MSR_i}} I_2 \tag{15.16}$$

where

$$I_1 = \frac{1}{r_{\max} - r_{\min}} \int_{r_{\min}}^{r_{\max}} \frac{\sigma^*_{MSR_f}}{r^*_{MSR_f} - r_f} \mathrm{d}r_f \tag{15.17}$$

$$I_1 = \frac{1}{r_{\max} - r_{\min}} \int_{r_{\min}}^{r_{\max}} \frac{\sigma^*_{MSR_f}\, r_f}{r^*_{MSR_f} - r_f} \mathrm{d}r_f. \tag{15.18}$$

Proposition 2 *The efficient portfolio* $i = (\sigma^*_{MSR_i}, r^*_{MSR_i})$ *that maximizes the cross-efficiency* CE_i*, in the interval* $[r_1, r_2]$*, is reached when*

$$r_i^* = r_{GMV}^* + \sigma_{GMV}^* \frac{\dfrac{r_{MSR_2}^* - r_2}{\sigma_{MSR_2}^*} - \dfrac{r_{MSR_1}^* - r_1}{\sigma_{MSR_1}^*}}{\ln\left(\dfrac{\dfrac{r_{MSR_2}^* - r_2}{\sigma_{MSR_2}^*} - \dfrac{r_{GMV}^* - r_2}{\sigma_{GMV}^*}}{\dfrac{r_{MSR_1}^* - r_1}{\sigma_{MSR_1}^*} - \dfrac{r_{GMV}^* - r_1}{\sigma_{GMV}^*}}\right)}. \tag{15.19}$$

Proof of Proposition 2. See the appendix. □

Corollary 1 *The maximum cross-efficiency (MCESR) portfolio in the interval* $[0, r_{GMV}^*]$ *is reached when*

$$r_i^* = r_{GMV}^* + \sigma_{GMV}^* \frac{\dfrac{r_{MSR_2}^* - r_{GMV}^*}{\sigma_{MSR_2}^*} - \dfrac{r_{TP}^*}{\sigma_{TP}^*}}{\ln\left(\dfrac{r_{MSR_2}^* - r_{GMV}^*}{\sigma_{MSR_2}^*}\right) - \ln\left(\dfrac{r_{TP}^*}{\sigma_{TP}^*} - \dfrac{r_{GMV}^*}{\sigma_{GMV}^*}\right)} \tag{15.20}$$

and, we can write the above expression as

$$r_i^* = r_{GMV}^* \left(1 - \frac{\dfrac{m_{ah}}{m_{GMV}} - \dfrac{m_{TP}}{m_{GMV}}}{\ln\left(\dfrac{m_{TP}}{m_{ah}} - \dfrac{m_{GMV}}{m_{ah}}\right)}\right), \tag{15.21}$$

where $m_{ah} = (r_{MSR_2}^* - r_{GMV}^*)/\sigma_{MSR_2}^*$ is the slope of the asymptote of W_ρ, $m_{TP} = r_{TP}^*/\sigma_{MSR_2}^*$ is the slope of the CML line when $r_f = 0$, i.e., the slope of the linear line from the origin to the tangent portfolio, and, $m_{GMV} = r_{GMV}^*/\sigma_{GMV}^*$ is the slope of the linear line from the origin to the global minimum variance portfolio (GMV), see Fig. 15.6.

Proof of Corollary 1. It follows from proposition 2. □

Proposition 3 *There exists a Pythagorean relationship between the slopes of the Tangent and Global Minimum portfolios and the slope of the asymptote of* W_ρ.

$$m_{TP}^2 = m_{ah}^2 + m_{GMV}^2. \tag{15.22}$$

Proof of Proposition 3. See the appendix. □

Corollary 2 *The maximum cross-efficiency (MCESR) portfolio in* $[0, r_{GMV}^*]$ *depends only on Minimal Global Variance and Tangent portfolios.*

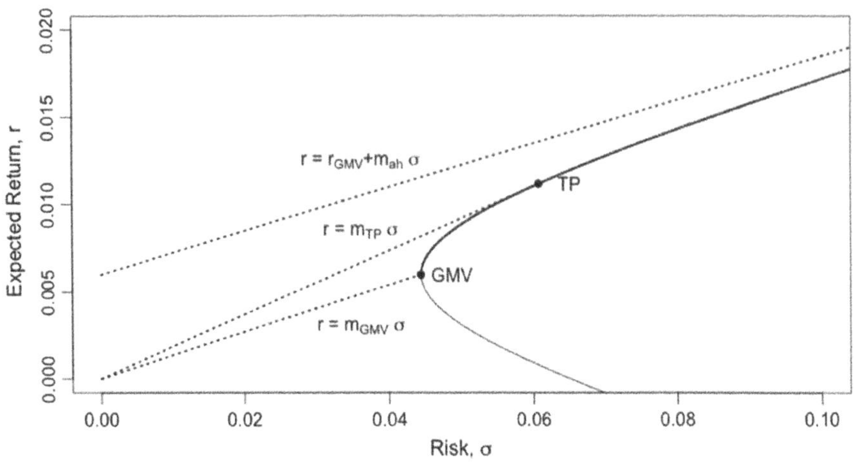

Fig. 15.6 Linear lines for the global minimum variance and tangent portfolios; and for the asymptote of the hyperbola

$$r_i^* = r_{GMV}^* \left(1 - \frac{\sqrt{\frac{r_{TP}^*}{r_{GMV}^*}} - \sqrt{\frac{r_{TP}^*}{r_{GMV}^*} - 1}}{\ln\left(\sqrt{\frac{r_{TP}^*}{r_{GMV}^*}} - 1\right) - \ln\left(\sqrt{\frac{r_{TP}^*}{r_{GMV}^*}} - 1\right)} \right). \tag{15.23}$$

Proof of Corollary 2. See the appendix. □

15.3.2 No Short-Sales Constraint

This constraint corresponds to the requirement that all asset weights be non-negative. If no short selling is allowed, then we need to add the non-negativity of each weight in vector w to the maximization Sharpe ratio problem (15.7),

$$\begin{aligned} w_{MSR}^* = \max_{w} & \frac{w^T(\mu - r_f)}{\sqrt{w^T \Sigma w}} \\ s.t. \quad & w^T 1_n = 1 \\ & w \geq 0. \end{aligned} \tag{15.24}$$

Markowitz's original formulation (15.1–15.3) included those constraints as an integral part of the portfolio optimization method. Note that the inclusion of these non-negativity constraints makes impossible to derive an analytical solution for the portfolio optimization problem (15.24). Model (15.24) is not a convex problem, so it is not easy to solve it. In Tütüncü (2003), R.H. Täutäuncäu present a convex quadratic programming problem equivalent to (15.24). This new formulation of the problem considers a higher dimensional space where the quadratic problem is convex when applying the lifting technique that follows.

It is easy to derive the equivalent problem of (15.24) as

$$\begin{aligned} \min \quad & x^T \Sigma x \\ s.t. \quad & x^T(\mu - r_f) = 1 \\ & x \geq 0 \end{aligned} \tag{15.25}$$

where the weight vector w of (15.24) is given by

$$w = \frac{x}{x^T 1_n}.$$

Note that problem (15.25) can be solved by using the well-known techniques for convex quadratic programming problems.

Although it is not possible to find a closed expression for the Maximum Cross-Efficiency (MCESR) portfolio, model (15.25) allows us to obtain an optimal portfolio, for the maximization Sharpe ratio problem, and for different values of the risk-free asset. We propose the following procedure to obtain an approximation to the Maximum Cross-Efficiency (MCESR) portfolio when no short-sales constraints are present.

1. Dividing the interval $[r_{\min}, r_{\max}]$ into n equal parts.
2. For (i = 1 to n + 1), solving (15.25) with $r_f = r_{\min} * (n - i + 1)/n + r_{\max} * (i - 1)/n$, and obtaining the efficient portfolio $i = (\sigma_{MSR_i}, r_{MSR_i})$, $\forall i = 1$ to n.
3. Computing the solution (u_i^*, v_i^*, u_{i0}^*) of problem (15.11) through expressions (15.12). Note that is not necessary to solve the problem (15.11), the solution of the problem is the tangent hyperplane to efficient curve W_ρ.
4. For (i = 1 to n + 1), calculating the cross-efficiency (CE_i) of portfolio i as the mean of the efficiency score of portfolio i by using the optimal weights of the remaining portfolios in the interval, i.e.,

$$CE_i = \frac{1}{n+1} \sum_{j=1}^{n+1} \frac{(r_{MSR_i}^* - r_f^j)/\sigma_{MSR_i}^*}{(r_{MSR_j}^* - r_f^j)/\sigma_{MSR_j}^*}. \tag{15.26}$$

5. Obtaining the efficient portfolio i that maximizes the cross-efficiency CE_i.

15.4 Numerical Example

We carried out two computational studies in order to illustrate the proposed approach. In the first one, we evaluate the goodness of the maximum cross-efficiency portfolio (MCESR) and draw some conclusions. The second part of the study allows us to compare the (MCESR) allocation depending on whether short sales are allowed or not.

The whole computational study was conducted on a MAC-OSX with a 2.5GHz Intel Core i5 and 4GB of RAM. We used the R-Studio, v0.97.551 with the library *stockPortfolio*, Diez and Christou (2020). In our computational study the required computational time did not exceed a few seconds; for this reason the times have not been reported.

15.4.1 Case 1. EUROSTOCK

In this section we compare the performance of our Maximum Cross-Efficiency Sharpe Ratio (MCESR) portfolio allocation with different Sharpe ratio allocations on a small example with real data. The set of assets that were chosen are listed in Table 15.1, and these were obtained from EUROSTOCK50. We selected the six Spanish assets in EUROSTOCK50.

Table 15.1 shows some descriptive statistics for the set of assets considered: return (average weekly returns), risk (standard deviation of weekly returns) and the Minimum and Maximum return. The first row- block corresponds to the period from 2009 to 2012 (in-sample or estimation period) and the second row-block to the period from 2013 to 2014 (out-sample or test period); being the last row-block the aggregate data from both periods. Figures 15.9, 15.8 and 15.7 show the accumulated weekly returns for the two periods considered and for the entire period.

In order to evaluate the performance of our solution, the Maximum Cross-Efficiency Sharpe Ratio (MCESR), we compare it with the global minimum variance (GMV), tangent (TP) and Maximum Sharpe Ratio (MSR) portfolios. Table 15.2 shows the different solutions evaluated in the in-sample period for the four portfolios considered. The risk-free asset in the interval (0, 0.003) was considered to evaluate the MCESR portfolio, and we chose the upper limit of the interval considered by the risk-free asset to evaluate the Maximum Sharpe Ration portfolio. Note that the optimal MCESR is obtained when r_f is 0.001773.

Table 15.3 shows the reaching value for each out-sample portfolio, i.e., in the test period. Note that we divided the out-sample period in three sub-periods of 25 weeks each in order to evaluate the evolution of the four allocations. In the first 25 weeks, all the portfolios decrease in value, being the minimum variance portfolio (GMV) the best hold. Returns increase in the next 25 weeks and in this case the portfolio with the higher volatility (MSR) obtains a better performance. Finally, for the entire period out-sample, the GMV portfolio is the only one that provides benefits, and

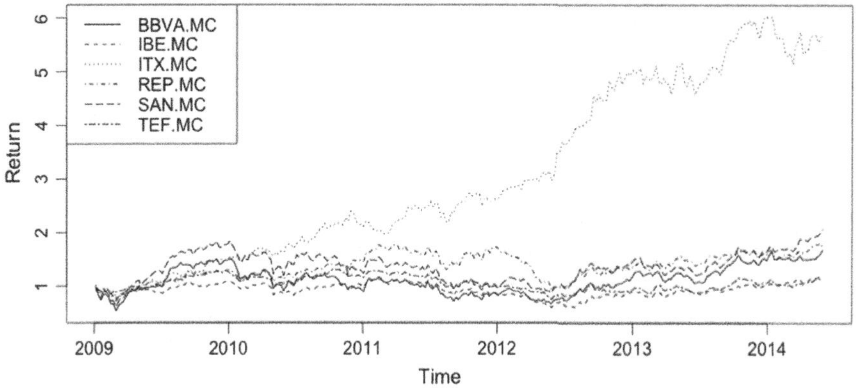

Fig. 15.7 Case 1. Returns for the total period, from 2009-01-01 to 2014-06-06

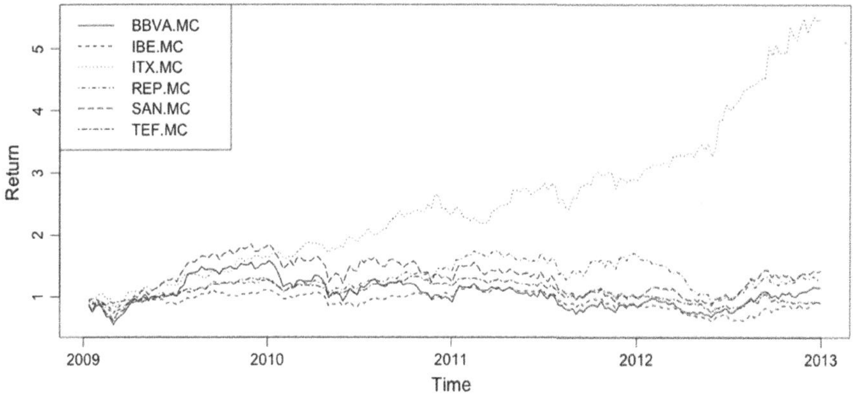

Fig. 15.8 Case 1. Returns for in-sample period, from 2009-01-01 to 2012-12-31

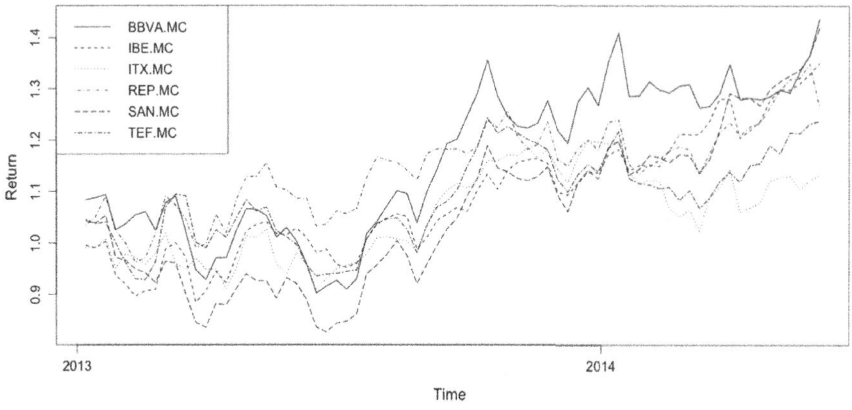

Fig. 15.9 Case 1. Returns for out-sample period, from 2013-01-01 to 2014-06-06

Table 15.1 Case 1. Weekly descriptive statistics returns for 6 EUROSTOCK assets

	BBVA.MC	IBE.MC	ITX.MC	REP.MC	SAN.MC	TEF.MC
	In-sample (Estimation) Period (from 2009-01-01 to 2012-12-31)					
Return	0.0026	0.00044	0.0092	0.0022	0.0034	0.00018
Risk	0.0623	0.04268	0.0381	0.0455	0.0587	0.03400
Minimum	−0.1916	−0.1483	−0.1558	−0.1496	−0.1760	−0.0977
Maximum	0.1838	0.1385	0.1292	0.1249	0.1916	0.1139
	Out-sample (Test) Period (from 2013-01-07 to 2014-06-02)					
Return	0.0056	0.0045	0.0021	0.0036	0.0053	0.0034
Risk	0.0377	0.0287	0.0297	0.0316	0.0335	0.0316
Minimum	−0.0882	−0.0909	−0.0723	−0.0697	−0.0773	−0.0826
Maximum	0.0943	0.0861	0.0669	0.0719	0.0901	0.1102
	Total period (from 2009-01-01 to 2014-06-02)					
Return	0.0035	0.0014	0.0069	0.0028	0.0040	0.0011
Risk	0.0567	0.0394	0.0365	0.0422	0.0530	0.0333
Minimum	−0.1916	−0.1483	−0.1558	−0.1496	−0.1760	−0.0977
Maximum	0.1838	0.1385	0.1292	0.1249	0.1916	0.1139

Table 15.2 Case 1. In-sample results for different portfolio solutions

GMV—Global Minimum Variance Portfolio	
Expected return = 0.00338	Risk = 0.02762
TP—Tangent Portfolio	
Expected return = 0.01612	Risk = 0.06025
MSR—Maximum Sharpe Ratio Portfolio ($r_f = 0.003$)	
Expected return = 0.1149	Risk = 0.4699
MCESR—Maximum Cross-Efficiency Sharpe Ratio Portfolio ($r_f = 0.001773$)	
Expected return = 0.02940	Risk = 0.1129

the worst benefit is obtained for the MSR portfolio where the losses outweigh the investment. Note that this last situation is possible because short sales are allowed.

Figure 15.10 shows the out-sample performance for the four strategies considered. We see the high volatility associated with the MSR ($r_f = 0.003$) portfolio. Note that although the MCESR portfolio is worse than GMV and TP portfolios, the MCESR provides greater benefits in good times and contains the losses in the bad ones .

Table 15.3 Case 1. Change in portfolio value for the out-sample period, from 2014-01-07 to 2014-06-02

Portfolio	First 25 weeks (%)	50 weeks (%)	75 weeks (%)
GMV	−6.1	5.9	15.1
TP	−11.1	8.6	−0.3
MSR ($r_f = 0.003$)	−49.3	29.0	−119.8
MCESR ($r_f = 0.001773$)	−16.2	11.3	−16.4

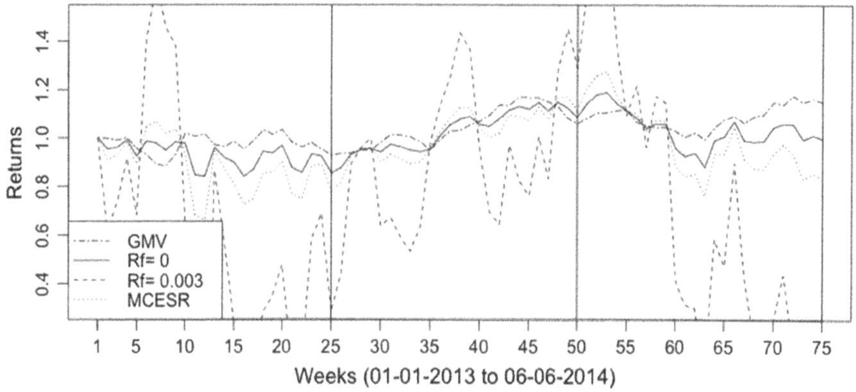

Fig. 15.10 Case 1. Out-sample expected returns from 2013-01-01 to 2013-12-31

Table 15.4 Dataset description

Case 2	N	In-sample (estimation) period	Out-sample (test) period
Ten industry portfolios	10	Jan 1963 to Dec 2012	Jan 2013 to Jul 2014
Monthly data		(600 observations)	(19 observations)

Source Ken French's Web Site

15.4.2 Case 2. USA Industry Portfolios

For the second numerical example, we selected 10 industry portfolios from the USA market. In the same way as in the case above, we considered two time periods, the first time period to estimate (in-sample period) and the second period (out-sample) to evaluate the performance of each strategy. The data source and the number of observations are shown in Table 15.4. In Table 15.5 we keep the same observations as in Table 15.1.

Table 15.6 shows the different portfolios in the in-sample period, and for each portfolio we report their allocation. The risk-free asset in the interval (0, 0.9) was considered in order to calculate the MCESR portfolio, which leads to obtaining the optimal MCESR r_f as 0.57103. If short sales are not allowed, the optimal MCESR is

Table 15.5 Case 2. Monthly descriptive statistics returns for 10 industry portfolios

	NoDur	Durbl	Manuf	Enrgy	HiTec	Telcm	Shops	Hlth	Utils	Other
	In-sample (estimation) Period (from Jan 1963 to Dec 2012)									
Return	1.1	0.86	0.97	1.1	0.96	0.86	1.0	1.1	0.83	0.92
Risk	4.3	6.36	4.98	5.4	6.59	4.67	5.2	4.9	4.03	5.35
Minimum	−21.03	−32.63	−27.33	−18.33	−26.01	−16.22	−28.25	−20.46	−12.65	−23.6
Maximum	18.88	42.62	17.51	24.56	20.75	21.34	25.85	29.52	18.84	20.22
	Out-sample (test) Period (Jan 2013 to Jul 2014)									
Return	1.4	2.3	1.7	1.7	2.0	1.9	1.5	2.4	1.4	1.9
Risk	3.3	4.0	3.2	3.7	2.5	2.8	3.5	3.4	3.8	3.2
Minimum	−5.71	−4.6	−4.33	−6.97	−2.84	−3.94	−6.65	−3.67	−6.96	−4.37
Maximum	5.21	9.9	6.03	7.71	5.97	5.62	5.97	8.11	5.51	6.87
	Total period (Jan 1963 to Jul 2014)									
Return	1.1	0.9	0.99	1.1	1.0	0.89	1.0	1.1	0.85	0.95
Risk	4.3	6.3	4.93	5.4	6.5	4.63	5.2	4.9	4.02	5.30
Minimum	−21.03	−32.63	−27.33	−18.33	−26.01	−16.22	−28.25	−20.46	−12.65	−23.6
Maximum	18.88	42.62	17.51	24.56	20.75	21.34	25.85	29.52	18.84	20.22

Table 15.6 Case 2. In-sample results for different portfolio solutions. 10 industry portfolios. Short Sales

	NoDur	Durbl	Manuf	Enrgy	HiTec	Telcm	Shops	Hlth	Utils	Other
	GMV—Global Minimum Variance portfolio									
	Expected return = 0.96			Risk = 3.45						
Allocation	0.29	0.00	0.09	0.11	0.02	0.26	0.08	0.15	0.45	−0.44
	TP—Tangent portfolio									
	Expected return = 1.09			Risk = 3.68						
Allocation	−0.70	−0.05	−0.05	0.30	0.03	0.18	0.14	0.17	0.17	−0.58
	MSR—Maximum Sharpe Ratio portfolio ($r_f = 0.9$)									
	Expected return = 3.08			Risk = 20.84						
Allocation	6.85	−0.77	−2.13	3.19	0.24	−1.04	1.01	0.53	−4.15	−2.72
	MCESR—Maximum Cross-Efficiency Sharpe Ratio portfolio ($r_f = 0.57103$)									
	Expected return = 1.29			Risk = 4.67						
Allocation	1.29	−0.12	−0.25	0.58	0.05	0.06	0.22	0.21	−0.26	−0.79

obtained when r_f is 0.576. See Table 15.7 for the same results as in Table 15.6 when short sales are not allowed. Figure 15.11 shows the efficient frontier of Malkowitz for both cases, with and without short sales.

In order to compare the performance of four strategies, we evaluated them in the out-sample period. The expected returns for each portfolio, with and without short sales, are shown in Table 15.8. Note that if short sales are allowed, the $MSR(r_f = 0.9)$ portfolio originates losses of 14.3%, while the same strategy portfolio causes a benefit of 33.5% if short sales are not allowed.

Table 15.7 Case 2. In-sample results for different portfolio solutions. 10 industry portfolios. No short sales

	NoDur	Durbl	Manuf	Enrgy	HiTec	Telcm	Shops	Hlth	Utils	Other
	GMV—Global Minimum Variance portfolio									
	Expected return = 0.93			Risk = 3.70						
Allocation	0.30	0.00	0.03	0.00	0.00	0.14	0.00	0.06	0.47	0.00
	TP—Tangent portfolio									
	Expected return = 1.03			Risk = 3.90						
Allocation	0.53	0.00	0.02	0.07	0.00	0.00	0.03	0.19	0.15	0.00
	MSR—Maximum Sharpe Ratio portfolio ($r_f = 0.9$)									
	Expected return = 1.076			Risk = 4.14						
Allocation	0.67	0.00	0.00	0.23	0.00	0.00	0.00	0.10	0.00	0.00
	MCESR—Maximum Cross-Efficiency Sharpe Ratio portfolio ($r_f = 0.576$)									
	Expected return = 1.07			Risk = 4.09						
Allocation	0.63	0.00	0.00	0.14	0.00	0.00	0.00	0.22	0.00	0.00

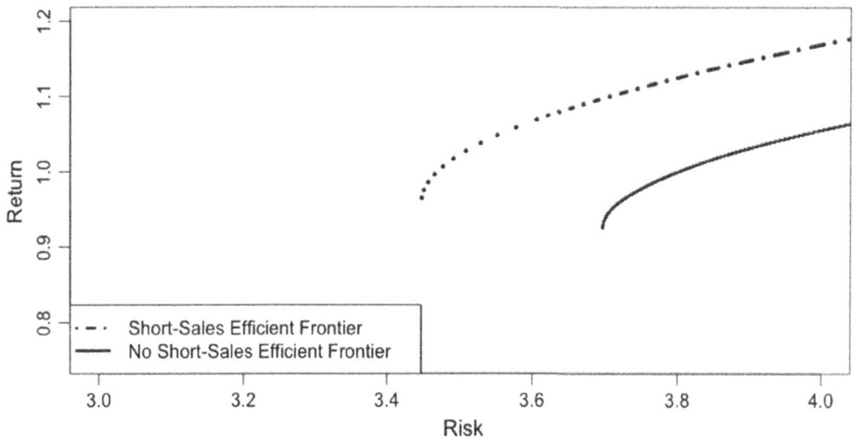

Fig. 15.11 Case 2. Efficient frontiers for 10 industry portfolios

The portfolio with less variation with or without short sales is the GMV portfolio. The MCESR portfolio provides a profit of 25.5% with short sales, and 36.1% without short sales, being this last profit the highest value for all portfolios considered in both situations.

Figures 15.12 and 15.13 show the out-sample performance for the four strategies considered. Note the high volatility of the MSR portfolio when short sales are allowed in front to the homogeneity of the rest. If short sales are not allowed, the four portfolios present practically the same curve, although in this case the MCESR provides the best performance.

Table 15.8 Case 2. Change in portfolio value for the period from 2014-01-07 to 2014-06-02

Portfolio	Short sales (%)	No short sales (%)
GMV	32.7	32.9
Tangent Portfolio	29.8	35.1
MSR ($r_f = 0.9$)	−14.3	33.5
MCESR	25.5	36.1

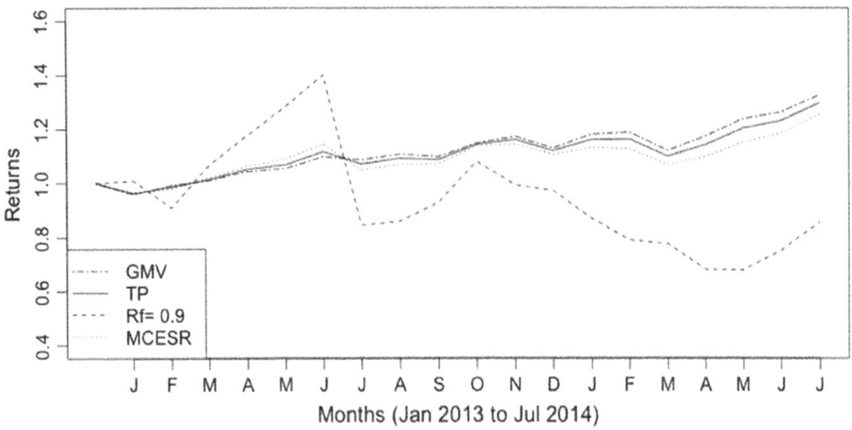

Fig. 15.12 Case 2. Returns from 2014-01-07 to 2014-06-02. With short sales

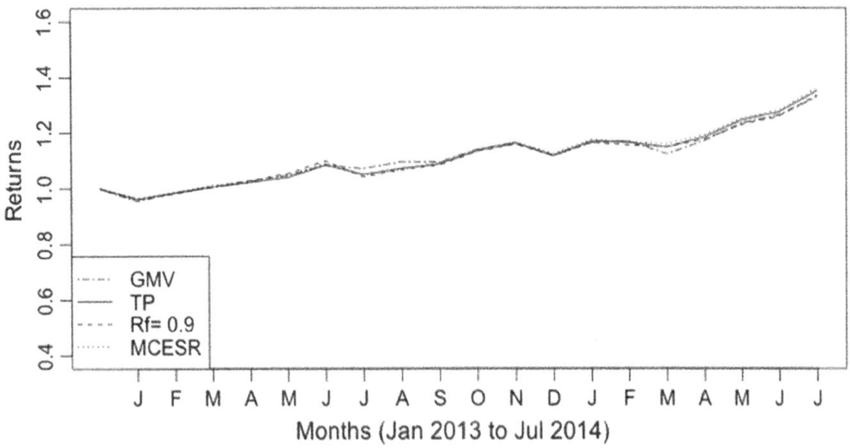

Fig. 15.13 Case 2. Returns from 2014-01-07 to 2014-06-02. Without short sales

15.5 Conclusions

This paper proposes a new portfolio selection strategy based on a cross-efficiency evaluation. We compare the new allocation with the classic global minimum and tangent portfolios through a numerical study. The results show that our allocation is comparable with the others in terms of performance in the out-sample period.

We have derived an explicit expression for the MCESR portfolio when short sales are allowed, and proposed procedures to obtain it when short sales are not allowed. We have also found a relationship between the slopes of the three portfolios considered and that between the MCESR portfolio with the expected returns of the GMV and TP portfolios.

For future research, we plan to apply this new portfolio solution (MCESR) to a large testbed in order to investigate their advantages over the rest.

Acknowledgements The authors thank the financial support from the Spanish Ministry for Economy and Competitiveness (Ministerio de Economa, Industria y Competitividad), the State Research Agency (Agencia Estatal de Investigacin) and the European Regional Development Fund (Fondo Europeo de Desarrollo Regional) under grant MTM2016-79765-P (AEI/FEDER, UE).

15.6 Appendix

Proof of Proposition 2. The efficient portfolio $i = (\sigma^*_{MSR_i}, r^*_{MSR_i})$ that maximizes the cross-efficiency CE_i, in the interval $[r_1, r_2]$, is reached when

$$r_i^* = r^*_{GMV} + \sigma^*_{GMV} \frac{\dfrac{r^*_{MSR_2} - r_2}{\sigma^*_{MSR_2}} - \dfrac{r^*_{MSR_1} - r_1}{\sigma^*_{MSR_1}}}{\ln\left(\dfrac{\dfrac{r^*_{MSR_2} - r_2}{\sigma^*_{MSR_2}} - \dfrac{r^*_{GMV} - r_2}{\sigma^*_{GMV}}}{\dfrac{r^*_{MSR_1} - r_1}{\sigma^*_{MSR_1}} - \dfrac{r^*_{GMV} - r_1}{\sigma^*_{GMV}}} \right)}.$$

The cross-efficiency of portfolio i, CE_i, depends of the risk-free rate, r_i, associated with the portfolio i. We can considered CE_i as a function of r_i, for $r_i \in [r_{\min}, r_{\max}]$. We can write $CE_i(r_i)$ as follows

$$CE_i(r_i) = \frac{r^*_{MSR_i}}{\sigma^*_{MSR_i}} \int_{r_{\min}}^{r_{\max}} \frac{\sigma^*_{MSR}}{r^*_{MSR} - r_f} \mathrm{d}r_f - \frac{1}{\sigma^*_{MSR_i}} \int_{r_{\min}}^{r_{\max}} \frac{\sigma^*_{MSR}\, r_f}{r^*_{MSR} - r} \mathrm{d}r_f .$$

From expressions (15.4), (15.9) and (15.10), using notation of (15.5), we can derivate the following identities for the expected return and variance of GMV and MSR portfolios:

$$r^*_{GMV} = \frac{c}{b}, \quad \sigma^*_{GMV} = \frac{1}{\sqrt{b}}, \quad \frac{r^*_{GMV} - r_f}{\sigma^{*2}_{GMV}} = c - b\ r_f, \quad \frac{r^*_{MSR}}{\sigma^*_{MSR}} = \frac{a - c\ r_f}{\sqrt{a - 2c\ r_f + b\ r_f^2}},$$

$$\frac{1}{\sigma^*_{MSR}} = \frac{c - b\ r_f}{\sqrt{a - 2c\ r_f + b\ r_f^2}}, \quad \text{and} \quad \frac{\sigma^*_{MSR}}{r^*_{MSR} - r_f} = \frac{1}{\sqrt{a - 2c\ r_f + b\ r_f^2}}, \tag{15.27}$$

and write the cross-efficiency, $CE_i(r_i)$, in terms of variable r_i.

$$CE_i(r_i) = \frac{a - c\ r_i}{\sqrt{a - 2c\ r_i + b\ r_i^2}} I_1 - \frac{c - b\ r_i}{\sqrt{a - 2c\ r_i + b\ r_i^2}} I_2$$

where

$$I_1 = \frac{1}{r_{\max} - r_{\min}} \int_{r_{\min}}^{r_{\max}} \frac{dr_f}{\sqrt{a - 2c\ r_f + b\ r_f^2}} \quad \text{and} \quad I_2 = \frac{1}{r_{\max} - r_{\min}} \int_{r_{\min}}^{r_{\max}} \frac{r_f\ dr_f}{\sqrt{a - 2c\ r_f + b\ r_f^2}}.$$

The function $CE_i(R_i)$ has first derivate

$$\begin{aligned} CE_i'(r_i) =& \frac{-c\sqrt{a - 2cr_i + br_i^2} - (a - cr_i)(br - c)/\sqrt{a - 2cr_i + br_i^2}}{\left(\sqrt{a - 2cr_i + br_i^2}\right)^2} I_1 - \\ &- \frac{-b\sqrt{a - 2cr_i + br_i^2} - (c - br_i)(br_i - c)/\sqrt{a - 2cr_i + br_i^2}}{\left(\sqrt{a - 2cr_i + br_i^2}\right)^2} I_2 \\ =& \frac{(c^2 r_i - abr_i)I_1 + (ba - c^2)I_2}{\left(\sqrt{a - 2cr_i + br_i^2}\right)^3}. \end{aligned}$$

It is left to show that $CE_i'(r_i) = 0$ for $r_i = I_2/I_1$, therefore, $r_i = I_2/I_1$ is a point with slope zero, and it is a candidate to a maximum in the interval $[r_{\min}, r_{\max}]$. The second derivate of the function $CE_i(r_i)$ is given by the following expression

$$CE_i''(r_i) = \frac{(c^2 - ab)I_1\left(\sqrt{a - 2cr_i + br_i^2}\right)^3}{\left(\sqrt{a - 2cr_i + br_i^2}\right)^6} - \frac{3\left((ba - c^2)I_2 + (c^2 - ab)I_1 r_i\right)(br_i - c)}{\left(\sqrt{a - 2cr_i + br_i^2}\right)^5} \tag{15.28}$$

and the second term of (15.28) is zero at $r_i = I_2/I_1$, and

$$CE''(I_2/I_1) = \frac{(c^2 - ab)I_1}{(a - 2c\, I_2/I_1 + b\, I_2^2/I_1^2)}.$$

Since Σ^{-1} is positive-definite matrix, then $(\mu - r)^T\Sigma^{-1}(\mu - r) = a - 2cr + br^2 > 0$, with discriminant $4(c^2 - ab) < 0$, then the second derivate at $r_i = I_2/I_1$, $CE''(I_2/I_1)$, is less to 0.

Next, we show the expression of I_2/I_1.

$$(r_{\max} - r_{\min})I_1 = \int_{r_{\min}}^{r_{\max}} \frac{dr_f}{\sqrt{a - 2cr_f + br_f^2}} = \left[\frac{1}{\sqrt{b}} \ln\left(\sqrt{b}\sqrt{a - 2cr_f + br_f^2} + br_f - c\right)\right]_{r_{\min}}^{r_{\max}}$$

$$(r_{\max} - r_{\min})I_2 = \int_{r_{\min}}^{r_{\max}} \frac{r_f\, dr_f}{\sqrt{a - 2cr_f + br_f^2}} =$$

$$= \left[\frac{c}{\sqrt{b^3}} \ln\left(\sqrt{b}\sqrt{a - 2cr_f + br_f^2} + br_f - c\right) + \frac{1}{b}\sqrt{a - 2cr_f + br_f^2}\right]_{r_{\min}}^{r_{\max}}.$$

Now, we can write the above expression using the identities of (15.27) as follows:

$$(r_{\max} - r_{\min})I_1 = \sigma^*_{GMV} \ln\left(\frac{\frac{r^*_{MSR_2} - r_2}{\sigma^*_{MSR_2}} - \frac{r^*_{GMV} - r_2}{\sigma^*_{GMV}}}{\frac{r^*_{MSR_1} - r_1}{\sigma^*_{MSR_1}} - \frac{r^*_{GMV} - r_1}{\sigma^*_{GMV}}}\right)$$

$$(r_{\max} - r_{\min})I_2 = r^*_{GMV}\sigma^*_{GMV} \ln\left(\frac{\frac{r^*_{MSR_2} - r_2}{\sigma^*_{MSR_2}} - \frac{r^*_{GMV} - r_2}{\sigma^*_{GMV}}}{\frac{r^*_{MSR_1} - r_1}{\sigma^*_{MSR_1}} - \frac{r^*_{GMV} - r_1}{\sigma^*_{GMV}}}\right) + \\ + \sigma^{*2}_{GMV}\left(\frac{r^*_{MSR_2} - r_2}{\sigma^*_{MSR_2}} - \frac{r^*_{MSR_1} - r_1}{\sigma^*_{MSR_1}}\right)$$

and, finally we can write the maximum r_i as

$$r^*_i = r^*_{GMV} + \sigma^*_{GMV} \frac{\frac{r^*_{MSR_2} - r_2}{\sigma^*_{MSR_2}} - \frac{r^*_{MSR_1} - r_1}{\sigma^*_{MSR_1}}}{\ln\left(\frac{\frac{r^*_{MSR_2} - r_2}{\sigma^*_{MSR_2}} - \frac{r^*_{GMV} - r_2}{\sigma^*_{GMV}}}{\frac{r^*_{MSR_1} - r_1}{\sigma^*_{MSR_1}} - \frac{r^*_{GMV} - r_1}{\sigma^*_{GMV}}}\right)}. \tag{15.29}$$

□

Proof of Proposition 3. There exists a Pythagorean relationship between the slopes of the Tangent and Global Minimum portfolios and the slope of the asymptote of W_ρ.

$$m^2_{TP} = m^2_{ah} + m^2_{GMV}. \tag{15.30}$$

From expressions (15.5), we can derivate the following identities for the m_{TP}, m_{ah} and m_{GMV} slopes:

$$m_{TP} = \sqrt{a}, \qquad m_{ah} = \sqrt{\frac{ab - c^2}{b}} \quad \text{and} \quad m_{GMV} = \frac{c}{\sqrt{b}}. \tag{15.31}$$

and now, we can derivate the relationship $m_{TP}^2 = m_{ah}^2 + m_{GMV}^2$,

$$m_{ah}^2 + m_{GMV}^2 = \frac{ab - c^2}{b} + \frac{c^2}{b} = a = m_{TP}^2$$

□

***Proof of Corollary* 2.** The maximum cross-efficiency (MCESR) portfolio in $[0, r_{GMV}^*]$ depends only of Minimal Global Variance and Tangent portfolios.

$$r_i^* = r_{GMV}^* \left(1 - \frac{\sqrt{\frac{r_{TP}^*}{r_{GMV}^*}} - \sqrt{\frac{r_{TP}^*}{r_{GMV}^*} - 1}}{\ln\left(\sqrt{\frac{r_{TP}^*}{r_{GMV}^*}} - 1\right) - \ln\left(\sqrt{\frac{r_{TP}^*}{r_{GMV}^*}} - 1\right)} \right). \tag{15.32}$$

From (15.27) and (15.31), we can derivate the following expressions:

$$\frac{m_{TP}^2}{m_{GMV}^2} = \frac{a}{c^2/\sqrt{b}^2} = \frac{a/c}{c/b} = \frac{r_{TP}^*}{r_{GMV}^*}, \text{ then } \frac{m_{TP}}{m_{GMV}} = \sqrt{\frac{r_{TP}^*}{r_{GMV}^*}}$$

$$\frac{m_{ah}^2}{m_{GMV}^2} = \frac{\frac{ab-c^2}{b}}{c^2/b} = \frac{ab - c^2}{c^2} = \frac{ab}{c^2} - 1 = \frac{r_{TP}^*}{r_{GMV}^*} - 1, \text{ then } \frac{m_{ah}}{m_{GMV}} = \sqrt{\frac{r_{TP}^*}{r_{GMV}^*} - 1}$$

$$\frac{m_{TP}^2}{m_{ah}^2} = \frac{a}{\frac{ab-c^2}{b}} = \frac{a/c}{a/c - c/b} = \frac{r_{TP}^*}{r_{TP}^* - r_{GMV}^*}, \text{ then } \frac{m_{TP}}{m_{ah}} = \sqrt{\frac{r_{TP}^*}{r_{TP}^* - r_{GMV}^*}}.$$

From the expression (15.21), it is left to show that (15.32) is true. □

References

Artzener, P., Delbaen, F., Eber, J. M., & Heth, D. (1999). Coherent measures of risk. *Mathematical Finance*, *3*, 203–228.

Ben-Tal, A., & Nemirovski, A. (1999). Robust solutions of uncertain linear programs. *Operations Research Letters*, *25*, 1–13.

Bertsimas, D., & Pachamanova, D. (2008). Robust multiperiod portfolio management in the presence of transaction cost. *Computers and Operations Research*, *35*, 3–17.

Chapados, N. (2011). *Portfolio choice problems: An introductory survey of single and multiperiod models*. Springer.

Charnes, A., Cooper, W.W., & Rhodes, E. (1978). Measuring the efficiency of decision-making units. *European Journal of Operational Research, 2*, 429–444.

DeMiguel, V., Garlappi, L., & Uppal, R. (2009). Optimal versus Naive diversification: How inefficient is the 1/N portfolio strategy? *Review of Financial Studies, 18*, 1219–1251.

Diez, D., & Christou, N. StockPortfolio package. http://cran.r-project.org/web/packages/stockPortfolio/stockPortfolio.pdf

Doyle, J. R., & Green, R. (1994). Efficiency and cross-efficiency in data envelopment analysis: Derivatives, meanings and uses. *Journal of the Operational Research Society, 45*, 567–578.

Galagedera, D. U. A. (2013). A new perspective of equity market performance. *Journal of International Financial Markets, Institutions and Money, 26*, 333–357.

Goldfarb, D., & Iyengar, G. (2003). Robust portfolio selection problems. *Mathematics of Operations Research, 28*, 1–38.

Jin, H., Markowitz, H. M., & Zhou, X. Y. (2006). A note on semi variance. *Mathematical Finance, 16*, 53–61.

Krokhmal, P., Palmquist, J., & Uryasev, S. (2002). Portfolio optimization with conditional value-at-risk objective and constraints. *The Journal of Risk, 4*, 11–27.

Lim, S., Oh, K. W., & Zhu, J. (2014). Use of DEA cross-eficiency evaluation in portfolio selection: An application to Korean stock market. *European Journal of Operational Research, 236*, 361–368.

Mansini, R., Ogryczak, W., & Grazia-Speranza, M. (2014). Twenty years of linear programming based portfolio optimization. *European Journal of Operations Research, 234*, 518–535.

Markowitz, H. M. (1952). Portfolio selection. *Journal of Finance, 7*, 77–91.

Markowitz, H. M. (1959). *Portfolio selection: Efficient diversification of investment*. New York, London, Sydney: Wiley.

Nawrocki, D. (1999). A brief history of downside risk measures. *Journal of Investing, 8*, 9–25.

Pätäri, E., Leivo, T., & Honkapuro, S. (2012). Enhancement of equity portfolio performance using data envelopment analysis. *European Journal of Operational Research, 220*, 786–797.

Sexton, T. R., Silkman, R. H., & Hogan, A. J. (1986). Data envelopment analysis: Critique and extensions. In R. H. Silkman (Ed.), *Measuring efficiency: An assessment of data envelopment analysis* (pp. 73–105). San Francisco, CA: Jossey-Bass.

Tütüncü, R. H. (2003). *Optimization in finance, Advance Lecture on Mathematical Science and Information Science*.

The manufacturer's authorised representative in the EU is Springer Nature Customer Service Centre GmbH, Europaplatz 3, 69115 Heidelberg, Germany. If you have any concerns regarding our products, please contact ProductSafety@springernature.com

Printed and bound by CPI Group (UK) Ltd, Croydon, CR0 4YY
22/09/2026
02221048-0008